# Seeing the Future

# Seeing the Future

*Theoretical Perspectives on Future-Oriented Mental Time Travel*

**EDITED BY KOURKEN MICHAELIAN**

**STANLEY B. KLEIN**

**KARL K. SZPUNAR**

OXFORD
UNIVERSITY PRESS

OXFORD
UNIVERSITY PRESS

Oxford University Press is a department of the University of Oxford. It furthers the University's objective of excellence in research, scholarship, and education by publishing worldwide. Oxford is a registered trade mark of Oxford University Press in the UK and certain other countries.

Published in the United States of America by Oxford University Press
198 Madison Avenue, New York, NY 10016, United States of America.



Library of Congress Cataloging-in-Publication Data
Names: Michaelian, Kourken, editor. | Klein, Stanley B, editor. | Szpunar, Karl K.
Title: Seeing the future : theoretical perspectives on future-oriented mental time travel / edited by Kourken Michaelian, Stanley B. Klein, Karl K. Szpunar.
Description: New York : Oxford University Press, [2016] | Includes index.
Identifiers: LCCN 2015042731 (print) | LCCN 2015046731 (ebook) | ISBN 9780190241537 (hardcover : alk. paper) | ISBN 9780190241544 (UPDF)
Subjects: LCSH: Space and time. | Time travel. | Episodic memory. | Memory (Philosophy)
Classification: LCC BF467 .S44 2016 (print) | LCC BF467 (ebook) | DDC 153—dc23
LC record available at http://lccn.loc.gov/2015042731

# CONTENTS

# CONTRIBUTORS

**Donna Rose Addis**
School of Psychology and Centre for Brain Research
University of Auckland

**Cristina M. Atance**
School of Psychology
University of Ottawa

**Melissa Brinums**
School of Psychology
University of Queensland

**Nicola S. Clayton**
Department of Psychology
University of Cambridge

**Michael C. Corballis**
School of Psychology
University of Auckland

**Gianfranco Dalla Barba**
Inserm, Départment de Neurologie
Université Pierre et Marie Curie-Paris 6;
Dipartimento di Scienze della Vita
Università degli Studi di Trieste

**Arnaud D'Argembeau**
Neurosciences Cognitives—Psychopathologie Cognitive
University de Liège

**Felipe De Brigard**
Department of Philosophy
Center for Cognitive Neuroscience;
Duke Institute for Brain Sciences
Duke University

**Dorothea Debus**
Department of Philosophy
University of York

**Aleea L. Devitt**
School of Psychology and Centre for Brain Research
University of Auckland

**Alexandra Ernst**
Laboratoire d'Etude de l'Apprentissage et du Développement
Université de Bourgogne

**Bryce S. Gessell**
Department of Philosophy
Duke University

**Christoph Hoerl**
Department of Philosophy
University of Warwick

**Kana Imuta**
School of Psychology
University of Queensland

**Muireann Irish**
School of Psychology
University of New South Wales;
Neuroscience Research Australia
Australian Research Council Centre of Excellence in Cognition and Its Disorders

**Stanley B. Klein**
Department of Psychological and Brain Sciences
University of California, Santa Barbara

**Caitlin E. V. Mahy**
Department of Psychology
Brock University

**Liliann Manning**
Cognitive Neuropsychology and Physiopathology of Schizophrenia
Université de Strasbourg

**Gema Martin-Ordas**
Centre for Behaviour and Evolution
Newcastle University

**Teresa McCormack**
School of Psychology
Queen's University Belfast

**Kourken Michaelian**
Department of Philosophy
University of Otago

**Denis Perrin**
UFR Sciences Humaines
Université Pierre-Mendès-France; Institut Universitaire de France

**Giovanni Pezzulo**
Institute of Cognitive Sciences and Technologies
National Research Council (Italy)

**Daniel L. Schacter**
Department of Psychology and Center for Brain Science
Harvard University

**R. Nathan Spreng**
Department of Human Development and Human Neuroscience Institute
Cornell University

**Chloe Steindam**
Department of Psychological and Brain Sciences
University of California, Santa Barbara

**Thomas Suddendorf**
School of Psychology
University of Queensland

**Karl K. Szpunar**
Department of Psychology
University of Illinois at Chicago

**James M. Thom**
Department of Philosophy
King's College London

1

# The Past, the Present, and the Future of Future-Oriented Mental Time Travel

*Editors' Introduction*

**KOURKEN MICHAELIAN, STANLEY B. KLEIN, AND KARL K. SZPUNAR** ■

This introductory chapter reviews research on future-oriented mental time travel to date (the past), provides an overview of the contents of the book (the present), and enumerates some possible research directions suggested by the latter (the future).

## 1. THE PAST

In the years since Tulving first argued for episodic memory as a distinct memory system (Tulving, 1983), episodic memory has become a major area of research in psychology. Initially viewed primarily as a distinct store for information deriving from experienced episodes, the definition of episodic memory shifted over time, with researchers increasingly viewing episodic memory as a form of mental time travel into the personal past (Suddendorf & Corballis, 1997), drawing on a range of recently evolved and late-developing cognitive capacities and characterized by a unique form of consciousness: autonoesis, or consciousness of the self in subjective time (Wheeler et al., 1997). At the same time, recognition has grown that mental time travel into the past cannot be understood independently of future-oriented mental time travel (FMTT) (Atance & O'Neill, 2001; Schacter & Addis, 2007; Suddendorf & Corballis, 1997, 2007; Szpunar, 2010); indeed, FMTT may be primary, with our capacity to remember the past being derivative of the more basic capacity to imagine the future (Klein, 2013).

Research on FMTT has exploded over the past decade, with over a hundred articles published in just the last five years. Much of this research has revealed that, despite various differences in the cognitive and neural mechanisms that characterize future- and past-oriented mental time travel (MTT) (for a review, see Schacter et al., 2007, 2012), there also exist striking similarities. For instance, people tend to think about future and past events that are temporally near, as opposed to temporally distant (Spreng & Levine, 2006); temporally near events are mentally represented in greater detail than temporally distant events, whether those events are oriented toward the future or the past (D'Argembeau & Van der Linden, 2004); and people who are more apt to engage in mental imagery in their daily lives are more likely to generate detailed mental representations of the personal future and past (D'Argembeau & Van der Linden, 2006).

Functional neuroimaging and neuropsychological studies have provided important insights into the close cognitive links between future- and past-oriented MTT. Specifically, functional neuroimaging studies have revealed that a core network of brain regions, sometimes referred to as the default network (Buckner et al., 2008; Raichle et al., 2001), become engaged when people think about their personal future and past (Addis et al., 2007; Okuda et al., 2003; Szpunar et al., 2007; for a recent review, see Benoit and Schacter, 2015). Moreover, neuropsychological observations of patients with damage to specific regions within this network have revealed concurrent deficits of future- and past-oriented MTT (Klein et al., 2002; Tulving, 1985). One striking example of such overlapping deficits comes from hippocampal amnesia (Hassabis, Kumaran, Vann, & Maguire, 2007). Patients with damage to the hippocampus suffer from debilitating deficits of memory, losing the ability to remember details about their personal past and to form new memories following brain damage (Eichenbaum & Cohen, 2001). Evidence for a concurrent deficit in future thinking in hippocampal amnesia suggests that memory, at least the kind of memory that is supported by the hippocampus, plays an important role in future thinking (for further details, see Addis & Schacter, 2012).

The many consistent observations arising from the cognitive neuroscience literature have led to the suggestion that an important adaptive function of human memory is to provide the building blocks that make up the contents of future-oriented cognition (Schacter & Addis, 2007), along with other modes of cognition that rely on memory-based processing (Buckner & Carroll, 2007; for related theoretical interpretations, see Hassabis & Maguire, 2007, 2009). In support of such claims, more recent work with healthy populations that possess underdeveloped or deteriorating episodic memory systems, such as younger children (Atance, 2008; Busby & Suddendorf, 2005; Russell et al., 2010) and older adults (Addis et al., 2008; Gaesser et al., 2011), has likewise shown that these individuals exhibit an impoverished ability to engage in future-oriented MTT. Moreover, populations that possess varying degrees of episodic memory impairment, such as those diagnosed with Alzheimer's disease (Addis et al., 2009), mild cognitive impairment (Gamboz et al., 2010), schizophrenia (D'Argembeau et al., 2008), and post-traumatic stress disorder (Brown et al., 2013), also encounter difficulty in thinking about the future.

Ongoing research in this emerging field is focusing on delineating the extent to which the human brain represents temporal features of future- and past-oriented MTT (e.g., Nyberg et al., 2010; de Vito et al., 2012a, 2012b), the extent to which non-human animals are able to engage in FMTT (Corballis, 2013; Suddendorf, 2013), the role of general world knowledge—also known as semantic memory—and its interactions with episodic memory in supporting future-oriented cognition (Klein, 2013; Irish & Piguet, 2013), and in further decomposing the nature of neural contributions to FMTT. For instance, recent advances in neuroimaging techniques applied to the study of FMTT have revealed that specific regions of the default network serve to represent details associated with specific components of simulated events (e.g., people, places, and scenarios; Hassabis et al., 2014; Szpunar, St. Jacques, et al., 2014). Lastly, it is important to keep in mind that various modes of thinking about the future can be subsumed under the overarching concept of FMTT. On a daily basis, people simulate events, make predictions about those events, form intentions, and plan for the future. Researchers are busily working to draw links and identify possible interactions among these various modes of future thinking.

Despite the wealth of data that has been amassed concerning FMTT and its relation to episodic memory, there nonetheless exists a lack of theoretical presuppositions and implications of this research. The present volume seeks to redress this imbalance by including chapters that foreground conceptual questions, drawing on the theoretical resources of a wide range of disciplines and speaking to readers based in these diverse fields.

## 2. THE PRESENT

The 19 contributions to the volume intersect in multiple ways, but we have grouped them into eight sections, determined by the following themes: the varieties of future-oriented cognition; relationships between FMTT and episodic memory; subjective temporality in FMTT; the self in FMTT; and functional, evolutionary and comparative, developmental, and clinical perspectives on FMTT.

### 2.1. Varieties of Future-Oriented Cognition

The bulk of the literature on FMTT so far has treated it primarily as a matter of *episodic* cognition, paying less attention to the wide variety of additional forms of future-oriented cognition of which we are capable. In Chapter 2, **Szpunar, Spreng, and Schacter** build on their recent work (Szpunar, Spreng, et al., 2014) to situate episodic FMTT with respect to the latter, developing a novel taxonomy of forms of future thinking and putting it to work in an exploration of interrelationships and interactions among them. Szpunar, Spreng, and Schacter see future thinking as falling into four basic categories: simulation, prediction, intention, and planning. They argue that these four forms of thought interact to support our capacity to think about the future from the initial stage of conceiving of possible future events to the final stage of bringing about the realization of a desired event. In keeping with a theme running through the rest of the volume, moreover, they argue that an exclusive focus on episodic thinking is bound to lead to an incomplete understanding of FMTT. Much of our future thinking is semantic in character, and they suggest that simulation, prediction, intention, and planning are sometimes episodic, sometimes semantic, and sometimes a hybrid of the two. Thus, for example, they review evidence suggesting that semantic simulation may be dissociable from episodic simulation, identifying future-oriented personal semantic cognition (Renoult et al., 2012) as a promising area for further investigation.

Overall, their taxonomy suggests that episodic forms of simulation, prediction, intention, and planning are better understood than their semantic and hybrid counterparts and that there is a need for additional work on the latter. Many of the remaining contributions to the volume take important steps in this direction.

### 2.2. Relationships Between Future-Oriented Mental Time Travel and Episodic Memory

While Szpunar, Spreng, and Schacter urge us to broaden our focus to include semantic and hybrid forms of future-oriented cognition, research on FMTT grows out of research on episodic memory, and the nature of the relationship between FMTT and episodic memory is still not fully understood. In Chapter 3, **Perrin** delineates two broad schools of thought about this matter. *Continuists* maintain that, despite the various differences between episodic memory and FMTT that have been observed at the neural, cognitive, and phenomenological levels, they are ultimately two instances of a single general capacity for MTT. *Discontinuists*, in contrast, maintain that the differences between them are sufficiently important for episodic memory and FMTT to constitute two distinct capacities.

Continuism is the dominant view in psychology and neuroscience, but Perrin himself favors discontinuism. Drawing on the philosophical literature on memory and mind, he identifies two apparent discontinuities between episodic memory and FMTT. The first is epistemological. Since the identities of the individuals involved in remembered episodes are fixed by the subject's past causal interactions with them, subjects engaged in episodic memory can commit errors of misidentification (Shoemaker, 1968) (i.e., they can make mistakes about the identities of the remembered individuals). In contrast, since the identities of the individuals involved in imagined future episodes are not fixed by past causal interactions, subjects engaged in FMTT cannot commit such errors. The second discontinuity is causal. In line with the causal theory of memory (Martin & Deutscher, 1966), Perrin argues that episodic memory presupposes a causal connection with past experiences, such that the subject who remembers an event must be numerically identical to the subject who experienced it, whereas FMTT presupposes neither causal connection nor identity. This second discontinuity is reflected in a subtle but important difference at the phenomenological level: whereas, in episodic memory, autonoesis is a necessary feature of the process and normally results from automatic, unconscious monitoring, in FMTT autonoesis need not occur and, when it does, it is the result of deliberate, conscious inference.

If discontinuism is right, it is a mistake to treat FMTT and episodic memory as differing only in their respective temporal orientations, and the concept of MTT would at best be misleading when applied to episodic memory. Defending a view of episodic memory as mental time travel (Michaelian, 2016), **Michaelian** in Chapter 4 responds to arguments for both *metaphysical* discontinuism, which posits a difference in kind between episodic memory and FMTT themselves, and *epistemological* discontinuism, which posits a difference in kind between the knowledge of past events produced by memory and the knowledge of future events produced by FMTT. Against Perrin's argument, he maintains that episodic memory, like FMTT, does not presuppose a causal connection with the relevant event. Against Debus's related argument (2014), according to which episodic memory involves a relationship to a particular past event, while FMTT does not involve a relationship to a particular future event, he defends a unified view of the objects of memory and FMTT. Finally, he maintains that any mechanism capable of producing autonoesis in episodic memory is likewise capable of producing autonoesis in FMTT, and vice versa, concluding, on grounds of parsimony, in favor of a unified account of the phenomenology of FMTT. Turning to epistemological discontinuism, Michaelian responds both to older arguments for quantitative and qualitative differences between our knowledge of past events and our knowledge of future events and to Perrin's claim that immunity to error through misidentification is characteristic of FMTT but not of episodic memory; on the alternative view developed by Michaelian, immunity to error through misidentification is essential neither to episodic memory nor to FMTT, with some but not all instances of episodic memory being vulnerable to error through misidentification and some but not all instances of FMTT being immune to such error.

Continuism may receive further support from **Devitt and Addis**'s investigation of bidirectional interactions between episodic memory and FMTT in Chapter 5. It is natural to focus on the ways in which imagining the future is shaped by remembering the past, but Devitt and Addis urge us to attend equally to the impact that imagining the future can have on remembering the past. Hence their chapter investigates both the ways in which memory contributes to the simulation of future events and the ways in which imagination influences memory. Under the former heading, they point out that, on both the constructive episodic simulation hypothesis (Schacter & Addis, 2007) and the scene construction hypothesis (Hassabis & Maguire, 2007), access to remembered information underwrites our capacity for phenomenologically rich simulation of both past and future events. Under the latter heading, they argue that the characteristics of imagined episodes influence both false memories of events and accurate memories of imagined events. The ability to remember the future is adaptive only given that imagined events are made available, through memory, for future use, and the encoding and retrieval of imagined events are themselves affected by the characteristics of those events, including vividness, plausibility, emotional valence, and field versus observer perspective. When imagined events rate highly on these characteristics, however, they may be mistaken for memories of real events; thus imagination may give rise to false memory, as in imagination inflation (Garry et al., 1996).

The research reviewed by Devitt and Addis supports a view on which information stored in memory makes symmetrical contributions to the process of remembering the past and the process of imagining the future, with the distinction between remembered and imagined events being an outcome of further processing by the subject. Such a view would appear to favor continuism, but the debate between continuists and discontinuists also depends on a number of more specific issues. One such issue is the role of autonoetic consciousness and subjective temporality more generally in episodic memory and FMTT; this is the focus of the following section.

## 2.3. Subjective Temporality in Future-Oriented Mental Time Travel

**Dalla Barba** in Chapter 6 builds on his earlier work on memory, consciousness, and temporality (Dalla Barba, 2002) to argue that information stored in memory, while providing the raw ingredients for both remembering the past and imagining the future, has no intrinsic temporal dimension; the temporal dimension of MTT is a further ingredient, supplied by temporal consciousness. In contrast to *knowing* consciousness, which grounds impersonal awareness of information about past and future events, *temporal* consciousness—closely related to Tulving's (1985) concept of autonoetic consciousness—enables the subject to orient himself to his personal past and future. The absence or distortion of temporal consciousness provides an elegant explanation of the patterns of confabulation observed in amnesic patients, with confabulation resulting from distortion of temporal consciousness (due to damage to the hippocampus), while outright loss of temporal consciousness (due to complete destruction of the hippocampus) does not result in confabulation, since patients with no temporal consciousness have no capacity to orient themselves with respect to subjective time.

Whereas Dalla Barba sees a largely symmetrical role for temporal consciousness in episodic memory and FMTT, **Klein and Steindam** (Chapter 7) see a potential asymmetry, arguing that FMTT involves a diverse range of capacities, only some of which depend on episodic memory and its characteristic form of subjective time. In line with Szpunar, Spreng, and Schacter's taxonomy, they argue against the assumption that FMTT stands in a privileged relationship to episodic memory, citing research that demonstrates the possibility of semantic memory-based FMTT. They argue, moreover, that this research demonstrates that FMTT is not characterized by a unitary form of subjective temporality. On their view, when FMTT does draw on episodic memory, its phenomenology is that of *lived time*: the subject has a pre-reflective sense that he is "pre-living" a possible future, a counterpart of the pre-reflective sense of reliving the past that is characteristic of episodic memory. When it draws on semantic memory, in contrast, the phenomenology of FMTT is that of *known time*, inferential rather than pre-reflective and experienced as impersonal rather than as pre-living (Klein et al., 2002). Lived time maps onto Dalla Barba's concept of temporal consciousness and Tulving's concept of autonoetic consciousness, but known time maps neither onto Dalla Barba's concept of knowing consciousness nor onto Tulving's concept of noetic consciousness, since both of the latter lack a strictly temporal dimension. It does, however, align with the philosophical distinction between the A-series, in which time is experienced as flowing from future to past, and the B-series, in which time is experienced as a fixed before-and-after relationship (McTaggart, 1908). While the distinction between the A-series and the B-series has informed little FMTT research to date, Klein and Steindam's contribution suggests that gains in conceptual clarity are to be had by taking both of these forms of subjective time into account.

While Dalla Barba, on the one hand, and Klein and Steindam, on the other hand, provide distinct accounts of subjective temporality in episodic memory and FMTT, they are in agreement on its centrality. **De Brigard and Gessell**, in contrast, call the centrality of subjective temporality into question in Chapter 8. Their starting point is the now-familiar observation that episodic memory and FMTT engage a common set of brain regions, the default network. A standard explanation for this observation is the *tensed content* view, according to which the default network is concerned with simulating episodes with explicit temporal content. Against the tensed content view, De Brigard and Gessell invoke evidence that the default network is engaged not only when imagining temporally located episodes but also when imagining episodes with no temporal location, as well as evidence that patients with hippocampal amnesia have an impaired capacity to imagine episodes without an explicit temporal location, despite remaining capable of other forms of temporal thought. Their alternative explanation of the involvement of the default network in episodic memory, FMTT, and other forms of episodic thought is the *dynamic structure* view, which relies on the distinction between intentional contents (how a mental state represents something) and representational vehicles (the brain structure that realizes a mental state with representational content). Contents can be temporal (i.e., the state can have tensed content), but so can vehicles (i.e., the state can itself have a dynamic character), and, on their view, it is the dynamic character of the vehicles of episodic simulation that accounts for the involvement of the default network in simulating both temporally located episodes and episodes with no determinate temporal location. Damage to the default network would thus account for amnesic patients' inability to imagine past, future, and non-temporal episodes, despite their preserved ability to engage in other forms of temporal thought.

While the dynamic structure view will require further investigation, De Brigard and Gessell suggest that it may provide a means of reconciling the constructive episodic simulation hypothesis and the scene construction hypothesis, a suggestion that harmonizes with Devitt and Addis's treatment of the relationship between these hypotheses.

## 2.4. The Self in Future-Oriented Mental Time Travel

The dynamic structure view is friendly to the idea that FMTT may ultimately depend as much on semantic memory as it does on episodic memory. The role of semantic memory in MTT is examined in detail in Chapter 9 by **Manning**, who further links it to the role of the self. Building on a rich body of empirical and theoretical literature, Manning explores the relationship between, on the one hand, episodic and semantic forms of MTT and, on the other hand, the *I-self*—the phenomenological, knowing self—and the *Me-self*—the conceptual, known self (James, 1890). In a departure from Tulving's (2005) influential view, which focuses on links between FMTT and the I-self and episodic memory, she foregrounds links between FMTT and the Me-self and semantic memory. The Me-self grounds the subject's capacity to think of his or her personal attributes and has a conceptual, semantic character. Specific, episodic memories may provide constraints on what the Me-self can be imagined to become in the future, but semanticized information about the self may also provide a critical input to simulations of future events that are compatible with the subject's self-knowledge. The I-self, in turn, grounds the subject's capacity to project himself or herself into possible past and futures in a phenomenologically rich manner. Ultimately, however, there may be a sense in which episodic memory is primary in MTT, since, while the Me-self and semantic memory provide inputs to simulations of future events, Manning emphasizes that the semanticized memories that feed into FMTT originate in episodic memories based on the I-self.

As the exchange between Perrin and Michaelian demonstrates, both the role of autonoetic consciousness and its determinants are controversial questions. **D'Argembeau**'s Chapter 10 is focused on the latter question, exploring the possibility that autonoesis depends not only on the subject's simulation of an event but also on contextualization of the event within knowledge of the self, with knowledge of personal goals, in particular, playing a central role. Paralleling autobiographical knowledge of the personal past, knowledge of the personal future consists of a variety of components, including the representations of possible selves described by Manning, as well as future life periods, general events, and cultural life scripts. D'Argembeau argues, first, that only when they are consistent with such knowledge are simulations of future events experienced as belonging to the personal future. He argues, second, that knowledge of personal goals plays an especially central role here. In support of the latter claim, he adduces evidence that activating knowledge of personal goals is a common first step in episodic future thinking, with personal goals being used to guide simulation of events, that memories of simulated future episodes are more easily accessed when they are relevant to personal goals, and that autobiographical knowledge plays a role in linking and organizing representations of specific future episodes, as well as neuroimaging evidence supporting the link between FMTT and goal processing. In support of the former claim, he reviews evidence that the importance of a simulated future event with respect to personal goals contributes to the involvement of autonoesis in FMTT.

In addition to reinforcing the plausibility of the claim that semantic memory makes critical contributions to FMTT, Manning's and D'Argembeau's discussions of the role of the self resonate with the suggestion—made in a number of contributions in subsequent sections—that FMTT may have much of its beneficial impact on future-directed decision-making via its interaction with the subject's personal goals.

## 2.5. Functional Perspectives

In one of several chapters to discuss the impact of FMTT on future-directed decision-making, **Debus**, in an exploration of the nature and the value of FMTT in Chapter 11, cautions that it may in fact have both benefits and costs. Beginning with the nature of FMTT, she echoes De Brigard and Gessell's observation that, in addition to imagining future events, subjects can imagine past events, present events, and events with no determinate temporal location, arguing that imagining future events differs from imagining events of these other types, first, in that the imagined event is subjectively located in the future (the "specific temporal orientation" claim) and, second, in that the future (unlike the past or the present) is, in an important sense, open (the "openness" claim). Rejecting interpretations of the specific temporal orientation claim on which future-orientation is a matter of temporal phenomenology, Debus argues, in line with D'Argembeau's approach, that it is determined by the occurrence of the event in the context of relevant future-directed *beliefs*. Rejecting both metaphysical and epistemological interpretations of the openness claim, Debus argues that the openness of the future is to be understood in agential terms, in the sense that the subject himself or herself is in principle able to affect the occurrence of the event in question. Turning from the nature of FMTT to its value, Debus argues that the value of imagining future events lies in the effect that doing so can have on their realization, in part by motivating the subject to bring them about. But she also notes that FMTT can in some cases have an opposite, demotivating effect.

In contrast to Debus's focus on the impact of FMTT itself, **Hoerl and McCormack** focus in Chapter 12 on the function of episodic memory in future-oriented decision-making. Though there is little consensus in the literature on the function of episodic memory, there is increasing agreement that its function is in some important sense future-oriented. Hoerl and McCormack's key insight is that the function of episodic memory may be future-oriented, not merely in the sense that it provides the raw ingredients for the simulation of future events, but also in the sense that it is a prerequisite for the ability to feel the emotion of *regret* (e.g., over how a given past event might have turned out differently had one chosen differently), hence shaping decisions (e.g., by leading one to choose differently on future occasions). Picking up on recent work on the role of regret in decision-making, they argue that regret may have an impact on future choices both by motivating one to choose differently when faced with similar choices and, more generally, by motivating one not to miss opportunities. Considering how episodic memory might underlie the ability to experience regret, Hoerl and McCormack argue that experiencing regret presupposes the ability to mentally project oneself into a particular past situation, as well as to understand that the situation might have unfolded differently than it did in fact. Considering how episodic memory might underlie the ability to anticipate regret, they argue that anticipating regret presupposes the ability to project oneself forward into a possible future situation, as well as the ability to imagine one's future self looking back on—remembering—one's present decision. That regret presupposes episodic memory in both of these ways is supported by the research on episodic counterfactual thinking discussed by De Brigard and Gessell, as well as research on the loss of ability to experience regret in episodic amnesia. Having made a case for the view that episodic memory is a prerequisite for the ability to feel regret, Hoerl and McCormack argue that regret is indeed involved in adaptive decision-making, since there is no simpler alternative mechanism that might play the same role.

Like Debus, **Pezzulo** focuses directly on FMTT in Chapter 13. His aim, however, is to situate FMTT in the context of a broader future-oriented perspective on cognition as a whole in order to shed light on the relationship between MTT and *predictive processing*. There is increasing consensus that the brain is functionally oriented toward the future, constantly generating short-term and longer-term predictions. As illustrated by research on internally generated sequences in the rodent hippocampus—which enable rodents not only to revisit spatial paths that they have actually taken but also to explore counterfactual and possible future paths—predictive processing requires the brain to self-generate information, as opposed to extracting it from sensory stimuli. Pezzulo speculates that such internally generated dynamics, which require an ability to detach from the agent's current context, might underlie more sophisticated forms of detached cognition, including past- and future-oriented MTT and episodic counterfactual thought. A link between simpler forms of predictive processing and more sophisticated forms of episodic cognition could shed light on the evolution of MTT by showing how the latter, highly complex capacity might have emerged on the basis of considerably simpler capacities. With respect to the adaptive benefits of phenomenologically rich FMTT, in particular, Pezzulo suggests that the sense of "pre-experiencing" that is characteristic of FMTT might enable the subject not only to anticipate the outcomes of future decisions but also to evaluate them in the here and now by anticipating his or her own future evaluative responses.

The three chapters in this section contribute to our increasingly sophisticated functional understanding of FMTT. As Pezzulo's chapter suggests, however, function can only be fully understood from an evolutionary perspective, and the chapters in the following section turn to the evolution of our capacity for FMTT.

## 2.6. Evolutionary and Comparative Perspectives

**Thom and Clayton** in Chapter 14 review the ongoing debate over MTT in animals. In the debate so far, the main challenge for partisans of animal MTT has been to identify clear counterexamples to the Bischof-Köhler hypothesis, according to which animals cannot anticipate future needs that differ from present needs (Suddendorf & Corballis, 2007). Identifying such counterexamples is problematic due to the difficulty of ruling out the possibility that apparently future-oriented behavior is in fact governed by current drive states. Seeking to resolve this difficulty, proponents of animal MTT have employed ingenious experiments to provide evidence of FMTT in great apes, other primates, and birds, but the evidence so far remains inconclusive. Rather than seeking further evidence of the same sort, Thom and Clayton suggest that the lack of firm conclusions in this area may in part reflect the limitations of the Bischof-Köhler hypothesis itself, arguing that consideration of the case of *human* FMTT shows that dissociating from current needs is neither necessary nor sufficient for engaging in FMTT. As opposed to looking at the evolutionary emergence of the capacity for MTT as a whole, they maintain, we would do better to focus on the emergence of the more specific cognitive capacities that give rise to it—if these, taken individually, have adaptive benefits, it becomes increasingly plausible to see animal and human MTT as continuous.

A similarly optimistic take on animal MTT is developed by **Martin-Ordas** in Chapter 15. Focusing on autonoesis, she points out that treating the presence of autonoetic phenomenology as a criterion for MTT makes it difficult to determine whether non-human animals, who cannot provide verbal reports on their phenomenology, are capable of MTT in the same sense as humans. Guided by the Bischof-Köhler hypothesis, researchers have therefore focused on criteria such as flexibility, learning from single experiences, and the absence of relevant stimuli in the present environment. Building on some of the same studies discussed by Thom and Clayton, Martin-Ordas similarly argues both that there is some evidence for MTT in animals and that the Bischof-Köhler hypothesis has important limitations. But whereas Thom and Clayton argue that dissociation from current needs is neither necessary nor sufficient for FMTT, Martin-Ordas argues that tests based on the hypothesis presuppose distinctions between episodic and semantic memory and between memory and foresight, which may ultimately be less than clear. In light of these points, standard criticisms of evidence for MTT in animals—especially great apes—begin to look less convincing. Martin-Ordas does not deny that human MTT is considerably more complex than animal MTT. But, adopting an anthropological perspective, she does suggest that this additional complexity might derive not from autonoesis but rather from semantic knowledge, since human planning, especially long-term planning, often depends less on episodic than on semantic knowledge. Much of the latter, in turn, is intergenerationally transmitted, opening up the possibility of cultural variability in FMTT. Thus Martin-Ordas suggests that, in addition to investigating similarities and differences between FMTT in animals and in humans, we investigate similarities and differences between FMTT in different human populations.

Like Martin-Ordas, **Corballis** (Chapter 16) is optimistic with respect to animal MTT, even while calling our attention to the unique features of human MTT. Drawing on some of the same research on the rat hippocampus cited by Pezzulo, Corballis suggests that at least basic forms of MTT may not be unique to humans, indeed that they may go far back in evolution. Human MTT, however, is clearly more complex than any form of MTT for which we have evidence in animals, in part because it is linked to more general forms of mind wandering, which greatly augment the powers of MTT, allowing us to anticipate not only likely future events but also a much broader range of possible events, thus preparing ourselves for the unexpected. Mind wandering overlaps with theory of mind, which in turn is a prerequisite for linguistic communication, since language use requires us to anticipate the mental states of others. And language, in turn, allows us to share the results of our mind wandering, which we do primarily through storytelling. It may ultimately be the link with storytelling that is the most distinctive feature of human MTT; indeed, the emergence of the capacity for storytelling may have marked a key step in human evolution (Scalise Sugiyama, 2011). Thus the overall message that emerges from Corballis's contribution, as well as the other contributions in this section, is that, while there may be precursors of MTT in other species, human MTT has important unique features.

## 2.7. Developmental Perspectives

Studying MTT in animals presents many of the same challenges as studying it in children. Bridging evolutionary and developmental perspectives, **Suddendorf, Brinums, and Imuta** in Chapter 17 investigate the role of FMTT in enabling the subject to shape his or her future self through the deliberate practice of motor skills. They point out that, while the improvement of such skills depends on changes to procedural systems, practice that results in changes can itself be triggered by imagining a future self with improved skills. There are controversies over the role of deliberate practice in enabling subjects to achieve the highest levels of performance, but it is clear that practice is an important factor in enabling us to achieve some level of improvement in our performance in various areas. Similarly, there are controversies over the ability of non-human animals to engage in FMTT, but it is clear that flexible, long-term foresight is restricted to humans. Hence the improvement of motor skills through deliberate practice is likely to be uniquely human. Indeed, Suddendorf, Brinums, and Imuta argue that deliberate practice, enabled by FMTT, might well have been a prerequisite for the emergence, in evolutionary time, of recognizably human culture, given that it is critical to the characteristically human ability to specialize in order to respond to shifting environmental demands (and to reshape the environment itself). Turning from evolution to development, Suddendorf, Brinums, and Imuta describe the gradual emergence of deliberate practice in children through the development of the multiple capacities that it presupposes, suggesting that the building blocks of deliberate practice are in place around the time children start formal schooling. Finally, they describe an initial study (Davis et al., in press) on the development of deliberate practice in children, which provides some confirming evidence for this suggestion.

A more general take on FMTT in children is provided in Chapter 18 by **Atance and Mahy**, who review both verbal and behavioral approaches to studying children's capacity to mentally project into the future. Verbal methods have provided evidence that, while they are able to anticipate the future, 3-year-olds have difficulty *accurately* anticipating the future. Due to concerns about the appropriateness of using verbal methods with children who have an imperfect grasp of the meanings of temporal terms, researchers have also investigated FMTT in children using behavioral methods—akin to those used to study MTT in non-human animals—inspired by Tulving's (2005) "spoon test," which requires the subject to act now in anticipation of a situation likely to be faced in the future. Such behavioral methods have provided evidence that memory limitations—the inability to remember relevant past information—may account for children's failures in some episodic future thinking tasks, and there is indeed evidence for a correlation between children's ability to think about the past and their ability to think about the future. In addition to episodic memory, however, the development of theory of mind and executive function may be important to explaining the development of FMTT in children. As there is little developmental work on the contributions of these factors, this remains an important area for future research.

## 2.8. Clinical Perspectives

Insights into the nature and mechanisms of FMTT can be gleaned not only from comparisons between humans and non-human animals and between adults and children, but also from comparisons between clinical and non-clinical populations. The final two contributions to the volume present clinical perspectives on FMTT.

Integrating findings on FMTT in clinical populations with functional neuroimagining studies of FMTT in healthy individuals, **Irish** in Chapter 19 examines the pivotal role played by semantic memory in FMTT. Going back to Tulving's foundational work (Tulving, 1972), semantic memory has been understood as lacking a conscious temporal dimension, and the standard view is that episodic memory, via autonoetic consciousness, provides the essential foundation for MTT into both past and future. There is, however, mounting evidence of significant interplay between episodic and semantic memory (e.g., overlap between the semantic memory system and the core network). Moreover, semantic representations, which are abstracted from multiple experiences, are more flexible and hence well suited to projection into future contexts; indeed, due to the lesser flexibility of episodic representations, it is plausible that semantic memory may not only facilitate FMTT but may even be essential to it. Irish therefore argues that we should expect to find that semantic memory makes important contributions to FMTT. After briefly reviewing relevant neuroimagining evidence, she turns to studies of amnesia and semantic dementia, which show that patients with episodic amnesia may be capable of temporal consciousness based on semantic memory, whereas patients with semantic dementia have impaired FMTT despite preserved episodic memory. These findings support the *semantic scaffolding* hypothesis, according to which semantic knowledge provides the framework enabling simulation of past and future events (Irish & Piguet, 2013). The semantic scaffolding hypothesis, in turn, fits with the taxonomy developed by Szpunar, Spreng, and Schacter in Chapter 2, which situates episodic and semantic forms of FMTT as points on a continuum.

In Chapter 20, **Ernst and Manning** review the findings of behavioral and neuroimaging studies of FMTT impairments in multiple sclerosis (MS) patients. MS is accompanied by a variety of cognitive impairments, and Ernst and Manning argue that there is indeed a neural and cognitive basis for FMTT impairment in MS, ruling out a psychological basis on the ground that similar patterns of impairment are observed in MS patients in both FMTT and autobiographical memory. A comparison of FMTT impairments in MS patients and FMTT impairments in other clinical populations leads them to suggest that MS primarily affects the early stages of the process of simulating future events, with patients having difficulty extracting and recombining details into novel and coherent events. At the structural level, they argue that the impairments result from changes to a widespread brain network, consistent with the diffuse nature of the brain lesions characteristic of MS. Finally, they turn to strategies for improving autobiographical memory functioning to assess the possibilities for cognitive rehabilitation of FMTT in MS patients, finding that the use of mental visual imagery provides a particularly promising means of improving not only autobiographical memory but also FMTT.

## 3. THE FUTURE

What is likely to come next in FMTT research? Collectively, the contributions to this volume identify a number of promising areas for future research.

- *The diversity of FMTT*: There are important differences and complex interactions among different forms of FMTT. Future research is unlikely to focus only on episodic forms of FMTT or to assume that episodic memory is necessarily involved in every form of FMTT. In particular, we are likely to see more research on the contribution of semantic memory to FMTT and on semantic and hybrid forms of FMTT.
- *The basis of MTT*: Restricting our attention to episodic MTT, there are arguments both in favor of the primacy of episodic memory and in favor of the primacy of FMTT. This question is particularly thorny, as it intersects in complex ways with comparative, developmental, and clinical questions.
- *Subjective temporality*: Multiple frameworks for understanding temporal phenomenology are on offer; it is unclear to what extent these can be reconciled. More broadly, it remains to be seen whether the temporal

dimension of MTT is to be understood primarily in terms of phenomenology, content, or representational vehicles.

- *The self and time*: Both the self and time are usually viewed as being essential to FMTT. It is becoming increasingly evident that there are multiple forms of self and multiple forms of time that might be involved in MTT, and there are arguments for viewing neither the self nor time as being essential to the basic capacity underlying MTT. Untangling the relations among the various forms of MTT, self, and time is likely to be an especially challenging theoretical task.
- *Function and evolution of MTT*: A range of costs and benefits of FMTT have been identified, and many competing explanations of the evolution of FMTT have been proposed; it remains to be seen to what extent these accounts can be reconciled with each other.
- *Components of FMTT*: The relationship of FMTT to more basic forms of future-oriented cognition is another area that will feed back into our evolutionary understanding of the capacity. Evolutionary, developmental, and clinical studies are also increasingly taking into account the way in which various more fine-grained cognitive capacities interact to give rise to the capacity for the kind of FMTT we find in adult humans. This might represent a way out of the impasse over MTT in non-human animals.

In our view, three general points should be kept in mind going forward. First, while FMTT has often been taken to be a general neurocognitive capacity whose different manifestations are largely a function of situational and motivational contingencies, there are conceptual and empirical difficulties for this view. The contributions to the volume suggest that FMTT has a number of distinct neurocognitive underpinnings and that the diversity of its empirical presentation is not merely superficial variation resulting from task demands; rather, it reflects the fact that there is not a single, uniform capacity for FMTT—there are different ways of imagining the future, underwritten by distinct mechanisms. Compare, for example, imagining what one will do tomorrow with imagining what will happen on a camping trip with imagining what one will be like in 10 years. These and other forms of FMTT may be only loosely related in virtue of their concern with possible futures. Research on FMTT has often assumed that episodic memory is the common factor that lends an underlying unity to this apparent diversity. But, as a number of the contributors argue, this assumption is increasingly ill supported. At this stage, a possibility deserving of serious consideration is that the treatment of FMTT as a unified psychological faculty is largely a matter of methodological convenience, together with a desire for theoretical parsimony. While parsimony is a laudable goal in theory construction, it may nevertheless turn out to be the case that the search for a common mechanism underlying the various forms of FMTT is futile.

Second, as a number of the chapters in this volume indicate, it is becoming increasingly clear that greater consideration is needed of those aspects of consciousness that enable retrieved mental content to be imagined and considered with respect to its implications for future behavior. It remains to be seen whether this is Tulving's autonoetic consciousness, Dalla Barba's temporal consciousness, or Klein and Steindam's lived temporality, but it is likely to be the case that no single form of subjective temporality can account for all FMTT in its various manifestations. The sooner we begin sustained investigation of the conscious processes that enable us to transcend the present, the sooner we will come to terms with exactly what it means to engage in the act of mental time travel.

Finally, we suggest caution with respect to the tendency to assume that empirical studies are sufficient by themselves to shed light on the nature of FMTT. The history of psychological science suggests that the mere accumulation of data, often in support of very local theories, has sometimes stood in for the development of more global understanding of the phenomena in question. If they are to be fully appreciated, the diverse phenomena constituting acts of FMTT need to be securely situated in a broader understanding of the workings of the mind. It is our hope that the present volume moves us significantly closer to this goal.

## REFERENCES

Addis, D. R., Sacchetti, D. C., Ally, B. A., Budson, A. E., & Schacter, D. L. (2009). Episodic simulation of future events is impaired in mild Alzheimer's disease. *Neuropsychologia, 47,* 2660–2671.

Addis, D. R., & Schacter, D. L. (2012). The hippocampus and imagining the future: where do we stand? *Frontiers in Human Neuroscience, 5,* 173.

Addis, D. R., Wong, A. T., & Schacter, D. L. (2007). Remembering the past and imagining the future: common and distinct neural substrates during event construction and elaboration. *Neuropsychologia, 45,* 1363–1377.

Addis, D. R., Wong, A. T., & Schacter, D. L. (2008). Age-related changes in the episodic simulation of future events. *Psychological Science, 19,* 33–41.

Atance, C. M. (2008). Future thinking in young children. *Current Directions in Psychological Science, 17,* 295–298.

Atance, C. M., & O'Neill, D. K. (2001). Episodic future thinking. *Trends in Cognitive Sciences, 5*(12), 533–539.

Benoit, R. G., & Schacter, D. L. (2015). Specifying the core network supporting episodic simulation and episodic memory by activation likelihood estimation. *Neuropsychologia, 75,* 450–457.

Brown, A. D., Root, J. C., Romano, T. A., Chang, L. J., Bryant, R. A., & Hirst, W. (2013). Overgeneralized autobiographical memory and future thinking in combat veterans with posttraumatic stress disorder. *Journal of Behavioral Therapy and Experimental Psychiatry, 44,* 129–134.

Buckner, R. L., Andrews-Hanna, J. R., & Schacter, D. L. (2008). The brains default network: anatomy, function, and relevance to disease. *Annals of the New York Academy of Sciences, 1124,* 1–38.

Buckner, R. L., & Carroll, D. C. (2007). Self-projection and the brain. *Trends in Cognitive Sciences, 11,* 49–57.

Busby, J., & Suddendorf, T. (2005). Recalling yesterday and predicting tomorrow. *Cognitive Development, 20,* 362–372.

Corballis, M. C. (2013). Mental time travel: a case for evolutionary continuity. *Trends in Cognitive Sciences, 17,* 5–6.

D'Argembeau, A., Raffard, S., & Van der Linden, M. (2008). Remembering the past and imagining the future in schizophrenia. *Journal of Abnormal Psychology, 117,* 247–251.

D'Argembeau, A., & Van der Linden, M. (2004). Phenomenal characteristics associated with projecting oneself back into the past and forward into the future: influence of valence and temporal distance. *Consciousness and Cognition*, *13*, 844–858.

D'Argembeau, A., & Van der Linden, M. (2006). Individual differences in the phenomenology of mental time travel: the effects of vivid imagery and emotion regulation strategies. *Consciousness and Cognition*, *15*, 342–350.

Dalla Barba, G. (2002). *Memory, consciousness, and temporality*. Boston: Kluwer.

Davis, J. T. M., Cullen, E., & Suddendorf, T. (In press). Understanding deliberate practice in preschool aged children. *The Quarterly Journal of Experimental Psychology*. DOI: 10.1080/17470218.205.1082140.

de Vito, S., Gamboz, N., & Brandimonte, M. A. (2012a). What differentiates episodic future thinking from complex scene imagery? *Consciousness and Cognition*, *21*, 813–823.

de Vito, S., Gamboz, N., Brandimonte, M. A., Barone, P., Amboni, M., & Della Sala, S. (2012b). Future thinking in Parkinson's disease: an executive function? *Neuropsychologia*, *50*, 1494–1501.

Debus, D. (2014). "Mental time travel": remembering the past, imagining the future, and the particularity of events. *Review of Philosophy and Psychology*, 5(3), 333–350.

Eichenbaum, H. E., & Cohen, N. J. (2001). *From conditioning to conscious recollection: memory systems of the brain*. New York: Oxford University Press.

Gaesser, B., Sacchetti, D. C., Addis, D. R., & Schacter, D. L. (2011). Characterizing age-related changes in remembering the past and imagining the future. *Psychology and Aging*, *26*, 80–84.

Gamboz, N., de Vito, S., Brandimonte, M. A., Pappalardo, S., Galeone, F., Iavarone, A., & Della Sala, S. (2010). Episodic future thinking in amnesic mild cognitive impairment. *Neuropsychologia*, *48*, 2091–2097.

Garry, M., Manning, C., Loftus, E., & Sherman, S. (1996). Imagination inflation: imagining a childhood event inflates confidence that it occurred. *Psychonomic Bulletin & Review*, *3*(2), 208–214.

Hassabis, D., Kumaran, D., Vann, S. D., & Maguire, E. A. (2007). Patients with hippocampal amnesia cannot imagine new experiences. *Proceedings of the National Academy of Sciences U S A*, *104*, 1726–1731.

Hassabis, D., & Maguire, E. A. (2007). Deconstructing episodic memory with construction. *Trends in Cognitive Sciences*, *11*(7), 299–306.

Hassabis, D., & Maguire, E. A. (2009). The construction system of the brain. *Philosophical Transactions of the Royal Society of London B: Biological Sciences*, *364*(1521), 1263–1271.

Hassabis, D., Spreng, R. N., Rusu, A. A., Robbins, C. A., Mar, R. A., & Schacter, D. L. (2014). Imagine all the people: how the brain creates and uses personality models to predict behavior. *Cerebral Cortex*, *24*, 1979–1987.

Irish, M., & Piguet, O. (2013). The pivotal role of semantic memory in remembering the past and imagining the future. *Frontiers in Behavioral Neuroscience*, *7*(27).

James, W. (1890). *The principles of psychology*. New York: Holt.

Klein, S. B. (2013). The temporal orientation of memory: it's time for a change of direction. *Journal of Applied Research in Memory and Cognition*, *2*(4), 222–234.

Klein, S. B., Loftus, J., & Kihlstrom, J. F. (2002). Memory and temporal experience: the effects of episodic memory loss on an amnesic patient's ability to remember the past and imagine the future. *Social Cognition*, *20*(5), 353–379.

Martin, C. B., & Deutscher, M. (1966). Remembering. *The Philosophical Review, 75*(2), 161–196.

McTaggart, J. E. (1908). The unreality of time. *Mind, 17*(68), 457–474.

Michaelian, K. (2016). *Mental time travel: episodic memory and our knowledge of the personal past.* Cambridge, MA: MIT Press.

Nyberg, L., Kim, A. S., Habib, R., Levine, B., & Tulving, E. (2010). Consciousness of subjective time in the brain. *Proceedings of the National Academy of Sciences U S A, 107*, 22356–22359.

Okuda, J., Fujii, T., Ohtake, H., Tsukiura, T., Tanji, K., Suzuki, K., Kawashima, R., Fukuda, H., Itoh, M., & Yamadori, A. (2003). Thinking of the future and past: the roles of the frontal pole and the medial temporal lobes. *NeuroImage, 19*, 1369–1380.

Raichle, M. E., MacLeod, A. M., Snyder, A. Z., Powers, W. J., Gusnard, D. A., & Shulman, G. L. (2001). A default mode of brain function. *Proceedings of the National Academy of Sciences U S A, 98*, 676–682.

Renoult, L., Davidson, P. S. R., Palombo, D. J., Moscovitch, M., & Levine, B. (2012). Personal semantics: at the crossroads of semantic and episodic memory. *Trends in Cognitive Sciences, 16*(11), 550–558.

Russell, J., Alexis, D., & Clayton, N. S. (2010). Episodic future thinking in 3- to 5-year-old children: the ability to think of what will be needed from a different point of view. *Cognition, 114*, 56–71.

Scalise Sugiyama, M. (2011). The forager oral tradition and the evolution of prolonged juvenility. *Frontiers in Psychology, 2*(133).

Schacter, D. L., & Addis, D. R. (2007). The cognitive neuroscience of constructive memory: remembering the past and imagining the future. *Philosophical Transactions of the Royal Society of London B: Biological Sciences, 362*(1481), 773–786.

Schacter, D. L., Addis, D. R., & Buckner, R. L. (2007). Remembering the past to imagine the future: the prospective brain. *Nature Reviews Neuroscience, 8*(9), 657–661.

Schacter, D. L., Addis, D. R., Hassabis, D., Martin, V. C., Spreng, R. N., & Szpunar, K. K. (2012). The future of memory: remembering, imagining, and the brain. *Neuron, 76*, 677–694.

Shoemaker, S. (1968). Self-reference and self-awareness. *Journal of Philosophy, 65*, 555–567.

Spreng, R. N., & Levine, B. (2006). The temporal distribution of past and future autobiographical events across the lifespan. *Memory & Cognition, 34*, 1644–1651.

Suddendorf, T. (2013). Mental time travel: continuities and discontinuities. *Trends in Cognitive Sciences, 17*, 151–152.

Suddendorf, T., & Corballis, M. C. (1997). Mental time travel and the evolution of the human mind. *Genetic, Social, and General Psychology Monographs, 123*(2), 133–167.

Suddendorf, T., & Corballis, M. C. (2007). The evolution of foresight: what is mental time travel, and is it unique to humans? *Behavioral and Brain Sciences, 30*, 299–313.

Szpunar, K. K. (2010). Episodic future thought: an emerging concept. *Perspectives on Psychological Science, 5*(2), 142–162.

Szpunar, K. K., Spreng, R. N., & Schacter, D. L. (2014). A taxonomy of prospection: introducing an organizational framework for future-oriented cognition. *Proceedings of the National Academy of Sciences, 111*(52), 18414–18421.

Szpunar, K. K., St. Jacques, P. L., Robbins, C. A., Wig, G. S., & Schacter, D. L. (2014). Repetition-related reductions in neural activity reveal component processes of mental simulation. *Social Cognitive and Affective Neuroscience, 9*, 712–722.

Szpunar, K. K., Watson, J. M., & McDermott, K. B. (2007). Neural substrates of envisioning the future. *Proceedings of the National Academy of Sciences U S A, 104*, 642–647.

Tulving, E. (1972). Episodic and semantic memory. In E. Tulving & W. Donaldson (eds.), *Organization of memory* (pp. 381–403). New York: Academic Press.

Tulving, E. (1983). *Elements of episodic memory.* Oxford: Oxford University Press.

Tulving, E. (1985). Memory and consciousness. *Canadian Psychology/Psychologie canadienne, 26*(1), 1–12.

Tulving, E. (2005). Episodic memory and autonoesis: uniquely human? In H. S. Terrace & -J. Metcalfe (eds.), *The missing link in cognition: origins of self-reflective consciousness* (pp. 3–56). Oxford: Oxford University Press.

Wheeler, M. A., Stuss, D. T., & Tulving, E. (1997). Toward a theory of episodic memory: the frontal lobes and autonoetic consciousness. *Psychological Bulletin, 121*(3), 331–354.

PART I

# Varieties of Future-Oriented Cognition

2

# Toward a Taxonomy of Future Thinking

KARL K. SZPUNAR, R. NATHAN SPRENG, AND DANIEL L. SCHACTER

The contents of the present volume clearly demonstrate that the topic of future-oriented mental time travel has made a significant impact in the fields of philosophy, psychology, and neuroscience. In this chapter, we evaluate future-oriented mental time travel in light of a related concept, namely prospection—the ability to represent what might happen in the future (Gilbert & Wilson, 2007). In so doing, our goal is to highlight that there are many ways in which people are able to think about or *mentally travel into* the future. On a daily basis, people think about and evaluate possible encounters with friends, colleagues, romantic partners, and even their future selves; form intentions to deliver messages, take prescribed medications, and pick up miscellaneous items at the grocery store; and plan daily routines, vacations, and savings strategies for retirement (see D'Argembeau, Renaud, & Van Der Linden, 2011). Despite the vast diversity in the ways in which people think about the future, we recently proposed that prospective cognition can be organized into four basic modes of future thinking (Szpunar, Spreng, & Schacter, 2014): *simulation* (construction of a detailed mental representation of the future); *prediction* (estimation of the likelihood of, and/or one's reaction to, a particular future outcome); *intention* (the mental act of setting a goal); and *planning* (the identification and organization of steps toward achieving a goal state). Historically, these modes of future thinking have been studied in relative isolation, and the main goal of our proposed framework is to bring these largely disparate lines of research and theory into meaningful discourse.

In addition to proposing that prospection can be parsed into four distinct categories, our framework also holds that these modes of future thinking can vary in terms of their representational contents. Specifically, the framework relates simulation, prediction, intention, and planning to two well-characterized types of memory or knowledge: episodic and semantic (Tulving, 1983, 2002). In this context, the term *episodic* is meant to refer to simulations, predictions, intentions, or plans in relation to specific autobiographical events that may take place in the future (e.g., thinking about an upcoming event that will take place next week). The term *semantic* is meant to refer to simulations, predictions, intentions, and plans that relate to more general or abstract states of the world that may arise in the future (e.g., thinking about what the environmental state of the world will be like 20 years from now; for a review, see Abraham & Bubic, 2015). Note, however, that not all instances of future thinking can be strictly classified as either episodic or semantic. Specifically, some instances of future thinking represent autobiographical states—general states of the world that are autobiographical in nature (e.g., imagining that one will attain one's career aspirations in the future). To accommodate such hybrid forms of future thinking, our framework conceptualizes the episodic-semantic distinction as a continuous variable. We provide relevant examples of hybrid forms of future thinking throughout.

The result of our approach to delineating prospection is a set of four basic modes of future-oriented cognition that may vary in the extent to which they draw on episodic and semantic knowledge structures (see Figure 2.1; see also Szpunar, Spreng, et al., 2014). In what follows, we (1) briefly elaborate on the various modes of future-oriented thinking that fall within the purview of our classification scheme, (2) demonstrate the manner in which our framework can be used to ask new questions about how different modes of future-oriented cognition interact with one another, and (3) discuss how our framework can be used to provide insights into deficits of future-thinking.

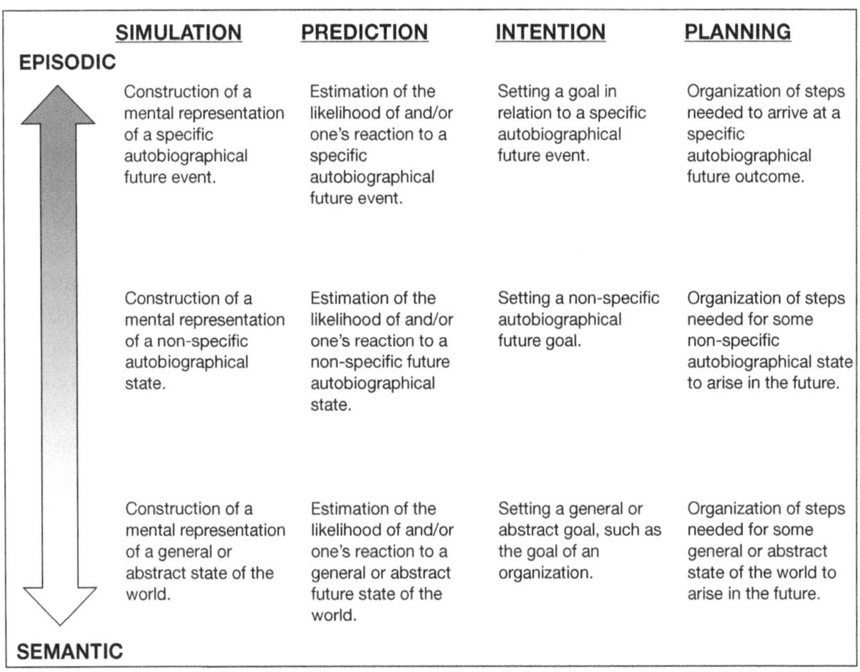


**Figure 2.1.** A taxonomy of prospective cognition.
SOURCE: Szpunar, K. K., Spreng, R. N., & Schacter, D. L. (2014, 18415). A taxonomy of prospection: introducing an organizational framework for future-oriented cognition. *Proceedings of the National Academy of Sciences U S A*, *111*, 18414–18421.

## 1. SIMULATION

The construct of simulation has been used to define the ability to cognitively represent various aspects of personal experience, including real or imagined events (Taylor & Schneider, 1989), the minds of others (Goldman, 2006), and person-environment interactions (Barsalou, 2003). Here, we take simulation to refer to the construction of hypothetical events, both specific and general.

### 1.1. Episodic Simulation

Episodic simulation is the construction of a detailed mental representation of a specific autobiographical future event (Schacter, Addis, & Buckner, 2008). Considerable work has focused on delineating the relation of episodic simulation to episodic memory and on identifying their common neural correlates. In a seminal case study, Tulving (1985) showed that an amnesic patient was unable to remember details about events from his past or to imagine details about possible events that might take place in his future. One limitation of this observation was that the patient had sustained diffuse brain damage following a motorcycle accident, and it was impossible to know what aspect(s) of the brain damage were related to his episodic memory and simulation deficits. More recently, Hassabis, Kumaran, Vann, and Maguire (2007) tested a set of patients with brain damage largely limited to the hippocampus, a region of the brain known to play an important role in memory for specific experiences, and found that those patients had deficits in constructing coherent mental images about the personal past and future. Although not all hippocampal amnesic patients demonstrate this pattern of results (e.g., Squire et al., 2010), there is good reason to believe that the hippocampus plays a pivotal role in memory for and simulation of specific autobiographical events (for a review, see Addis & Schacter, 2012).

Taking a broader approach, functional neuroimaging studies of episodic simulation and episodic memory provide insights into the neural underpinnings of these related abilities across the entire brain. For instance, using positron emission topography, a neuroimaging technique that requires researchers to inject a non-radioactive tracer into the bloodstream and to track the uptake of the tracer by blood vessels in the brain, Okuda et al. (2003) found that long periods of remembering and future thinking were characterized by similar patterns of brain activity in various frontal and temporal regions of the brain. The subsequent development of event-related functional magnetic resonance imagining (fMRI) (Dale & Buckner, 1997) enabled researchers to associate estimates of neural activity to specific memories and simulated events. Studies using event-related fMRI have demonstrated a striking overlap in medial and lateral regions of frontal, parietal, and temporal cortex as participants remember and imagine events from their past and future (Addis, Wong, & Schacter, 2007; Szpunar, Watson, & McDermott, 2007; for reviews, see Benoit & Schacter, 2015; Schacter et al., 2012). More recently, researchers have been able to parse the contributions of these regions to episodic simulation. Repetition suppression (also called neural adaptation) refers to a phenomenon whereby regions of the brain responsible for processing specific stimuli will show reduced patterns of responding following repeated presentations of those stimuli (Grill-Spector,

Henson, & Martin 2006). Capitalizing on this phenomenon, Szpunar, St. Jacques, Robbins, Wig, & Schacter (2014) asked participants to repeatedly simulate various aspects of future events (e.g., people and locations) and were able to pinpoint the contributions of stimulus-specific regions of the brain to simulation (e.g., parahippocampal cortex and retrosplenial cortex, regions known to be involved in spatial processing, were particularly sensitive to repeated simulations of simulated locations; see also Hassabis et al., 2014).

The collection of neuropsychological and neuroimaging findings relating episodic simulation to episodic memory have sparked various theoretical considerations. Schacter and Addis (2007) proposed the *constructive episodic simulation hypothesis*, which postulates that a constructive memory system that is prone to errors may nonetheless be adaptive in that it enables the individual to draw flexibly upon elements of past experiences in the service of simulating novel future events. Others have focused on more broadly relating episodic memory and simulation to other constructs. For instance, Buckner and Carroll (2007) identified episodic simulation and episodic memory as forms of self-projection, which, along with theory of mind (i.e., the ability to take the mental perspective of another person; Goldman, 2006) and spatial navigation (i.e., the ability to imagine moving through a familiar spatial location; Ekstrom et al., 2003), allow the individual to experience mental states that are removed from the immediate environment. Still others have focused on illuminating specific cognitive mechanisms involved in episodic memory, simulation, and related constructs such as spatial navigation. For instance, Hassabis and Maguire (2007; see also, Mullally & Maguire, 2014) argue that scene construction—the process of mentally generating and maintaining a complex and coherent scene or event—is a key cognitive mechanism that involves retrieving and integrating disparate details stored in modality-specific regions, a process believed to be mediated by the hippocampus.

## 1.2. Semantic Simulation

Semantic simulation is the construction of a detailed mental representation of a general or abstract state of the world. Although semantic simulation has not received as much research attention as episodic simulation, evidence suggests that the two may be, at least to some extent, dissociable from one another. For instance, Klein, Loftus, and Kihlstrom (2002) found that an episodic amnesic patient who was unable to reliably express what he might do in the future was nonetheless able to think about problems that might face the world in the future (e.g., global warming). Race, Keane, and Verfaellie (2013) similarly showed that patients with medial temporal lobe damage were able to generate possible issues that the world would face in the future, but also found that their patients were impaired in their ability to elaborate on those issues. This latter finding suggests that episodic amnesic patients may possess more fine-grained deficits in semantic simulation. Notably, Manning, Denkova, and Unterberger (2013) reported preserved episodic simulation but impaired semantic simulation in a patient whose left temporal lobe, a region known to play an important role in representing general or semantic knowledge (e.g., De Renzi, Liotti, & Nichelli, 1987), was resected following epileptic seizures localized to that portion of his brain. The available data suggest that damage to the hippocampus and more anterior portions of the temporal lobe may respectively lead to dissociable deficits of episodic and semantic simulation. At the same time, it is important to note that episodic and semantic knowledge may interact in the context of simulation. For instance, Irish and colleagues (Irish, Addis, Hodges, & Piguet, 2012; Irish & Piguet, 2013) have demonstrated that episodic simulation is also impaired with damage to the anterior temporal lobes, a finding that has led to the proposal that semantic knowledge may serve as a scaffold for episodic simulation.

Some forms of simulation cannot be neatly classified as either episodic or semantic, but rather represent hybrids of episodic and semantic simulation. These hybrid forms of simulation may take the form of personal semantic knowledge (e.g., one may have an interest in business) that is projected into the future (e.g., one may envision playing an important role in business in the future). Such simulations are autobiographical but are not related to specific future episodes. Whereas considerable research has been devoted to investigating the relation of personal semantic knowledge to episodic and semantic memory (Renoult, Davidson, Palombo, Moscovitch, & Levine, 2012), to our knowledge no research has specifically examined the relation of hybrid forms of simulation to episodic and semantic simulation. Prior work has shown that the extent to which personal semantic knowledge is characterized by patterns of neural activity that are similar to episodic and semantic memory can depend on the nature of the information under investigation (e.g., autobiographical facts versus repeated events; Renoult et al., 2012). The extent to which such factors may influence the relation of hybrid forms of simulation to episodic and semantic simulation should provide insights into not only the nature of simulation, but also relations between episodic, semantic, and personal semantic knowledge.

## 2. PREDICTION

Prediction is a fundamental task of the brain (Bar, 2009; Friston & Kiebel, 2009). Predictions about the future include short-term predictions about what object may appear next in a scene (Bar, 2007) and prediction errors concerning expected rewards that are crucial to learning (Schultz, Dayan, & Montague, 1997). Our focus is on longer-term predictions about specific events and general or abstract states of the world that may arise in the future and the manner in which such predictions may interact with other modes of future thinking.

### 2.1. Episodic Prediction

Episodic prediction is the estimation of the likelihood of, and/or one's reaction to, a specific autobiographical future event. Here, we focus specifically on predictions about reactions to events (for further detail regarding predictions of event likelihood, see the following section on Semantic Prediction; see also Szpunar, Spreng, et al., 2014). Perhaps the most clear-cut example of episodic prediction of reactions to events comes from work on affective forecasting, wherein social psychologists have attempted to highlight the shortcomings of the ability to predict our reactions to the outcomes of future events (Wilson & Gilbert, 2003, 2005). Gilbert and

Wilson (2007) argued that people base their predictions on episodic simulations of the future and succinctly summarized three common limitations of simulations that lead to errors in prediction. Specifically, the authors highlighted that (1) simulations are often based on easily accessible but unrepresentative memories of similar experiences (e.g., remembering a particularly negative experience of having interacted with a particular individual when predicting one's emotional reaction to having to interact with that person in the future); (2) simulations often focus on essential details (e.g., the joys of parenthood) while omitting inessential details that can influence future happiness (e.g., how it will feel to change diapers or go to work not having slept the previous night); and (3) simulations are often abbreviated and tend to focus on the initial aspects of future experience while omitting consideration of how circumstances may change over time (e.g., imagining the positive feeling associated with getting a promotion but failing to consider the extra work that would come with it). Recent studies have shown that people are able to generate more detailed episodic simulations of the future when they have been trained previously to report on episodic details in the context of a cognitive interview about a recently experienced event (Madore, Gaesser, & Schacter, 2014; for review and discussion of induction studies, see Schacter & Madore, in press). Whether such inductions of specificity could be used to enhance predictive accuracy awaits future research. For instance, it may be the case that inductions help to bring to mind non-essential details of simulations that are typically omitted but that could serve to enhance predictive accuracy. Later, we will consider the potentially important implications that episodic prediction might have on other modes of future thinking (i.e., intention and planning).

## 2.2. Semantic Prediction

Semantic prediction is the estimation of the likelihood of, and/or one's reaction to, a general or abstract state of the world. Semantic prediction represents an important type of future thinking that is highly valued in disciplines such as politics and economics. Research within the domain of affective forecasting has considered not only the manner in which people believe they will react to future events, but also how they may react to future states of the world, such as those brought about in the aftermath of elections. In these contexts, the shortcomings that hamper episodic simulations appear to similarly limit semantic predictions. For instance, Gilbert, Pinel, Wilson, Blumberg, and Wheatley (1998) showed that people overestimated the duration of their emotional reactions to the end of personal relationships and the outcomes of elections. Although we focus on predictions of reactions to events and changes in the state of the world, predictions about the likelihood of the emergence of future events and states of the world also represent an important aspect of prediction. Indeed, interesting questions about individual differences in predictions related to personal events and general states of the world remain to be answered. For instance, are people who are good at predicting the occurrence of personal experiences also good at predicting the occurrence of changes in general states of the world? Such questions can be further extended to the relation of episodic and semantic prediction to predictions about autobiographical states. For instance, how well are people able to predict success/failure in their pursuit of vocational goals?

Moreover, do predictions for future autobiographical states involve similar or different mechanisms as compared to those that are involved in predicting personal events or general states of the world?

## 3. INTENTION

Understanding the conscious determinants of human action has represented a central fixture in psychological research since the cognitive revolution (Ryan, 1970). Considerable research has been devoted to illuminating the nature of the underlying intentions that guide the behavior of the individual (Ajzen, 1991) and also the more general goal setting that is the driving force behind the growth and development of organizations (Locke & Latham, 2002). Next, we provide a brief overview of specific and more general or abstract intentions and identify possible links to other modes of future-oriented cognition.

### 3.1. Episodic Intention

Episodic intention is the mental act of setting a goal in relation to a specific autobiographical future event. Once goals are contemplated and set, whether or not those goals are actualized represents an important obstacle to adaptive behavior. Hence, one important aspect of setting intentions is remembering that those intentions have been set in the first place. Consider the role of setting and actualizing goals in the context of prospective memory (e.g., I need to remember to pick up bread on my way home from work; for a review, see Kliegel, McDaniel, & Einstein, 2008). Studies of prospective memory have long investigated the extent to which the quality of encoding of episodic intentions predicts the success with which those intentions are carried out in the future. Early research on implementation intentions found that explicitly stating when and where an intention will be carried out (e.g., when *X* occurs, I will perform *Y*) enhanced prospective memory performance (for a review, see Gollwitzer, 1999). Notably, many studies employing implementation intention instructions required participants to generate episodic simulations that revolved around themselves carrying out the future task at hand (Chen et al., 2015), and recent studies that explicitly relate episodic simulation to prospective memory performance have reported similar benefits (e.g., Neroni, Gamboz, & Brandimonte, 2014). It is important to note that implementation intentions may be effective in part because they involve formulating a plan of action. However, as we will highlight later, most conceptualizations of planning ability extend beyond simple if-then statements characteristic of implementation intentions to include the organization of complex action sequences.

### 3.2. Semantic Intention

Semantic intention is the mental act of setting a general or abstract goal, such as the goal of an organization. To our knowledge, no work has been conducted in the future thinking literature to identify the cognitive determinants that underlie the formation of a general or abstract intention for a particular organization (for example, setting a target of 10% industry growth). The development of research programs aimed at identifying the possible overlapping and non-overlapping cognitive and neural mechanisms that give rise to episodic and semantic intentions should have considerable implications for various fields of study, including psychology, business, and economics. As with simulation and prediction, not all instances of intention can be classified as episodic or semantic. For instance, one may possess an inherent interest in business (personal semantics; Renoult et al., 2012) and project that information forward by forming the intention to pursue a relevant career path. The extent to which such hybrid intentions rely on similar or different mechanisms compared with more clear-cut examples of episodic and semantic intention awaits future work. For instance, to what extent might similar mechanisms underlie the acts of forming intentions to pick up bread from the grocery store and to pursue a particular career?

## 4. PLANNING

For intended behaviors to be carried out in an effective manner, plans are often necessary. Although various scholars have defined the concept of planning (Hayes-Roth & Hayes-Roth, 1979; Mumford, Schultz, & Van Doorn, 2001; Ward & Morris, 2005), most definitions commonly conform to the notion of a plan as a "predetermination of a course of action aimed at achieving some goal" (Hayes-Roth & Hayes-Roth, 1979). Here, we focus on the nature of plans that are aimed at achieving goals in relation to specific autobiographical and more general contexts, and the extent to which other modes of future-oriented cognition may factor into the planning process.

### 4.1. Episodic Planning

Episodic planning is the identification and organization of steps needed to arrive at a specific autobiographical future event or goal state. One notable aspect of research on episodic planning is that tasks that are used to gauge planning ability can vary considerably in terms of how well they approximate real-life planning. For instance, in the Tower of London test, participants are typically presented with two arrays of beads organized on two separate sets of pegs, with the goal of reorganizing the beads presented in the first array to match the arrangement of the second, or goal, array. The measure of interest is the number of steps taken to achieve the goal state (Shallice, 1982). Other tests strive to more closely mimic real-world planning. For instance, the Six Elements Test (SET) requires participants to carry out a series of laboratory tasks in a specified time frame and order. The Multiple Errands Test is similar to the SET but involves completing daily tasks in a real-world setting (see Shallice & Burgess, 1991). Much of the research on planning has been conducted in the context of testing patients with frontal lobe damage. The fact that frontal lobe patients have trouble with each of the above-noted tasks highlights that processes subserved by the frontal lobes (e.g., executive control) play an important role in planning and future thinking more generally. More recent neuroimaging studies (Gerlach, Spreng, Madore, & Schacter, 2014; Spreng, Gerlach, Turner, & Schacter, 2015 Spreng & Schacter, 2012; Spreng, Stevens, Chamberlain, Gilmore, & Schacter, 2010) have examined the neural underpinnings of episodic or autobiographical planning, using a task in which participants mentally construct personal plans containing specific steps in order to achieve particular personal goals (e.g., academic success, getting out of debt). Results indicate that episodic planning is associated with activity in the same core network of brain regions that has been linked previously to episodic simulation (the default network; Andrews-Hanna, Smallwood, & Spreng, 2014; Buckner, Andrews-Hanna, & Schacter, 2008; Spreng, Mar, & Kim, 2009). Moreover, activity in the default network during episodic planning was closely coupled with activity in a distinct frontoparietal control network (Niendam et al., 2012; Vincent, Kahn, Snyder, Raichle, & Buckner, 2008) that has been linked to executive control and goal-directed cognition, and that also supports planning performance on the Tower of London task (Spreng et al., 2010). Finally, real-world episodic planning tasks have further illuminated the extent to which episodic planning draws upon other modes of future thinking. For instance, one study showed that participant descriptions of strategies for completing a pseudo shopping-planning task included simulation, prediction, and intention formation (Simons, Coates, & Channon, 2005). We address this point in further detail in the next section.

## 4.2. Semantic Planning

Semantic planning is the identification and organization of steps needed for some general or abstract goal state in the world to arise in the future. Semantic planning is perhaps best represented in the context of strategic (Blatstein, 2012) and urban planning (Rydin et al., 2012), tasks during which steps required to achieve a particular goal state for an organization or community are explicitly mapped out. Although there exists a paucity of research related to the cognitive determinants of semantic planning, research with frontal lobe patients has provided some initial data. For instance, patients with frontal lobe damage exhibit difficulty in formulating both episodic plans (see previous section) and financial plans for others (Goel, Grafman, Tajik, Gana, & Danto, 1997). It is important to point out that financial planning, and possibly other forms of semantic planning, may incorporate episodic knowledge (e.g., people may use their own experiences to formulate plans for others; see Goel et al., 1997, 1882). Nonetheless, the results of this study suggest that cognitive functions subserved by the frontal lobes may be important for episodic and semantic planning. To our knowledge, however, no study has directly compared episodic and semantic planning deficits in frontal lobe patients, and so more work is needed to understand the extent to which these forms of planning are supported by similar and different mechanisms. Indeed, the broader literature on planning does not generally distinguish between episodic and semantic planning (for a review, see Mumford et al., 2001) or the extent to which the two may work together in particular contexts (e.g., retirement planning; Wang & Shi, 2014), and research along these lines has the potential to answer novel questions regarding individual differences in planning ability. For example, are people generally good planners, or can someone be a good semantic planner but a poor episodic planner, and vice versa?

## 5. INTERACTIONS BETWEEN MODES OF FUTURE THINKING

We have already alluded to the fact that simulated representations of the future can be utilized in the context of prediction, intention, and planning. However, the extent to which the level of detail associated with a simulation influences performance in other domains remains to be investigated, particularly in relation to prediction and planning. For example, could specificity inductions (Madore et al., 2014) be used to improve the quality of plans? In a similar vein, next to nothing is currently known about the extent to which predictions associated with the likelihood of the occurrence of a future event might influence the formation and retention of intentions or the quality of plans. For instance, is an intention less likely to influence behavior if the perceived probability of occurrence is less than certain? The development of research programs that strive to answer these and related questions will represent an important step forward for the study of future-oriented cognition.

## 6. APPLICATIONS

The organizational framework discussed here provides a benchmark against which clinical populations with deficits in future thinking can be assessed in order to develop a profile of their future-thinking abilities. For instance, some episodic amnesic patients are able to think about the future in more general terms (Klein et al., 2002), although they may have difficulty generating semantic details of future events (Race et al., 2013). Nonetheless, more work is needed to further understand the extent to which such patients are able to engage in semantic prediction and semantic planning. Similarly, although it is well known that frontal lobe patients exhibit considerable deficits in episodic and semantic planning, much less is known about the extent to which such patients are able to generate useful simulations and predictions of episodic and semantic future events (Berryhill, Picasso, Arnold, Drowos, & Olson, 2010). Finally, studies of mood and anxiety-related disorders have historically focused on the fluency with which individuals with depression and anxiety think about positive and negative events that may occur in the future (MacLeod, Tata, Kentish, & Jacobsen, 1997). Although this work has provided important insights into future thinking in these populations, relatively little is known about the extent to which these individuals are able to engage in episodic and semantic forms of simulation, prediction, intention, and planning (for relevant work on episodic predictions, see Marroquín & Nolan-Hoeksema, 2015; Martin & Quirk, 2015; Wenze, Gunthert, & German, 2012). Development of research programs that consider the role of these various modes of future thinking could enhance our understanding of the ability of individuals afflicted with various mood and anxiety disorders to engage in adaptive behavior (e.g., how well are individuals with depression or anxiety able to predict their reactions to future events? How well are they able to formulate episodic or semantic plans for the future?).

## 7. CONCLUSIONS

In this chapter we have summarized a framework for organizing key cognitions involved in thinking about or mentally traveling into the future. We suggest that decomposing prospective cognition in terms of episodic, semantic, and hybrid forms of simulation, prediction, intention, and planning provides a useful framework for discriminating, and developing connections among, various forms of future thinking. Nonetheless, additional cognitive, emotional, and motor processes may contribute to and support future-oriented behavior (see Szpunar, Spreng, et al., 2014), which will likely lead to an expanded and refined taxonomy of prospective cognition. For instance, cognitive operations such as mental contrasting (e.g., Oettigen, 2012) may play an important role in decision-making about the future in contexts wherein predictions about the future are compared with considerations of present states (e.g., delay of gratification, Metcalfe & Mischel, 1999; temporal discounting, Benoit, Gilbert, & Burgess, 2011). For now, we believe that our organizational framework can encourage the cross-fertilization of research and theory across various domains of prospective, clinical, developmental, and comparative psychology, and thereby both broaden and deepen our understanding of the processes that support mental time travel into the future.

## REFERENCES

Abraham, A., & Bubic, A. (2015). Semantic memory as the root of imagination. *Frontiers in Psychology, 6*, Article 325.

Addis, D. R., Wong, A. T., & Schacter, D. L. (2007). Remembering the past and imagining the future: common and distinct neural substrates during event construction and elaboration. *Neuropsychologia, 45*, 1363–1377.

Addis, D. R., & Schacter, D. L. (2012). The hippocampus and imagining the future: where do we stand? *Frontiers in Human Neuroscience, 5*, 173.

Ajzen, I. (1991). The theory of planned behavior. *Organizational Behavior and Human Decision Processes, 50*, 179–211.

Andrews-Hanna, J. R., Smallwood, J., & Spreng, R. N. (2014). The default network and self-generated thought: Component processes, dynamic control, and clinical relevance. *Annals of the New York Academy of Sciences, 1316*, 29–52.

Bar, M. (2007). The proactive brain: using analogies and associations to generate predictions. *Trends in Cognitive Sciences, 11*, 280–289.

Bar, M. (2009). The proactive brain: memory for predictions. *Philosophical Transactions of the Royal Society B: Biological Sciences, 364*, 1235–1243.

Barsalou, L.W. (2003). Situated simulation in the human conceptual system. *Language and Cognitive Processes, 18*, 513–562.

Benoit, R. G., Gilbert, S. J., & Burgess, P. W. (2011). A neural mechanism mediating the impact of episodic prospection on farsighted decisions. *Journal of Neuroscience, 31*, 6771–6779.

Benoit, R. G., & Schacter, D. L. (2015). Specifying the core network supporing episodic simulation and episodic memory by activation likelihood estimation. *Neuropsychologia, 75*, 450–457.

Berryhill, M. E., Picasso, L., Arnold, R., Drowos, D., & Olson, I. R. (2010). Similarities and differences between parietal and frontal patients in autobiographical and constructed experience tasks. *Neuropsychologia, 48*, 1385–1393.

Blatstein, I. M. (2012). Strategic planning: predicting or shaping the future? *Organization Development Journal, 30*, 31–38.

Buckner, R. L., Andrews-Hanna, J. R., & Schacter, D. L. (2008). The brain's default network: anatomy, function, and relevance to disease. *Annals of the New York Academy of Sciences, 1124*, 1–38.

Buckner, R. L., & Carroll, D. C. (2007). Self-projection and the brain. *Trends in Cognitive Sciences, 11*, 49–57.

Burgess, P., Simons, J. S., Coates, L. M. A., & Channon, S. (2005). The search for specific planning processes. In R. Morris & G. Ward (eds.), *The cognitive psychology of planning* (pp. 199–227). New York: Psychology Press.

Chen, X., Wang, Y., Liu, L., Cui, J., Gan, M., Shum, D. H. K., et al. (2015). The effect of implementation intention on prospective memory: a systematic and meta-analytic review. *Psychiatry Research, 226*, 14–22.

Dale, A. M., & Buckner, R. L. (1997). Selective averaging of rapidly presented individual trails using fMRI. *Human Brain Mapping, 5*, 329–340.

D'Argembeau, A., Renaud, O., & Van der Linden, M. (2011) Frequency, characteristics, and functions of future-oriented thoughts in daily life. *Applied Cognitive Psychology, 35*, 96–103.

De Renzi, E., Liotti, M., & Nichelli, P. (1987). Semantic amnesia with preservation of autobiographical memory: a case report. *Cortex, 23*, 575–597.

Ekstrom, A. D., Kahana, M. J., Caplan, J. B., Fields, T. A., Isham, E. A., Newman, E. L., et al. (2003). Cellular networks underlying human spatial navigation. *Nature, 425*, 184–188.

Friston, K., & Kiebel, S. (2009). Predictive coding under the free-energy principle. *Philosophical Transactions of the Royal Society B: Biological Sciences, 364*, 1211–1221.

Gerlach, K. D., Spreng, R. N., Madore, K. P., & Schacter, D. L. (2014). Future planning: default network activity couples with frontoparietal control network and reward-processing regions during process and outcome simulations. *Social Cognitive and Affective Neuroscience, 9*, 1942–1951.

Gilbert, D. T., Pinel, E. C., Wilson, T. D., Blumberg, S. J., & Wheatley, T. P. (1998). Immune neglect: a source of durability bias in affective forecasting. *Journal of Personality and Social Psychology, 75*, 617–638.

Gilbert, D. T., & Wilson, T. D. (2007). Prospection: experiencing the future. *Science, 317*, 1351–1354.

Goel, V., Grafman, J., Tajik, J., Gana, S., & Danto, D. (1997). A study of the performance of patients with frontal lobe lesions in a financial planning task. *Brain, 120*, 1805–1822.

Goldman, A.I. (2006). *Simulating minds: the philosophy, psychology, and neuroscience of mindreading.* New York: Oxford University Press.

Gollwitzer, P. M. (1999). Implementation intentions: strong effects of simple plans. *American Psychologist, 54*, 493–503.

Grill-Spector, K., Henson, R., & Martin, A. (2006). Repetition and the brain: neural models of stimulus-specific effects. *Trends in Cognitive Sciences, 10*, 14–23.

Hassabis, D., Kumaran, D., Vann, S. D., & Maguire, E. A. (2007). Patients with hippocampal amnesia cannot imagine new experiences. *Proceedings of the National Academy of Sciences U S A, 104*, 1726–1731.

Hassabis, D., & Maguire, E. A. (2007). Deconstructing episodic memory with construction. *Trends in Cognitive Sciences, 11*, 299–306.

Hassabis, D., Spreng, R. N., Rusu, A. A., Robbins, C. A., Mar, R. A., & Schacter, D. L. (2014). Imagine all the people: how the brain creates and uses personality models to predict behavior. *Cerebral Cortex, 24*, 1979–1987.

Hayes-Roth, B., & Hayes-Roth, F. (1979). A cognitive model of planning. *Cognitive Science, 3*, 275–310.

Irish, M., Addis, D. R., Hodges, J. R., & Piguet, O. (2012). Considering the role of semantic memory in episodic future thinking: evidence from semantic dementia. *Brain, 119*, 2178–2191.

Irish, M., & Piguet, O. (2013). The pivotal role of semantic memory in remembering the past and imagining the future. *Frontiers in Behavioral Neuroscience, 7*, 27.

Klein, S. B., Loftus, J., & Kihlstrom, J. F. (2002). Memory and temporal experience: the effects of episodic memory loss on an amnesic patient's ability to remember the past and imagine the future. *Social Cognition, 20*, 353–379.

Kliegel, M., McDaniel, M. A., & Einstein, G. O. (2008). *Prospective memory: cognitive, neuroscience, developmental, and applied perspectives.* Mahwah, NJ: Lawrence Erlbaum Associates.

Locke, E. A., & Latham, G. P. (2002). Building a practically useful theory of goal setting and task motivation: a 35-year odyssey. *American Psychologist, 57*, 705–717.

MacLeod, A. K., Tata, P., Kentish, J., Jacobsen, H. (1997). Retrospective and prospective cognitions in anxiety and depression. *Cognition and Emotion, 11*, 467–479.

Madore, K. P., Gaesser, B., & Schacter, D. L. (2014). Constructive episodic simulation: dissociable effects of a specificity induction on remembering, imagining, and describing in young and older adults. *Journal of Experimental Psychology: Learning, Memory, and Cognition, 40*, 609–622.

Manning, L., Denkova, E., & Unterberger, L. (2013). Autobiographical significance in past and future public semantic memory: a case-study. *Cortex, 49*, 2007–2020.

Marroquín, B., & Noeln-Hoeksema, S. (2015). Event prediction and affective forecasting in depressive cognition: using emotion as information about the future. *Journal of Social and Clinical Psychology, 34*, 117–134.

Martin, S. M., & Quirk, S. W. (2015). Social anxiety and the accuracy of predicted affect. *Cognition and Emotion, 29*, 51–63.

Metcalfe, J., & Mischel, W. (1999). A hot/cool-system analysis of delay of gratification: dynamics of willpower. *Psychological Review, 106*, 3–19.

Mullally, S. L., & Maguire, E. A. (2014). Memory, imagination, and predicting the future: a common brain mechanism? *Neuroscientist, 20*, 220–234.

Mumford, M. D., Schultz, R. A., & Van Doorn, J. R. (2001). Performance in planning: processes, requirements, and errors. *Review of General Psychology, 5*, 213–240.

Neroni, M. A., Gamboz, N., & Brandimonte, M. A. (2014). Does episodic future thinking improve prospective remembering. *Consciousness and Cognition, 23*, 53–62.

Niendam, T. A., Laird, A. R., Ray, K. L., Dean, Y. M., Glahn, D. C., & Carter, C. S. (2012). Meta-analytic evidence for a superordinate cognitive control network subserving diverse executive functions. *Cognitive Affective Behavioral Neuroscience, 12*, 241–268.

Oettingen, G. (2012). Future thought and behavior change. *European Review of Social Psychology, 23*, 1–63.

Okuda, J., Fujii, T., Ohtake, H., Tsukiura, T., Tanjii, K., Suzuki, K., et al. (2003). Thinking of the future and past: the roles of the frontal pole and the medial temporal lobes. *NeuroImage, 19,* 1369–1380.

Race, E., Keane, M. M., & Verfaellie, M. (2013). Losing sight of the future: impaired semantic prospection following medial temporal lobe lesions. *Hippocampus, 23,* 268–277.

Renoult, L., Davidson, P. S. R., Palombo, D. J., Moscovitch, M., & Levine, B. (2012). Personal semantics: at the crossroads of sematic and episodic memory. *Trends in Cognitive Sciences, 16,* 550–558.

Ryan, T. A. (1970). *Intentional behavior.* New York: Ronald Press.

Rydin, Y., Bleahu, A., Davies, M., Davila, J. D., Friel, S., De Grandis, G., et al. (2012). Shaping cities for health: complexity and the planning of urban environments in the 21st century. *Lancet, 379,* 2079–2108.

Schacter, D. L., & Addis, D. R. (2007). The cognitive neuroscience of constructive memory: remembering the past and imagining the future. *Philosophical Transactions of the Royal Society B: Biological Sciences, 362,* 773–786.

Schacter, D. L, Addis, D. R., & Buckner, R. L. (2008). Episodic simulation of future events: concepts, data, and applications. *The Year in Cognitive Neuroscience, Annals of the New York Academy of Sciences, 1124,* 39–60.

Schacter, D. L., Addis, D. R., Hassabis, D., Martin, V. C., Spreng, R. N., & Szpunar, K. K. (2012). The future of memory: Remembering, imagining, and the brain. *Neuron, 76,* 677–694.

Schacter, D. L., & Madore, K. P. (in press). Remembering the past and imagining the future: identifying and enhancing the contribution of episodic memory. *Memory Studies.*

Schultz, W., Dayan, P., & Montague, P. R. (1997). A neural substrate of prediction and reward. *Science, 275,* 1593–1599.

Shallice, T. (1982). Specific impairments of planning. *Philosophical Transactions of the Royal Society B: Biological Sciences, 298,* 199–209.

Shallice, T., & Burgess, P. (1991). Deficits in strategy application following frontal lobe damage in man. *Brain, 114,* 727–741.

Spreng, R. N., Gerlach, K. D., Turner, G. R., & Schacter, D. L. (2015). Autobiographical planning and the brain: activation and its modulation by qualitative features. *Journal of Cognitive Neuroscience, 27,* 2147–2157.

Spreng, R. N., Mar, R. A., & Kim, A. S. (2009). The common neural basis of autobiographical memory, prospection, navigation, theory of mind and the default mode: a quantitative meta-analysis. *Journal of Cognitive Neuroscience, 21,* 489–510.

Spreng, R. N., & Schacter, D. L. (2012). Default network modulation and large-scale network interactivity in healthy young and old adults. *Cerebral Cortex, 22,* 2610–2621.

Spreng, R. N., Stevens, W. D., Chamberlain, J. P., Gilmore, A. W., & Schacter, D. L. (2010). Default network activity, coupled with the frontoparietal control network, supports goal-directed cognition. *NeuroImage, 31,* 303–317.

Squire, L. R., van der Horst, A. S., McDuff, S. G. R., Frascino, J. C., Hopkins, R. O., & Mauldin, K. N. (2010). Role of the hippocampus in remembering the past and imagining the future. *Proceedings of the National Academy of Sciences U S A, 107,* 19044–19048.

Szpunar, K. K., Spreng, R. N., & Schacter, D. L. (2014). A taxonomy of prospection: introducing an organizational framework for future-oriented cognition. *Proceedings of the National Academy of Sciences U S A, 111,* 18414–18421.

Szpunar, K. K., St. Jacques, P. L., Robbins, C. A., Wig, G. S., & Schacter, D. L. (2014). Repetition-related reductions in neural activity reveal component processes of mental simulation. *Social Cognitive and Affective Neuroscience, 9*, 712–722.

Szpunar, K. K., Watson, J. M., & McDermott, K. B. (2007). Neural substrates of envisioning the future. *Proceedings of the National Academy of Sciences U S A, 104*, 642–647.

Taylor, S. E. & Schneider, S. K. (1989). Coping and the simulation of events. *Social Cognition, 7*, 174–194.

Tulving, E. (1983). *Elements of episodic memory*. New York: Oxford University Press.

Tulving, E. (1985). Memory and consciousness. *Canadian Psychology, 26*, 1–12.

Tulving, E. (2002). Episodic memory: from mind to brain. *Annual Review of Psychology, 53*, 1–25.

Vincent, J. L., Kahn, I., Snyder, A. Z., Raichle, M. E., & Buckner, R. L. (2008). Evidence for a frontoparietal control system revealed by intrinsic functional connectivity. *Journal of Neurophysiology, 100*, 3328–3342.

Wang, M., & Shi, J. (2014). Psychological research on retirement. *Annual Review of Psychology, 65*, 209–233.

Ward, R., & Morris, G. (2005). *The cognitive psychology of planning*. New York: Psychology Press.

Wenze, S. J., Gunthert, K. C., & German, R. E. (2012). Biases in affective forecasting and recall in individuals with depression and anxiety symptoms. *Personality and Social Psychology Bulletin, 38*, 895–906.

Wilson, T. D., & Gilbert, D. T. (2003). Affective forecasting. *Advances in Experimental Social Psychology, 35*, 345–411.

Wilson, T. D., & Gilbert, D. T. (2005). Affective forecasting: knowing what to want. *Current Directions in Psychological Science, 14*, 131–134.

**PART II**

# Relationships Between Future-Oriented Mental Time Travel and Episodic Memory

3

# Asymmetries in Subjective Time

DENIS PERRIN

This chapter tackles the issue of the relationship between backward and forward mental time travel (MTT). It makes a case for a moderate version of discontinuism with regard to the question of the processes involved in both forms of MTT. To do so, it provides a philosophical argument against the continuist position on episodic thought that is widely held by psychologists. More specifically, it points out and elaborates on the asymmetries in the epistemological and causal process that are intrinsic to the subjectivity feature—or so-called autonoeticity—of episodic thought.

## 1. CHARTING THE DEBATE BETWEEN CONTINUISM AND DISCONTINUISM

"Episodic future thought" (Atance & O'Neill, 2001)—hereafter referred to as "episodic anticipation"—designates the neurocognitive ability to mentally pre-experience personal particular events. It involves the ability to project oneself through subjective time toward an experiential perspective that is different from the present. It should be distinguished, on the one hand, from other forms of anticipatory thoughts—such as planning or prospective memory, which generally do not involve sensory imagination—and, on the other hand, from mere sensory imagination, which involves neither personal relevance nor temporal location (Szpunar, 2010; Szpunar & Tulving, 2011). Due to its similarity to the ability of episodic memory to re-experience a personal particular event in the past, Endel Tulving (1985) considers episodic future thought and episodic remembering as two sides of the same ability to mentally travel through subjective time. A major and still intensely debated issue is how both kinds of mental time travel are specifically related. This raises the question of the functional dependence of the simulation of the future on the remembering of the past. According to one—albeit questionable (Juskaneite et al., 2014; Klein, 2015; Schacter et al., 2012)—leading view, episodic memory is the neurocognitive system that provides raw material from the past to flexibly construct future episodes (Addis et al., 2007; Schacter & Addis, 2007; Szpunar et al., 2007). According to the "constructive episodic simulation hypothesis," re-experiences in episodic memory are constructed rather than replayed, and their fundamental processes of construction are precisely the ones engaged for pre-experienced personal scenarios. The latter is thought to emerge from a flexible recombination of episodic bits of mnemonic information. Whether past or future, personal particular events are considered products of the same processes of construction. In this respect, both backward and forward MTT are essentially symmetrical. "The continuity of mental time travel into the past and future" (Suddendorf & Corballis, 2009, 1318) would be a striking and basic neurocognitive feature. While this broadly symmetrical view of the relationship between episodic memory and episodic future thought commands fairly unanimous acceptance, evidence for some asymmetries between them has also been found (Schacter et al., 2012). But these differences are rarely seen as pertaining to the kinds of processes at work in each form of mental time travel.[1] In this chapter, I discuss the implication link that is usually taken for granted, from the shared constructive character of episodic memory and episodic anticipation to the overall identity of their respective processes of construction, and make a claim for the existence of different kinds of processes.

More specifically, I will expand upon the aforementioned debate between continuity and discontinuity. Against the widely held continuist[2] view, I favor discontinuism and will provide a philosophical argument for this. My argument aims to point out the differences in processing that distinguish episodic memory and episodic anticipation. It will demonstrate that discontinuism is the preferable view with regard to a pivotal aspect of mental time travel, namely its specific subjectivity. I believe that this way of tackling the debate is relevant because subjectivity is a feature that researchers generally consider as being unquestionably similar in both forms of mental travel. Most of them admit that Tulving's notion of autonoetic consciousness (1985, 2002a, b, 2005) properly captures this feature. In both forms of mental travel, the subject is thought to be aware of itself as having an experiential life extending across time and to shift from its present perspective to other perspectives, either past or future. I argue that this account is unsatisfactory.

In doing so, I have two primary goals in mind, which determine the specific theoretical perspective of this chapter. First, I intend to provide an analysis belonging to the field of the philosophy of psychology. I side with a naturalistic approach whereby any satisfactory philosophical account of mental life has to seriously take psychological studies into account. I thus aim to provide a critical discussion that draws on the literature on mental time travel and which is strongly grounded in empirical research. However, in employing philosophical tools, I would also like to adopt a critical stance that suggests reframing some significant tenets of studies in mental time travel. This chapter thus offers an example of speculative psychology—but speculation that is supported empirically. In the rest of this section, I will chart the continuism/discontinuism debate with a view to drafting a rough outline of the perspective I advocate.

I would first like to give a broad characterization of the positions I call *continuism* and *discontinuism* before refining them. According to continuism,

> Except for the respective temporal orientations of backward and forward mental time travel, the neurocognitive systems and/or processes at work in mental time travel are identical.

Both backward and forward MTT, in other words, are symmetrical with regard to their neurocognitive nature. On the contrary, according to discontinuism:

> In addition to the respective temporal orientations of backward and forward mental time travel, the neurocognitive systems and/or processes at play in mental time travel are different.

In other words, both backward and forward MTT are asymmetrical.

As I said, each of these views comes across in different ways. To begin with, continuism can assume one of two forms. As for strong continuism, except for the respective temporal directions of backward and forward mental time travel, there is no difference between the two. They are strictly symmetrical. One can find this view in the early literature devoted to mental time travel (e.g., Suddendorf & Corballis, 1997, 149). But the more common—and now more widespread—form of continuism is more moderate. To introduce it, the classical distinction between differences in degree and nature is useful. On the one hand, for two entities to be different in *degree* means that they are of the same nature and thus possess the same features, although in different quantitative measures. For example, regarding intuitive analysis, two people can feel the same toothache in their respective mouths but with different degrees of intensity. They describe the same types of pain (e.g., throbbing, piercing, or general pain throughout the whole mouth) but can indicate different levels of pain on a scale of 1–10. More generally, differing degrees are properly described in terms of a quantitative numerical variation. On the other hand, two entities can be distinguished by a difference in their respective *natures*. For example, regarding intuitive analysis, two people have two mental experiences that are different in nature if one entertains the image of a landscape and the other a feeling of sadness. Although the degree/nature difference ultimately proves tricky when pressed further, this rough formulation will suffice for my argument.

According to moderate continuism, both backward and forward MTT are indeed different, but only by virtue of differences of degrees, such as the degree of vividness of the images that occur or the degree of the positive emotional valence (see review later in this chapter). On the contrary (and this is what separates discontinuism from continuism), discontinuism holds that there exist differences of nature between the two forms of temporal mental travel. Before developing what I mean by *nature*, it is worth noting that discontinuism can also assume different forms. Strong discontinuism rejects the existence of something such as one genus of the episodic thought that would allow one to mentally travel toward the past and the future. According to this, the alleged forms of mental time travel are in fact different in nature, one being memory and the other imagination—two mental faculties usually taken to be distinct. For instance, a replay view of memory is committed to such a form of discontinuism, since it argues that remembering reactivates a representational trace deposited in the brain, while episodic future thought consists of the activity of construction. What underlies the current literature on mental time travel is the rejection of strong discontinuism. Indeed, against the latter, moderate continuism claims that there are sufficient neurocognitive commonalities between the re- and pre-experience of personal events—in particular, their shared constructive character—to speak of *episodic thought* in both cases. But there exists another form of discontinuism, which I will introduce here.

In this chapter, I would like to explore and advocate a different stand from the ones mentioned. While I certainly accept the constructivist approach to episodic memory and the idea that episodic future thinking draws on episodic memory, I reject the idea that both backward and forward MTT are symmetrical. In my view, the constructivist idea whereby re- and pre-experiences are constructed mental scenarios has led to the mistaken idea that the processes at work for each cognitive achievement are generally similar. I thus favor a moderate form of discontinuism that is midway between strong discontinuism and moderate continuism. Moderate discontinuism argues that there are important differences between the two forms of mental time travel, which cannot be reduced to differences of degree, but that such differences do not suffice when it comes to blocking the claim that there is a unique genre of the episodic thought. To grasp the point of moderate discontinuism, one has to adopt the process view—acknowledged by most researchers in the field of mental time travel—according to which episodic future thinking and episodic remembering should be conceived as processes. Along these lines, I will conceive of the difference of the natures of backward and forward mental time travels as differences in *kinds* of underlying processes. My point is therefore to advance arguments that show that backward and forward mental time travel each rely on different kinds of processes—particularly regarding their allegedly shared type of subjectivity—against the claim of process identity upon which most researchers agree. What I mean by *process* should thus be specified.

*Process* is an oft-used but little defined term in psychology. To provide a brief definition, it refers to a neurocognitive activity that causally underlies cognitive functions. For instance, within the framework of Tulving's General Abstracting Processing System (1983, 1984), the function—or the cognitive achievement of the episodic memory system (namely, episodic remembering)—emerges from the ordered, interrelated series of the processes of a specific encoding, a specific recoding, an ecphory, and a conversion, which results in the mental state of remembering. As this example suggests, the notion should be refined in addition to the distinction between process/function, and an overall process should be distinguished from its component processes. Such components form a dynamically ordered and causally interrelated series from which emerges the cognitive achievement.[3] For instance, according to the GAPS, the ecphory component occurs after encoding and recoding, and it is causally dependent upon them. To summarize:

> ***Process*** is a series of dynamically ordered and causally interrelated neurocognitive components that make a subject able to carry out cognitive functions.

Provided that mental notions such as memory are conceived in terms of processes, the operating components involved and the way their respective operations are interrelated are definitional of such notions. Thus, series involving different components that are related in different ways arguably define different kinds of processes.

Now how does such a processing view bear on the continuism/discontinuism debate? Regarding the moderate continuism advocated by the constructive episodic simulation hypothesis, the two forms of mental time travel are on the whole identical when it comes to processing. As for my contention against such a claim, the two forms of MTT involve different components, and the components they share are arranged in a different causal and sequential order from other components.

According to the definition of process given earlier, this allows me to claim that both backward and forward MTT are *different when it comes to processing*—according to the classical idiom, they are different in nature. More specifically, my central claim focuses on the aspect of subjectivity and functions as follows. Autonoetic consciousness is certainly involved in both forms of mental time travel. However, it does not come into the picture the same way in both cases. Subjectivity is different in each case with regard to processing. Subjectivity in backward mental time travel is provided from the outset by autonoetic awareness, which is triggered through a process of detection.[4] For this reason, episodic remembering is essentially personal. In contrast, subjectivity in forward mental time travel is not automatically personal. While it has a first-person perspectival character, the self whose experience is simulated is not necessarily the self of the anticipating subject. Only a process of contingent assignment perhaps adds the personal feature. The two asymmetries I will eventually point out are intended to provide an argument for the claim of a difference in process and give the opportunity to explore the aforementioned differences. If I am on the right track, then I will manage to show a difference in kind regarding the subjectivity of backward and forward mental time travel and thus make a case for moderate discontinuism.

With a view to supporting the diagnosis that most researchers in the field of mental time travel strongly favor moderate continuism, I provide in this section a brief review of the similarities/dissimilarities put forward.

By now, well-known symmetries have been pinpointed in the literature on the subject. They have strengthened the claim that there is some legitimacy in using the same phrase *mental time travel*, regardless of the temporal orientation. They can be grouped as follows. First, phenomenological commonalities have been identified, such as the dependence of the wealth of context and detail on the assessed temporal distance of the represented event. Second, important symmetries have also occurred in cases of neurological and cognitive impairment. It has been shown that deficits in the ability to remember one's own past strongly correlate with the impaired ability to imagine one's own personal future (Hassabis et al., 2007; Klein et al., 2002; Robenbaum et al., 2005). Moreover, neuroimaging of healthy subjects confirms the claim that widely overlapping regions of the brain are solicited by both forms of mental time travel (Addis et al., 2007), which gives rise to the idea of a "core network" (or "default network") as the neural basis of mental time travel in general. Third, further symmetries are provided by developmental investigations. They show that the capacities to remember and imagine future personal scenarios emerge at roughly the same age (Atance, 2008; Perner et al., 2010; Viard et al., 2012). Fourth, both backward and forward mental time travel have been shown to display the same type of self-relatedness, both at the phenomenological level—allegedly encompassed by Tulving's concept of autonoeticity—and at the organization level, since it has been shown that the same life narrative structures organize autobiographical memories and autobiographical future events (Rathbone, Conway, & Moulin, 2011). Last but not least, the same processes are said to be involved in both cases (Race et al., 2011), in particular the linking processes that carry out the binding of episodic elements in a single represented episode (Schacter & Addis, 2007; Schacter et al., 2012), even if researchers have not reached an agreement with regard to the kinds of processes involved (see, for instance, the debate between Buckner & Carroll, 2007 and Hassabis & Maguire, 2007).

However, the neat picture of a symmetrical duplication with regard to the future of what episodic memory achieves with regard to the past has increasingly been qualified over the past few years (see Schacter et al., 2012). Does this imply a shift toward discontinuism? I would like to briefly review some key asymmetries here. Beyond the commonalities I have just mentioned, phenomenological discrepancies have been discovered. Remembered past events were found to be associated with richer and more vivid sensory and contextual details than imagined future events (D'Argembeau & Van Linden, 2004, 2006). A similar asymmetry marks the status of emotional valence, such that imagined personal future events display a positivity bias relative to remembered personal events (Berntsen & Bohn, 2010; Rasmussen & Berntsen, 2013). Moreover, neuroimaging studies have revealed that the construction of imagined scenarios are more cognitively demanding and solicit regions of the brain that are not solicited by remembering (Addis et al., 2007; Schacter & Addis, 2007, 781–782; Szpunar et al., 2007). This is generally interpreted as reflecting the additional effort of constructing imagery due to the element of novelty in the combination that emerges in a future scenario. Based on the aforementioned differences, functional asymmetries have been invoked. On the one hand, remembering and imagining personal events apparently fulfill different functions with respect to the self, emotion, and behavioral regulation (Rasmussen & Berntsen, 2013; Shao et al., 2010); from the internal point of view of how the episodic memory system functions, it is proposed that remembering has the functional significance of providing bits of information, the flexible recombination of which in simulated scenarios apparently allows human beings to optimally tune their behavior to the infinitely diverse situations that are likely to happen.

The striking feature of the aforementioned asymmetries is that they are almost never conceived as differences pertaining to the very nature of the cognitive processes involved in each of the two forms of mental time travel. For instance, phenomenological differences are scored on similar scales, according to vividness or emotional valence. Similarly, the asymmetry in cognitive effort is often couched in terms of "the more intensive constructive processes required by imagining future events relative to retrieving past events" (Schacter & Addis, 2007, 782), which suggests that the asymmetry under consideration merely consists of a higher degree of intensity of one and the same type of cognitive operation. This striking feature motivates the repeated claim for process identity in mental time travel studies. The following are a few examples (among many): there is a "considerable overlap in the psychological and neural processes involved in remembering the past and imagining the future" (Schacter & Addis, 2007, 774); "it should be possible to document a direct link between processes underlying memory distortion and those underlying mental simulations of the future" (782); "in view of the generative aspect of episodic memory, it seems reasonable to suppose that basically the same mechanisms might be involved in imagining the future as in constructing the past" (Suddendorf & Corballis, 1997, 149); "our memories for episodes are made up of combinations of people, actions, objects and places, along with qualities such as time of day, season, emotional states, and so forth. Imagined future events are similarly constructed" (Suddendorf & Corballis, 2009, 1320); and "episodic remembering and episodic future thinking are supported by the same neurocognitive processes" (Berntsen & Bohn, 2010, 265). On the whole, moderate continuism is therefore favored. Contrary to this position, I advocate a different one that is occasionally expressed, such as when Friedman says, "In my view, it remains an open question to what extent common processes underlie episodic memory and future-directed thinking in general" (Friedman, 2007, 323, concerning Suddendorf and Corballis's claim for continuity). Other researchers side with the notion of two forms of MTT that differ with regard to processing (Irish et al., 2012; Juskaneite et al., 2014; Klein, 2015; Klein & Steindam, Chapter of this volume; Rasmussen & Berntsen, 2013; and Debus, 2014, on the philosophical side). Expanding on a view that is clearly in the minority, I would like to provide a philosophical argument in favor of moderate discontinuism.

I will proceed as follows. First, I will delineate the target of the critical part of my argument, which I call the Continuity Thesis, regarding the aspect of subjectivity in episodic thought. I will then explain the strategy of my argument. Drawing on what I call the Actuality Claim, I propose to show that episodic thought is actuality sensitive such that the actual (respectively, inactual) nature of its object has causal and epistemological bearings on the kind of process that grounds the subjectivity feature of episodic memory and anticipation. In order to carry out this strategy, I will point out two asymmetries between both forms of mental time travel: one epistemological and the other causal. I will then proceed to explain that the pinpointed asymmetries arise from asymmetries of process and suggest that they are empirically plausible.

## 2. THE CONTINUITY THESIS AND THE ACTUALITY CLAIM

My target in this chapter is one of the main theses defended by moderate continuism, which I call the Continuity Thesis. It concerns the subjectivity involved in mental time travel. According to a common assumption of the literature inherited from Tulving (1985, 2002a, b, 2005), autonoeticity is a form of consciousness that allows one to project oneself toward both the past and the future. Being autonoetically aware of an event consists of the ability to mentally relate oneself to the event in an experiential way. Instead of simply knowing that an event happened in the past or will happen in the future, one mentally relives or pre-lives the event in question. In this way, one is not merely aware of an event experienced in the past or to be experienced in the future, but is aware of it as having been (or to be) experienced by oneself. In the remembering or anticipatory mode, one can typically see the occurrence of an event from a sensory and first-person embodied perspective. Autonoeticity thus provides a specific way for a piece of information to be subjective. Subjectivity does not consist here of the object of information, but of the specific way—sometimes called "flavor"—in which the information is accessed. It thus forms one specific component of what is nowadays known as complex and multilevel autobiographical memory (Conway & Pleydell-Pearce, 2000; Conway et al., 2004; Conway, 2009). Using more recent analyses (Szpunar, 2011; Tulving, 2005), Tulving distinguishes within his initial notion of autonoeticity (1985) between the capacity of autonoesis in the strict sense of the term and the capacity of chronesthesia (2002b). While the latter refers to the capacity of being aware of the subjective *time* in which one's conscious lives are extended, the former refers to the capacity of being aware of one's *self* as temporally extended. And the self who should be emphasized must be conceived in the experiencing way I described earlier.

According to Tulving, autonoeticity depends on the system of episodic memory. Provided episodic memory is defined as the neurocognitive system that "makes possible mental time travel through subjective time—past, present, and future" (2005, 9), autonoeticity contributes to the specific subjectivity of such an achievement. As Tulving puts it, "This mental time travel allows one, as an 'owner' of episodic memory ('self'), through the medium of autonoetic awareness, to remember one's own previous 'thought-about' experiences, as well as to 'think about' one's own possible future experiences." In short, autonoeticity provides the specific experiential sense of "mineness" of relived and pre-lived episodes. Suddendorf and Corballis take up this very idea when they say that, through the episodic memory operation, an episode is "related to the self in subjectively sensed time" (2009, 1318). Therefore, mental time travel, conceived as the achievement of the episodic memory system, centrally involves the process component of autonoeticity. The latter constitutes a necessary condition for one to be aware of the event as experienced, whether re- or pre-experienced. This view is summarized in the following thesis:

*The Continuity Thesis (CT)*

Whatever the temporal direction (backward or forward), mental time travel operations draw on the same capacity of autonoeticity in order to simulate and thereby mentally experience personal episodes.[5]

I would like to articulate the various elements of CT. First, the definition of autonoeticity it involves is as follows:

(*CT1*) Autonoeticity is the ability to be aware of a mentally experienced episode as belonging to the thinking subject's temporally extended self.

A corollary of CT1 that is crucial to my argument is the claim that the past/future self is the same as the self that exercises its autonoetic capacity. This is what I call the "personal" nature of episodes in CT. The corollary is the following:

(*CT2*) Autonoeticity implies the identity of the self whose experience is simulated with the simulating subject.

Drawing on CT1 and CT2, CT adopts the following position on the nature of the subjectivity of episodic thinking:

(*CT3*) The subjective character of any episodic thought is due to the capacity for autonoeticity.

Due to CT1 through CT3, any form of episodic thought implies the identity of the self whose personal episode is mentally experienced with the self mentally experiencing this episode. According to this characterization, CT thus puts one and the same constraint of the personal identity of the subject of the mentally experienced episode for backward and forward mental time travel. This sameness is what I call the *continuity feature* of mental time travel expressed by CT. My argument thus consists of discarding CT3.

To do so, I would first like to introduce some conceptual and theoretically rather neutral tools that will be useful for building my argument.[6] Within a cognitive process and its resulting mental state, I distinguish between the mode, the content, and the object. The mode specifies the type of cognitive activity or operation carried out by a process, such as memory, perception, reasoning, and so on. The content specifies the information transmitted by the process, such as the semantic memory information that Paris is the capital of France or the perceptual information that there is a tree in front of me. The object specifies the part of reality to which the content refers (e.g., Paris and France or the tree in front of me). Once these notions are provided, how do the elements to which they refer relate to each other? A much neglected though highly important issue, in my view, concerns the way the object's ontological status determines the content and the mode. Arguably, the mode and the content of the state could very well be different, depending on whether the object of a mental state is an actual fact or a mere possibility. According to the main stream of the existing literature on mental time travel, the actual/inactual contrast between the respective objects of episodic memory and episodic anticipation does not imply deep differences in the respective grounding processes of these cognitive achievements.[7] I object to this view, based on the following notion:

> *The Actuality Claim* (*AC*): A cognitive process is *actuality-* or *inactuality-sensitive* if, and only if, it covaries in kind with the actual or inactual character of the object of the cognitive function it grounds.

The respective objects of episodic anticipation and episodic memory have two distinct modal ontological statuses, as expressed by the following:

(*IF*) The object of episodic anticipation, as of any thought about the future, is *inactual* (*possible*).

(*AP*) The object of an episodic memory, as of any thought about the past, is *actual*.

The notion of actuality is a notoriously tricky and complex one. I do not claim to define it. As it is used here, it can be minimally developed in the following way. The property of being *actual* endows an entity with the capacity for bearing two kinds of relations.[8] On the one hand, it confers upon it the possibility of bearing efficient causal relations. A past event can have present effects, while a future one cannot. The *idea* of such an event could certainly have such causal efficiency. For instance, the idea of my next talk leads me to prepare my presentation. But the event itself, being merely possible, cannot have such efficiency—it will possibly have causal effects as it will become present, but as future it cannot. On the contrary, the occurrence of a past talk that has been recorded has as a present causal effect the existence of a recording, while no such record is conceivable for a future talk.[9] On the other hand, being *actual* endows an entity with the possibility of being experientially known. One can certainly say in a loose way that one knows that such a future event will take place. But for the sake of my argument, following the objection formulated by Dorothea Debus (2014) will suffice. According to her, experiential knowledge is a factive relation. It cannot occur without the existence of the alleged object of knowledge. In such a case, one is faced with a mere illusion of experiential knowledge. Episodic remembering can be said to be the experiential knowledge of a particular event, since it consists of reliving the past encounter of an actual event. But since events as future are events deprived of any occurrence (otherwise, they would be past or present), a relation of experiential knowledge cannot be projected onto them.[10]

With this characterization of the notion of actuality in mind, I claim that episodic thought processes turn out to be sensitive to the actual/inactual contrast pointed out by AP and IF. If this is the case, then episodic memory and episodic anticipation are grounded by two different kinds of processes, as desired by moderate discontinuism. This line of reasoning sets up the strategic role of the two asymmetries I will introduce. They provide an argument for the following contention. Regarding the subjectivity feature of episodic thought, the kinds of processes—a notion that should be conceived in terms of the aforementioned definition of the latter term—that ground episodic memory and anticipation covary with the actual/inactual contrast. Consequently, the Continuity Thesis must be rejected. Inasmuch as moderate continuism rests on this thesis, it must be rejected as well. Thus, moderate discontinuism should be favored. I would now like to move on to the asymmetries before expanding upon how they ground the contention of differences in the kinds of process.

## 3. FIRST ASYMMETRY: THE EPISTEMIC CRITERION

The first asymmetry claims that episodic anticipation and episodic memory are *de se* thoughts with different epistemic features. I call *de se* thought any thought that a subject is able to have about itself as about itself. For instance, to imagine oneself doing such and such a thing, to see oneself in a mirror, or to self-attribute a belief to oneself are instances of this kind of thought. Arguably, one must master a certain self-concept in order to be able to mentally travel through time, as it has been documented (Hoerl & McCormack, 1999; Moore & Lemmon, 2001; Suddendorf & Corballis, 1997). Minimally, one should master a concept according to which the self is a temporally extended entity. More specifically, this temporal extendedness should be conceived as a succession of experiential perspectives through which the self passes and between which it can shift by re- and pre-experiencing them mentally.[11] I will take this as a given here. In order to make my point about the first asymmetry, I need a further notion. The error of attributing to a person or a thing an identity that differs from the person's or the thing's actual identity is known as *error through misidentification*. For instance, if I see Jacques in the distance raking leaves in his garden and think, "Jean is raking leaves in his garden," I am the prey to an error through misidentification. I attribute to Jacques the identity of Jean. Such an error can also be made regarding oneself. In a famous anecdote, Ernst Mach recounts that as he saw himself in the mirror of a bus he was boarding, he thought about himself without realizing he was entertaining a thought about himself, "What a shabby professor!" His *de re* thought about himself was affected by an error through misidentification. Obviously, some *de se* thoughts can be the prey to error through misidentification, while others are immune to such an error.[12] For instance, I can confuse the arm of someone sitting close to me with my own arm, thus misidentifying the person to whom the arm belongs. In contrast, one can hardly misidentify the person experiencing pain when one feels pain in one's own arm.[13] To which category does episodic thought thus belong? I propose a distinct answer for thoughts in episodic memory and episodic anticipation.

Arguably, *de se* errors through misidentification are conceivable with regard to episodic memories. A famous thought experiment suggests imagining that an episodic memory can be transplanted from *A*'s mind to *B*'s mind unbeknownst to *B*.[14] *B* can then episodically remember *A*'s perceptions of past scenes while believing that he, *B*, episodically remembers some of his own past experiences. The episodic *de se* thought he entertains thus misidentifies the past subject whose perceptions it represents. In a possible world in which such misattributed episodic memories exist, a proper account of the content of episodic memories would include an identification component.[15] For instance, *B*'s episodic thought could be the following:

*s* broke his arm while he was jogging two days ago & *s* = me

Here, "*s* = me" is the (mistaken) identification component. Given what the actual world is like, the content of episodic memories does not have to include such a component. But this is due to the contingent fact that our world does not include transplantation technologies similar to the aforementioned one. Misattributed episodic memories are actually excluded but could potentially occur. The identification component actually secured by the conformation of the world might have to migrate into the content of our episodic memories. Therefore, the content of one's memory could be wrong about the identity of the subject whose past experience one re-experiences while remembering. Given that knowledge is minimally defined as a belief whose content about the fact it represents is true, it should be maintained that the content of a true (or false) episodic memory provides knowledge (or no knowledge) about the identity of the subject whose past experience is re-experienced, however guaranteed this knowledge may be. Thus, it should be said, one is *epistemologically* related to one's past self through episodic memory.

Is it similarly the case that *de se* errors through misidentification are (at least theoretically) possible in episodic anticipations? In other words, is the identity of the subject whose experience one anticipates episodically something about which one could be wrong? My answer is no. To prove this, I draw an epistemological distinction regarding the respective content of episodic memory and anticipation. To introduce this distinction I will use an example that is not *de se* but that permits us to set the distinction in a general manner. *De se* thoughts will then fall under the general distinction as a particular case. Let's consider the following report of an act of episodic remembering:

Frances comes closer to me and tips a full bucket of ice-cold water over me. I have one second of shock, long enough to see her turn and walk away. Then I feel the chill. (reported in Conway, 2008)

Let's also consider that this report corresponds to an event that the rememberer actually experienced. Suppose now that Frances has an identical twin sister, Jacqueline, and that the rememberer was actually assaulted by Jacqueline and not Frances. Because of the suggested mistake, the report admits two versions, an intensional one and an extensional one (Quine, 1956; Williams, 1973),[16] defined as follows:

> [*intensional*] If *s* remembers$_e$ *Frances* φ-ing, *Frances* belongs to the episodic memory content.
>
> [*extensional*] If *s* remembers$_e$ *Frances* φ-ing, whereas it was *Jacqueline*, *Jacqueline* belongs to the episodic memory content.

There can be a discrepancy between what one believes one is remembering and what one is actually remembering. For this reason, the rememberer could be compelled to acknowledge that the person he or she sees in his or her episodic remembering is Jacqueline and not Frances, as he or she thought in the first place. In other words, there can be a discrepancy between what he or she thinks his or her mnemonic content is about and what the latter is actually about and presents him or her with. Here the identity of the (real) object determines the identity of the content. The situation is different in the case of episodic anticipation. Consider the initial report as the description of a personal event the subject fears it will have to experience. In such a case:

> [*intensional*] If *s* anticipates$_e$ *Frances* φ-ing, *Frances* belongs to the episodic anticipation content.

But could there be an extensional counterpart of this intensional version, as in the previous case of episodic remembering? If so, it would read as follows:

> [*extensional*] If *s* anticipates$_e$ *Frances* φ-ing, whereas it will be *Jacqueline*, *Jacqueline* belongs to the episodic anticipation content.

Note first that the extensional version of the episodic anticipation makes sense to a certain extent. It could actually occur that some time after the occurrence of the episodic anticipation, the anticipated event happens exactly as it has been anticipated, except that it is Jacqueline, not Frances, who carries out the liquid assault. Yet it is dubious that one would accept, as the (quasi-exactly) anticipated event happens to occur, that the fact that Jacqueline φ-s was in any way part of the content of the anticipation. One would not be prepared to say, as the subject is anticipating, that it is anticipating Jacqueline φ-ing in spite of its belief that it is anticipating Frances φ-ing. Here the identity of the (real) object does not determine the identity of the content. The absence of a discrepancy in possible extensional-intensional versions makes an argument for the claim that there is no experiential knowledge relation between the subject and the object it anticipates through the content. Indeed, there is nothing in it about which the subject could conceivably be wrong, and therefore nothing it could be said to know. What the anticipation content is about is only what the subject decides it is about. Couched in terms sometimes used in philosophical literature, episodic anticipation is not *world-involving*, in contrast with episodic memory (Martin, 2001; Recanati, 2007). *World-involvingness* is a property of the contents of mental states. It consists of the possibility for the identity of content to be determined by the actual world independently from what the subject puts in the content and believes about it, and it thereby confers upon the content its epistemic significance of experiential knowledge of the world. It is minimally captured by the extensional-intensional discrepancy. Thus, as I have shown, episodic memory is world-involving, while episodic anticipation is not. As the subject anticipates the liquid assault, it cannot be wrong about the identity of the person whose experience it anticipates. It "stipulates" (Recanati, 2007) this identity, or this identity is fixed by the subject's "imaginative project" (Williams, 1973), which decides what does and does not figure in the content of the episodic anticipation. The contrast regarding world-involvingness is a general property of the respective contents of episodic remembering and episodic anticipation. It therefore concerns in particular the identity of the subject whose experience is re-, or pre-experienced. Thus *de se* immunity to error through misidentification is a necessary feature of the latter, while it is contingent as far as episodic memory is concerned.[17] On the whole, our diachronic relation to ourselves through episodic memory is epistemological, while it is not through episodic anticipation, in which it is *stipulative*.[18]

## 4. SECOND ASYMMETRY: THE CAUSAL CRITERION

The second asymmetry claims that episodic memory and episodic anticipation do not depend in the same way on the personal identity of the subject whose experience is mentally experienced.[19] To put it briefly, episodic memory mode causally depends on such an identity such that a mnemonic mental state must be directly caused by one of the rememberer's past experiences for it to be an episodic memory.[20] In contrast, the episodic anticipation mode does not display such a causal dependence. Indeed, provided one admits that a future experience can cause its anticipation—which is controversial (see earlier discussion)—the episodic anticipation mode remains the same, even though the subject whose experience is pre-experienced is different from the anticipating subject.

To elaborate on this point, I need to refine the notion of a thought about oneself. Some authors have proposed to distinguish between three main types of experiential thoughts about oneself (Recanati, 2007; Vendler, 1979; Williams, 1973).

i. ***De re* thoughts**: these are thoughts not accompanied by the subject's awareness that they are about it. They include in their content the self whose experience is represented—although they misidentify it. Mach's anecdote illustrates this first category.
ii. **Explicit *de se* thoughts**: these are *de se* thoughts accompanied by the subject's awareness that they are about it. Moreover, they include in their explicit content the self whose experience is represented. This category is illustrated by a thought of which the content is adequately expressed by "I imagine *myself* giving a talk."

iii. **Implicit *de se* thoughts**: these are *de se* thoughts accompanied by the subject's awareness that they are about it. In contrast with the previous category, however, the identity of the subject whose experience is represented is not included in the content of such thoughts. This category is illustrated by a thought of which the content is adequately expressed by "I imagine giving a talk."

A key feature of these kinds of thoughts is that they all imply the identity of the thinking self and the self that is thought about. Arguably, episodic memory and episodic anticipation belong in category *ii* or *iii*. But it would be worth introducing an additional kind of thought. Suppose you engage in the project of imagining being Napoleon. Such cognitive achievement could be said to differ, on the one hand, from imagining *oneself* being Napoleon, which would obviously amount to a metaphysical impossibility. It also differs, on the other hand, from the project of imagining oneself in the circumstances in which Napoleon happened to be, since such a project would not involve him. Preferably, one should say that the achievement under consideration consists of simulating Napoleon's experiential point of view. Now this achievement does not imply any reference to oneself on the part of the simulating subject, despite the fact that it is an achievement carried out by the self. Thus, just as transplanted episodic memories are not real *de se* thoughts, imagining being Napoleon is not a real *de se* thought, although it offers the possibility of mentally experiencing someone else's *de se* thought—by transplantation in one case and by simulation in the other. Hence the proposed name of this additional category (Recanati, 2007):

iv. **Quasi-*de se* thoughts**: these are thoughts simulating the *de se* thought of a different subject from the simulator. This category is illustrated by a thought of which the content is adequately expressed by "I imagine what it is like to be Napoleon and fighting on the Waterloo battlefield," or "I imagine what it is like to give a talk."

How do episodic memory and episodic anticipation behave with respect to this fourth category of thought? I argue that they behave in sharply contrasting ways.

It seems that an episodic memory cannot be quasi-*de se*. Suppose I believe I have an episodic memory of a scene *e* from my childhood. For instance, my parents repeatedly told me for years that I had experienced *e*. In fact, it was not me but my brother who experienced *e*. Nonetheless, I have vivid and detailed imagery about *e* accompanied by the awareness of having myself experienced *e*—my awareness of *e* being autonoetic. Two features of this type of mental state are worth noting. First, for the appearance of episodic memory to occur, I must have the belief that I am the subject whose past experience I represent. This identity belief is a condition of the episodic appearance. This suggests that the identity of selves is constitutive of episodic memory. Second, as suggested, such a mental state is a mere illusion of episodic memory, or a false episodic memory. Something is missing for it to be a real episodic memory. This something is a causal link between the current mental state and the represented past experience. Such a link is specific. It has to be direct or firsthand such that the current mental state must not derive from others' testimony or from inferences carried out by the subject, and it has to link to experiential stages of one and the same self. I would like to call the requirement of such a link the *causal constraint* proper to episodic memory.[21] For instance, imagine *A* experiences *e*, then recounts it to *B* before forgetting it. Imagine now that *B* recounts *e* to *A* later on, and *A* comes to believe (rightly) that she experienced *e*. If *A* has vivid images of *e* and therefore believes she episodically remembers *e*, according to the causal constraint she is mistaken. On the whole, for the mode of a mental state to be episodic memory, it is a necessary condition that the corresponding content causally derives in a direct way from a past experience of the remembering subject. By contraposition, if the content does not have this feature of causal derivation, the corresponding mode cannot be episodic memory. In the case of episodic memory, switching the self switches the mode. A quasi-*de se* episodic memory is *not* an episodic memory. In other words, the episodic memory mode is causally dependent on the personal identity of the subject whose experience it represents.

On the contrary, I argue, no such dependence occurs in the case of episodic anticipation. It is free from the causal constraint. To make my point, I will examine two types of cases that ground my claim. I call them the *uncertainty case* and the *self-substitution case*. Suppose first that someone from the board at your company will have to tell Jean that he is fired. You belong to this board. Not knowing whether and fearing that you will have to carry out this unpleasant task, you anticipate by imagining what this event will be like. The content of your sensory imaginative anticipation could be expressed as follows:

> I imagine what it must be like to tell Jean that he is fired, but I do not know whether I will have to do it.

In such a case, the personal identity of the person whose unpleasant experience you anticipate is arguably indeterminate. The following is as an example of a case of self-substitution. Your daughter will have to take an exam. You are very worried about it and want to help her. You thus imagine what your daughter will have to endure. The content could be expressed as follows:

> I imagine what it must be like to take the exam, but not as something I will have to endure myself.

In this case, the personal identity of the person whose future experience you anticipate is determinate. But it is not yours. Thus, both of these examples display cases of the mental pre-experience of future particular events, even though the anticipating self and the self whose future experience is anticipated differ from each other. Note that, at some point when you anticipate the firing of Jean, you could suppose that you will have to carry out the task, before supposing that Jacques will do it. Shifting from a case of identical to a case of different selves does not alter the mode of the mental state. From one case to the other, you continue to episodically anticipate a particular experiential event. Thus, in episodic anticipation, switching the self does *not* switch the mode, in sharp contrast to episodic memory. A plausible explanation is that episodic remembering is an epistemological mental state and can be considered as such because it is about actual events. Therefore, in order to provide the knowledge it seems to provide (the experiential firsthand knowledge of a particular event), it has to comply with the causal constraint. On the contrary, episodic anticipation is about possible particular events and thus cannot provide the experiential firsthand knowledge of such an event. For this reason, it does not have to comply with the causal constraint. Thus, the episodic anticipation mode does not depend on the personal identity of the future self whose experience is anticipated.

## 5. ARGUING FOR THE AUTONOETICITY/ EPISODICITY DISTINCTION

Both asymmetries have motivated the claim that the actuality of its object has specific epistemological and causal bearings on episodic memory and, conversely, that the inactuality of its object has specific epistemological and causal bearings on episodic anticipation.[22] As explained, these are different bearings. The role of this final section is to draw out a certain contention from the two asymmetries, namely that the respective subjectivity features of episodic memory and anticipation are grounded in different kinds of processes. I will also argue that this contention can be grounded on empirical data.

Episodic memory and episodic anticipation are reputedly driven by constructive processes. In particular, the sense of mineness with which they are infused results from such processes. Moreover, it is held that to a far extent the construction of future scenarios draws on bits of imagery information provided by episodic memory. Regarding these two tenets, moderate discontinuism sides with the constructivist framework favored by the main stream of the mental time travel literature. But it crucially diverges in two ways.

First, it claims that drawing on mnemonic episodic bits of information does *not* imply the importing of autonoetic awareness into the imagined future scenario. Indeed, what is autonoetically experienced is a past event belonging to one's autobiography, namely a specifically articulated cluster of episodic details. Once severed from this whole, an element—for instance, the visual image of a particular object—plays no part in such an event and can simply be a mere element of perspective imagery. Many such details are accessible in this way without being autonoetically conscious. Thus, the flexible recombination of such elements does not imply autonoetic consciousness, though it emerges in mental scenarios that possess the subjective character of experienced scenes—in particular, they involve a perspective character and quasi-sensory details. For instance, when you were young, you could have taken an exam very similar to your daughter's and still have vivid memories of that experience. Your use of such mnemonic elements allows you to episodically anticipate someone else's future, meaning in a non-autonoetic way. Provided such scenarios are deprived from autonoetic phenomenology, should they be categorized as non-episodic? Two alternatives would suggest a positive answer. On the one hand, one could propose to categorize them as merely episodic-like (Griffith et al., 1999), since their content includes a particular location, a particular moment, and a particular event. But as the uncertainty case shows, except for the identity of the self, nothing changes in the content of the anticipation when one shifts from the case in which one assigns the imagined future scene to one's own personal future to the case in which one assigns it to someone else's future. Thus, the episodic-like alternative will not do. On the other hand, one could propose that the self-substitution case is merely semantic imagination. But, here again, the uncertainty case poses a problem. If the alternative were right, it would imply that, while hesitating about the identity of the person who will have to endure the imagined unpleasant experience, the subject oscillates between a scene constructed by means of episodic bits of information and a scene constructed by drawing on semantic knowledge. Such a complete cognitive shift is highly implausible. I thus propose instead that the imagined scenes in the cases of both uncertainty and self-substitution are episodic without being autonoetic. More generally, in my view, the problematic cases show that episodic subjectivity is not equivalent to and not dependent on autonoetic subjectivity. It displays non-personal subjectivity in the sense that it does not imply the identity of the self who is episodically thinking with the self whose experience is being thought about. Thus, episodic anticipation does not necessarily consist of pre-experiencing personal scenarios. If this is right, then the Continuity Thesis is in trouble, since its CT3 defining component has to be discarded.

The contingency of the autonoetic character of episodic anticipation leads moderate discontinuism to a second diverging claim. Indeed, as suggested by the first diverging claim, *when* episodic anticipation happens to be autonoetic, this sense of mineness does not come into the picture, as it does in the case of episodic memory. On the one hand, as I have said, episodic memory involves the causal constraint. According to a converging cluster of psychological data, this constraint has cognitive reality. By this, I mean that the direct causal link of a mental state with its past cause is a procedural feature of the construction process of mnemonic representation. The subpersonal property of, for instance, the fluency with which the construction of the scene is carried out makes the mental scene feel like it is mine and already experienced, meaning that it represents one of my own past experiences (Cerles & Rousset, 2012; Dokic, 2014; Gomez et al., 2009; Jacoby et al., 1989). Along these lines, autonoetic consciousness enters into the construction process of episodic memory in a specific way:

i. It is triggered by a detecting or monitoring operation applying to procedural features of the construction process of the mental state content. Arguably, the binding of the episode repeats, at least partially, a past perceptual binding. This can be detected while episodically remembering, but not in the episodic anticipation that carries out a new binding.[23]
ii. Autonoeticity is a feature that is automatically attributed to the content, rather than a property that would emerge from an inferential deliberation carried out by the subject on the ground of the information transmitted by the content. This automatic character of the introduction of autonoeticity is reflected by its impenetrability to beliefs. Indeed, even if one comes to believe that one's apparent memory is illusory and represents an event that one did not experience, one will likely still have the same autonoetic impression in mind.
iii. As shown by the fact that its withdrawal switches the episodic memory mode, autonoetic awareness is a necessary feature of episodic memory.

To summarize, the introduction of autonoeticity in episodic memory should be described in terms of the following process:

> ***Detection process.*** In episodic memory, at the phenomenological level of content, the necessary autonoetic sense of mineness reflects the automatic detection of procedural properties intrinsic to the construction of the mnemonic content.

On the other hand, the introduction of autonoetic consciousness in episodic anticipation features strikingly different properties. Let's return to the example of firing Jean. Suppose that, when thinking about the particular task that has to be carried out, you reach the conclusion that you will have to do it. Suppose, for instance, you are rather close to Jean, and you know the specific content of the professional mistakes he made. By detecting some properties of the content (i.e., who Jean is and the reasons that should be invoked to justify the decision of firing him), you are arguably led to give yourself the role of the agent of the imagined future scene. But beyond the *prima facie* similarity, there are sharp differences between this case and the previous one. Indeed,

i. the autonoetic sense of mineness of the imagined firing of Jean does not result from a subpersonal and automatic attribution occurring during the construction process of the content. It instead emerges from an inferential process drawing on some elements of semantic knowledge concerning your professional relationship with Jean. And it supposes that the content is already constructed in order to occur, since you must have it at your disposal, as it were, to carry out your inferential activity.
ii. Being inferential, it is not automatically attributed and is penetrable to belief. For instance, if you suddenly remember that Jacques has a closer professional relationship with Jean than you, maybe will you believe that Jacques will have to do the job, and your episodic anticipation will consequently no longer display the sense of mineness feature.
iii. As confirmed by the second asymmetry, the withdrawal of the autonoeticity feature does not switch the mode of episodic anticipation. This feature is thus a contingent feature of episodic anticipation.

The introduction of autonoeticity in episodic anticipation should thus be described in terms of the following process:

> ***Assignment process.*** In episodic anticipation, in the phenomenology of content, the sense of mineness reflects a deliberate and contingent self-assignment carried out on the basis of the already constructed content.

To summarize: according to the two diverging claims, on the one hand, autonoetic consciousness is not necessarily required for episodic anticipation to have first-person perspectival subjectivity; on the other hand, when episodic anticipation is autonoetic, autonoetic consciousness does not enter into the construction process in the same way as in episodic memory. When recalling the aforementioned definition of a kind of process (see section 1), one should say that the processes that confer subjectivity to episodic memory or episodic anticipation differ from each other in kind. This is due to the sensitivity of episodic thought to actuality (respectively, inactuality). CT3 must therefore be rejected along with CT. Thus, moderate discontinuism should be preferred to moderate continuism regarding the subjectivity issue. This strongly suggests that it should be more generally preferred.

## ACKNOWLEDGMENTS

I greatly benefited from comments by Christoph Hoerl, Stan Klein, Kourken Michaelian, and Stéphane Rousset on a first draft of this chapter. I warmly thank them for that. Thanks also to Dorothea Debus for exchanges on the topic of the chapter.

## NOTES

1. See my review later in this chapter, which supports this claim.
2. As will become clear in what follows, by "continuity" I mean the neurocognitive identity of backward and forward mental time travels.
3. See Brent (1984, 18–19) and Toth & Hunt (1999, 255–256) for a similar definition, despite differing distinctions between function and process. The definition I propose is intended to be neutral with regard to the debate on process-based versus structure-based approaches, which I will leave aside here (on this debate, see Foster & Jelicic, 1999).
4. See section 5 for more details.
5. Wheeler et al. (1997, 331); Suddendorf & Corballis (1997, 137, 149); Tulving (2002b, 315); Tulving (2005, 9); Szpunar (2010, 144).
6. Recanati (2007) and Dokic (2014), for instance, make similar use of this threefold distinction.
7. See the aforementioned review.
8. On actuality in this context, see Williams (1973), Perner and Kloo (2007), and Debus (2014).
9. Note that the same argument would hold couched in terms of B-relations. It would say that events that come earlier can have causal bearings on events that come later, but not the reverse. Actuality would then be an epistemologically relative notion.
10. Here again my argument could be epistemologically relativized without loss.
11. I refine the notion of *de se* thought in section 4.
12. For recent developments and classical references, see Possner and Recanati (2012).
13. For tricky cases such as thought insertion, see Campbell (1999) and Gallagher (2000). On the rubber-hand illusion, see Blanke & Metzinger (2009).
14. For quasi-memories, see, in particular, Shoemaker (1970), Evans (1982), and Coliva (2006).
15. This notion is based on Evans (1982, 180).
16. "*s*" means "a subject," "remembers$_e$" means "episodically remembers," "anticipates$_e$" means "episodically anticipates," and "φ-ing" is an abbreviation of the expression of the action mentioned in the episodic memory report.
17. The following situation could create a complication. Suppose that after having confused himself with someone else, Mach initiates a train of sensory imaginative

thoughts concerning the future of the person he saw in the mirror. It could be argued that he would thus carry out *de se* error through misidentification in his imaginative achievement. To override this objection, it suffices to restrict the immunity claim to strictly imaginative content. Indeed, the objection involves a modality that differs from imagination, namely perception. With this restriction, the claim holds (Recanati, 2007).

18. There exists one other criticism of the alleged symmetrical content/object relation in episodic memory and episodic anticipation in the philosophical literature. It is related to Debus's and Campbell's relationalist account (Campbell, 2002, and Debus, 2014; see also Martin, 2001) and claims there is a direct epistemic relation to past events through episodic memory, while such contact is impossible for future events. Roughly speaking, only general concepts could be employed to think of future particular events. In my view, relationalism is a strong theoretical commitment requiring a case of its own. To this idea of a direct epistemic link with the past event, which is controversial and has a strong theoretical commitment, I prefer the less theoretically committed idea of the individuation of the content by the object.
19. A special thank to Christoph Hoerl for his comments and suggestions on this section.
20. In cases of memory transplants, the apparent episodic memory mode would thus only be an apparent (*quasi*) episodic memory mode. More should be said on this, which I must renounce for reasons of space.
21. This is a much noted feature of episodic memory; see, for instance, Campbell (1997).
22. As noted earlier, I do not intend thereby to commit myself to any thesis about the ontology of time. My point would remain the same if couched in terms of an epistemologically relative notion of actuality.
23. If pressed further, I would say—along these proceduralist lines—that (*pace* Tulving) autonoeticity is not a neurocognitive ability, but an epiphenomenon deriving from the detection of certain procedural features of scene reconstruction.

## REFERENCES


Addis, D. R., et al. (2007). Remembering the past and imagining the future: common and distinct neural substrates during event construction and elaboration. *Neuropsychologia*, *45*, 1363–1377.

Atance, C. M. (2008). Future thinking in young children. *Current Directions in Psychological Science*, *17*(4), 295–298.

Atance, C. M., & O'Neill, D. K. (2011). Episodic future thinking. *Trends in Cognitive Sciences*, *5*(12), 533–539.

Berntsen, D., & Bohn, A. (2010). Remembering and forecasting: the relation between autobiographical memory and episodic future thinking. *Memory and Cognition*, *38*(3), 265–278.

Blanke, O., & Metzinger, T. (2009). Full-body illusions and minimal phenomenal selfhood. *Trends in Cognitive Sciences*, *13*(1), 7–13.

Brent, S. B. (1984). *Psychological and social structures*. Hillsdale, NJ: Lawrence Erlbaum.

Buckner, R. L., & Carroll, D. C. (2007). Self-projection and the brain. *Trends in Cognitive Sciences, 11*(2), 49–57.
Campbell, J. (1997). The structure of time in autobiographical memory. *European Journal of Philosophy, 5*(2): 105–118.
Campbell, J. (1999). Immunity to Error Through Misidentification and the Meaning of a Referring Term. *Philosophical Topics, 6*(1-2), 89–104.
Campbell, J. (2002). *Reference and consciousness*. Oxford: Oxford University Press.
Cerles, M., & Rousset, S. (2012). Bias in self-motion perceived speed can enhance episodic memory. *Cognitive Process, 13*(Suppl 1), 121–124.
Coliva, A. (2006). Error through misidentification: some varieties. *The Journal of Philosophy, 103*(8), 403–425.
Conway, M. (2008). Exploring episodic memory, In E. Dere, et al. (eds.), *Handbook of episodic memory* (vol. 18, pp. 19–29). Amsterdam; Oxford: Elsevier.
Conway, M. (2009). Episodic memories. *Neuropsychologia, 47*, 2305–2313.
Conway, M., & Pleydell-Pearce, C. (2000). The construction of autobiographical memories in the self-memory system. *Psychological Review, 107*, 261–288.
Conway, M., et al. (2004). The self and autobiographical memory: correspondence and coherence. *Social Cognition, 22*(5), 491–529.
D'Argembeau, A., & Van Linden, M. (2004). Phenomenal characteristics associated with projecting oneself back into the past and forward into the future: influence of valence and temporal distance. *Consciousness and Cognition, 13*, 844–858.
D'Argembeau, A., & Van Linden, M. (2006). Individual differences in the phenomenology of mental time travel: The effect of vivid imagery and emotion regulation strategies. *Consciousness and Cognition, 15*, 342–350.
Debus, D. (2014). 'Mental time travel': remembering the past, imagining the future, and the particularity of events. *Review of Philosophy and Psychology, 5*(3), 333–350.
Dokic, J. (2014). Feeling the past: a two-tiered account of episodic memory. *Review of Philosophy and Psychology, 5*(3), 413–426.
Evans, G. (1982). *The varieties of reference*. Oxford: Oxford University Press.
Foster, J. K., & Jelicic, M. (eds.). (1999). *Memory: systems, process, or function?* Oxford: Oxford University Press.
Friedman, W. J. (2007). The meaning of 'time' in episodic memory and mental time travel. *Commentary*, Suddendorf and Corballis: "The evolution of foresight." *Behavioural and Brain Sciences, 30*(3), 323.
Gallagher, S. (2000). Philosophical conceptions of the self: implications for cognitive science. *Trends in Cognitive Sciences, 4*(1), 14–21.
Gomez, A., et al. (2009). Egocentric-updating during navigation facilitates episodic memory retrieval. *Acta Psychologica, 132*(3), 221–227.
Griffith, D., et al. (1999). Episodic memory: what can animals remember about their past? *Trends in Cognitive Sciences, 3*(2), 74–80.
Hassabis, D., et al. (2007). Patients with hippocampal amnesia cannot imagine new experiences. *Proceedings of the National Academy of Science U S A, 104*, 1726–1731.
Hassabis, D., & Maguire, E. A. (2007). Deconstructing episodic memory with construction. *Trends in Cognitive Sciences, 11*(7), 299–306.
Hoerl, C., & McCormack, T. (1999). Memory and temporal perspective: the role of temporal frameworks in memory development. *Developmental Review, 19*, 154–182.
Irish, M., et al. (2012). Considering the role of semantic memory in episodic future thinking: evidence from semantic dementia. *Brain, 135*, 2178–2191.

Jacoby, L. L., et al. (1989). Memory attributions. In H. L. Roediger & F. I. M. Craig (eds.), *Varieties of memory and consciousness: essays in honour of Endel Tulving* (pp. 391–422). Hillsdale, NJ: Lawrence Erlbaum.

Juskaneite, A., et al. (2014). Mental simulation of future scenarios in transient global amnesia. *Neuropsychologia, 63*, 1–9.

Klein, S. B. (2015). Autonoetic consciousness: Reconsidering the role episodic memory in future-oriented self-projection. *The Quaterly Journal of Experimental Psychology*, 1–21.

Klein, S. B., et al. (2002). Memory and temporal experience: the effects of episodic memory loss on an amnesic patient's ability to remember the past and imagine the future. *Social Cognition, 20*, 353–379.

Martin, M. G. F. (2001). Out of the past: episodic recall as retained acquaintance. In C. Hoerl & T. McCormack (eds.), *Time and memory: issues in philosophy and psychology* (pp. 257–284). Oxford: Clarendon Press.

Moore, C., & Lemmon, K. (eds.). (2001). *The self in time: developmental perspectives.* Hillsdale, NJ: Lawrence Erlbaum.

Perner, J., et al. (2007). Episodic memory development: theory of mind is part of re-experiencing experienced events. *Infant and Child Development, 16*, 471–490.

Perner, J., et al. (2010). Retro- and prospection for mental time travel: emergence of episodic remembering and mental rotation in 5- to 8-year old children. *Consciousness and Cognition, 19*, 802–815.

Possner, S., & Recanati, F. (eds.). (2012). *Immunity to error through misidentification.* Cambridge: Cambridge University Press.

Quine, W. V. (1956). Quantifiers and propositional attitudes. *The Journal of Philosophy, 53*(5), 177–187.

Race, E., et al. (2011). Medial temporal lobe damage causes deficits in episodic memory and episodic future thinking not attributable to deficits in narrative construction. *Journal of Neurosciences, 31*, 10262–10269.

Rasmussen, A. S., & Berntsen, D. (2013). The reality of the past versus the ideality of the future: emotional valence and functional differences between past and future mental time travel. *Memory and Cognition, 41*, 187–200.

Rathbone, C. J., et al. (2011). Remembering and imagining: the role of the self. *Consciousness and Cognition, 20*, 1175–1182.

Recanati, F. (2007). *Perspectival thought.* Oxford: Oxford University Press.

Robenbaum, R. S., et al. (2005). Theory of mind is independent of episodic memory. *Science, 318*, 1257.

Schacter, D. L., & Addis, D. R. (2007). The cognitive neuroscience of constructive memory: remembering the past and imagining the future. *Philosophical Transactions of the Royal Society, Biological Sciences, 362*, 773–786.

Schacter, D. L., et al. (2012). The future of memory: remembering, imagining, and the brain. *Neuron, 76*, 677–694.

Shao, Y., et al. (2010). Does the self drive mental time travel? *Memory, 18*(8), 855–862.

Shoemaker, S. (1970). Persons and their past. *American Philosophical Quaterly, 7*, 269–285.

Suddendorf, T., & Corballis, M. (1997). Mental time travel and the evolution of the human mind. *Genetic, Social, and General Psychology Monographs, 123*(2), 133–167.

Suddendorf, T., & Corballis, M. (2009). Mental time travel and the shaping of the human mind. *Philosophical Transactions of the Royal Society, Biological Sciences, 364*, 1317–1324.

Szpunar, K. K. (2010). Episodic future thought: an emerging concept. *Perspectives on Psychological Science, 5*(2), 142–162.

Szpunar, K. K. (2011). On subjective time. *Cortex, 47*, 409–411.

Szpunar, K. K., et al. (2007). Neural substrates of envisioning the future. *Proceedings of the National Academy of Science U S A, 104*, 642–647.

Szpunar, K. K., & Tulving, E. (2011). Varieties of future experience. In M. Bar (ed.), *Predictions in the brain: using our past to generate a future* (pp. 3–12). Oxford: Oxford University Press.

Toth, J. P., & Hunt, R. R. (1999). Not one versus many, but zero versus any: structure and function in the context of the multiple memory systems debate. In J. K. Foster & M. Jelicic (eds.), *Memory: systems, process, or function?* (pp. 232–272). Oxford: Oxford University Press.

Tulving, E. (1983). *Elements of episodic memory.* Oxford: Clarendon Press.

Tulving, E. (1984). Précis of *Elements of episodic memory. The Behavioral and Brain Sciences, 7*, 223–268.

Tulving, E. (1985). Memory and consciousness. *Canadian Psychology, 26*, 1–12.

Tulving, E. (2002a). Episodic memory: from mind to brain. *Annual Review of Psychology, 53*, 1–25.

Tulving, E. (2002b). Chronesthesia: conscious awareness of subjective time. In D. T. Stuss & R. C. Knight (eds.), *Principles of frontal lobe function* (pp. 311–325). New York: Oxford University Press.

Tulving E. (2005). Episodic memory and autonoesis: uniquely human? In H. S. Terrace & J. Metcalfe (eds.), *The missing link in cognition* (pp. 4–56). New York: Oxford University Press.

Vendler, Z. (1979). Vicarious experience. *Revue de Métaphysique et de Morale*, 161–173.

Viard, A., et al. (2012). Factors affecting medial temporal lobe engagement for past and future episodic events: an ALE meta-analysis of neuroimaging studies. *Brain and Cognition, 80*, 111–125.

Wheeler, M. A., et al. (1997). Toward a theory of episodic memory: the frontal lobes and autonoetic consciousness. *Psychological Bulletin, 121*(3), 331–354.

Williams, B. (1973). Imagination and the self. In his *Problems of the Self* (pp. 26–45). Cambridge: Cambridge University Press.

4

# Against Discontinuism

## *Mental Time Travel and Our Knowledge of Past and Future Events*

**KOURKEN MICHAELIAN**

### 1. CONTINUISM AND DISCONTINUISM ABOUT MENTAL TIME TRAVEL

Over the last several decades, the psychology of memory has undergone a major reorientation away from earlier conceptions of episodic memory as a specialized system dedicated to the storage and retrieval of information about the "what, when, and where" of past events and toward a conception of episodic remembering as a form of constructive mental time travel (MTT). Reinforced by impressive brain imaging evidence and extensive research on representational and phenomenological overlap between remembering the past and imagining the future (for recent reviews, see Klein, 2013; Schacter et al. 2012; Szpunar 2010), this new conception emphasizes the deep similarities between episodic memory, future-oriented mental time travel (FMTT), and, increasingly, processes such as episodic counterfactual thinking (De Brigard, 2014; De Brigard & Gessell, Chapter 8 of this volume; Pezzulo, Chapter 13 of this volume), in which the subject imagines alternatives to past events. Taking the new conception to its logical conclusion, many have argued that, rather than distinct faculties of episodic remembering and imagining, what we in fact have is "a single general faculty of mental time travel" (Suddendorf & Corballis, 2007). On this view, remembering the past and imagining the future are strictly continuous.

Philosophers have recently begun to take note of the turn toward MTT in psychology and, in line with older and more general philosophical suspicions about the ability of imagination to give us knowledge of future events, to push back against the associated "continuist" view. "Discontinuism," as we may refer to this emerging position, maintains that, notwithstanding the strength of the empirical support for continuism, there are fundamental metaphysical and/or epistemological differences between episodic memory and FMTT, differences in virtue of which it would be a mistake to view our capacities to remember the past and imagine the future as being merely special cases of a general capacity for MTT. This chapter aims to meet discontinuists on their own ground, providing a philosophical defense of continuism against discontinuist arguments. The chapter argues that, while treating episodic memory as a form of MTT not distinct in kind from FMTT requires us to abandon certain common philosophical views about memory and imagination, there are good philosophical reasons to abandon these views and therefore to reject the view that there are fundamental discontinuities between episodic memory and future-oriented MTT. The picture that emerges of a fully symmetrical ability to mentally travel backward and forward in time suggests that philosophers would do well to follow the lead of psychologists, attending more closely than they have historically done to our capacity for future-directed episodic thought and to the ways in which that capacity gives rise to knowledge of future events.

## 1.1. Forms of Continuism and Discontinuism

Let us begin by fixing terms. *Discontinuists* maintain that, in addition to their distinct temporal orientations, there are further important differences between episodic memory and FMTT and that, in virtue of these further differences, episodic memory and FMTT amount to qualitatively different processes or states. As Debus puts it, "[episodic memories] of past events and [episodic imaginations] of future events are ultimately mental occurrences of two different kinds" (2014, 337).[1] On the discontinuist view, it is a mistake to see MTT as a single, unified capacity that can be directed at either the past or the future. *Continuists*, in contrast, maintain that MTT is precisely such a unified capacity. On the continuist view, temporal orientation is the only important difference between episodic memory and FMTT.

Three further distinctions—between extreme and moderate, narrow and broad, and metaphysical and epistemological forms of (dis)continuism—will help to structure the following discussion.

### 1.1.1. Extreme and Moderate Discontinuism

Perrin (who introduced the terms *continuism* and *discontinuism* in Chapter 3 of this volume) distinguishes between extreme and moderate forms of (dis)continuism. *Extreme continuism* denies the existence of any difference, beyond subjective temporal orientation, between episodic memory and FMTT. *Moderate continuism*, in turn, acknowledges that there may be limited differences between them, but denies that these are sufficiently important to give rise to a difference in kind between the processes. *Extreme discontinuism* maintains that the differences between episodic memory and FMTT are so dramatic that they represent completely different kinds of process, incapable of being subsumed under a common superordinate kind. *Moderate discontinuism*, in turn, acknowledges that they may have enough features in common for us to classify them as belonging to a common kind, but insists that the differences between them are sufficiently important for them to constitute distinct kinds of process within that broader kind.

We are thus faced with a spectrum of possible positions, ranging from extreme continuism to extreme discontinuism. Which of these positions should we take seriously? As Perrin suggests, someone who holds a simple "replay" view of memory, combined with a view of FMTT as a constructive or simulational process, might endorse extreme discontinuism. It is unlikely that anyone now holds such a combination of views, and we may thus set extreme discontinuism aside. Similarly, disregarding occasional rhetorical excesses, it is unlikely that anyone subscribes to extreme continuism, since the fact that there are a variety of differences, at the neural, representational, and phenomenological levels, between remembering the past and imagining the future, is now well established; we may thus set extreme continuism aside. The real debate, then, is between moderate continuism and moderate discontinuism: everyone will agree that there are some important similarities and some important dissimilarities between episodic memory and FMTT; the question is whether the dissimilarities are merely a matter of degree, or whether they reflect a qualitative difference between the two processes.

### 1.1.2. Narrow and Broad Discontinuism

The debate between continuists and discontinuists focuses on the contrast between FMTT and episodic memory, but episodic memory is only one form of past-oriented mental time travel (PMTT), which also includes processes such as episodic counterfactual thought. Thus we can distinguish between narrow and broad forms of (moderate) discontinuism. *Narrow discontinuism* maintains that, in addition to their distinct temporal orientations, there are further qualitative differences between FMTT and episodic memory, remaining silent about other forms of PMTT. *Broad discontinuism* maintains that, in addition to their distinct temporal orientations, there are further qualitative differences between FMTT and PMTT as a whole, including, but not limited to, episodic memory. (Continuism is always broad: given that we admit that episodic memory is continuous with FMTT, there would seem to be no reason to maintain that other forms of PMTT are not continuous with FMTT.)

The explicitly discontinuist arguments that have been offered in the literature are best understood as arguments for narrow discontinuism. As we will see, this is an important weak point, shared by a number of otherwise distinct arguments. Discontinuists single out episodic memory and argue for a distinction in kind between it and FMTT. But what are we to say about other forms of PMTT? The discontinuist is faced with a choice. Either he groups other forms of PMTT with episodic memory, or he groups them with FMTT. If they are grouped with episodic memory, narrow discontinuism collapses into broad discontinuism. If they are grouped with FMTT, narrow discontinuism maintains not only that there are qualitative discontinuities between episodic memory and FMTT but also that there are qualitative discontinuities within PMTT, between episodic memory and other forms of PMTT. Neither of these positions is easy to maintain. In the case of the latter position, while it is somewhat plausible to contrast episodic memory with FMTT, it is much less plausible to contrast episodic memory with episodic counterfactual thought and other forms of PMTT; narrow discontinuism tends to lead to broad discontinuism. The former position (broad continuism), however, is itself an unstable position, as it is not particularly plausible to contrast PMTT as a whole with FMTT; grouping other forms of PMTT with episodic memory thus tends to lead us simply to adopt full-blown (moderate) continuism.

### 1.1.3. Metaphysical and Epistemological Discontinuism

A final distinction is between metaphysical and epistemological forms of (dis)continuism. The distinction between metaphysics and epistemology aligns roughly with the difference between descriptive questions about the mind (studied by philosophers of mind) and normative questions about knowledge (studied by epistemologists). Thus *metaphysical discontinuism* accepts (and metaphysical continuism rejects) the existence of descriptive differences in kind between episodic memory and FMTT. For example, a view on which episodic memory has concrete past events as its objects, whereas FMTT has abstract types of possible future events as its objects, would amount to a form of metaphysical discontinuism (see section 2.2). Correspondingly, *epistemological discontinuism* accepts (and epistemological continuism rejects) the existence of differences in kind between the knowledge of past events that is provided by episodic memory and the knowledge of future events that is provided by FMTT. For example, a view on which certain kinds of misrepresentation are possible in remembering the past but not in imagining the future would amount to a form of epistemological discontinuism (see section 3.3).

Section 2 of this chapter discusses metaphysical continuism; section 3 discusses epistemological continuism. Which combinations of these metaphysical and epistemological views are coherent? Epistemological discontinuism presupposes metaphysical discontinuism, in the sense that, if metaphysical discontinuism turns out to be incorrect, it is unclear what motivation might remain for epistemological discontinuism. Thus, if we could be certain that the arguments given in section 2 were successful, we could in principle omit any discussion of the arguments of section 3. But given that the former arguments are not decisive, it is worth exploring independent arguments for epistemological discontinuism. While epistemological discontinuism seems to presuppose metaphysical discontinuism, metaphysical discontinuism does not entail epistemological discontinuism—there might in principle be qualitative differences between episodic memory and FMTT at the descriptive level that fail to translate into differences at the normative level. Thus one might in principle endorse metaphysical discontinuism while also endorsing epistemological continuism.

Before proceeding, a note on where this chapter fits into the wider literature: the strategy of the chapter is primarily negative, in that it defends continuism by responding to arguments for discontinuism. The positive empirical case for metaphysical continuism has been made in the psychological literature discussed earlier; a positive philosophical case for metaphysical continuism is made in Michaelian (2016). Supposing that the main arguments of this chapter—negative arguments against metaphysical and epistemological discontinuism—are successful, what is missing at this stage is a positive case for epistemological continuism; the final section of the chapter briefly sketches such a case, but this remains an important area for future work.

### 1.2. Empirical Evidence and A Priori Arguments

There is a danger that philosophers and psychologists debating the continuity or discontinuity of episodic memory and FMTT will talk past each other. Psychologists endorse continuism primarily on the basis of empirical evidence for similarities between episodic memory and other forms of MTT. Philosophers endorse discontinuism primarily on the basis of arguments of a largely conceptual character. Can empirical evidence decide between continuism and discontinuism? Are purely conceptual arguments relevant to a debate over distinctions among kinds of mental processes?

### 1.2.1. The Relevance of Empirical Evidence

The general thrust of MTT research—which has demonstrated extensive similarities between episodic memory and FMTT—clearly favors continuism. As Perrin points out, however, the MTT literature also includes a number of empirical findings concerning differences between episodic memory and FMTT at both the neural and the representational levels.[2]

A discontinuist might be tempted to invoke these findings in support of her view, but there are two difficulties involved in doing so. The first difficulty is that if the observed differences were to suggest deep or extensive asymmetries between episodic memory and FMTT, they might in principle support discontinuism, but in practice they amount to differences of degree and are therefore compatible with continuism. At the neural level, the discontinuist might cite the fact that brain imaging has revealed the involvement in FMTT of brain regions that are not involved, or are involved to a lesser extent, in episodic memory (see, e.g., Szpunar et al., 2007). Philosophers with no special expertise on brain imaging ought to defer to psychologists and neuroscientists on the interpretation of this sort of evidence, and this chapter will not discuss these findings in any detail. It is sufficient here to note that the consensus among psychologists and neuroscientists (e.g., Schacter & Addis, 2007) is that the findings in question suggest differences of degree, rather than differences in kind, between episodic memory and FMTT, perhaps reflecting the greater processing demands of the latter. They thus provide little support for discontinuism.

At the representational level, one typical finding—echoing Hume's (1739) argument that memory is distinguished from imagination by its greater vivacity—is that episodic remembering produces representations containing greater sensory and contextual detail than is contained by those produced by FMTT (e.g., D'Argembeau & Van der Linden, 2004). These representational-level differences might seem to provide a more promising basis for drawing a qualitative distinction between episodic memory and FMTT, but here, too, the differences ultimately seem to be a matter of degree, rather than kind. Reid (1764)—responding to Hume—remarked that he "would gladly know . . . how one degree of vivacity fixes the existence of the object to the present moment; another carries it back to time past; a third, taking a contrary direction, carries it into futurity; and a fourth carries it out of existence altogether." Translated into contemporary terms,[3] Reid's point is, roughly, that the level of detail of representations of both future and past events is highly variable. Both episodic memory and FMTT may produce more or less detailed representations, and, while the representations produced by episodic memory may, on average, be more detailed than those produced by FMTT, the difference holds only on average. This means that there can be no clear cutoff point above which representations are sufficiently detailed to count as memory representations and below which they are too generic to count as memory representations. Thus the observed differences between episodic memory and FMTT at the representational level provide little support for discontinuism.

The second difficulty is posed by the fact that episodic memory is not the whole of PMTT, which also includes processes such as episodic counterfactual thought.

Episodic counterfactual thought, like FMTT, calls on slightly different brain regions than does episodic memory (e.g., De Brigard et al., 2013) and might display differences at the representational level as well. If differences of this sort were to support a qualitative distinction between episodic memory and FMTT, they would presumably likewise support a similar distinction between episodic memory and other forms of PMTT. The discontinuist, then, can either treat the observed differences among episodic memory, FMTT, and episodic counterfactual thought as irrelevant, or can opt for a form of narrow discontinuism that explicitly claims that episodic memory is qualitatively distinct not only from FMTT but also from other forms of PMTT, grouping episodic counterfactual thought with FMTT under the broader heading of episodic imagination. If the discontinuist takes the former route, he must provide alternative reasons in support of discontinuism. If he takes the latter route, he owes us a reason for thinking that episodic memory, in particular, should be treated as special—since the various forms of episodic cognition all call on slightly different brain regions, we need some additional reason for singling out episodic memory while treating the other processes in question as being merely different forms of episodic imagination. Either way, the discontinuist must fall back on conceptual arguments.

### 1.2.2. The Status of A Priori Arguments

The chief arguments for discontinuism are indeed largely conceptual or a priori in character. Those of a naturalistic methodological bent—this includes many philosophers, in addition to most psychologists—might therefore doubt that the threat from discontinuism is particularly serious, suggesting that we base our understanding of the relationship between remembering the past and imagining the future on a demonstrably productive empirical research program, rather than on ungrounded a priori intuitions. Indeed, since discontinuists do not deny the existence of important empirical symmetries between PMTT and FMTT, arguing, in many cases, for metaphysical and epistemological asymmetries that might obtain even if there were no empirical asymmetries whatsoever,[4] naturalists might suggest that the asymmetries between PMTT and FMTT with which discontinuists are concerned are not ones with which we should be concerned, even in principle.

Discontinuism should, however, be of interest even to those who are skeptical of conceptual arguments. First, while conceptual arguments go beyond empirical MTT research, they ought to be consistent with the main findings of that research. Second, as the editors point out in their Introduction to this volume, MTT research has proceeded at a breakneck pace that has left little time to develop the concepts needed to come to grips with its accumulating findings, and one potential benefit of engaging with philosophical resistance to the MTT framework is increased conceptual clarity: when we say that remembering the past and imagining the future are both forms of mental time travel, alike aside from their respective temporal orientations, what, exactly, do (or should) we mean?

## 2. AGAINST METAPHYSICAL DISCONTINUISM

This section considers arguments for metaphysical discontinuism, discussing the implications of apparent causal (section 2.1), intentional (section 2.2), and phenomenological (section 2.3) discontinuities between episodic memory and FMTT.[5]

## 2.1. Causal Discontinuities

In addition to the quantitative representational-level differences noted earlier, an argument for discontinuism might be based on more fundamental apparent intentional differences between episodic memory and FMTT. One such argument is developed by Debus (2014), who sees intentional and causal discontinuities as intertwined. Debus's case for discontinuism rests on the view that, in episodic memory, the subject is experientially aware of a particular past event, whereas, in FMTT, the subject is not (and could not be) experientially aware of a particular future event.

She summarizes the argument as follows.

1. When a subject [episodically remembers] a past event, the subject is experientially aware of a past particular event.
2. When a subject [episodically imagines] a future event, the subject could not possibly be experientially aware of a future particular event.
3. Being experientially aware of a particular event and being in an experiential state in which one could not possibly be aware of a particular event are two different kinds of mental state.
4. Therefore, [episodic memories of] past events and [episodic imaginations] of future events are ultimately mental occurrences of two different kinds [i.e., metaphysical discontinuism is true].

Granting premise 3, this argument succeeds in establishing metaphysical discontinuism unless either premise 1 or premise 2 is false. Premise 1 has two components. First, it says that, in episodic memory, the subject enjoys *experiential awareness* of past events. Call this the positive relation claim. Second, it says that, in episodic memory, the subject is aware of *particular* past events. Call this the positive object claim. Premise 2 likewise has two aspects. First, it says that, in FMTT, the subject does not enjoy *experiential awareness* of future events. Call this the negative relation claim. Second, it says that, in FMTT, the subject is not aware of *particular* future events. Call this the negative object claim. This section discusses the relation claims (the causal aspect of the argument); section 2.2 discusses the object claims (the intentional aspect).[6]

### 2.1.1. The Negative Relation Claim

In support of the negative relation claim, Debus argues, first, that in order for a subject to be experientially aware of an event at a given time, the event must be "actual" "from the subject's own spatio-temporal perspective at the time" and, second, that future events are not actual. At first glance, this seems plausible enough. If experiential awareness is the kind of awareness that one has one when perceives an event, and if what it is for an event to be actual from the subject's spatiotemporal perspective is for it to occur at the relevant time, then premise 2 is uncontroversially right: one cannot be experientially aware of future events simply because one is not contemporaneous with them—they have not yet occurred. But if this is how we understand experiential awareness, premise 1 is likewise uncontroversially wrong: one cannot be experientially aware of past events simply because one is not contemporaneous with them—they have already occurred. Thus what Debus needs is a notion of experiential awareness on which a subject may be experientially aware of an event with which she is not contemporaneous, but only when the event in question is located in the past, not when it is located in the future.

### 2.1.2. The Positive Relation Claim

The discontinuist therefore must understand experiential awareness in a looser, less literal sense. And indeed, Debus ultimately suggests that what it is for one to be aware of a past event, in the relevant sense, is for one to have experienced the event when it occurred and for one's current thought of the event to have the right sort of *causal* connection to one's original experience of it.[7] If this is what "experiential awareness" amounts to, then, while the label is misleading, a subject may be aware of an event that does not occur at the time at which he represents it, as long as he experienced it at the time at which it did occur. And if enjoying this sort of awareness is definitive of episodic memory, then it does mark a qualitative asymmetry between remembering past events and imagining future events, since, even if one will eventually experience an event, one's current representation of an event cannot be caused by one's future experience of it. The tenability of continuism thus rests on the tenability of the denial of the claim that remembering requires a causal connection between the subject's current representation of a past event and his previous experience of that event, a condition that is at the heart of the influential causal theory of memory (Martin and Deutscher, 1966).

Before discussing the causal condition, some background is in order. Philosophers generally take it for granted that we can meaningfully distinguish between genuine and merely apparent memory. As it is standardly drawn, this distinction presupposes two claims. The first is the semantic claim that genuine remembering is "factive," where factivity means, in the case of semantic memory, that the content of the apparent memory is true or correct and, in the case of episodic memory, that the apparently remembered event actually occurred. If a subject seems to remember a proposition but the proposition is false, the apparent memory is said to be merely apparent. If a subject seems to remember an event but the event did not actually occur, the apparent memory is likewise said to be merely apparent. Factivity will be discussed further in section 2.2. The second claim is that, since an apparent memory might be accurate for the wrong sort of reason (e.g., due to blind luck), something more than simple correspondence between the subject's current representation and the event must be required to distinguish genuine memory from merely apparent memory (i.e., factivity is not enough for genuine memory). Different philosophical theories propose different accounts of the nature of this "something more." The causal theory, in particular, says that the subject's current representation must be caused (in the right way)[8] by her experience of the event itself.

Though it has been influential in philosophy, and though something like it may occasionally be presupposed in discussions of false memory in psychology, there is considerable empirical evidence against the causal theory, and hence the theory in turn can provide little support for metaphysical discontinuism. When arguing from empirical memory research to the conclusion that the causal theory is incorrect, there are two different paths that we might take.[9] First, we might start from the research on mental time travel discussed earlier. In outline—see Michaelian (2016) for a more detailed version of this argument—the strategy here would be to argue that the overlap that has been demonstrated between remembering the past, imagining the future, episodic counterfactual thought, and other forms of episodic cognition shows that, intuitions to the contrary notwithstanding, there is nothing special about episodic memory—episodic memory is just one form of episodic imagination among others. Like any other form of episodic imagination, episodic remembering is a simulational process that draws on information originating in a variety of different sources to construct a representation of a target episode. Just as other forms of episodic imagination of necessity do not draw on any information originating in the episodes they target (since the subject has not experienced those episodes), episodic memory need not draw on any information originating in the episodes it targets. In many cases, information originating in the subject's experience of the target episode will in fact end up being incorporated into the subject's memory representation of the event. In such cases, memory may ensure an appropriate causal connection between the memory representation and the original experience. But even in such cases, information originating in experience of the target episode may represent only a small fraction of the total content of the memory representation; there may thus be equally strong or stronger connections to another episode or episodes, making the relevance of the causal connection to the target episode unclear. And in many other cases, *no* information originating in the subject's experience of the target episode ends up being incorporated into the subject's memory representation of the event. In such cases, memory involves no causal connection with the original experience whatsoever.

If this picture of episodic memory as one form of episodic imagination among others is right, the causal condition does not provide an adequate account of the "something else" distinguishing genuine from merely apparent remembering,[10] and discontinuists may not appeal to the causal theory to ground the claim that there is a qualitative metaphysical difference between episodic memory and FMTT. Discontinuists might in principle respond to this argument by drawing a distinction within episodic memory, arguing that only representations of events from the personal past that are produced by processes that do satisfy the causal condition qualify as genuine memories. This response, however, sits ill with the workings of the episodic construction system. There is overwhelming evidence that, while the system aims, within certain limits, at producing *accurate* representations of past episodes, it is simply not designed to ensure *causal connections* with past episodes. Remembering regularly fails to involve causal connections, and there is no difference, at the neurocognitive level, between (apparent) memories that do involve a causal connection and (apparent) memories that do not. Given that we want our philosophical account of remembering to correspond to a process that unfolds at the neurocognitive level, there is simply no justification for drawing this sort of distinction within episodic memory.

Situating episodic memory as a form of PMTT further undermines the causal condition. In addition to producing representations of episodes from the actual personal past, the episodic construction system frequently aims at producing representations of counterfactual episodes from the personal past. And while it may be intuitively plausible to claim that episodic memory is distinguished from FMTT by the fact that the former necessarily involves a causal connection with the target episode, it is considerably less plausible to claim that it is distinguished from episodic counterfactual thought in this manner. The boundary between episodic memory and episodic counterfactual thought is fluid, in the sense that we are able to shift freely back and forth between (attempting to) remember events as they actually occurred and (attempting to) imagine them occurring in more or less different ways. In fact, given that episodic remembering need not be fully accurate to count as remembering, so that it is compatible with a certain amount of modification of the remembered event, and given that much episodic counterfactual thinking consists not in imagining wholly novel events but rather in modifying aspects of past events, in some cases it may be indeterminate whether the subject is engaged in episodic remembering or episodic counterfactual thinking. Episodic counterfactual thought, of course, does not presuppose a causal connection between the subject's current representation and his previous experience of the event, simply because counterfactual events (by definition) did not occur, or did not occur as represented. Given the fluidity of the boundary between episodic memory and episodic counterfactual thought, it becomes difficult to justify imposing the causal condition on remembering.

Now, this first path to the rejection of the causal theory risks being question-begging in the current context, since it requires us to grant that there is no qualitative distinction between episodic memory and other forms of episodic imagination, which is precisely the matter at issue in the debate between continuists and discontinuists. The second path from empirical memory research to the rejection of the causal theory avoids this difficulty (see Sutton and Michaelian [forthcoming] for a more detailed version of this argument). Both continuists and discontinuists ought to accept the basic lesson of constructive memory research: remembering is never a matter of simply retrieving a stored "copy" of an experience but is always to some extent reconstructive, where reconstruction implies, inter alia, the inclusion, in the "retrieved" memory representation, of information not originating in the subject's experience of the target episode. Call the inclusion of such information *supplementation*. Is the occurrence of supplementation compatible with satisfaction of the causal condition? One might argue that the causal condition can be satisfied in cases of supplementation, as long as the retrieved representation does not include too much information originating in sources other than the subject's experience of the target episode: some degree of supplementation is compatible with genuine remembering; only in cases where supplementation results in the inclusion of too much information originating in such sources does remembering fail to occur (Michaelian, 2011). One obvious difficulty with this position is its vagueness: How much supplementation is too much? Setting this difficulty aside, though, the position attempts to draw a distinction where none is to be found. The process that is at work when the subject, in an attempt to remember a given past episode, constructs a representation that includes a large fraction of information originating in her experience of the episode is not distinct, at the neurocognitive level, from the process that is at work when she constructs a representation that includes only a small fraction of information originating in her experience of the episode; nor is it distinct from the process that is at work when she constructs a representation that includes no information whatsoever originating in her experience of the episode. If no information originating in her experience of the episode is preserved, the causal condition is not satisfied. Thus the constructive character of remembering rules out the claim that genuine, as opposed to merely apparent, remembering presupposes satisfaction of the causal condition. Again, the causal condition does not provide an adequate account of the "something else" distinguishing genuine from merely apparent memory, and discontinuists should not appeal to the causal theory to ground the claim that there is a qualitative metaphysical difference between episodic memory and FMTT.[11]

## 2.2. Intentional Discontinuities

In short, while there is much more that might be said both for and against the causal theory, the ground on which the theory stands at this point is arguably too shaky for it to provide significant support for metaphysical discontinuism. Since it is at best unclear whether genuinely remembering a past event does require a causal connection with the remembered event, we should look elsewhere for qualitative descriptive differences between episodic memory and FMTT. Debus maintains both that episodic memory involves causal connection to events (the positive relation claim) and that it involves awareness of particular events (the positive object claim). She maintains, further, that FMTT does not involve causal connection to events (the negative relation claim) and that it does not involve awareness of particular events (the negative object claim). While the negative relation claim is unproblematic, we have seen that there is reason to doubt the positive relation claim. Nevertheless, Debus's argument for metaphysical discontinuism will go through if both object claims are correct. There are reasons to doubt each of these claims.

### 2.2.1. The Negative Object Claim

If we set aside the negative relation claim, the negative object claim collapses into the claim that it is impossible for subjects to represent particular future events. By way of argument for this claim, Debus begins by sketching a two-step account of what is involved in imagining future events. On her account, imagining a future event (e.g., having lunch with a friend next week) involves, first, imagining a general type of event (having lunch with a friend) and, second, imaginatively projecting that event into the future (next week). She then argues that, given this account, imagining a future event never amounts to imagining a *particular* future event. There are two problems with this strategy. First, it does not show that we cannot imagine particular future events by employing some other procedure. Second, it does not convincingly show that we cannot imagine particular future events by employing the two-step procedure.

As far as the first problem goes, it is by no means clear that the procedure described by Debus is in fact employed in all cases of FMTT, and other ways of engaging in FMTT may enable us to imagine particular future events. For example, it would seem to be possible to imagine a future event by, first, remembering a particular past event and, second, imaginatively projecting that event into the future. I might, for example, imagine having lunch with a friend next week by remembering having lunch with him last week and then imagining that event occurring next week rather than last week. If this procedure is employed, then, assuming that the representation produced by remembering qualifies as a representation of a particular past event, it should arguably not cease to count as a representation of a particular event simply because it is projected into the future.

The discontinuist might suggest that the representation does indeed cease to count as a representation of a particular event when it is projected into the future on the ground that this sort of projection involves imaginative modification of an important feature of the event, namely, its temporal location. Taking the counterfactual dimension of PMTT into account, however, undermines this suggestion. Thinking of a particular past event is compatible with imaginatively modifying various feature of the event. I might, for example, remember having lunch with a friend last week but imagine eating slightly different food. In this case, I am presumably still thinking of a particular event. I might go further, remembering having lunch with a friend but imagining eating in a different restaurant, modifying the spatial location of the event. In this case, I may or may not count as being engaged in remembering, but I am presumably still thinking of a particular event. If imaginatively modifying their spatial locations is compatible with thinking of particular events, presumably imaginatively modifying their temporal location is as well. Nor does it seem to matter whether I modify the temporal location of the event by imagining it occurring at a different time in the past or by imagining it occurring in the future. I might remember having lunch with a friend last week but imagine the lunch occurring somewhat later—closer to the present—than it did. If I then shift from episodic counterfactual thought to FMTT and project the event forward into the future—say, tomorrow, rather than yesterday—nothing would seem to change, as far as the particularity of the event that I am imagining is concerned.

Conceding this point, the discontinuist might argue that the possibility of imagining a particular future event by remembering a particular past event and projecting it into the future is compatible with the spirit of discontinuism, since one's imagination of a particular future event in such cases is parasitic on one's memory of a particular past event, which might be enough to secure a metaphysical asymmetry between episodic memory and FMTT. The second problem for Debus's strategy, however, undermines this move. The problem is that, even in cases in which imagining future events employs the original two-step procedure—thinking of a general type of event and projecting it into the future—it is unclear whether the agent necessarily fails to imagine a particular future event. As Debus admits, whether one can, by means of this procedure, imagine a particular event may depend on how we individuate events. If events are individuated in a thin manner, simply by reference to spatiotemporal location, the procedure clearly enables subjects to think of particular future events. There is no barrier to imagining a general type of event occurring at a given spatiotemporal location.

Debus's argument thus presupposes a thicker way of individuating events. If events are individuated in a slightly thicker manner, say, by reference to the particular agents they involve, in addition to spatiotemporal location, the procedure still enables subjects to think of particular future events. There is no barrier to imagining a general type of event occurring at a given spatiotemporal location and involving given people. We might suspect that, if events are individuated in a maximally thick manner, as fully determinate, concrete entities, the procedure will not enable subjects to think of particular future events, since imagination of a general type of event will necessarily fail to capture the full richness of the real, concrete event. At this point, however, the question of event individuation starts to look like a red herring. Regardless of how events are individuated, one need not, in order to think of a particular past event, recall it in every detail. In many cases, one's memory for past events may in fact be quite schematic. Analogously, one need not, in order to think of a particular future event, imagine it in every detail. Of course, if events are individuated in a maximally thick manner, detailed representations of future events will very often be somewhat inaccurate, simply because there are limits to our powers of anticipation. But the same thing goes for detailed representations of past events, which are likewise often inaccurate—there are limits to our powers of recollection, just as there are limits to our powers of anticipation. At least within certain limits, inaccuracy need not prevent us from thinking of particular past events, or from thinking of particular future events.

The suspicion that the two-step procedure may suffice to allow us to imagine particular future events is reinforced by the observation that, in many cases, the very same procedure is employed in episodic memory: in many cases, remembering a past event amounts precisely to imagining a general type of event and imaginatively projecting it into the past. Consider memory for frequently repeated events. In many such cases, one retains little information from any particular repetition of the event, but one does retain a general representation of the event type. Remembering a particular repetition of the event may amount to forming a representation of the event type and imaginatively projecting it back to the relevant past time. If remembering can put us in contact with particular events when it employs this procedure, then so, presumably, can imagining. Of course, a discontinuist might suggest that, in cases where remembering employs this procedure, it does not count as genuinely episodic remembering (i.e., that it does not in fact involve representing a particular past event). But this suggestion is difficult to reconcile with the reconstructive character of episodic memory, which routinely draws both on information originating in experience of particular events and on generic knowledge, including scripts for common event types. In fact, as several of the contributions to this volume make clear (Szpunar, Spreng, & Schacter, Chapter 2; Klein & Steindam, Chapter 7; Irish, Chapter 19), both PMTT and FMTT may in some cases depend exclusively on semantic knowledge. Alternatively, the discontinuist might grant that remembering sometimes employs the two-step procedure but suggest that it counts as genuinely episodic remembering only if there is a shift from projecting a general type of event to a past time to thinking of a particular event, strictly understood. But any plausible explanation of what such a shift might consist in is likely to invoke the causal criterion, and, as we have seen, there are independent problems for that criterion.

In light of these two problems for Debus's strategy, the discontinuist might look for another means of motivating the negative object claim. Rather than motivating it by appealing to the nature of *representations* of future events, the discontinuist might attempt to motivate it by appealing to the nature of future events themselves, arguing that we cannot represent particular future events simply because there is no such thing as a future event. This alternative strategy attempts to base a metaphysical asymmetry between episodic memory and FMTT on a metaphysical asymmetry between the past and the future, invoking an indeterminist metaphysics according to which, while there is a matter of fact about what happened in the past, there is no determinate matter of fact about what will happen in the future. While such an approach would provide no guarantee that we are able to think of particular past events, it would guarantee that we are unable to think of particular future events and so would constitute a step toward metaphysical discontinuism.

This alternative strategy is not a particularly promising means of motivating the negative object claim. First, the metaphysical openness of the future is a controversial matter (Debus, Chapter 11 of this volume); the strategy is thus risky, as it depends on the outcome of an ongoing debate in the metaphysics of time. Second, the strategy seems to miss the point. What matters, in the debate between continuists and discontinuists, is not the *metaphysical* openness of the future but rather its *cognitive* openness. Discontinuism becomes no less plausible if we suppose that metaphysical determinism is true: even if the future is metaphysically closed, the discontinuist can maintain that it is cognitively open and that we therefore cannot enter into cognitive contact with particular future events. By the same token, the continuist can maintain that, even if the past is metaphysically closed, it is cognitively open (since our knowledge of the past is limited) and that we therefore cannot enter into cognitive contact with particular past events. Invoking a metaphysical asymmetry between the past and the future is thus unlikely to provide adequate support for a metaphysical asymmetry between past- and future-oriented MTT.

### 2.2.2. The Positive Object Claim

While the considerations raised in the previous section cast doubt on the negative object claim, according to which we are unable to imagine particular future events, they do not establish that we are in fact able to imagine particular future events. If it should turn out that we are unable to imagine particular future events, the discontinuist might fall back on the positive object claim, according to which we are able to remember particular past events, to motivate metaphysical discontinuism. Thus our assessment of the case for metaphysical discontinuism requires an examination of the case for the positive object claim. This section argues that, while the discussion so far has taken for granted that we are able to remember particular past events, there is a real possibility that the positive object claim is false (i.e., that we are unable to imagine particular past events).

In order to make a case for the claim that episodic memory puts the subject in contact with particular past events, Debus appeals to her relational account of memory (Debus, 2008). In addition to the positive relation claim (discussed earlier), the relational account includes the claim that, when a subject remembers, the remembered event itself is (part of) the content of her memory. The relational account thus amounts to a form of direct realism about the objects of memory (hence Debus cites Reid 1764 as a precursor of her view).[12] In view of the serious problems facing direct realism, the relational account can provide little support for the positive object claim.

The distinction between direct and indirect (or representative) realism has been most extensively discussed in the philosophy of perception. The direct realist about perception claims that, in perceiving, what the subject is related to in the first instance is an external object. The indirect realist, in contrast, claims that what the subject is related to in the first instance is an internal representation; the subject counts as seeing the external object when his internal representation is related to it in the right way. Because it posits internal representations that function as intermediaries between the perceiving subject and the perceived object, the story told by indirect realism about the nature of perception is more complicated than that told by direct realism. But there is a powerful motivation for the introduction of this additional complexity, in the form of the argument from illusion, which motivates the introduction of internal representations by pointing out that perception is not always veridical.

In cases of hallucination, for example, the subject "sees" something that is not in fact there. Veridical perception seems to have something important in common with hallucination; in principle, after all, hallucination might be indistinguishable, from the subject's perspective, from veridical perception. Direct realism, since it views perception in terms of a direct relation between subject and object, has difficulty accounting for this fact. Indirect realism, in contrast, can accommodate it without difficulty, since it can point to the internal representation as the factor common to veridical perception and hallucination. The upshot is that direct realists are forced to adopt a "disjunctive" view of perception: veridical perception and hallucination are not two kinds of perception; only veridical perception is perception; hallucination is another kind of state entirely.

The direct realist about memory claims that, in remembering, what the subject is related to in the first instance is a past event. The indirect realist, in contrast, claims that what the subject is related to in the first instance is an internal representation; the subject counts as remembering a past event when his internal representation is related to it in the right way. Just as indirect realism about perception can be motivated by an argument from perceptual illusion, indirect realism about memory can be motivated by an argument from memory illusions—distorted or inaccurate memories. In the case of "memory hallucinations" (i.e., wholly false memories), the subject "remembers" an event that did not in fact occur. Accurate memory seems to have something important in common with memory hallucination; in principle, memory hallucination might be indistinguishable, from the subject's perspective, from accurate memory. And just as direct realism about perception has difficulty accommodating perceptual hallucination, direct realism about memory has difficulty accommodating memory hallucination. Indirect realism about memory, in contrast, can accommodate it without difficulty, since it can point to the memory representation as the factor common to accurate memory and memory hallucination. The upshot is that, just as direct realists about perception are forced to adopt a disjunctive view of perception, direct realists about memory are forced to adopt a disjunctive view of memory: accurate memory and memory hallucination are not two kinds of memory; only accurate memory is memory; memory hallucination is another sort of state entirely.

Though it has its defenders, disjunctivism is implausible not only from an intuitive but also from a naturalistic perspective. From a naturalistic perspective, we should not draw distinctions between mental states or processes where none is to be found at the neurocognitive level; this is precisely what disjunctivism requires us to do. One way to see this is to note that, just as direct realism implies disjunctivism, it implies factivity: as Debus puts it, "a subject could not possibly have an [episodic memory] of an event that did not take place" (2008, 410). As noted earlier, it is a commonplace in philosophy that memory is factive, but factivity is empirically untenable. At the neurocognitive level, there need be no qualitative difference between the process at work when the episodic construction system produces a wholly inaccurate representation of a past event and the process at work when the system produces an accurate representation of a past event. It is tempting to suppose that there must be at least some quantitative difference, such as greater fluency in the case of accurate memory than in the case of memory hallucination (Perrin, Chapter 3 of this volume), but there need not be even a quantitative difference. A counterfactual event that has been frequently imagined, for example, might be remembered just as fluently as an experienced event (Loftus, 1996). Factivity—and disjunctivism—should thus be rejected.[13]

Since direct realism about perception entails disjunctivism, it encounters an analogous objection. But direct realism about memory also faces a problem that is not shared by direct realism about perception. The objects of perception are present (i.e., they are contemporaneous with the perceiving subject); hence, setting aside disjunctivism, it is relatively easy to understand how they might form part of the content of the subject's perceptual states. The objects of episodic memory, in contrast, are in the past; hence it is much more difficult to understand how they might form part of the content of the subject's memories. The memory exists now; the event does not; how, then, might the latter be part of the content of the former? One possibility here is to invoke mnemic causation (Russell, 1921), the idea that past events can be direct causes of our present memories of them. Were the notion of mnemic causation legitimate, it might provide the sort of direct connection between present representations and past events needed to allow the latter to constitute part of the former. The problem, of course, is that the notion of causation at a temporal distance is borderline unintelligible (Bernecker, 2008), and hence relying on it to save direct realism would be a desperate move.

While this brief discussion cannot provide a decisive case against the disjunctivism entailed by direct realism, we have at least good prima facie reason to reject the positive object claim, if it is understood along the direct realist lines suggested by Debus. The alternative is to understand it along indirect realist lines, as saying that remembering puts subjects in contact with particular past events by means of internal representations that are appropriately related to those events. As we saw earlier, however, if we understand the negative object claim in the same way, there is little reason to think that imagination does not put subjects in contact with particular future events. In sum, neither Debus's object claims nor her relation claims go through, and her argument does not establish metaphysical discontinuism.

## 2.3. Phenomenological Discontinuities

If the argument so far is on track, neither causal nor intentional considerations establish metaphysical discontinuism. The possibility remains that there are qualitative phenomenological discontinuities between episodic memory and FMTT. Perrin (Chapter 3 of this volume) claims to identify two such discontinuities, one concerning the role of autonoesis, the form of consciousness characteristic of MTT, and one concerning the mechanisms respectively responsible for the production of autonoesis in episodic memory and FMTT. This section discusses these alleged phenomenological discontinuities in turn.

### 2.3.1. The Role of Autonoesis

Continuists maintain that autonoesis plays essentially the same role in FMTT as it does in episodic memory—conferring a sense of "mineness," or ownership, on an episodic representation, a sense that the represented event belongs to the subject's own personal past or personal future (as opposed to the life of some other subject or possibly of no one in particular). Perrin argues that this is a mistake. While he grants that FMTT, like episodic memory, draws on content originating in the subject's own experiences, he argues that the fact that FMTT draws on experiential content does not guarantee the involvement of autonoetic consciousness in the process. He points out, in particular, that a subject might, while drawing on the very same experiential content, switch back and forth between imagining his own future experience and imagining a future experience belonging to another subject. Only in the former case will the subject enjoy autonoetic consciousness; the latter case will involve a different sort of phenomenology. Thus Perrin seeks to drive a wedge between autonoetic consciousness and mere episodic subjectivity: the latter, unlike the former, "does not imply the identity of the self who is episodically thinking with the self whose experience is being thought about" (Perrin, Chapter 3 of this volume). If this is right, then autonoetic consciousness is not a necessary component of FMTT, whereas it is a necessary component of episodic memory.[14]

Given that autonoesis is distinct from chronesthesia—with autonoesis referring specifically to consciousness of the self in subjective time and chronesthesia referring to consciousness of subjective time more broadly (Tulving, 2002)—the continuist should admit that autonoesis is not intrinsic to episodic cognition. The involvement of experiential content in FMTT does not imply the involvement of autonoesis, and for precisely the reason that Perrin identifies: FMTT enables the subject to imagine not only his own future experiences, but also those of other subjects. Situating episodic memory as one form of PMTT, however, it is clear that the same thing goes for PMTT. The involvement of experiential content in PMTT does not imply the involvement of autonoesis, for precisely the same reason: PMTT enables the subject to imagine not only his own past experiences (i.e., to remember), but also to imagine those of other subjects.

At this point, the discontinuist might retreat from broad to narrow discontinuism, restricting the scope of the claim about the role of autonoesis to episodic memory in particular, as opposed to PMTT as a whole, to argue that episodic memory necessarily involves autonoesis, while FMTT and other forms of PMTT do not. But there are two problems with this move. First, even disregarding the question of the status of animal "episodic-like" memory (on which see, e.g., Thom & Clayton, Chapter 14 of this volume; Martin-Ordas, Chapter 15 of this volume; Corballis, Chapter 16 of this volume), it is not in fact clear whether episodic memory necessarily involves autonoesis. There are, for example, cases of subjects who have intact episodic memory, in the sense that they are able to construct detailed representations of events from the personal past, despite suffering from impaired autonoesis, in the sense that they do not feel as if the events in question belong to them (e.g., Klein & Nichols 2012).

This does not provide an open-and-shut case against the move; for example, the discontinuist might maintain that autonoesis is necessarily involved in ordinary, non-pathological cases of episodic memory, but the second problem comes into play at this point. Even if we grant that episodic memory does necessarily involve autonoesis, Perrin's argument asks us to compare apples and oranges—or rather to compare a single apple selected from basket of varied apples to a whole basket of varied oranges. There are multiple forms of FMTT and multiple forms of PMTT.

Thus it is illegitimate to select one form of PMTT (namely, episodic memory), point out that it has a certain phenomenological feature, note that some forms of FMTT lack that phenomenological feature, and argue on that basis for an asymmetry between episodic memory and FMTT. Since there are multiple forms of FMTT and PMTT, what the discontinuist needs to show is that the particular form of FMTT that corresponds to the particular form of PMTT in question—episodic memory—lacks or can lack the phenomenology characteristic of that form of PMTT. And there is no reason to think that it does or can: when I imagine an event of my personal future, the event seems to belong to me, just as remembered events do. There may be subtle phenomenological differences between MTT into the personal future and episodic memory, stemming from the greater uncertainty of the future. But the fact that I cannot be sure whether an anticipated event will actually occur as represented does not imply that I cannot be sure that I am the person at issue in my representation of the event, just as the fact that I cannot be sure whether a remembered event actually occurred as represented does not imply that I cannot be sure that I am the person at issue in my representation of the event.[15] Moreover, episodic memory is not the only form of PMTT to involve autonoetic consciousness, since the latter is also involved in episodic counterfactual thought; and any phenomenological difference between MTT into the personal future is unlikely to be greater than the phenomenological difference between episodic memory and episodic counterfactual thought.

### 2.3.2. Mechanisms of Autonoesis

The second apparent discontinuity identified by Perrin concerns not the role but rather the production of autonoesis, specifically, the mechanism that brings autonoetic consciousness into play in a given episodic constructive process. He argues, first, that, in episodic memory, autonoesis is the result of an automatic, subpersonal-level monitoring process (roughly, a type 1 process), suggesting that it derives from the detection of certain features of the first-order episodic simulation process, perhaps its fluency. Due to its automatic, subpersonal-level character, autonoesis is impenetrable to belief in cases of episodic memory: even if one reflectively concludes that the represented event does not in fact belong to one's personal past, one will continue to feel as if it does. He argues, second, that, in FMTT, autonoesis (when it occurs) is the result of a deliberate, personal-level process (roughly, a type 2 process); for example, one might anticipate the occurrence of a certain event and infer, on the basis of semantic knowledge, that one will participate in the event in a given role. Due to its deliberate, personal-level character, autonoesis *is* penetrable to beliefs in cases of FMTT: if one concludes that the represented event does not in fact belong to one's personal future, one will cease to feel as if it does.

The first part of this argument is plausible. Given that other forms of PMTT can be just as fluent as episodic memory, the monitoring process is likely to rely on multiple factors, rather than fluency alone, but the production of autonoesis in episodic remembering might indeed result from a subpersonal-level monitoring process. The second part of the argument is much less plausible. On the one hand, if it were indeed possible for a personal-level process to generate autonoesis in the case of FMTT, the same process could in principle be responsible for generating autonoesis in the case of PMTT, including episodic memory. It is unclear, however, how a personal-level process might, in either case, result in the sort of primitive phenomenal state that autonoesis is generally taken to be. On the other hand, it is by no means clear that a subpersonal-level monitoring process could not be responsible for generating autonoesis in the case of FMTT. Whatever factors such a monitoring process might draw on in the case of episodic memory will also be available in the case of FMTT. Fluency alone will not discriminate perfectly between representations of one's own future experiences and those of other subjects, but neither does it discriminate perfectly between representations of one's own past experiences and those of other subjects.[16] Moreover, an account that posits two distinct mechanisms, one responsible for generating autonoesis in the case of PMTT, the other responsible for generating autonoesis in the case of FMTT, is unparsimonious; if an account positing a single mechanism is available, it is to be preferred. On the whole, then, it is possible that a subpersonal-level mechanism generates autonoesis in the case of episodic memory, while a personal-level mechanism generates autonoesis in the case of FMTT, but the more likely possibility is that a subpersonal-level mechanism is at work in both cases.

In short, neither of the apparent phenomenological discontinuities between episodic memory and FMTT identified by Perrin appears to be bona fide. Overall, the case for metaphysical discontinuism is weak.

## 3. AGAINST EPISTEMOLOGICAL DISCONTINUISM

If metaphysical discontinuism is incorrect, epistemological discontinuism is unlikely to be correct. Nevertheless, it is worth looking separately, if more briefly, at arguments for epistemological discontinuism. Even if metaphysical discontinuism is correct, epistemological discontinuism might turn out to be incorrect, since metaphysical discontinuities do not necessarily entail epistemological discontinuities. A discontinuist unconvinced by the considerations raised against metaphysical discontinuism in the foregoing thus might still be persuaded to reject epistemological discontinuism.

This section therefore discusses possible epistemological discontinuities between episodic memory and FMTT, looking at the epistemic openness of the future and the past (section 3.1), the directness or indirectness of our knowledge of past and future events (section 3.2), and immunity to error through misidentification—or the lack thereof—in episodic memory and FMTT (section 3.3).

### 3.1. The Epistemic Openness of the Future and the Past

A natural first thought to have about potential epistemological discontinuities between episodic memory and FMTT is that, if FMTT can provide us with any knowledge of the future at all, that knowledge is bound to be dramatically less secure than the knowledge of the past with which we are provided by episodic memory. As noted in section 2, the future is cognitively or epistemically open, whether or not it is metaphysically open: even if there is a determinate matter of fact about what will happen, we cannot be *certain* about what will happen. We might go further: even if there is a determinate matter of fact about what will happen, we cannot be certain that *anything at all* will happen—no matter how detailed our projections, it is entirely consistent with our epistemic situation vis-à-vis the future that the entire world will blink out of existence five minutes from now. In contrast, one might maintain, even if we cannot be fully certain that the details of our memories of past events are accurate, we can at least be certain that the world did not blink into existence five minutes ago. The contrast between our inability to rule out the unreality of the future and our ability to rule out the unreality of the past might thus seem to underwrite a qualitative asymmetry between our knowledge of future events and our knowledge of past events.

While this is a natural thought to have, it is mistaken. As Russell (1921) notoriously pointed out, we are not in fact able to rule out the unreality of the past. No matter how detailed our memories, no matter how subjectively convincing they are, it is entirely consistent with our epistemic situation vis-à-vis the past that the entire world blinked into existence five minutes ago, complete with our apparent memories of past events. This is, of course, just a particular instance of a general point about skepticism in epistemology: if we raise the standards for knowledge sufficiently high, requiring not just reliability but certainty, we deprive ourselves of virtually all knowledge, including knowledge of the very existence of the external world; if certainty is a prerequisite for knowledge, we have knowledge neither of past events, nor of future events, nor even of present events. Non-skeptics—and that includes virtually all of us—should thus set the standards for knowledge well below the level of certainty. And if knowledge presupposes only some lower level of reliability, we have no reason (so far) to think that there is a qualitative asymmetry between our knowledge of past events and our knowledge of future events. Our beliefs about future events may well be somewhat less reliable than our beliefs about past events, but this gives us a merely quantitative epistemological difference between episodic memory and FMTT.

## 3.2. Direct and Indirect Knowledge

If a qualitative epistemological difference between episodic memory and FMTT cannot be located at the level of reliability, perhaps it can be located elsewhere, either in the directness or indirectness of our knowledge of past and future events, or in the extent of our indirect knowledge.

### 3.2.1. Indirect Knowledge

As far as indirect or inferential knowledge goes, our beliefs about the future appear to be on a par with our beliefs about the past. Future events are not given but inferred, and our knowledge of them thus depends on the reliability of the inferences that we perform, on the basis of the evidence available to us in the present, about what is going to happen. But the very same thing goes for our knowledge of past events. As Kneale (1971, 11) puts it,

> We may . . . be misled by a truism, that what is past has happened and that what is future has not, into the belief that the past is necessarily more accessible for knowledge than the future. After all, it is *there* and we should be able to find out about it. But this does not follow. Past events have happened, but

> whether they can be known depends upon whether they are connected with present events by laws which allow of backward inference, and whether present observation, memory, and records provide us with sufficient information to apply such laws.

In general, past events are no more given than are future events, and our knowledge of them depends on the reliability of the inferences that we perform, on the basis of the evidence available to us in the present, about what must have happened. This point holds both for systematic, scientific inquiries into the past and the future and for episodic simulation of past and future events.

Why, if there is no qualitative difference between our indirect knowledge of past events and our indirect knowledge of future events, do we ordinarily take ourselves to know much more about the past than we do about the future? Kneale suggests that this habit is explained by two broad psychological tendencies. On the one hand, when we think about the past, we tend to focus on what we *do* know. Her examples are primarily historical, but this goes for episodic memory as well: there are whole periods of one's life about which one remembers little or nothing, but one is normally only dimly aware of these, preferring to focus on well-remembered events ("meeting my wife," "the birth of my child", . . .), as it is the latter that tend to be especially significant. On the other hand, when we think about the future, we tend to focus on what we *do not* know. Again, Kneale's examples are primarily historical, but this goes for FMTT as well: there are many future events that one can anticipate in great detail, but one normally focuses on especially significant events ("my child leaving home," "my retirement," . . .), and these are often difficult to envision in any detail.

This explanation might itself be taken to provide an alternative route to an epistemological discontinuity between episodic memory and FMTT. Our knowledge of the past, the discontinuist might argue, includes a great deal of specific information about events, whereas our knowledge of the future consists mostly of information about generic, repeated features of events—this is why we know more about the personally important events of the past than we do about the personally important events of the future. If the claim is merely that we know more about particular past events than we do about particular future events, it is probably right but does not establish a qualitative difference between episodic memory and FMTT.[17] If the claim is, more strongly, that, while we can have knowledge of particular past events, we cannot have knowledge of particular future events, it is vulnerable to the considerations raised in section 2.2 against Debus's positive and negative object claims. Thus the suggested route does not look promising.

### 3.2.2. Direct Knowledge

With respect to indirect or inferential knowledge, Kneale argues for continuity between our knowledge of the future and our knowledge of the past. With respect to direct or non-inferential knowledge, however, she grants that there is a fundamental discontinuity between them, in that only memory provides us with direct knowledge of events. Indeed, she treats "the capacity to have non-inferential self-certifying beliefs about particular matters of fact in the past" as definitive of memory and argues that "we have no such self-certifying beliefs about the future" (1971, 1–2). In this, she is in agreement with a number of other philosophers. Swinburne, for example, argues that "men have memory but not foreknowledge. Their non-inferential knowledge is of the past but not of the future" (1966, 167).

While this view has been popular among philosophers, it is not immediately clear what it might mean to say that memory is a direct source of knowledge. Kneale makes clear that her point is not about reliability: she acknowledges that memory is imperfectly reliable and suggests that, even if anticipation were just as reliable as memory, it would not count as a direct source of knowledge of the future. Ultimately, her point boils down to the claim that it is part of the very concept of memory that "the memory event [i.e., the occurrence of remembering] should have as a part-cause the occurrence of the event recollected" (Kneale, 1971, 2). Given this assumption, as well as the innocuous assumption that causes necessarily precede their effects, it follows that we are necessarily unable to "remember the future" (i.e., that we are necessarily unable to spontaneously anticipate future events, in parallel to the way we spontaneously remember past events). The apparent epistemological difference between our knowledge of the past and our knowledge of the future thus collapses into the apparent metaphysical difference that provides the basis for the causal theory of memory. And, as we saw in section 2.1, while the causal theory is intuitively plausible, there is good reason to reject it.

Kneale's discussion of "direct" memory knowledge is nevertheless instructive, in that it provides an important clue to the source of the appeal of the thought that there must be a fundamental epistemological asymmetry between episodic memory and FMTT. Consider her discussion of alleged cases of "pre-cognition" (1971, 2):

> The pre-cognitions investigated by the Society for Psychical Research do not correspond to memories, because the investigator at least, and generally the percipient too, regards them as subject to confirmation, and no serious scientist or historian would think of such "precognitions" as providing premises for further inferences. . . . We have no faculty for recognizing truths about the future analogous to memory.

The point is not that precognition does not exist. The point—and this is independent of the causal theory—is rather that, even if it were to exist, we would be reluctant to regard it as a source of knowledge in the same way that we regard memory as a source of knowledge, because it would always be subject to further confirmation. Similarly, we can observe that, whereas we regularly treat memory, for example in legal contexts, as a more or less decisive source of evidence about what happened, we do not have any equivalent practice of treating FMTT as a source of evidence about what will happen. More generally, when someone claims to remember the occurrence of an event, we often accept his claim at face value, but when someone anticipates the occurrence of an event, we normally want additional confirming evidence.

There is thus an asymmetry between our epistemic practices with respect to episodic memory and our epistemic practices with respect to FMTT. Whether this difference at the level of epistemic *practice* reflects an underlying difference of epistemic *status* is another question. There is good reason to think that it does not. Paradoxically, the practical difference seems to arise because confirming or disconfirming evidence is often available, at least in principle, at the time at which a subject remembers an event, whereas confirming evidence cannot be available at the time at which a subject anticipates an event. The fact that a memory can be confirmed at the time at which it occurs reassures us; we are therefore less inclined to ask for confirmation. The fact that a prediction cannot be confirmed at the time at which it occurs disquiets us; we are therefore more inclined to ask for confirmation. But as far as the epistemic status of the representations respectively produced by episodic memory and FMTT are concerned, there may yet be no qualitative difference between them.

Within a reliabilist framework (e.g., Goldman, 2012)—analogous points can be made within most other epistemological frameworks—the epistemic status of a representation is determined simply by the reliability of the process that produced it. If episodic memory and FMTT are both sufficiently reliable for knowledge, they both produce knowledge, even if we have available processes for *confirming* that we have knowledge in the case of episodic memory (e.g., seeking corroborating physical evidence) and no such process available for confirming that we have knowledge in the case of FMTT. The point is that one can know without being in a position to know that one knows. When, seeming to remember an event, I ask myself whether I know that the event occurred, I often have available some means of answering the question. In contrast, when, anticipating an event, I ask myself whether I know that the event will occur, I normally have available no means of answering the question. But the fact that I cannot know whether I know does not imply that I do not know. Our epistemic practices with respect to epistemic memory and FMTT thus suggest at most a difference between our knowledge of our knowledge of the past and our knowledge of our knowledge of the future, but this does not entail a difference between our knowledge of the past itself and our knowledge of the future itself.

## 3.3. Immunity to Error Through Misidentification

Perrin (Chapter 3 of this volume) argues for a final, subtler epistemological discontinuity between episodic memory and FMTT: on his view, FMTT displays immunity to error through misidentification, and episodic memory does not. Some background is required in order to understand this claim. First, we need the notion of *error through misidentification*. A thought involves an error through misidentification when it is incorrect due to the fact that it attributes an incorrect identity to something or someone. For example, suppose that I hear the phone ring and think, "my wife is calling," when, in reality, it is my colleague who is calling. In this case, my thought correctly describes an event—someone really is calling—but it is nonetheless incorrect overall because I am mistaken about the identity of one of the people involved in the event. Second, we need the notion of *immunity* to error through misidentification. A thought manifests immunity to error through misidentification when it is impossible for it to be incorrect due to the fact that it attributes an incorrect identity to something or someone. My thought that my wife is calling is, obviously, not immune to error through misidentification. Even thoughts about oneself need not be immune to error through misidentification. For example, if I mistake my reflection in a mirror for that of a colleague and think "my colleague is wearing a sport coat" when in fact I am the one wearing a sport coat, I commit an error through misidentification. But certain thoughts about oneself are arguably immune to error through misidentification. For example, if I introspect and think "I feel hungry right now," I arguably cannot be mistaken about the identity of the person who feels hunger.

The first part of Perrin's claim, then, is that when a subject episodically remembers, his thoughts are *not* immune to error through misidentification: it is possible for them to be mistaken due to the fact that they attribute an incorrect identity to something or someone. Errors through misidentification in episodic memory may result from attributions of incorrect identity to oneself,[18] but they may also result from attributions of incorrect identity to other subjects. For example, I might remember talking to a specific colleague but misidentify him, perhaps because I mistook him at the time of the conversation for another colleague. The second part of Perrin's claim is that, when a subject imagines a future episode, her thoughts *are* immune to error through misidentification: it is impossible for them to be mistaken due to the fact that they attribute an incorrect identity to something or someone. The thought here is that, when, for example, I imagine talking to a colleague tomorrow, I cannot attribute an incorrect identity to the colleague in question. One cannot be mistaken about the identities of the entities involved in imagined events, simply because the identities of the entities one imagines are wholly determined by one's intentions. In contrast, one can be mistaken about the identities of the entities involved in remembered events, because the identities of the entities one remembers are not wholly determined by one's intentions. This is due to the fact that episodic memory is "world-involving" (Recanati, 2007): the content of one's memory is partly determined by one's interactions with the world.

Faced with this argument, the epistemological continuist has three options: first, he might maintain that *neither* episodic memory nor FMTT is immune to error through misidentification; second, he might maintain that *both* episodic memory and FMTT are immune to error through misidentification; third, he might maintain that *some* instances of episodic memory and *some* instances of FMTT are immune to error through misidentification, while others are not. Given the characterizations of FMTT and episodic memory developed in previous sections, the last option seems likely to be correct.

Might some instances of FMTT fail to be immune to error through misidentification? We have already seen why it is plausible to hold that error through misidentification is possible in episodic memory: one's past interactions with the entities involved in the events that one remembers affect the content of one's memories, with the consequence that one's own judgement about the identities of the entities involved in the events that one remembers may be mistaken. The same line of thought, however, suggests that error through misidentification is possible in FMTT. As noted in section 2, one may imagine a future event by remembering a past event and imaginatively projecting that event into the future. In such cases, one clearly imagines events involving entities with which one has previously interacted. Even when one imagines more novel events, one often imagines events involving entities with which one has previously interacted, and if previous interactions suffice to fix identity in cases of episodic memory, they presumably do so as well in cases of FMTT. Consider again a case in which I imagine talking to a specific colleague tomorrow. While it is natural to suppose that my intentions determine the identity of the colleague in question (that I effectively stipulate his identity), this need not be the case. Episodic memory is world-involving, if it is, because the subject has causally interacted with the entities involved in his memories; those past causal interactions can arguably partly determine the content of his memory representations. But the same mechanism allows FMTT to be world-involving: if the subject imagines a conversation with a person with whom she has interacted in the past, the past causal interaction arguably determines the identity of the person of whom she is thinking. Thus when I imagine a conversation with a specific colleague, I might be mistaken about the identity of the colleague in question, perhaps because I mistook him during our past interactions for another colleague. If this is right, then in at least some cases FMTT is not immune to error through misidentification.[19]

This argument only shows that *some* instances of FMTT are not immune to error through misidentification; if other instances of FMTT are immune to error through misidentification and no instances of episodic memory are immune to error through misidentification, the discontinuist can maintain that there is an important epistemological asymmetry between episodic memory and FMTT. But might some instances of episodic memory be immune to error through misidentification? Perrin's argument tacitly assumes that episodic memory always satisfies the factivity constraint: genuine memory is always memory of events that actually occurred. But we saw in section 2 that memory of events that did not actually occur has just as much claim to be genuine memory as memory of events that actually occurred (i.e., that episodic memory does not always satisfy the factivity constraint). In cases where the factivity constraint is not satisfied, the subject need not have interacted with the entities involved in the event he remembers. Thus there are no causal interactions available to fix the identities of those entities, and their identities are presumably fixed by the subject's intentions: just as, when I imagine a wholly novel future event, my intentions determine the identities of the things and persons involved, when I remember a wholly novel past event, my intentions determine the identities of the things and persons involved.

Thus it appears that both parts of Perrin's claim are to be rejected. It is not the case that, when a subject episodically remembers, his thoughts are never immune to error through misidentification; in some cases, they may not be, but in other cases they may be. And it is not the case that, when a subject imagines a future episode, his thoughts are always immune to error through misidentification; in some cases, they may be, but in other cases they may not be. In short, episodic memory and FMTT appear to be on a par, as far as immunity to error through misidentification is concerned.

## 4. TOWARD A POSITIVE CASE FOR EPISTEMOLOGICAL CONTINUISM

This chapter has largely been devoted to responding to the available arguments for metaphysical and epistemological continuism. Supposing that the response has been successful in undermining those arguments, should we move to metaphysical and epistemological continuism? The focus of positive continuist arguments thus far has been on metaphysical continuism, with psychologists and philosophers appealing to MTT research to support a picture of remembering the past and imagining the future as qualitatively similar constructive processes. What remains to be provided, then, is a positive case for epistemological continuism, the view that our knowledge of future events is qualitatively similar to our knowledge of past events.

Such a case may be less difficult to develop than one might initially suspect. Against the background of metaphysical continuism, Michaelian (2016) develops a reliabilist account of our knowledge of past events, arguing that episodic memory provides reliable access to the personal past despite its simulational character. The account involves two core ideas: first, that mental simulation of past events is governed by reliable heuristics; second, that the threat of unreliability introduced by the simulational character of remembering is further reduced by reliable metacognitive monitoring. It may be possible to extend this account to our knowledge of future events in a straightforward manner. Once we give up the idea that remembering necessarily involves a causal connection to the relevant past event (i.e., the causal theory) and the idea that remembering puts us in direct contact with the relevant past event (i.e., direct realism), it becomes clear that memory knowledge, in line with other forms of indirect, inferential knowledge, is underwritten entirely by the reliability of the process of remembering. FMTT, of course, involves neither a causal connection to nor direct contact with the relevant future event, which suggests that here, too, knowledge is underwritten entirely by the reliability of the process of imagining. Thus our knowledge of the personal future need pose no special epistemological challenge: we acquire knowledge of future events, just as we acquire knowledge of past events, through reliable simulation and reliable metacognitive monitoring.

Such a reliabilist account of our knowledge of the personal future dovetails nicely with recent responses to the charge that imagination is "epistemically irrelevant" (Kind, forthcoming) (cf. Balcerak Jackson, forthcoming). As Kind points out, philosophers such as Sartre (1948) and Wittgenstein (1980) have argued that the imagination is fundamentally epistemically irrelevant: while it can play an incidental role in the generation of knowledge, for example by leading us to formulate hypotheses which can then be verified by other means, it cannot itself generate new knowledge. Kind identifies three assumptions that have been used to ground this conclusion: first, imagination is typically under our voluntary control; second, it is not world-sensitive, in the sense that "its content is determined by the imaginer, not by the world"; third, imagination is uninformative, in the sense that "an act of imagining can provide us with no new information."

Kind grants the first two assumptions, but argues that they do not imply that imagination cannot aim at the truth. The epistemological continuist might also challenge the first two assumptions more directly. As far as the first assumption is concerned, much FMTT is spontaneous or involuntary (Berntsen & Jacobsen, 2008). As far as the second assumption is concerned, this amounts to a version of the claim that FMTT is not world-involving, and we have already seen that this claim can be challenged. But Kind focuses her resistance to the charge of epistemic irrelevance on the third assumption, arguing that imagination can indeed provide the imaginer with new information. The thought behind the third assumption is that a representation produced by imagination does not contain anything that the imaginer did not put into it, which suggests that it cannot provide him with new information. Kind points out that this is simply mistaken. A computer program, for example, contains nothing that the programmer did not put into it, and yet it can of course provide new information when it is run. The same thing goes for imagination: in episodic imagination, one simulates an event, but the simulation can provide one with new information. Consider, for example, the case of mind reading or theory of mind: when one simulates the mental states of another subject, the simulation contains nothing that one did not put into it, but can nevertheless, assuming that the simulation process is reliable, provide one with new information about the target's mental states. Similarly, when one simulates one's own possible future experience, the simulation will contain nothing that one did not put into it, but can nevertheless, assuming that the simulation process is reliable, provide one with new information about those experiences.

This is only a sketch of a positive account, but it is enough to suggest that the reliabilist explanation of how episodic memory provides us with knowledge of past events can be extended to explain how FMTT provides us with knowledge of future events.[20] The future of continuism about our knowledge of past and future events thus looks bright.

## ACKNOWLEDGMENTS

Thanks to Dorothea Debus, Denis Perrin, André Sant'Anna, and audiences at the University of Auckland and the University of Waikato for comments.

## NOTES

1. Debus formulates her view in terms of "recollective memory" and "sensory imagination of future events," but these terms map onto "episodic memory" and "future-oriented mental time travel."
2. Similarities and differences at the phenomenological level have also been important in MTT research; possible phenomenological discontinuities between episodic memory and FMTT are discussed at length in section 2.3.
3. And ignoring Reid's direct realism; see section 2.2 on direct versus indirect theories of memory.
4. Debus (2014), for example, argues that her discontinuist view is compatible with essentially all empirical MTT research.
5. One might also argue for a functional discontinuity, suggesting that the function of episodic memory is to provide information that can be recombined to permit the simulation of possible future events, the latter being the function of FMTT. If this view of the respective functions of episodic memory and FMTT were accurate, it would indeed support a difference in kind between them. But the argument depends on an equivocation, treating episodic memory as a *store*—a source of raw materials for the simulation of episodes—and FMTT as a *process* that draws on the materials provided by episodic memory to simulate events. Bearing in mind the distinction between episodic *memory* (understood as a store) and episodic *remembering* (understood as a process), it is equally plausible to suggest that the function of episodic memory is to provide information that can be recombined to permit the simulation of past events, either counterfactual past events, in episodic counterfactual thinking, or actual past events, in episodic remembering. As far as functional considerations go, then, the fact that episodic memory provides raw

materials for the simulation of possible future events does not support an asymmetry between episodic remembering and FMTT. There is, moreover, a case to be made against the view that there is a dedicated store of episodic information. Klein (2015), in particular, argues that what makes a form of cognition episodic is a feature at the level of phenomenology, rather than information or content, namely, the involvement of autonoetic consciousness. If there is no dedicated episodic store, the functional argument does not get going in the first place.

6. The initial stages of the discussion employ the language of representations; for reasons given later in this chapter, Debus herself rejects this language.
7. Debus claims that experiential awareness *supervenes* on these factors. As philosophers use the notion of supervenience, experiential awareness will supervene on previous experience and causal connection just in case there can be no difference at the level of experiential awareness without a corresponding difference at the level of previous experience and causal connection. Debus resists the claim that experiential awareness *reduces* to these factors. Whether the latter claim is actually stronger than the former—i.e., whether supervenience is sufficient for reduction—is a technical question with which we need not be concerned here.
8. The qualification that causation must occur "in the right way" is required to cope with deviant causal chains, a problem that need not concern us here; see Michaelian (2011) for background.
9. Both paths assume that the causal link supposed to be characteristic of genuine remembering is an informational link, that is, that the causal connection between the retrieved memory representation and the original experience goes via stored information deriving from the experience. In principle, the causal theorist might deny this assumption, but it is not clear that a non-informational characterization of non-deviant causal chains can be given.
10. What sort of account is available, if this picture is right? One possibility worth exploring is that genuine remembering is distinguished from merely apparent remembering simply by the fact that the subject's episodic construction system takes an episode from the subject's actual personal past, as opposed to the personal future or the counterfactual personal past, as its target. See Michaelian (2016) for discussion of this possibility. Another possibility is that genuine remembering is distinguished from merely apparent remembering by the subject's personal-level intentions. See Hoerl (2014) for discussion of this possibility. The remainder of the argument of this chapter does not depend on a particular alternative to the causal condition as a means of distinguishing between genuine and merely apparent memory.
11. If the causal theory is incorrect, how can the reliability of remembering be explained? This question is taken up in section 4.
12. As noted earlier, this means that it is misleading to describe Debus's view using the language of memory *representations*, as we have been doing so far; but nothing in the foregoing argument depends on this.
13. As far as the distinction between genuine and merely apparent memory is concerned, the rejection of factivity implies that the distinction rests entirely on the second sort of difference between genuine and merely apparent memory that rival theories of remembering attempt to capture. Thus it might rest, for example, entirely on the aims of the episodic construction system or the subject's personal-level intentions.

14. Klein has made a related point; see Klein (2016), Klein & Steindam (Chapter 7 of this volume).
15. Perrin might grant this point, arguing that the difference concerns the source of the certainty: FMTT is, and episodic memory is not, immune to error through misidentification. This argument is discussed in section 3.
16. See Michaelian (2012; 2016) for suggestions as to other potential factors that might be involved in metacognitive monitoring of episodic construction.
17. Even if it is right, there may be a chicken-and-egg problem here: To what extent do we remember events because they are important to us, and to what extent are events important to us because we remember them?
18. As in hypothetical cases of "quasi-memory," in which the memories of one subject are "transplanted" into another subject (Shoemaker, 1970).
19. The discontinuist might object that the world-involvingness of FMTT is parasitic on that of episodic memory, since it depends on traces laid down during past causal interactions—that is, on episodic memories. But this objection depends on ignoring the distinction (noted earlier in connection with alleged functional disanalogies between episodic memory and FMTT) between memory as a store and remembering as a process. As we have seen, the same stored content is employed both in episodic remembering and in FMTT; there is no convincing reason to assign any priority to the former process, rather than the latter.
20. An alternative strategy for securing the epistemological continuity of episodic memory and FMTT is to grant the epistemic irrelevance of FMTT but to argue that episodic memory is also epistemically irrelevant. Hopkins, for example, argues that "[i]f in memory we represent the past to ourselves as being a certain way, and if representing it so is an action of ours, it seems we can only get the past right if we in some way already know how it was. Remembering cannot be the source of knowledge of how things were, since it presupposes such knowledge" (2014, 321); this amounts to the suggestion that episodic memory cannot provide us with new information. Those who acknowledge the generative, simulational character of remembering will not be tempted to avail themselves of this argument.

## REFERENCES


Balcerak Jackson, M. (Forthcoming). Justification by imagination. In F. Dorsch & F. Macpherson (eds.), *Perceptual memory and perceptual imagination*. Oxford: Oxford University Press.

Bernecker, S. (2008). *The metaphysics of memory*. Dordrecht: Springer.

Berntsen, D., & Jacobsen, A. S. (2008). Involuntary (spontaneous) mental time travel into the past and future. *Consciousness and Cognition*, *17*(4), 1093–1104.

D'Argembeau, A., & Van der Linden, M. (2004). Phenomenal characteristics associated with projecting oneself back into the past and forward into the future: influence of valence and temporal distance. *Consciousness and Cognition*, *13*(4), 844–858.

De Brigard, F. (2014). Is memory for remembering? Recollection as a form of episodic hypothetical thinking. *Synthese*, *191*(2), 155–185.

De Brigard, F., Addis, D., Ford, J., Schacter, D., & Giovanello, K. (2013). Remembering what could have happened: neural correlates of episodic counterfactual thinking. *Neuropsychologia*, *51*(12), 2401–2414.

Debus, D. (2008). Experiencing the past: a relational account of recollective memory. *Dialectica*, *62*(4), 405–432.

Debus, D. (2014). "Mental time travel": remembering the past, imagining the future, and the particularity of events. *Review of Philosophy and Psychology*, *5*(3), 333–350.

Goldman, A. I. (2012). *Reliabilism and contemporary epistemology: essays*. Oxford: Oxford University Press.

Hoerl, C. (2014). Remembering events and remembering looks. *Review of Philosophy and Psychology*, *5*(3), 351–372.

Hopkins, R. (2014). Episodic memory as representing the past to oneself. *Review of Philosophy and Psychology*, *5*(3), 313–331.

Hume, D. (1739). *A treatise of human nature*. London: John Noon.

Kind, A. (Forthcoming). How imagination gives rise to knowledge. In F. Dorsch & F. Macpherson (eds.), *Perceptual memory and perceptual imagination*. Oxford: Oxford University Press.

Klein, S. B. (2013). The complex act of projecting oneself into the future. *Wiley Interdisciplinary Reviews: Cognitive Science*, *4*(1), 63–79.

Klein, S. B. (2015). What memory is. *Wiley Interdisciplinary Reviews: Cognitive Science*, *6*(1), 1–38.

Klein, S. B. (2016). Autonoetic awareness: reconsidering the role of episodic memory in future-oriented mental time travel. *Quarterly Journal of Experimental Psychology*, *69*(2), 381–401.

Klein, S. B., & Nichols, S. (2012). Memory and the sense of personal identity. *Mind*, *121*(483), 677–702.

Kneale, M. (1971). Our knowledge of the past and of the future. *Proceedings of the Aristotelian Society*, *72*, 1–12.

Loftus, E. (1996). *Eyewitness Testimony*. 2nd ed. Cambridge, MA: Harvard University Press.

Martin, C. B., & Deutscher, M. (1966). Remembering. *Philosophical Review*, *75*(2), 161–196.

Michaelian, K. (2011). Generative memory. *Philosophical Psychology*, *24*(3), 323–342.

Michaelian, K. (2012). (Social) metacognition and (self-)trust. *Review of Philosophy and Psychology*, *3*(4), 481–514.

Michaelian, K. (2016). *Mental time travel: episodic memory and our knowledge of the personal past*. Cambridge, MA: MIT Press.

Recanati, F. (2007). *Perspectival thought: a plea for (moderate) relativism*. Oxford: Oxford University Press.

Reid, T. (1764). *An inquiry into the human mind*. Dublin: Printed for Alexander Ewing.

Russell, B. (1921). *The analysis of mind*. London: George Allen & Unwin.

Sartre, J.-P. (1948). *The psychology of imagination*. New York: Philosophical Library.

Schacter, D. L., & Addis, D. R. (2007). The cognitive neuroscience of constructive memory: remembering the past and imagining the future. *Philosophical Transactions of the Royal Society of London B: Biological Sciences*, *362*(1481), 773–786.

Schacter, D. L., Addis, D. R., Hassabis, D., Martin, V. C., Spreng, R. N., & Szpunar, K. K. (2012). The future of memory: remembering, imagining, and the brain. *Neuron*, *76*(4), 677–694.

Shoemaker, S. (1970). Persons and their pasts. *American Philosophical Quarterly*, *7*(4), 269–285.

Suddendorf, T., & Corballis, M. C. (2007). The evolution of foresight: what is mental time travel, and is it unique to humans? *Behavioral and Brain Sciences*, *30*, 299–313.

Sutton, J., & Michaelian, K. (Forthcoming). Memory. In E. N. Zalta (ed.), *Stanford Encyclopedia of Philosophy*.

Swinburne, R. G. (1966). Knowledge of past and future. *Analysis*, *26*(5), 166–172.

Szpunar, K. K. (2010). Episodic future thought: an emerging concept. *Perspectives on Psychological Science*, *5*(2), 142–162.

Szpunar, K. K., Watson, J. M., & McDermott, K. B. (2007). Neural substrates of envisioning the future. *Proceedings of the National Academy of Sciences*, *104*(2), 642–647.

Tulving, E. (2002). Chronesthesia: conscious awareness of subjective time. In R. T. Knight (ed.), *Principles of frontal lobe function* (pp. 311–325). Oxford: Oxford University Press.

Wittgenstein, L. (1980). *Remarks on the philosophy of psychology.* Chicago: University of Chicago Press.

5

# Bidirectional Interactions Between Memory and Imagination

ALEEA L. DEVITT AND DONNA ROSE ADDIS

Our inner world is invariably self-focussed and autobiographical. We reflect on our experiences, situating these episodes and our reactions within our view of ourselves and the narrative of our lives (Prebble, Addis, & Tippett, 2013). If a particular experience resulted in less than favorable outcomes, we might mentally rework the event to imagine new endings and glean lessons for the future (cf. counterfactual thinking; De Brigard, Addis, Ford, Schacter, & Giovanello, 2013). Also with a view to enhancing future outcomes, we can simulate events that are yet to occur, an ability that enables us to strategize, predict consequences, and make effective decisions (Benoit, Gilbert, & Burgess, 2011; Boyer, 2008; Gerlach, Dornblaser, & Schacter, 2014; Peters & Büchel, 2010; Schacter, Guerin, & St Jacques, 2011; Schacter, 2012).

While the cognitive and neural processes underlying autobiographical memory have been researched for decades, it was only in 2007 that the field shifted to consider the role that memory may play in broader, non-mnemonic functions such as future thinking. Following up on an early report that an amnesic patient struggled to think about his future (Tulving, 1985), the initial focus of the field has been to document parallel deficits of remembering and imagining in various clinical populations with varying degrees of memory impairments. Such evidence, along with neuroimaging findings that memory and imagination engage similar neural networks, strongly suggests that memory plays an important role in future thinking. Indeed, we have even described representations of future events as "memory-based simulations" (Addis & Schacter, 2013). What is becoming increasingly apparent, however, is that the relation between memory and imagination is not simply unidirectional. Although memory is critical for imagination, by providing the details that comprise a simulation, in turn the process of imagining events can influence the content, phenomenology, and veracity of memories.

In this chapter, we first consider various lines of evidence that memory contributes to the process of simulating future events, as well as more recent work suggesting that memory quality can influence the phenomenology of simulations. Second, we consider research suggesting that imagination can influence both the accurate encoding of imagined representations and the formation of false memories. Like autobiographical memories, simulations of the past and future can be episodic in nature—when one imagines a specific event that involves the self in a particular spatiotemporal context (Tulving, 1972, 1985)—and in this chapter, we focus primarily on episodic memory and its relation with episodic simulations. Nevertheless, we also consider semantic memory and semantic future thinking where relevant. Moreover, our focus here is primarily on future episodic simulations, given the adaptive function of imagining the future (Boyer, 2008; Klein, 2013; Schacter & Addis, 2007; Schacter, 2012; Suddendorf & Corballis, 1997, 2007; Tulving, 2005), although the same processes are presumably applicable to simulations of imaginary past and atemporal events.

## 1. CONTRIBUTIONS OF MEMORY TO IMAGINATION

> To build a future, you have to know the past.
>
> —Otto Frank *(Mazer, 1967)*

### 1.1. Episodic Memory Is Critical for Imagination

As mentioned earlier, a major focus of research on future thinking has been to establish its link with episodic memory for past events. Neuropsychological studies have provided much evidence for the overlap between past and future, with the most striking findings from studies of severely amnesic patients. Such patients have little or no ability to remember the past as a result of damage to critical memory structures such as the medial temporal lobe (MTL) and prefrontal cortex (PFC). Moreover, they also experience significant difficulties in imagining the future; either their mind draws a blank (e.g., Patient K. C.; Tulving, 1985), or the simulations they do produce are fragmented and devoid of specific episodic detail (Andelman, Hoofien, Goldberg, Aizenstein, & Neufeld, 2010; Hassabis, Kumaran, Vann, & Maguire, 2007; Kwan, Carson, Addis, & Rosenbaum, 2010; Race, Keane, & Verfaellie, 2011). Impairments in future simulations have also been documented in other clinical populations with memory deficits ranging from mild (e.g., healthy aging: Addis, Musicaro, Pan, & Schacter, 2010; Addis, Wong, & Schacter, 2008; Cole, Morrison, & Conway, 2012; De Beni et al., 2013; Gaesser, Sacchetti, Addis, & Schacter, 2011; Rendell et al., 2012; mild cognitive impairment: Gamboz et al., 2010) to severe (e.g., Alzheimer's disease: Addis, Sacchetti, Ally, Budson, & Schacter, 2009; Irish, Addis, Hodges, & Piguet, 2012; depression: Williams et al., 1996; and transient global amnesia: Juskenaite et al., 2014). For instance, with advancing age, individuals exhibit a decline in the number of episodic details comprising remembered past and imagined future events (Addis et al., 2010, 2008; Gaesser et al., 2011).

It is possible, however, that parallel changes in memory and imagination could reflect the decline of some other function common to both memory and imagination, such as the ability to describe mental representations in detail. To assess the contribution of narrative construction ability, a number of studies have utilized a picture description task (Gaesser et al., 2011) in which individuals are asked to describe the story depicted in a visual scene. Interestingly, participants with episodic amnesia who are impaired at past and future event tasks show no deficit on the picture description task (Race et al., 2011). Moreover, although the number of details provided in the picture narratives is significantly associated with the number of episodic details comprising future simulations, memory ability (i.e., the number of episodic details recalled in past events) still makes a significant contribution over and above narrative ability (Gaesser et al., 2011). This observation further supports the notion that being able to access details in episodic memory is critical to imagining the future.

Numerous imaging studies have established that there is striking overlap between the neural regions engaged during memory and imagination (see Figure 5.1; for reviews see Schacter, Addis, & Buckner, 2007; Schacter et al., 2012). Specifically, the regions comprise a whole-brain network of medial and lateral prefrontal, temporal, and parietal regions (cf. autobiographical memory retrieval network; Maguire, 2001; Spreng, Mar, & Kim, 2009); it encompasses the default mode network (Andrews-Hanna, 2012; Buckner, Andrews-Hanna, & Schacter, 2008) and connected structures such as the hippocampus (Ward et al., 2014). The network of overlapping functional activations is so compelling it has been called the "core network" due to its common role in various forms of simulation (including memory, future imagination, navigation, and theory of mind; Buckner & Carroll, 2007; Spreng, Mar, & Kim, 2009). However, it should be noted that memory and imagination do engage some of the regions in this network to differing degrees. For example, future simulation engages the frontal polar cortex and anterior hippocampus more than retrieval (Addis, Pan, Vu, Laiser, & Schacter, 2009; Addis, Wong, & Schacter, 2007; Addis & Schacter, 2008), while retrieval is more reliant on posterior regions associated with visuospatial detail (Addis et al., 2009a; Gilmore, Nelson, & McDermott, 2014). Nevertheless, it seems likely that the overlapping neural correlates reflect the core processes required by the internal representation of remembered or imagined episodes, irrespective of content (e.g., integration of details, Schacter & Addis, 2007; spatial processing, Hassabis & Maguire, 2007;

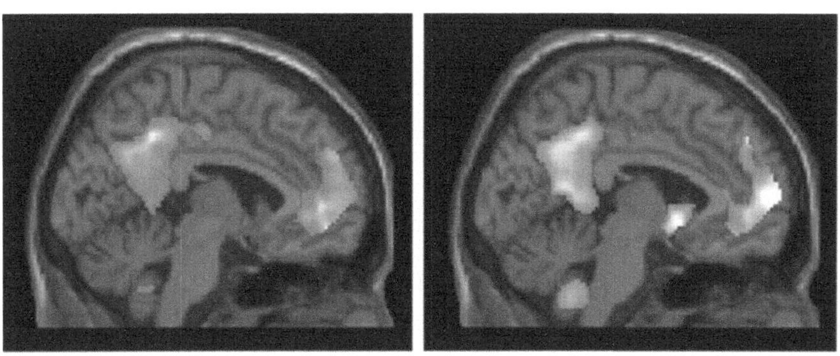

**Figure 5.1.** Activation of the core network during the retrieval of past events (*left*) and simulation of future events (*right*) relative to a control task. These medial slices illustrate the striking overlap in left medial prefrontal and parietal activity. (See color insert). Adapted from Addis et al. (2007).

self-projection, Buckner & Carroll, 2007), while the differences relate to task- or content-specific processes (e.g., intentional thinking and heightened constructive demands of future simulation, Addis et al., 2007; the increased visuospatial content of memories, Addis et al., 2009a).

One hypothesis strongly motivated by the overlap evident in the neuropsychological and neuroimaging studies is that memory provides something critical to the process of imagination. In the constructive episodic simulation hypothesis (Schacter & Addis, 2007), we have argued that access to details stored in episodic memory is necessary for both the recapitulation of episodic memories (Johnson & Rugg, 2007; Woodruff, Johnson, Uncapher, & Rugg, 2005) and the construction of novel simulations (van Mulukom, Schacter, Corballis, & Addis, 2013). Imagining personally relevant episodes requires the retrieval of autobiographical details, including people known to the individual and places they have experienced. That is, a scenario is built by integrating agents acting on items within a spatial context (cf. the binding of items and contexts model; Ranganath, 2010). Similarly, in their scene construction hypothesis, Maguire and colleagues argue that access to mnemonic information—specifically spatial details—is the critical framework on which one builds a coherent imagined scenario (Hassabis & Maguire, 2007; Maguire & Mullally, 2013).

There is some evidence that appears to speak against this interpretation: reports of patients with amnesia who can imagine the future without difficulty. That is, despite their memory impairments, they can construct detailed imaginings of the future. However, on closer inspection, it is apparent that these patients have a temporally graded amnesia affecting only recent memory. For instance, Squire et al.'s (2010) group of patients could retrieve events from more than a year ago, and showed only a non-significant deficit for events in the past year. Similarly, a patient with mammillary body damage and retrograde amnesia as a result of Korsakoff's syndrome was able to imagine the future (Tedder, Miller, & Lah, 2011). However, again the amnesia was temporally graded and limited to recent events. Thus it appears that an inability to access recent episodic memories is not sufficient to disrupt future simulation if the patients can draw on more remote episodic memories as fodder for simulations. Therefore, these cases provide further evidence for the notion that access to episodic memories—recent and/or remote—is a necessary condition for imagination.

### 1.2. The Quality of Episodic Memory Influences the Phenomenology of Imagination

Like episodic memories, a defining feature of episodic simulations is that they are associated with a subjective experience. We can pre-experience an imagined event with an intimate connection to the people, objects, and spatiotemporal context, seeing the event—yet to occur in reality—unfold in our mind's eye in vivid detail. However, simulations can also lack vivid detail and protracted plots, instead being just a static image of some future situation, and approximately half of future-oriented thoughts in daily life are conceptual in nature (D'Argembeau, Renaud, & Van der Linden, 2011; see Section 1.3 for more on semantic future thinking).

A number of factors have been shown to influence the phenomenology of imagined future events. The abilities of an individual can influence the nature of his or her simulations. For example, D'Argembeau and Van der Linden (2006) found that those with heightened visual imagery abilities imagine events that contain greater sensory details. In contrast, individuals who regulate their emotional experience via suppression tended to generate future events with less emotional detail; interestingly, this effect extends to sensory and contextual detail. This finding is broadly consistent with studies finding that depressed individuals typically generate future events that are reduced in specificity (Williams et al., 2007) and episodic detail (Murray, Hach, Tippett, & Addis, 2014) relative to non-depressed controls.

Being able to draw upon and use memory in a creative fashion may also influence the vividness of future events. In a recent study, we explored whether divergent thinking—the ability to generate ideas by combining disparate forms of information in new ways—is associated with future simulation, given that it requires the extraction and integration of episodic details to create novel scenarios. Indeed, we found that performance on the Alternate Uses Test, a standard measure of divergent thinking, was associated with the generation of future simulations that were richer in episodic detail (Addis, Pan, Musicaro, & Schacter, 2016). Interestingly, this effect did not extend to imagining past events, suggesting that the open-ended nature of future events allow for such individual differences to emerge.

One of the most influential factors on subjective experience is the temporal distance of an imagined event from the present. The majority of future thoughts relate to happenings within the next month (D'Argembeau et al., 2011; Spreng & Levine, 2006), and close future events are more specific in nature, containing significantly more sensorial and contextual details that more distant events (D'Argembeau & Van der Linden, 2004, 2012), and experienced from a first-person perspective more often (D'Argembeau & Van der Linden, 2012). It is likely that this temporal pattern relates to the fact that this trend is mirrored in past events: the majority of retrieved events are from the past month (Spreng & Levine, 2006) and more recent events are more vividly detailed (D'Argembeau & Van der Linden, 2004). Therefore, when contemplating current concerns or goals, more recent memory details are most relevant, and the majority of these are vivid and highly accessible—and the resulting simulations take on this vivid nature. Indeed, D'Argembeau and Van der Linden (2012) report that the effects of temporal distance on the vividness of imagined events are mediated by the vivid nature of the details used to construct the simulation.

Thus, if episodic details are the constituent elements of an episodic simulation, then the nature of the details extracted and recombined into an imagined event will influence the quality of the imagined representation itself. We recently investigated this hypothesis directly by gathering detailed phenomenological ratings on episodic details that are later used to construct simulations (McLelland, Devitt, Schacter, & Addis, 2015). Using an adapted version of the experimental recombination paradigm (Addis, Pan, Vu, Laiser, & Schacter, 2009; see Figure 5.2), participants first retrieved 110 episodic memories of recent events (occurring in the past 5 years). Importantly, they identified a person, place, and object featuring within each event, and then rated each of these episodic details for familiarity (i.e., how well a particular detail was known), emotionality (i.e., the intensity of emotion elicited by the detail), and personal significance (i.e., how important the detail is within

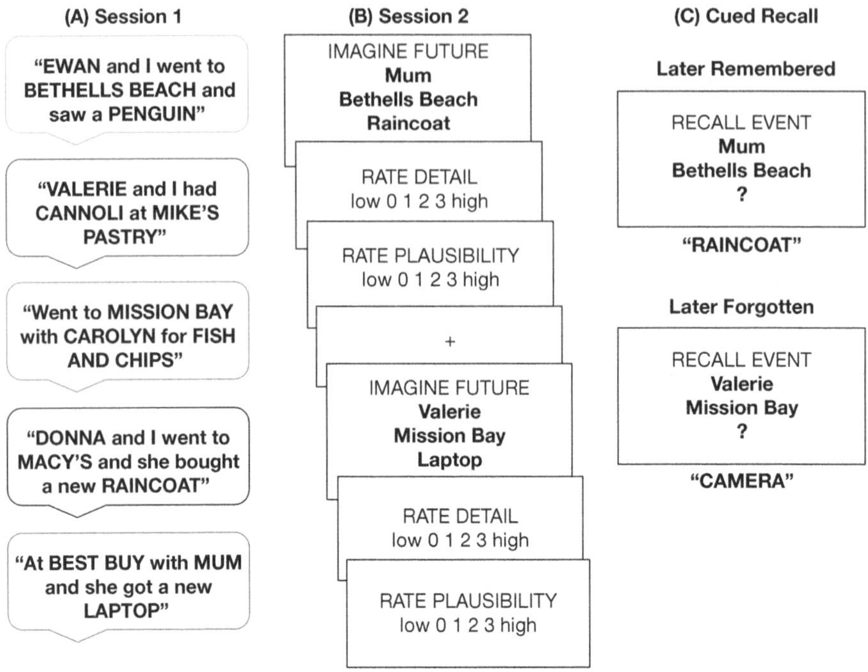


Figure 5.2. A schematic diagram of the experimental recombination paradigm where episodic details collected during session 1 (A) are recombined and presented during session 2 to elicit future simulations (B). Successful encoding of future simulations is then assessed using a cued-recall test (C).
Adapted from McLelland et al. (2015).

the participant's life). A week later, participants came back into the laboratory for a simulation session, in which we presented random recombinations of these episodic details (i.e., a person, place, and object taken from different memories) that would then form the components of an imagined future event. Simulated events were then rated for vividness (i.e., the amount of detail generated) and plausibility (i.e., how possible the event was, given personal circumstances).

Although the main focus of this study was on the factors that influence the encoding of future events (see Section 2.1), we also examined the factors affecting simulation phenomenology (McLelland et al., 2015). To capitalize on this rich trial-by-trial information, we utilized hierarchical linear modeling (HLM), which, instead of aggregating data into participant means, allowed us to directly test whether the quality of the particular event components presented on any given trial could predict the phenomenology of the simulation generated (as well as whether the simulation was encoded; for more discussion of the use of HLM in the analysis of data with an inherently nested structure, such as autobiographical memory, see Wright, 1998). We found that the mean familiarity of the components (averaged across the three components presented on a trial) significantly predicted the amount of detail in the imagined future event; average emotionality and significance ratings did not. This finding is consistent with that of D'Argembeau and Van der Linden (2012), who also found that the familiarity of event components predicts simulation vividness. However, it is possible that the familiarity of a particular component type is most important for driving the phenomenology of the resulting simulation. We constructed another model to test this question, and found that, in fact, the familiarity of all three components contributed significantly to the overall vividness of that simulation.[1]

In addition to the quality of the event components themselves, it is likely that the particular combination of details used in a given simulation will affect the quality of the imagined event. For instance, if episodic details need to be recombined and integrated to create a coherent simulation (Addis & Schacter, 2012; Schacter & Addis, 2007), then details that are incongruent may affect the extent to which this can occur. To explore this possibility, we modified the experimental recombination to manipulate the degree of congruence between the details to be integrated to form a future simulation. Specifically, instead of obtaining event components by having participants recall events and identify details, we had participants generate lists of people, places, and objects that are associated with distinct social spheres in their own lives (participants used the Facebook App "Friends Visual Map" to assist with determining their distinct social spheres). The premise is that details within a sphere (e.g., Brother, Mum's House, Dad's car) will be more congruent and easily recombined into a scenario than will details from across spheres (e.g., Mum, Prof. Schacter's office, soccer ball). Participant ratings confirmed that within-sphere details in the congruent condition were significantly more likely to co-occur in everyday life than across-sphere details.

The imagined future events generated using these detail sets differed on important dimensions (Addis et al., 2014; van Mulukom et al., 2013). Simulations in the incongruent condition were rated as significantly more novel, less plausible, and more difficult to construct than those in the congruent condition. Importantly, the congruence of the details from which a simulation was constructed was associated with differential levels of vividness: future events in the congruent condition were significantly more detailed than those in the incongruent condition. Preliminary fMRI results indicate that relative to the incongruent condition, the core network is more intensively activated during the construction of simulations in the congruent condition (see Figure 5.3; Addis et al., 2014). Together, these results appear to indicate that when the details are interrelated, one is able to fully engage in the simulation process and construct an elaborate representation of a plausible event. It is likely that the interrelatedness of the details provides a basic scaffold that assists the generation of an event that makes sense given the constraints of reality. Indeed, others have reported that plausible future events tend to be more detailed (Szpunar & Schacter, 2013; Weiler, Suchan, & Daum, 2010). However, when the details fail to provide this scaffold, the extent to which one can draw on prior experience is limited, and some framework is needed to guide the simulation process. We have argued that semantic memory is central to this scaffolding process.

### 1.3. The Role of Semantic Memory in Imagination

Often, when considering the future, a schematic representation will suffice. Semantic memory is sufficient for the construction of such schematic future representations (Addis & Schacter, 2013; Atance & O'Neill, 2001), as evident in patient

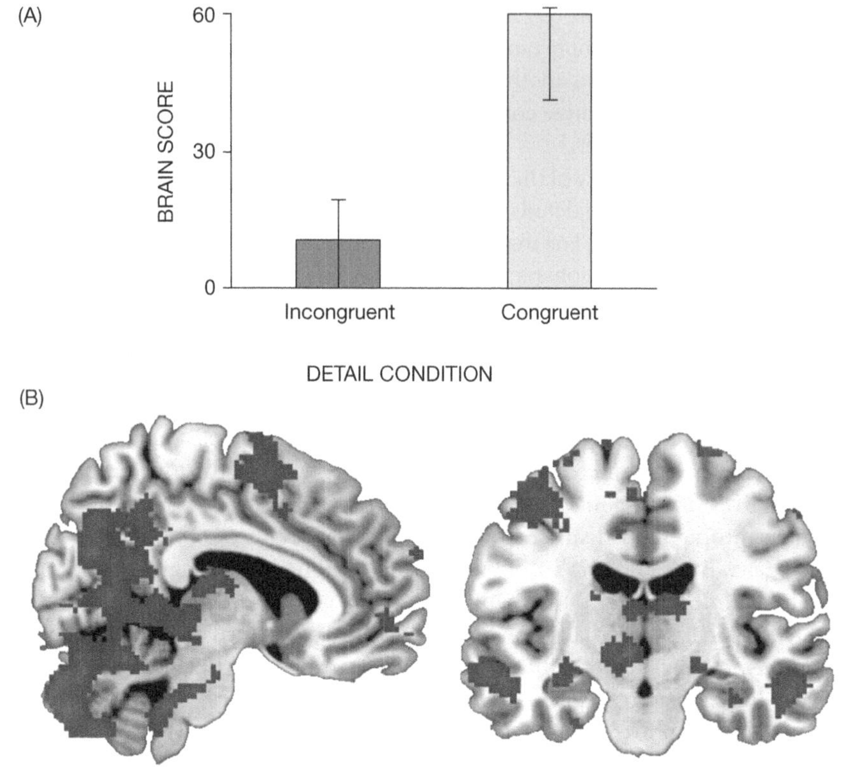


**Figure 5.3.** (A) Brain scores (weighted average of activity across all voxels) and (B) maps of brain activity during the simulation of future events constructed from congruent and incongruent detail combinations. (See color insert).
Adapted from Addis et al. (2014).

studies. For instance, patient D. B. (Klein, Loftus, & Kihlstrom, 2002) experienced amnesia for episodic memory as a result of hypoxia, and is unable to imagine future episodes. However, he is still able to generate semantic representations about the future and anticipate issues in the public domain (e.g., environmental issues, community issues), in line with his relatively preserved semantic memory. While Patient K. C. was not able to imagine future episodes, he had a semantic concept of time and could engage in future-oriented thinking, such as anticipating regret (Craver, Kwan, Steindam, & Rosenbaum, 2014). Moreover, many patients with episodic amnesia can imagine scenes (e.g., Cooper, Vargha-Khadem, Gadian, & Maguire, 2011; Hurley, Maguire, & Vargha-Khadem, 2011; Maguire, Vargha-Khadem, & Hassabis, 2010), likely on the basis of conceptual information about these contexts (Aminoff, Gronau, & Bar, 2007; Bar & Aminoff, 2003). One does not have to draw on a specific episodic memory of being at the beach to know that a beach scene typically includes white sand, waves, chairs, beach umbrellas, and relaxed people. Such content is either abstracted from previous experiences (Binder & Desai, 2011) or is learned in a non-experiential fashion (e.g., learning about a previously unexperienced context such as the jungle via books, Internet, or television). However, episodic memory can provide experiential detail to create richly detailed future simulations that are also flexible, such that one can explore different plots and outcomes. Indeed, scenarios based on semantic information (e.g., non-experienced contexts such as the jungle) tend to be less detailed and do not engage the core network to the same extent as events based on episodic memory of experienced contexts (Arnold, McDermott, & Szpunar, 2011; Szpunar, Chan, & McDermott, 2009).

Nevertheless, a number of lines of evidence suggest that semantic memory may be involved in episodic simulations. First, a ubiquitous finding in neuropsychological studies is that both autobiographical memories and future simulations comprise not only episodic details but also conceptual information memory (Addis et al., 2008; Levine, Svoboda, Hay, Winocur, & Moscovitch, 2002). If, in the context of episodic memory deficits, semantic memory is relatively unaffected, the amount of semantic content comprising such event descriptions increases significantly (e.g., in older adults; Schacter, Gaesser & Addis, 2010). This tendency to "flesh out" future-oriented thoughts with semantic information is not evident in patients where episodic memory loss is accompanied by semantic deficits (e.g., Alzheimer's disease; Addis et al., 2009b). Second, some regions of the core network associated with autobiographical memory retrieval and future simulation overlap with the semantic memory network (Binder, Desai, Graves, & Conant, 2009; Burianova, McIntosh, & Grady, 2010). For instance, neuroimaging studies of future simulation consistently report the activation of lateral temporal regions thought to mediate semantic memory (for a review, see Schacter et al., 2012). Interestingly, lateral temporal cortex activity appears to be up-regulated in populations that exhibit an over-generation of conceptual detail in their event representations, such as older adults (Addis, Roberts, & Schacter, 2011).

Although these findings suggest that semantic content is evident in episodic simulations, this evidence does not confirm that it is a critical component. However, studies with neuropsychological patients suggest that access to semantic memory may be necessary to imagine a specific future episode. Irish and colleagues (2012) examined simulation in semantic dementia patients, who exhibited deficits in semantic memory but no changes in the episodic content of remembered past events. Despite being able to access episodic details, the ability to imagine the future was significantly reduced relative to control participants. This deficit in future episodic detail was correlated with semantic memory performance and anterior lateral temporal lobe atrophy in the patients. Importantly, this finding suggests that semantic memory plays an important role in future simulation, possibly providing an organizational framework or script for an imaginary scenario that guides the simulation process. It is likely that once generated, this framework is then populated with details from episodic memory. Indeed, behavioral research in healthy adults has shown that during the simulation process, conceptual information is accessed before episodic details (D'Argembeau & Mathy, 2011).

The importance of semantic information when constructing mental representations of episodes was recently highlighted by Ranganath and Ritchey (2012). In a review of neural evidence, they propose that a posterior medial network mediates contextual and sensory (i.e., episodic) detail, while an anterior temporal network supports both the conceptual gist of the scenario and semantic information about the featured entities (such as people and objects). Being able to draw on such conceptual information facilitates the formation of item-item and item-context associations that are necessary to create a coherent simulation. Future research should explore whether these two networks contribute to the semantic and episodic aspects of the simulation process.

## 2. HOW IMAGINATION INFLUENCES MEMORY

> Inconsistency, at least in the way people remember their own lives, may be the price people pay for being imaginative.
>
> —Garry & Polaschek, *2000*

### 2.1. The Phenomenology of Imagination Influences Memory Encoding

The flexible nature of memory is generally believed to be adaptive; such a system can be employed to imagine novel scenarios, which allows us to creatively solve problems, anticipate upcoming events, and learn from the alternate ways in which past events may have transpired (Howe, Garner, Charlesworth, & Knott, 2011; Roese, 1997). However, this ability to simulate episodic events is only valuable if these imagined scenarios are maintained in memory, and recalled where appropriate to benefit behavior (see Szpunar, Addis, McLelland, & Schacter, 2013). The phenomenological characteristics of a simulated scenario, as well as the particular way in which the scenario is constructed, can influence the successful encoding and subsequent retrieval of that simulation. In other words, what and how we imagine determines its memorability.

We recently employed HLM to explore a range of potential phenomenological characteristics that may impact imagination memorability, at both the level of individual event components and the simulation as a whole (McLelland, Devitt, Schacter, & Addis, in press). In this study, participants generated future scenarios comprising novel arrangements of people, locations, and objects originating from autobiographical memories. Familiarity, emotionality, and personal significance ratings were made for these individual event components, and vividness and plausibility ratings were made for the simulation as a whole. Encoding of each simulation was tested after a short delay, by providing two of the event components and asking the participant to recall the third (see Figure 5.2). We found that those scenarios imagined with high vividness were more likely to be successfully encoded, replicating previous results (Martin, Schacter, Corballis, & Addis, 2011). Additionally, the plausibility of the simulation was also associated with higher subsequent recall. As discussed in section 1.2, more plausible events can be imagined in greater detail, and may provide a framework for further elaboration and integration with existing memories. Indeed, increased ratings of plausibility have been correlated with increases in ease of simulation and arousal (Szpunar & Schacter, 2013; though see Anderson, 2012), which may further contribute to the memorability of a simulated event (Szpunar, Addis, & Schacter, 2012). Familiarity of the person component was a third predictor of subsequent memory. It may be that familiarity aids the integration of simulation components with memories of past experiences, as well as enhancing simulation vividness (D'Argembeau & Van der Linden, 2012), which provides a scaffold for the encoding of these novel simulations as a whole (Poppenk & Norman, 2012). However, because the person component is typically more central, and thus integral, to an event (Dijkstra & Misirlisoy, 2009), further research is needed to disentangle the effect of component familiarity from that of component centrality.

A similar paradigm was employed to delineate the role of emotional valence on simulation memorability (Szpunar et al., 2012). Participants imagined positive, negative, and neutral future scenarios involving recombined autobiographical event components. Memory for these scenarios was tested after either a short (10 minute) or a long (one day) delay. At a short delay, emotional simulations (both positive and negative) were better remembered than neutral simulations. However, at a longer delay, positive and neutral events were more likely to be successfully recalled, suggesting that negative events fade more quickly in memory over time. It is motivationally beneficial to be optimistic about the future, and to remember positive events that can help obtain future rewards (Weinstein, 1980). It has also been proposed that emotional arousal aids in the integration of event components (Mather & Sutherland, 2011). Therefore, negative simulations may be more rapidly forgotten, as they are less beneficial to future behavior, and this fading over time may be compounded by the deterioration of component associations as affect diminishes (also see Holmes, 1970). Similar findings have been reported for the imagination of past events; when engaging in counterfactual thinking (where alternative outcomes for past events are simulated), positive events were both rated as more vivid (De Brigard et al., 2013) and subsequently better remembered than negative events (Gerlach et al., 2013).

A further influence on imagination encoding is the perspective with which a simulation is viewed: from either a field (first-person) perspective, or an observer (third-person) perspective (Nigro & Neisser, 1983). Manipulating the vantage point from which an imagined scenario is viewed can impact the strength with which the scenario can be visualized (Vella & Moulds, 2013), and subsequently its memorability. Participants who were instructed to recall memories and to imagine future scenarios using an alternative perspective than they naturally employ reported reduced vividness of those simulations, regardless of the original perspective used. While subsequent memory was not tested in this study, events that are imagined with greater vividness have been previously shown to be more memorable (Martin et al., 2011).

In order for an imagined event to be beneficial for future behavior, the simulation must be successfully encoded and retrieved at a later date. Simulations that are familiar and high in vividness, plausibility, and positive affect are more likely to be subsequently remembered. However, a unique problem occurs during retrieval when an imagined scenario is particularly high in these phenomenological characteristics—it may be misattributed as an authentic occurrence, resulting in the creation of a false memory.

## 2.2. The Influence of Imagination on Memory Veracity

A constructive memory system that allows us to imagine the past in different ways and anticipate the future may be adaptive, yet this system also leaves us vulnerable to memory distortions. The "lost in the mall" study was one of the first systematic attempts to create entirely false memories using imagination (Loftus & Pickrell, 1995). Participants were falsely informed that they were lost in a mall at some point in their childhood, and were encouraged to use imagination as a means of recalling "dormant" memories of this event. Crucially, the participant's parents were contacted prior to the study, to verify that this event never actually happened in the participant's past. After a series of imagination exercises, 25% of participants came to form at least a partial memory of being lost in a mall. This phenomenon, whereby imagining a past event subsequently increases confidence that the event truly took place, has been termed *imagination inflation* (Garry, Manning, Loftus, & Sherman, 1996). A multitude of studies have since demonstrated an increased belief in the veracity of a false event following imagination, from memories of simple actions (Goff & Roediger, 1998; Nash, Wade, & Lindsay, 2009), to childhood memories of a hot air balloon ride (Garry & Wade, 2005) or an improbable medical procedure (Mazzoni & Memon, 2003).

One theory put forward to account for this imagination inflation effect is that imagination increases the recollective quality of the simulation. Memories from different origins tend to have differing phenomenological characteristics, which typically allow an accurate source judgment to be made about a retrieved mental representation (cf. the Source Monitoring Framework; Johnson, Hashtroudi, & Lindsay, 1993; Johnson & Raye, 1981). For example, memories of authentic occurrences tend to be rated higher in perceptual, emotional, temporal, and spatial detail than memories for imagined events (Johnson, Foley, Suengas, & Raye, 1988; Justice, Morrison, & Conway, 2013; Porter, Yuille, & Lehman, 1999). However, the imagination process can enhance the phenomenological characteristics of a simulated event (Heaps & Nash, 2001; Lampinen, Odegard, & Bullington, 2003), making the event more memory-like in quality, and thus more difficult to distinguish from authentic memories. In support of this theory, the imagination inflation effect has been shown to be more effective when participants are encouraged to elaborate upon their imaginings by the inclusion of sensory information (Thomas, Bulevich, & Loftus, 2003). Furthermore, Marsh, Pezdek, and Lam (2014) demonstrate that recent events imagined from a field perspective result in a greater imagination inflation effect than those imagined from an observer perspective, with an inverse pattern observed for childhood events. As it is more common for recent memories to be seen from a field perspective, and older memories from an observer perspective (Nigro & Neisser, 1983), these findings are in line with the idea that imagined scenarios with similar properties to authentic memories are more likely to be falsely accepted as real.

Neuroimaging results provide further support for the imagination inflation effect being at least partially driven by an increase in recollective quality. Gonsalves et al. (2004) presented participants with object words, followed by either a pictorial representation of that word, or an instruction to form a mental image of the object, while monitoring brain activity using fMRI. On a subsequent source test, participants were presented with spoken words of all items shown, and had to determine whether they had previously been shown a picture with that word. Endorsement of seeing word-only items as a picture was associated with activity in regions associated with visual imagery and spatial attention (precuneus, inferior parietal cortex, and anterior cingulate), which may be reflective of greater perceptual detail accompanying the mental image for the object, contributing to later source confusion.

Imagination can also increase the subjective plausibility of the event in question (Szpunar & Schacter, 2013; though see De Brigard, Szpunar, & Schacter, 2013, for alternative findings). Plausibility, belief, and memory are believed to be nested constructs (Scoboria, Mazzoni, Kirsch, & Relyea, 2004), whereby events must first be perceived as plausible before they can be considered to have actually taken place (see also Hyman, Gilstrap, Decker, & Wilkinson, 1998; Mazzoni, 2007; Scoboria, Mazzoni, Jarry, & Shapero, 2012). Enhanced plausibility is associated with increases in vividness, arousal, and ease of imagining (Szpunar & Schacter, 2013), all of which can contribute to the source misattribution of an imagined event. Interestingly, the relationship between imagination and plausibility appears to be reciprocal; not only does imagination increase the subjective plausibility of an event, inflation is more effective for events that are first perceived as plausible (Pezdek, Blandon-Gitlin, & Gabbay, 2006). For example, positing a scenario as personally plausible, as opposed to more generally possible, not only inflates belief in that false event (such as contracting a childhood illness after eating spoiled peach yogurt), but also has measurable influence on behavior (avoiding eating peach yogurt) for up to one month following imagination (Scoboria, Mazzoni, Jarry, & Bernstein, 2012). A more persistent search is thought to ensue through memory for elaborative details for those events considered plausible (Anderson, 2012), resulting in a phenomenologically richer, and thus more convincing, simulation.

Yet it has been argued that the imagination inflation effect is not a result of imagination at all, but rather a consequence of increased processing fluency due to simple exposure to the fabricated scenario. The enhanced ease of processing may be misattributed as a sign that the event truly occurred in the past (Jacoby, Kelley, & Dywan, 1989; Sharman, Garry, & Beuke, 2004; Whittlesea, 1993). Hyman and Pentland (1996) were among the first to demonstrate the power of explicit imagination over and above simple exposure to a fabricated scenario. Participants were either prompted to imagine a childhood event in which they supposedly spilled punch on the parents of the bride at a wedding. A quarter of the participants who imagined this event subsequently formed a false memory, compared to only 9% of participants who were instructed to merely think about the event happening. A number of studies have further demonstrated imagination inflation over and above the effect of increased fluency alone, providing evidence that this inflation occurs due to mechanisms inherent to the imagination process (Mazzoni & Memon, 2003; Thomas et al., 2003; von Glahn, Otani, Migita, Langford, & Hillard, 2012). However, it is important to note that the source monitoring and processing fluency accounts of inflation need not be mutually exclusive; imagination may enhance both the recollective detail, and facilitate the fluency of later retrieval, both of which can be misattributed as indicators of event authenticity.

We recently explored the influence of both processing fluency and phenomenological quality on the formation of conjunction errors in autobiographical memory (Devitt, Monk-Fromont, Schacter, & Addis, 2016), using an amalgamation of the recombination and conjunction error paradigms (Burt, Kemp, & Conway, 2004; Odegard & Lampinen, 2004). Conjunction errors are a type of memory distortion that occurs when components from one memory are unknowingly incorporated into another. In this study, conjunction lures were constructed by recombining person, location, and object details taken from autobiographical memories. These lures were divided between imagination and association conditions; in the imagination condition, participants imagined a novel past event composed of the details, while in the association condition, participants ranked the details based on subjective pleasantness, as a control for exposure to the component combination. On a subsequent source-monitoring test, conjunction errors occurred when conjunction lures were incorrectly judged as depicting real memories.

It was found that conjunction errors were more likely to arise in the imagination condition than the association condition, suggesting that something other than increased processing fluency inflates conjunction error rates for imagined events. To further explore this effect, authentic memories, correctly identified imagined events, and conjunction errors were rated for phenomenological qualities following the source test, and short event descriptions were obtained using the Autobiographical Interview procedure (Levine et al., 2002). Conjunction errors were rated intermediary between real and imagined events in vividness and emotionality, thereby having a more memory-like subjective experience than events correctly identified as imagined (see Figure 5.4). Furthermore, conjunction errors and real memories were also more likely to be viewed from a field perspective than imagined events. Objective scoring of the event descriptions identified perceptual detail as playing an important role in conjunction error formation. Together, these results support the perspective that the imagination inflation effect is due to misattribution of memory-like qualities.

While repeated imagining can increase the likelihood of false memory formation, (Thomas et al., 2003), there may be a boundary to this inflation effect. Von Glahn et al. (2012) examined the impact of imagination richness by having participants generate no, three, or six details about a fabricated event; the greatest inflation effect was seen in the condition where three details were generated. These results may be explained through a mental workload hypothesis (Schwartz et al., 1991), whereby easy tasks confer high processing fluency and a sense of previous exposure to the event, but difficult tasks (such as having to generate more details) are associated with low fluency, and may further provide access to information about the cognitive operations underlying the imagination process, making the true source

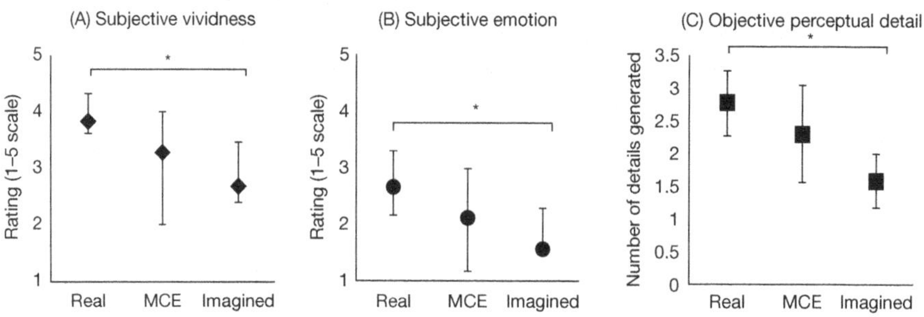


**Figure 5.4.** Median subjective ratings of (A) vividness and (B) emotion, and (C) mean objective scoring of perceptual detail for real memories, memory conjunction errors (MCE), and correctly identified imagined events. Errors bars for (A) and (B) represent interquartile range, error bars for (C) represent standard error.
Adapted from Devitt et al. (2016).

of the scenario more distinct (see also Johnson et al., 1993; Schwarz, 2012; Tversky & Kahneman, 1973). Results from Devitt et al. (2016) also support this workload hypothesis, as detail recombinations that were partially recombined (where only one of the three details was altered) were twice as likely to be falsely attributed as belonging to a real event than full recombinations (where all three details originate from three separate memories), even when controlling for the vividness and plausibility of the simulation. Constructing a scenario with three potentially incongruent details is more difficult than with only one incongruent detail, and this increased mental workload may later provide a cue as to the actual origin of the imagination.

## 3. CONCLUSIONS

Neuropsychological and neuroimaging studies reveal a striking overlap in the neural networks recruited during memory and imagination, suggesting an interplay between these two processes. It has been hypothesized that episodic memory contributes the necessary ingredients for the simulation of novel events, while semantic memory provides an organizational framework upon which an imagined event is constructed. However, while it is clear that imagination draws heavily on memory, this relationship appears to be a bidirectional one. The nature of an imagined scenario has an influence on its subsequent memorability, and can also contribute to the formation of distortions in memory. In order to understand how we imagine the future, we must look toward the way in which we construct the past, while bearing in mind that the past may in turn be a product of our imagination.

## NOTE

1. The coefficient and statistics for this HLM model predicting the detail rating for a given future simulation are as follows: person familiarity ($\beta = .188$; SE = .021; $t(20) = 9.004, p < .001$), location familiarity ($\beta = .071$; SE = .019; $t(20) = 3.795, p < .001$), and object familiarity ($\beta = .048$; SE = .018; $t(20) = 2.625, p < .01$).

## REFERENCES

Addis, D. R., Musicaro, R., Pan, L., & Schacter, D. L. (2010). Episodic simulation of past and future events in older adults: evidence from an experimental recombination task. *Psychology and Aging, 25*(2), 369–376.

Addis, D. R., Pan, L., Musicaro, R., & Schacter, D. L. (2016). Divergent thinking and constructing episodic simulations. *Memory, 24*, 89–97. http://doi.org/10.1080/09658211.2014.985591.

Addis, D. R., Pan, L., Vu, M., Laiser, N., & Schacter, D. L. (2009a). Constructive episodic simulation of the future and the past: distinct subsystems of a core brain network mediate imagining and remembering. *Neuropsychologia, 47*, 222–238.

Addis, D. R., Roberts, R. P., & Schacter, D. L. (2011). Age-related neural changes in autobiographical remembering and imagining. *Neuropsychologia, 49*(13), 3656–3669. http://doi.org/http://dx.doi.org/10.1016/j.neuropsychologia.2011.09.021.

Addis, D. R., Roberts, R. P., van Mulukom, V., Sumner, R., Grady, C. L., Schacter, D. L. (2014). *Disparateness of episodic details modulates default network activity during future simulation.* Poster presented at Society for Neuroscience Annual Meeting, Washington D.C., November 2014.

Addis, D. R., Sacchetti, D. C., Ally, B. A., Budson, A. E., & Schacter, D. L. (2009b). Episodic simulation of future events is impaired in mild Alzheimer's disease. *Neuropsychologia, 47*(12), 2660–2671. http://doi.org/http://dx.doi.org/10.1016/j.neuropsychologia.2009.05.018.

Addis, D. R., & Schacter, D. L. (2008). Constructive episodic simulation: temporal distance and detail of past and future events modulate hippocampal engagement. *Hippocampus, 18*(2), 227–237. http://doi.org/10.1002/hipo.20405.

Addis, D. R., & Schacter, D. L. (2012). The hippocampus and imagining the future: where do we stand? *Frontiers in Human Neuroscience, 5*, 1–15.

Addis, D. R., & Schacter, D. L. (2013). Future-oriented simulations: the role of episodic memory. *Journal of Applied Research in Memory and Cognition, 2*, 248–250.

Addis, D. R., Wong, A. T., & Schacter, D. L. (2007). Remembering the past and imagining the future: common and distinct neural substrates during event construction and elaboration. *Neuropsychologia, 45*, 1363–1377.

Addis, D. R., Wong, A. T., & Schacter, D. L. (2008). Age-related changes in the episodic simulation of future events. *Psychological Science, 19*, 33–41.

Aminoff, E., Gronau, N., & Bar, M. (2007). The parahippocampal cortex mediates spatial and nonspatial associations. *Cerebral Cortex, 17*(7), 1493–1503. http://doi.org/10.1093/cercor/bhl078.

Andelman, F., Hoofien, D., Goldberg, I., Aizenstein, O., & Neufeld, M. Y. (2010). Bilateral hippocampal lesion and a selective impairment of the ability for mental time travel. *Neurocase, 16*(5), 426–435. http://doi.org/10.1080/13554791003623318.

Anderson, R. J. (2012). Imagining novel futures: the roles of event plausibility and familiarity. *Memory, 20*(5), 443–451.

Andrews-Hanna, J. R. (2012). The brain's default network and its adaptive role in internal mentation. *Neuroscientist, 18*(3), 251–270.

Arnold, K., McDermott, K. B., & Szpunar, K. K. (2011). Imagining the near and far future: the role of location familiarity. *Memory & Cognition, 39*(6), 954–967. http://doi.org/10.3758/s13421-011-0076-1.

Atance, C. M., & O'Neill, D. K. (2001). Episodic future thinking. *Trends in Cognitive Sciences, 5*(12), 533–539. http://doi.org/http://dx.doi.org/10.1016/S1364-6613(00)01804-0.

Bar, M., & Aminoff, E. (2003). Cortical analysis of visual context. *Neuron, 38*(2), 347–358. http://doi.org/http://dx.doi.org/10.1016/S0896-6273(03)00167-3.

Benoit, R., Gilbert, S., & Burgess, P. (2011). A neural mechanism mediating the impact of episodic prospection on farsighted decisions. *Journal of Neuroscience, 31*, 6771–6779.

Binder, J. R., & Desai, R. H. (2011). The neurobiology of semantic memory. *Trends in Cognitive Sciences, 15*(11), 527–536. http://doi.org/http://dx.doi.org/10.1016/j.tics.2011.10.001.

Binder, J. R., Desai, R. H., Graves, W. W., & Conant, L. L. (2009). Where is the semantic system? A critical review and meta-analysis of 120 functional neuroimaging studies. *Cerebral Cortex, 19*(12), 2767–2796. http://doi.org/10.1093/cercor/bhp055.

Boyer, P. (2008). Evolutionary economics of mental time travel? *Trends in Cognitive Sciences, 12*(6), 219–224. http://doi.org/http://dx.doi.org/10.1016/j.tics.2008.03.003.

Buckner, R. L., Andrews-Hanna, J. R., & Schacter, D. L. (2008). The brain's default network: anatomy, function, and relevance to disease. *Annuals of the New York Academy of Sciences, 1124*, 1–38.

Buckner, R. L., & Carroll, D. C. (2007). Self-projection and the brain. *Trends in Cognitive Sciences, 11*(2), 49–57. http://doi.org/http://dx.doi.org/10.1016/j.tics.2006.11.004.

Burianova, H., McIntosh, A. R., & Grady, C. L. (2010). A common functional brain network for autobiographical, episodic, and semantic memory retrieval. *NeuroImage, 49*(1), 865–874. http://doi.org/http://dx.doi.org/10.1016/j.neuroimage.2009.08.066.

Burt, C. D. B., Kemp, S., & Conway, M. A. (2004). Memory for true and false autobiographical event descriptions. *Memory, 12*, 545–552.

Cole, S. N., Morrison, C. M., & Conway, M. A. (2013). Episodic future thinking: linking neuropsychological performance with episodic detail in young and old adults. *The Quarterly Journal of Experimental Psychology, 66*(9), 1687–1706. http://doi.org/10.1080/17470218.2012.758157.

Cooper, J. M., Vargha-Khadem, F., Gadian, D. G., & Maguire, E. A. (2011). The effect of hippocampal damage in children on recalling the past and imagining new experiences. *Neuropsychologia, 49*(7), 1843–1850. http://doi.org/http://dx.doi.org/10.1016/j.neuropsychologia.2011.03.008.

Craver, C. F., Kwan, D., Steindam, C., & Rosenbaum, R. S. (2014). Individuals with episodic amnesia are not stuck in time. *Neuropsychologia, 57*(0), 191–195. http://doi.org/http://dx.doi.org/10.1016/j.neuropsychologia.2014.03.004.

D'Argembeau, A., & Mathy, A. (2011). Tracking the construction of episodic future thoughts. *Journal of Experimental Psychology: General, 140*(2), 258–271. http://doi.org/10.1037/a0022581.

D'Argembeau, A., Renaud, O., & Van der Linden, M. (2011). Frequency, characteristics and functions of future-oriented thoughts in daily life. *Applied Cognitive Psychology, 25*(1), 96–103. http://doi.org/10.1002/acp.1647.

D'Argembeau, A., & Van der Linden, M. (2004). Phenomenal characteristics associated with projecting oneself back into the past and forward into the future: influence of valence and temporal distance. *Consciousness and Cognition, 13*(4), 844–858. http://doi.org/http://dx.doi.org/10.1016/j.concog.2004.07.007.

D'Argembeau, A., & Van der Linden, M. (2006). Individual differences in the phenomenology of mental time travel: the effect of vivid visual imagery and emotion regulation strategies. *Consciousness and Cognition, 15*(2), 342–350. http://doi.org/http://dx.doi.org/10.1016/j.concog.2005.09.001.

D'Argembeau, A., & Van der Linden, M. (2012). Predicting the phenomenology of episodic future thoughts. *Consciousness and Cognition, 21*(3), 1198–1206. http://doi.org/http://dx.doi.org/10.1016/j.concog.2012.05.004.

De Beni, R., Borella, E., Carretti, B., Zavagnin, M., Lazzarini, L., & Milojevi, G. (2013). Remembering the past and imagining the future: age-related differences between young, young-old and old-old. *Aging Clinical and Experimental Research, 25*(1), 89–97. http://doi.org/10.1007/s40520-013-0003-3.

De Brigard, F., Addis, D. R., Ford, J. H., Schacter, D. L., & Giovanello, K. S. (2013). Remembering what could have happened: neural correlates of episodic counterfactual thinking. *Neuropsychologia, 51*(12), 2401–2414. http://doi.org/10.1016/j.neuropsychologia.2013.01.015.

De Brigard, F., Szpunar, K. K., & Schacter, D. L. (2013). Coming to grips with the past: Effect of repeated simulation on the perceived plausibility of episodic counterfactual thoughts. *Psychological Science*, *24*(7), 1329–1334.

Devitt, A. L., Monk-Fromont, E., Schacter, D. L., & Addis, D. R. (2016). Factors that influence the generation of autobiographical memory conjunction errors. *Memory*, *24*, 204–222.

Dijkstra, K., & Misirlisoy, M. (2009). Recognition accuracy for original and altered verbal memory reports in older adults. *The Quarterly Journal of Experimental Psychology*, *62*, 248–256.

Gaesser, B., Sacchetti, D. C., Addis, D. R., & Schacter, D. L. (2011). Characterizing age-related changes in remembering the past and imagining the future. *Psychology and Aging*, *26*, 80–84.

Gamboz, N., De Vito, S., Brandimonte, M. A., Pappalardo, S., Galeone, F., Iavarone, A., & Sala, S. Della. (2010). Episodic future thinking in amnesic mild cognitive impairment. *Neuropsychologia*, *48*(7), 2091–2097. http://doi.org/http://dx.doi.org/10.1016/j.neuropsychologia.2010.03.030.

Garry, M., Manning, C. G., Loftus, E. F., & Sherman, S. J. (1996). Imagination inflation: imagining a childhood event inflates confidence that it occurred. *Psychonomic Bulletin & Review*, *3*(2), 208–214. http://doi.org/10.3758/BF03212420.

Garry, M., & Wade, K. A. (2005). Actually, a picture is worth less than 45 words: narratives produce more false memories than photographs do. *Psychonomic Bulletin & Review*, *12*(2), 359–366.

Gerlach, K. D., Dornblaser, D. W., & Schacter, D. L. (2014). Adaptive constructive processes and memory accuracy: consequences of counterfactual simulations in young and older adults. *Memory*, *22*(1), 145–162. http://doi.org/10.1080/09658211.2013.779381.

Gilmore, A. W., Nelson, S. M., & McDermott, K. B. (2014). The contextual association network activates more for remembered than for imagined events. *Cerebral Cortex*. http://doi.org/10.1093/cercor/bhu223.

Goff, L. M., & Roediger, H. L. (1998). Imagination inflation for action events: repeated imaginings lead to illusory recollections. *Memory and Cognition*, *26*, 20–33.

Gonsalves, B. D., Reber, P. J., Gitelman, D. R., Parrish, T. B., Mesulam, M. M., & Paller, K. A. (2004). Neural evidence that vivid imagining can lead to false remembering. *Psychological Science*, *15*, 655–660.

Hassabis, D., Kumaran, D., Vann, S. D., & Maguire, E. A. (2007). Patients with hippocampal amnesia cannot imagine new experiences. *Proceedings of the National Academy of Sciences*, *104*, 1726–1731.

Hassabis, D., & Maguire, E. A. (2007). Deconstructing episodic memory with construction. *Trends in Cognitive Sciences*, *11*(7), 299–306. http://doi.org/http://dx.doi.org/10.1016/j.tics.2007.05.001.

Heaps, C. M., & Nash, M. (2001). Comparing recollective experience in true and false autobiographical memories. *Journal of Experimental Psychology: Learning, Memory and Cognition*, *27*, 920–930.

Holmes, D. S. (1970). Differential change in affective intensity and the forgetting of unpleasant personal experiences. *Journal of Personality and Social Psychology*, *15*, 234–239.

Howe, M. L., Garner, S. R., Charlesworth, M., & Knott, L. (2011). A brighter side to memory illusions: false memories prime children's and adult's insight-based problem solving. *Journal of Experimental Child Psychology*, *108*, 383–393.

Hurley, N. C., Maguire, E. A., & Vargha-Khadem, F. (2011). Patient HC with developmental amnesia can construct future scenarios. *Neuropsychologia, 49*(13), 3620–3628. http://doi.org/http://dx.doi.org/10.1016/j.neuropsychologia.2011.09.015.

Hyman, I. E., Gilstrap, L. L., Decker, K., & Wilkinson, C. (1998). Manipulating remember and know judgements of autobiographical memories: an investiation of false memory creation. *Applied Cognitive Psychology, 12*, 371–386.

Hyman, I. E., & Pentland, J. (1996). The role of mental imagery in the creation of false childhood memories. *Journal of Memory and Language, 35*, 101–117.

Irish, M., Addis, D. R., Hodges, J. R., & Piguet, O. (2012). Considering the role of semantic memory in episodic future thinking: evidence from semantic dementia. *Brain, 135*(7), 2178–2191.

Jacoby, L. L., Kelley, C. M., & Dywan, J. (1989). Memory attributions. In H. L. Roediger & F. I. M. Craik (eds.), *Varieties of memory and consciousness: Essays in honor of Endel Tulving* (pp. 391–422). Hillsdale, NJ: Lawrence Erlbaum.

Johnson, J. D., & Rugg, M. D. (2007). Recollection and the reinstatement of encoding-related cortical activity. *Cerebral Cortex, 17*(11), 2507–2515. http://doi.org/10.1093/cercor/bhl156.

Johnson, M. K., Foley, M. A., Suengas, A. G., & Raye, C. L. (1988). Phenomenal characteristics of memories for perceived and imagined autobiographical events. *Journal of Experimental Psychology: General, 117*(4), 371–376.

Johnson, M. K., Hashtroudi, S., & Lindsay, D. S. (1993). Source monitoring. *Psychological Bulletin, 114*, 3–28.

Johnson, M. K., & Raye, C. L. (1981). Reality monitoring. *Psychological Review, 94*, 37–64.

Juskenaite, A., Quinette, P., Desgranges, B., Sayette, V. de La, Viader, F., & Eustache, F. (2014). Mental simulation of future scenarios in transient global amnesia. *Neuropsychologia, 63*(0), 1–9. http://doi.org/http://dx.doi.org/10.1016/j.neuropsychologia.2014.08.002.

Justice, L. V., Morrison, C. M., & Conway, M. A. (2013). True and intentionally fabricated memories. *The Quarterly Journal of Experimental Psychology, 66*,1196–1203.

Klein, S. B. (2013). The temporal orientation of memory: it's time for a change of direction. *Journal of Applied Research in Memory and Cognition, 2*(4), 222–234. http://doi.org/http://dx.doi.org/10.1016/j.jarmac.2013.08.001.

Klein, S. B., Loftus, J., & Kihlstrom, J. F. (2002). Memory and temporal experience: the effects of episodic memory loss on an amnesic patient's ability to remember the past and imagine the future. *Social Cognition, 20*(5), 353–379. http://doi.org/10.1521/soco.20.5.353.21125.

Kwan, D., Carson, N., Addis, D. R., & Rosenbaum, R. S. (2010). Deficits in past remembering extend to future imagining in a case of developmental amnesia. *Neuropsychologia, 48*(11), 3179–3186. http://doi.org/http://dx.doi.org/10.1016/j.neuropsychologia.2010.06.011.

Lampinen, J. M., Odegard, T. N., & Bullington, J. L. (2003). Qualities of memories for performed and imagined actions. *Applied Cognitive Psychology, 12*, 881–893.

Levine, B., Svoboda, E., Hay, J., Winocur, G., & Moscovitch, M. (2002). Aging and autobiographical memory: dissociating episodic from semantic retrieval. *Psychology and Aging, 17*, 677–689.

Loftus, E. F., & Pickrell, J. E. (1995). The formation of false memories. *Psychiatric Annuals, 25*, 720–725.

Maguire, E. A. (2001). Neuroimaging studies of autobiographical event memory. *Philosophical Transactions of the Royal Society (B), 356*, 1441–1451.

Maguire, E. A., & Mullally, S. L. (2013). The hippocampus: a manifesto for change. *Journal of Experimental Psychology: General, 142*(4), 1180–1189. http://doi.org/10.1037/a0033650.

Maguire, E. A., Vargha-Khadem, F., & Hassabis, D. (2010). Imagining fictitious and future experiences: evidence from developmental amnesia. *Neuropsychologia, 48*(11), 3187–3192. http://doi.org/http://dx.doi.org/10.1016/j.neuropsychologia.2010.06.037.

Marsh, B. U., Pezdek, K., & Lam, S. T. (2014). Imagination perspective affects ratings of the likelihood of occurrence of autobiographical memories. *Acta Psychologica, 150*, 114–119.

Martin, V. C., Schacter, D. L., Corballis, M. C., & Addis, D. R. (2011). A role for the hippocampus in encoding simulations of future events. *Proceedings of the National Academy of Sciences, 108*(33), 13858–13863. http://doi.org/10.1073/pnas.1105816108.

Mather, M., & Sutherland, M. R. (2011). Arousal-biased competition in perception and memory. *Perspectives on Psychological Science, 6*, 114–133.

Mazer, V. (Writer). (1967). The Legacy of Anne Frank [Television series episode]. In Executive Producer, Krents, M. E. *The Eternal Light.* New York, NY: National Broadcasting Corporation.

Mazzoni, G. A. (2007). Did you witness demonic possession?. A response time analysis of the relationship between event plausibility and autobiographical beliefs. *Psychonomic Bulletin & Review, 14*, 277–281.

Mazzoni, G. A., & Memon, A. (2003). Imagination can create false autobiographical memories. *Psychological Science, 14*, 186–188.

McLelland, V. C., Devitt, A. L., Schacter, D. L., & Addis, D. R. (2015). Making the future memorable: the phenomenology of remembered future events. *Memory, 23*, 1255–1263.

Murray, C., Hach, S., Tippett, L. J., & Addis, D. R. (2014, August). *Remembering the past and imagining the future in depression.* Poster presented at Australasian Winter Brain Research Conference, Queenstown, New Zealand.

Nash, R. A., Wade, K. A., & Lindsay, D. S. (2009). Digitally manipulating memory: effects of doctored videos and imagination in distorting beliefs and memories. *Memory and Cognition, 37*, 414–424.

Nigro, G., & Neisser, U. (1983). Point of view in personal memories. *Cognitive Psychology, 15*(4), 467–482. http://doi.org/http://dx.doi.org/10.1016/0010-0285(83)90016-6.

Odegard, T. N., & Lampinen, J. M. (2004). Memory conjunction errors for autobiographical events: more than just familiarity. *Memory, 12*, 288–300.

Peters, J., & Büchel, C. (2010). Episodic future thinking reduces reward delay discounting through an enhancement of prefrontal-mediotemporal interactions. *Neuron, 66*(1), 138–148. http://doi.org/http://dx.doi.org/10.1016/j.neuron.2010.03.026.

Pezdek, K., Blandon-Gitlin, I., & Gabbay, P. (2006). Imagination and memory: does imagining implausible events lead to false autobiographical memories? *Psychonomic Bulletin & Review, 13*, 764–769.

Poppenk, J., & Norman, K. A. (2012). Mechanisms supporting superior source memory for familiar items: a multi-voxel pattern analysis study. *Neuropsychologia, 50*(13), 3015–3026. http://doi.org/http://dx.doi.org/10.1016/j.neuropsychologia.2012.07.010.

Porter, S., Yuille, J. C., & Lehman, D. R. (1999). The nature of real, implanted, and fabricated memories for emotional childhood events: implications for the recovered memory debate. *Law and Human Behavior, 23*, 517–537.

Prebble, S. C., Addis, D. R., & Tippett, L. J. (2013). Autobiographical memory and sense of self. *Psychological Bulletin, 139*(4), 815–840.

Race, E. A., Keane, M. M., & Verfaellie, M. (2011). Medial temporal lobe damage causes deficits in episodic memory and episodic future thinking not attributable to deficits in narrative construction. *Journal of Neuroscience*, *31*(28), 10262–10269.

Ranganath, C. (2010). Binding items and contexts: the cognitive neuroscience of episodic memory. *Current Directions in Psychological Science*, *19* (3), 131–137. Retrieved from http://cdp.sagepub.com/content/19/3/131.abstract.

Ranganath, C., & Ritchey, M. (2012). Two cortical systems for memory-guided behaviour. *Nature Reviews Neuroscience*, *13*(10), 713–726. http://dx.doi.org/10.1038/nrn3338.

Rendell, P. G., Bailey, P. E., Henry, J. D., Phillips, L. H., Gaskin, S., & Kliegel, M. (2012). Older adults have greater difficulty imagining future rather than atemporal experiences. *Psychology and Aging*, *27*(4), 1089–1098. http://doi.org/10.1037/a0029748.

Roese, N. J. (1997). Counterfactual thinking. *Psychological Bulletin*, *121*(1), 133–148.

Schacter, D. L. (2012). Adaptive constructive processes and the future of memory. *American Psychologist*, *67*(8), 603–613. http://doi.org/10.1037/a0029869.

Schacter, D. L., & Addis, D. R. (2007). The cognitive neuroscience of constructive memory: Remembering the past and imagining the future. *Philosophical Transactions of the Royal Society (B)*, *362*, 773–786.

Schacter, D. L., Addis, D. R., & Buckner, R. L. (2007). Remembering the past to remember the future: the prospective brain. *Nature Reviews Neuroscience*, *8*, 657–661.

Schacter, D. L., Addis, D. R., Hassabis, D., Martin, V. C., Spreng, R. N., & Szpunar, K. K. (2012). The future of memory: remembering, imagining, and the brain. *Neuron*, *76*(4), 677–694. http://doi.org/http://dx.doi.org/10.1016/j.neuron.2012.11.001.

Schacter D. L., Gaesser B., & Addis D. R. (2010). Age-related changes in the episodic simulation of past and future events. In A. S. Benjamin (ed.), *Successful remembering and successful forgetting: A festschrift in honor of Robert A. Bjork* (pp. 505–525). New York: Psychology Press.

Schacter, D. L., Guerin, S. A., & St Jacques, P. L. (2011). Memory distortion: an adaptive perspective. *Trends in Cognitive Science*, *15*, 467–474.

Schwartz, N., Bless, H., Strack, F., Klumpp, G., Rittenauer-Schatka, H., & Simons, A. (1991). Ease of retrieval as information: another look at the avaliability heuristic. *Journal of Personality and Social Psychology*, *61*, 195–202.

Schwarz, N. (2012). Feelings-as-information theory. In P. A. M. Van Lange, A. W. Kruglanski, & E. T. Higgins (eds.), *Handbook of theories of social psychology (Vol. 1)* (pp. 289–308). Thousand Oaks, CA: Sage Publications.

Scoboria, A., Mazzoni, G. A., Jarry, J. L., & Bernstein, D. M. (2012). Personalized and not general suggestion produces false autobiographical memories and suggestion-consistent behavior. *Acta Psychologica*, *139*, 225–232.

Scoboria, A., Mazzoni, G. A., Jarry, J. L., & Shapero, D. (2012). Implausibility inhibits but does not eliminate false autobiographical beliefs. *Canadian Journal of Experimental Psychology*, *66*, 259–267.

Scoboria, A., Mazzoni, G. A., Kirsch, I., & Relyea, M. (2004). Plausibility and belief in autobiographical memory. *Applied Cognitive Psychology*, *18*, 791–807.

Sharman, S. J., Garry, M., & Beuke, C. J. (2004). Imagination or exposure causes imagination inflation. *American Journal of Psychology*, *117*, 157–168.

Spreng, R. N., & Levine, B. (2006). The temporal distribution of past and future autobiographical events across the lifespan. *Memory & Cognition*, *34*(8), 1644–1651. http://doi.org/10.3758/BF03195927.

Spreng, R. N., Mar, R. A., & Kim, A. S. N. (2009). The common neural basis of autobiographical memory, prospection, navigation, theory of mind, and the default mode: a quantitative meta-analysis. *Journal of Cognitive Neuroscience, 21*(3), 489–510. http://doi.org/10.1162/jocn.2008.21029.

Squire, L. R., van der Horst, A. S., McDuff, S. G. R., Frascino, J. C., Hopkins, R. O., & Mauldin, K. N. (2010). Role of the hippocampus in remembering the past and imagining the future. *Proceedings of the National Academy of Sciences, 107* (44), 19044–19048. http://doi.org/10.1073/pnas.1014391107.

Suddendorf, T., & Corballis, M. C. (1997). Mental time travel and the evolution of the human mind. *Genetic, Social, and General Psychological Monographs, 123*, 133–167.

Suddendorf, T., & Corballis, M. C. (2007). The evolution of foresight: what is mental time travel, and is it unique to humans? *Behavioral and Brain Sciences, 30*(03), 299–313.

Szpunar, K. K., Addis, D. R., McLelland, V. C., & Schacter, D. L. (2013). Memories of the future: new insights into the adaptive value of episodic memory. *Frontiers in Behavioral Neuroscience, 7*, 47. http://doi.org/10.3389/fnbeh.2013.00047.

Szpunar, K. K., Addis, D. R., & Schacter, D. L. (2012). Memory for emotional simulations: Remembering a rosy future. *Psychological Science, 23*, 24–29.

Szpunar, K. K., Chan, J. C. K., & McDermott, K. B. (2009). Contextual processing in episodic future thought. *Cerebral Cortex, 19* (7), 1539–1548. http://doi.org/10.1093/cercor/bhn191.

Szpunar, K. K., & Schacter, D. L. (2013). Get real: effects of repeated simulation and emotion on the perceived plausibility of future experiences. *Journal of Experimental Psychology: General, 142*, 323–327.

Tedder, J., Miller, L., & Lah, S. (2011, August). *Episodic future thinking: is recall of past personal experiences necessary?* Paper presented at the 5th International Conference on Memory, York, UK.

Thomas, A. K., Bulevich, J. B., & Loftus, E. F. (2003). Exploring the role of repetition and sensory elaboration in the imagination inflation effect. *Memory & Cognition, 31*, 630–640.

Tulving, E. (1972). Episodic and semantic memory. In E. Tulving & W. Donaldson (eds.), *Organization of memory* (pp. 381–402). New York: Academic Press.

Tulving, E. (1985). Memory and consciousness. *Canadian Psychologist, 26*, 1–12.

Tulving, E. (2005). Episodic memory and autonoesis: Uniquely human? In H. Terrace & J. Metcalfe (eds.), *The missing link in cognition: origins of self-reflective consciousness* (pp. 3–56). New York: Oxford University Press.

Tversky, A., & Kahneman, D. (1973). Availability: a heuristic for judging frequency and probability. *Cognitive Psychology, 5*(2), 207–232. http://doi.org/http://dx.doi.org/10.1016/0010-0285(73)90033-9.

Van Mulukom, V., Schacter, D. L., Corballis, M. C., & Addis, D. R. (2013). Re-imagining the future: repetition decreases hippocampal involvement in future simulation. *PLoS ONE, 8*(7), e69596. http://doi.org/10.1371/journal.pone.0069596.

Vella, N. C., & Moulds, M. L. (2013). The impact of shifting vantage perspective when recalling and imagining positive events. *Memory, 22*(3), 256–264. http://doi.org/10.1080/09658211.2013.778292.

Von Glahn, N. R., Otani, H., Migita, M., Langford, S. J., & Hillard, E. E. (2012). What is the cause of confidence inflation in the life events inventory (LEI) paradigm? *The Journal of General Psychology, 139*, 134–154.

Ward, A. M., Schultz, A. P., Huijbers, W., Van Dijk, K. R. A., Hedden, T., & Sperling, R. A. (2014). The parahippocampal gyrus links the default-mode cortical network with the medial temporal lobe memory system. *Human Brain Mapping*, *35*(3), 1061–1073. http://doi.org/10.1002/hbm.22234.

Weiler, J. A., Suchan, B., & Daum, I. (2010). Foreseeing the future: occurrence probability of imagined future events modulates hippocampal activation. *Hippocampus*, *20*(6), 685–690.

Weinstein, N. D. (1980). Unrealistic optimism about future life events. *Journal of Personality and Social Psychology*, *39*, 806–820.

Whittlesea, B. W. A. (1993). Illusions of familiarity. *Journal of Experimental Psychology: Learning, Memory and Cognition*, *19*, 1235–1253.

Williams, J. M. G., Barnhofer, T., Crane, C., Herman, D., Raes, F., Watkins, E., & Dalgleish, T. (2007). Autobiographical memory specificity and emotional disorder. *Psychological Bulletin*, *133*(1), 122–148. http://doi.org/10.1037/0033-2909.133.1.122.

Williams, J. M. G., Ellis, N. C., Tyers, C., Healy, H., Rose, G., & MacLeod, A. K. (1996). The specificity of autobiographical memory and imageability of the future. *Memory & Cognition*, *24*(1), 116–125.

Woodruff, C. C., Johnson, J. D., Uncapher, M. R., & Rugg, M. D. (2005). Content-specificity of the neural correlates of recollection. *Neuropsychologia*, *43*(7), 1022–1032. http://doi.org/http://dx.doi.org/10.1016/j.neuropsychologia.2004.10.013.

Wright, D. B. (1998). Modelling clustered data in autobiographical memory research: the multilevel approach. *Applied Cognitive Psychology*, *12*(4), 339–357.

# PART III

# Subjective Temporality in Future-Oriented Mental Time Travel

6

# Temporal Consciousness and Confabulation

## *When Mental Time Travel Takes the Wrong Track*

GIANFRANCO DALLA BARBA

## 1. INTRODUCTION

Mental time travel (MTT), including future mental time travel, does not exist *stricto sensu*. This might appear a provocative statement since "MTT" is largely used in the literature, but the term is actually misused. To travel means to move from one place to another place. Pierre, for example, travels from Paris to Venice. When Pierre arrives in Venice, he is no longer in Paris, but in Venice. He then plans to go to Florence, but he is not in Florence, he is still in Venice. In this example, Paris is Pierre's past and Florence his future. He remembers having been in Paris and he is waiting to go to Florence, but his present consists in being in Venice. When Pierre thinks about having been in Paris, he is not traveling to Paris, as he is not traveling to Florence when he thinks that he will go to Florence in the near future.

This seems trivial, but it is not if we consider the use of "time travel" in the literature, including the subtitle of this book, "Theoretical Perspectives on Future-Oriented Mental Time Travel." While "time travel" is well represented in science fiction, it is not part of the concrete human experience. Individuals can remember their past, can be aware of their present, and can foresee their future, but this happens *in the present*, without any need or possibility to "travel" in the past and in the future.

As soon I had finished writing a book on memory and temporality (Dalla Barba, 2002), I found out that what I wrote in about 250 pages was said in few sentences more than 16 centuries ago:

> One fact is limpid and clear: neither the future nor the past exist. It is inexact to say that there are three kinds of time: past, present and future. Maybe it would be more exact to say that time can be conceived of in three ways: the present of the past, the present of the present, and the present of the future. . . .

> The present of the past is memory, the present of the present is vision, the present of the future is wait. Le Confessioni di Sant Agostino, Einaudi: Torino, 1984, p. 333

This was Augustine in *The Confessions of Saint Augustine.*

As one can easily evict from Augustine's words, and from Dalla Barba (2002), neither the past, nor the present, nor the future exists by itself. They are subordinate structures of the personal temporality of individuals who are firmly anchored in their present, without any possibility of traveling in the past or in the future. In normal conditions, they can remember their past, can be present to their present world, and can foresee their future without traveling—without moving, physically or mentally, an inch from the present in which they are anchored.

In the following sections we will first discuss the phenomenological ontology of personal temporality, showing that the concept of memory traces is paradoxical since it is logically inconceivable that the past is contained in memory traces. Personal temporality and impersonal temporality are phenomenologically different; they are expressions of two different types of consciousness, namely temporal consciousness (TC) and knowing consciousness (KC), and there is no ontological or causal priority of personal past on personal future. We will then describe TC within the framework of the Memory, Consciousness, and Temporality Theory (Dalla Barba, 2002) and discuss confabulation and amnesia within this framework, showing that amnesia and confabulation reflect loss (the first) and distortion (the latter) of TC. Finally, we will discuss the neural correlates of TC.

## 2. THE PHENOMENOLOGICAL ONTOLOGY OF TIME

### 2.1. What Is Mental Time Travel?

If one remembers having had dinner at a restaurant last night, this memory happens in the present. The memory "having-had-dinner-last-night-at-the-restaurant" is a past event, but its being *past* is possible only because one individual takes the trouble to make it a past event (i.e., a memory belonging to the individual's present). Looking at the literature, terms such as "reviviscence" of a past event, or "re-experiencing" past events, are employed. In the same vein, some authors think that when foreseeing a future event, people pre-live a future event. These assumptions are synthetized in the concept of MTT. People can travel at ease in their past and future. However, this is not the case. Pierre had a car accident last week, his ankle was injured, and he was taken to the hospital. His injury was extremely painful. Now he is well. When, now, he remembers that episode, he is no longer in pain. He is remembering the accident and the pain that he suffered, but he is not suffering again. There is no "reviviscence," or "re-experience" of that event. There is "remembering" of that event. He remembers the pain he suffered, the ambulance taking him to the hospital, the doctors, and so on. But certainly, and hopefully, he has no "reviviscence," nor is he "re-experiencing" that event. In the same line of reasoning, when Pierre thinks about his future staying in Florence, he is not "pre-experiencing" his being in Florence. This is to say that people have a specific type of consciousness, TC (see later discussion), which allows them to have a personal past, present, and future.

A flourishing literature is now devoted to MTT, including to future-oriented time travel. Several concepts and terms are employed, including memory of the future (Ingvar, 1985), planning (Dalla Barba, Cappelletti, Signorini, & Denes, 1997) or imagining the future (Klein, Loftus, & Kihlstrom, 2002; Schacter, Addis, & Buckner, 2007), mental time travel (Suddendorf & Corballis, 2007), and chronesthesia (Tulving, 2002). However, these formulations do not explicitly describe what the relationship between past, present, and future is. We have shown elswhere (Dalla Barba, 2002) that past, present, and future are subordinate structures of personal temporality. This point has been largely neglected in recent neurocognitive litterature. "Reviviscence" of past experience and "pre-experiencing" of future events (see the preceding paragraph) are examples of this.

## 2.2. The Paradox of the Memory Trace

If personal temporality is not considered as a whole, in which past and future are past and future *of a present*, namely of a present consciousness that takes the trouble to consider objects and events *as past* or *as future*, then theories of memory and of MTT need to postulate that past and future events are represented in the brain and in the cognitive system. This way of thinking has produced the notion of *memory trace*, on which most theories of memory and of MTT are based. According to these theories, if I can remember a past event, this happens because that event is stored *as past* in a memory trace. If I can predict a future event, this happens because, although theories of MTT are not specific on this point, there is a sort of propedeutical role of the past on the future. In other words, one needs to have memory traces of past events in order to predict the future. It is easy to see that, however reformulated, any theory that bases the ability to recollect on the preservation of an event inside a trace contains a paradox, the fruit of a misleading assumption, which is the belief that time can exist in things. But it should be clear that "things" as such are not temporal. Objects of the world are neither present, nor past, nor future, but they acquire a temporal dimension only in the presence of a person who goes to the trouble of making them temporal. Time cannot even be found in a "thing" like the brain. This erroneous assumption, on which theories of memory are founded, is directly reflected in the paradox to which these theories fall victim. We have called this *the paradox of the memory trace* and we have discussed it in detail elsewhere (e.g., Dalla Barba, 2002). We will summarize it here because of its importance for what will be discussed in the following sections. The paradox of the memory trace consists in seeing memory, that is, what is given to consciousness as past, as originating in elements borrowed from the present. Let us see why.

The event that I now perceive, for example, the glass that is on the table in front of me, is without doubt a present event. This event of "glass-on-the-table" determines a modification in the equilibrium of a system, the nervous system, or the cognitive system, which I call *memory trace*. What is the temporal nature of this modification, namely of the memory trace? Without doubt, it is present. The "glass-on-the-table" that I now perceive is present, and if one accepts that this event produces a modification in my brain, or in my cognitive system, one will have to accept that said modification will be present, and that the event represented by that modification will also be present. What happens when that event contained in the trace is recollected in memory? When the event of "glass-on-the-table" is recollected, it happens in the present, that is, as the result of the reactivation of the modification that the event caused on a physiological, biochemical, neuroanatomical, or functional level. And so it is not at all clear how memory, whose basic characteristic is being memory of the past, can stem from a combination of present phenomena, perception, the preservation of the trace, and recollection. Activation of the memory trace should, if anything, coincide with a new perception, a *reviviscence* of the event contained within the trace, not with its memory, since the event contained in the trace was present as it ended up in the trace and continued to be so for as long as it remained enclosed in the trace. The paradox of the memory trace that current theories on memory overlook had already been clearly reported over 60 years ago. Merleau-Ponty (1945) wrote,

> … our very best reason for rejecting physiological preservation of the past is also a reason for rejecting 'psychological preservation,' and this reason is that no preservation, no physiological or psychic "trace" of the past can enable comprehension of consciousness of the past. This table carries traces of my past life, I have carved my initials, I have left ink stains. But, alone, these traces do not refer back to the past: they are present; and, if I find there signs of some 'previous' event, I find them because, by other means, I have a sense of the past, because I carry in me this meaning.… a preserved perception (in a 'physical' or 'psychic' trace) is a perception, it continues to exist, it is always in the present, it does not open behind us that dimension of escape and of absence that is the past. If, on the other hand, I recognize that particular event as past, this happens because I attribute a precise meaning to it, that of being past, a meaning which by definition cannot be contained in the trace since, in every moment of its existence, it has never ceased to be present." Merleau-Ponty, La fenomemorogia della percezione, Il Saggiatore, Milano, 1965, pp. 527–528

This does not mean that events do not cause modifications in our brain or in our cognitive system, but these modifications cannot be used to explain recollection. Besides, any object in the universe undergoes modification on account of events. Any object carries with it the marks of the past, and yet, if not metaphorically, no one would think of attributing the possibility of recollection to objects. The signs that events have left on objects acquire the meaning of *past* only by virtue of a consciousness that attributes it to them. Even the impression on the cushion on the sofa in front of me bears witness to past events, but in itself it is present and has never ceased to be so in any single instant of its existence. If I find in it signs of some past event, if I remember the thousands of times that my guests and I have sat on that sofa, it is because I attribute a meaning to those marks, that of being past.

If the possibility of remembering the past and projecting the future cannot rely on the existence of memory traces (i.e., on the preservation of the past in the individual who is remembering and on the use of this preservation to travel in the future), how is it possible that, still, we can remember our past and project our future? We have said that, in order for an event to become past or future, we need the present of a person who takes the trouble to *temporalize* these events, making them past or future events. This possibility is provided by temporal consciousness (TC), that is, a specific form of consciousness that allows individuals to locate objects, events, and information according to the three subordinate structures, past, present and future, of their subjective temporality (Dalla Barba, 2002). As we will discuss in detail later, complete loss of TC results in deep amnesia. Patients with deep amnesia not only are unable to remember their past (i.e., to have phenomenological experience of specific events of their past), but are also disoriented in their present world and do not have phenomenological experience of events in their future. Using another terminology, MTT in these patients has been abolished. Conversely, in patients who confabulate (i.e., patients who produce false memories), TC is present, but malfunctioning. The result is that these patients make errors not only when remembering, but also in their orientation in the present and in predicting future personal events. They travel in time along a wrong track.

Before discussing the role of TC in confabulation and amnesia, we need to describe TC within the framework of the Memory, Consciousness, and Temporality Theory (MCTT).

## 3. THE MEMORY, CONSCIOUSNESS, AND TEMPORALITY THEORY

In line with the continental phenomenological tradition (Brentano, 1874; Husserl, 1950; Merleau-Ponty, 1945; Sartre, 1943), the MCTT considers that consciousness is not an aspecific entity, but is intentionally projected toward its object, being always consciousness of *something*. The consequence is that consciousness addresses its object in different ways, further implying that different types, or modes, of consciousness exist. For example, regarding the pen in front of me on the desk, I can perceive it; if I close my eyes, I can imagine it; I can hate it or like it; I can know it (e.g., know that it is a pen and not a sailboat); and I can remember it (e.g., remember where and when I bought it). All these different relationships between consciousness and its object are *original*, because each one differs from each other, and *irreducible*, because they are not the final result of a causal or ontological cascade. The aim of this work is not to detail a taxonomy of different types of consciousness, but rather to use the distinction made by Dalla Barba (2002) between TC and knowing consciousness (KC).

KC is defined as a specific form of consciousness allowing individuals to be aware of past, present, and future *impersonal* knowledge and information. KC concerns, for example, knowing that George W. Bush was formerly president of the United States, that Barack Obama is currently in that position, and that in the next presidential elections Obama will not be allowed to run for a third term. KC is usually relatively preserved in both confabulating and non-confabulating amnesiacs (Dalla Barba & Boissé, 2010; Dalla Barba, Cappelletti, et al., 1997; S. B. Klein et al., 2002; La Corte, Pradat, & Dalla Barba, 2011; La Corte, Serra, Boissé, & Dalla Barba, 2010). Patients who have no phenomenological experience of remembering their personal past and of predicting their personal future not only are able to retrieve impersonal past information (i.e., semantic memories), but also are able to predict the impersonal future. For example, they have no difficulties in answering questions like "What is likely to be an important breakthrough in the medical domain in the next 10 years?" (Klein et al., 2002; La Corte et al., 2010; La Corte et al., 2011). They also have preserved "personal semantics" (i.e., they have access to personal past and present factual information). They can tell, for example, their date of birth, and that they went to school and then graduated. They can also use this type of information to make inferences about their future. Yet in none of these cases do they have the phenomenological experience of remembering specific episodes from their personal past and of predicting specific episodes in their personal future.

TC is a specific form of consciousness that allows individuals to remember their personal past, to be oriented in their present world, and to predict their personal future (Dalla Barba, 2002; Dalla Barba & La Corte, 2013). It is this type of consciousness that defines individuals as temporal entities with a personal past, present, and future. Personal temporality, as expressed by TC, is different from impersonal temporality, as expressed by KC. Patients without TC, due to bilateral hippocampal damage (see later discussion) still have impersonal temporality (i.e., they can access and use impersonal past, present, and future information), but they have lost the personal dimension of time. They can learn and *know* things about their past, as they can *know* things about their future. They can even learn their entire biography, like patient RM (Dalla Barba, Mantovan, Ferruzza, & Denes, 1997), but they have no phenomenological experience of remembering their past and of projecting themselves in specific future situations.

In normal conditions, TC addresses the object's *Uniqueness* (U), whereas KC addresses the object' *Multiplicity* (M). Let's consider the following example. This pen on the desk reveals both a U and an M, according to how my consciousness addresses it. It represents a U if I consider it *that-specific-pen-I-bought-last-week-and-that-I-will-be-using-this-afternoon-to-sign-a-check*. It represents an M if I consider it *an-indeterminate-pen-belonging-to-the-category-of-pens*. In short, U means *this specific pen* and not another one, whereas M means *a pen*, an object belonging to the multiplicity of objects of the same category. According to how consciousness addresses its object, the pen in this example, the object reveals a U, or an M. Anticipating what will be discussed later on, in normal conditions, TC's object represents U, whereas KC's object represents M.

In the next section we will see how what we have discussed so far is relevant to the interpretation of confabulation.

## 4. TEMPORAL CONSCIOUSNESS, CONFABULATION, AND AMNESIA

Confabulation is a kind of memory distortion, which, at a general level, can be defined as the production of statements or actions that are unintentionally incongruous to the subject's history, background, present, and future situation (Dalla Barba, 1993a; Dalla Barba, Cappelletti, et al., 1997). Classically described in Korsakoff's syndrome, following lesions of the mammillary bodies and the dorsomedial nucleus of the thalamus, and often present after lesions to the orbitofrontal cortex, confabulation is observed in several conditions affecting the nervous system and follows lesions located in more than 20 anterior and posterior brain areas (Dalla Barba and Boissé, 2010). According to our model (Dalla Barba & La Corte,

2013), confabulation reflects a distortion of TC, whereas amnesia due to complete, bilateral hippocampal damage reflects a loss of TC.

In confabulators, TC is present, but it is malfunctioning, because these patients confabulate when questioned about their past, present, and future. Conversely, non-confabulating amnesiacs, who have lost TC, have no phenomenological experience of remembering their personal past or of predicting their personal future. An increasing number of studies have addressed the question of the neurocognitive relationship between episodic memory and individuals' ability to predict their personal future. What is referred to as memory of the future (Ingvar, 1985), planning (Dalla Barba, Cappelletti, et al., 1997), or imagining the future (e.g., Klein et al., 2002; Schacter et al., 2007), mental time travel (e.g., Suddendorf & Corballis, 2007), episodic future thinking, and future-oriented cognition and chronesthesia (e.g., Szpunar, Addis, McLelland, & Schacter, 2013; Szpunar, Spreng, & Schacter, 2014; Tulving, 2002) are aspects of TC, as described in the MCTT (Dalla Barba, 2002). Schacter and colleagues (2007) propose that remembering the past is necessary in order to imagine the future. Although remembering the past and imagining the future depend on much the same neural machinery, namely the medial temporal lobe, there is, however, no ontological priority of remembering versus predicting the future. In other words, remembering is not a prerequisite to predict the future. It is not because I remember that I had a cup of tea this morning that I am able to predict having sushi for dinner tonight. This point has been extensively discussed by Klein (2015), who argues that episodic memory per se is not a determinative factor enabling individuals to predict their personal future.

As reported elsewhere by Dalla Barba and coworkers (Dalla Barba & Boissé, 2010; Dalla Barba & La Corte, 2013), aspects of the MCTT relevant to the interpretation of confabulation and amnesia are summarized in the following list.

1. Events produce atemporal and aspecific patterns of modifications in the brain. These modifications are atemporal in the sense that they do not contain any information concerning time. They do not represent the past, the present, or the future, nor are they organized according to the order of succession (i.e., there is nothing in Y, for example, that tells that Y comes before Z and after X). They are aspecific in the sense that they do not contain any information specifying that they are representing episodes, meanings, rules, procedures, or algorithms.
2. The patterns of modifications in the brain can be more or less stable and more or less vulnerable, depending on a number of variables. These variables include, among others, attention at encoding, emotional value of the event, depth of encoding, rehearsal, and repeated experience of the same or of a similar event.
3. TC is a specific form of consciousness that allows individuals to remember their personal past, to be oriented in their present world, and to predict their personal future. It is specific because it cannot be confounded with other forms of consciousness, such as, for example, perception and imagination. If, for instance, the average person perceives a tiger in front of him, he will be scared, but if he remembers or imagines such an event, he may not be scared at all.

5. TC is experimentally measurable and dissociable from impersonal temporality. Using the Confabulation Battery, which includes 11 dependent variables (Dalla Barba, 1993a; Dalla Barba & Decaix, 2009), it has been demonstrated that confabulators and amnesiacs either confabulate or have no phenomenological experience of remembering their personal past or of predicting their personal future, whereas they are able to answer questions about impersonal past, present, and future (e.g., what happened to Princess Diana, who the president of the United States is, and what is likely to be one of the most important breakthroughs in the medical domain in the next 10 years; La Corte et al., 2011; Klein et al., 2002).
6. TC is lost in amnesia following complete bilateral hippocampal damage and malfunctioning in confabulation, because it receives distorted information from more than 20 damaged, predominantly orbitofrontal, brain areas.
7. The object of consciousness represents a determination and an indeterminateness, which we have called U and M. U refers to unique events, whereas M refers to repeated events. TC addresses the object's U, whereas KC addresses its M.

In normal conditions, TC interacts with less stable patterns of modification of the brain in order to seize the object's U, past, present or future, whereas KC interacts with more stable patterns of modification of the brain in order to seize the object's M. The interaction between TC and less stable patterns of modification of the brain allows individuals to identify the "pen" as a U, that is, as an object belonging to a personal temporality—I have used this pen yesterday to sign a check, it is now in front of me, just some inches beside the computer's keyboard, and I can predict using it tomorrow to sign another check for the plumber. In contrast, the interaction of KC with more stable patterns of modification of the brain allows people to identify the "pen" as an M, that is, as a specific object, which is different from other objects—this pen in front of me is different from the computer's keyboard, although they share similar functions.

In confabulation, TC is still present, but it is not interacting with less stable patterns of modification of the brain, because these modifications are abolished or inaccessible in the mode of TC. In this condition, TC interacts with more stable patterns of modifications of the brain, and the result is that repeated events, habits, and over-learned information (in short, the object's multiplicity) are seized as unique events, past, present, or future. It is clinically well known, for instance, that hospitalized confabulators, when directly questioned on what they have done the previous day, usually report routine activities from their life before the accident. For example, they may say that the previous day they went to work or that they had dinner at home "as usual." In this case, irretrievable episodic memories (i.e., events that occurred in a unique and specific temporo-spatial context) are replaced by routines (i.e., multiple, repeated events that did not occur in a unique and specific temporo-spatial context). Therefore we can say that M (i.e., routines and repeated events) is mistaken for U (i.e., a specific unique event that occurred in a specific, unique temporospatial context, such as the previous day). This clinically well-known observation has been experimentally demonstrated for the first time in a recent work from the Dalla Barba group (Serra et al., 2014). In order to measure the ability to discriminate unique from repeated events, the authors used four runs of a recognition memory task, in which some items were seen only once at study, whereas others were seen four times. Confabulators, but not non-confabulating amnesiacs, considered repeated items as unique, thus mistaking M for U. The authors suggested that a crucial mechanism involved in the production of confabulations is thus the confusion between unique and repeated events.

In contrast to confabulation, in amnesia, due to complete hippocampal damage, TC is lost. Patients with classical amnesia are unable to temporalize objects. They cannot remember their past, they are disoriented in the present world, and they are unable to prospect their future. Since in these patients TC is lost, no interaction is possible between TC and more or less stable patterns of modification of the brain. In contrast, in these patients, KC is relatively preserved and interacts normally with more stable patterns of modification of the brain. Therefore they can access and use impersonal information concerning the past, the present, and the future. They can say, for example, that Kennedy was killed, that France is a republic whereas the UK is a kingdom, and that the candidate for the Democrats in the next US presidential elections will not be Barack Obama.

## 5. NEURAL CORRELATES OF TEMPORAL CONSCIOUSNESS

So far we have seen that what distinguishes confabulators from non-confabulating amnesiacs is a distorted TC, in the first, and a loss of TC, in the latter. We have also seen that available data indicate that the integrity, at least partial or unilateral, of the hippocampus seems to be a necessary condition in order for individuals to confabulate, whereas its complete damage not only is not associated with confabulation, but also results in a loss of TC, and consequently in deep amnesia.

It is known that some patients with bilateral lesions in the hippocampus confabulate (e.g., Kartsounis & de Silva, 2011; Kikuchi et al., 1999; Nahum, Ptak, Leemann, Lalive, & Schnider, 2010) In these patients, however, hippocampal destruction is not complete. Overall, there is overwhelming evidence supporting the conclusion that at least partially preserved hippocampus is a necessary condition for confabulation.

At variance with patients with hippocampal amnesia, patients with diencephalic amnesia have distorted TC, as defined in this and previous work from Dalla Barba and coworkers. Confabulation, which is the hallmark of a distorted TC, is a pathognomonic sign of Korsakoff's syndrome (e.g., Benson et al., 1996; Bonhoeffer, 1904; Dalla Barba, Cipolotti, & Denes, 1990; Korsakoff, 1889; Mercer, Wapner, Gardner, & Benson, 1977; Schnider, von Däniken, & Gutbrod, 1996; Talland, 1961; Wyke & Warrington, 1960), which is a diencephalic amnesia. Patients with non-Korsakoff thalamic lesions (e.g., Hodges & McCarthy, 1993), with orbitofrontal lesions (Dalla Barba, Cappelletti, et al., 1997; Kopelman, 1987; Moscovitch, 1995; Schnider et al., 1996) show deep anterograde, more variably, retrograde amnesia, and, invariably, various types of memory distortions (i.e., distorted TC, including confabulations).

Taken together, these observations show that hippocampal amnesia (complete bilateral destruction of the hippocampus) produces negative signs and symptoms (i.e., the failure to retrieve the desired information in TC), whereas non-hippocampal amnesia (diencephalic and frontal) produces positive signs such as memory distortions. This strongly suggests that the hippocampus is the neural correlate of TC (Dalla Barba & La Corte, 2013) and is supported by an increasing number of recent neuropsychological studies (Hassabis, Kumaran, Vann, & Maguire, 2007; Klein et al., 2002; Kwan, Carson, Addis, & Rosenbaum, 2010; Rosenbaum, Wincour, Grady, Ziegler, & Moscovitch, 2007) and neuroimaging studies (Addis, Cheng, Roberts, & Schacter, 2011; Botzung, Denkova, & Manning, 2008; Martin, 2001; Nyberg, Kim, Habib, & Tulving, 2010; Schacter & Addis, 2007) confirming that the hippocampus is a core structure within a network involved in individuals' temporal existence (i.e., their having phenomenological experience of a personal past, present, and future).

Hippocampal anatomy, physiology, and connectivity are all suggestive of a crucial function of this neural structure in associating experienced events in order to remember specific episodes from one's own past and to adapt to ongoing and future reality (Henke, 2010). The hippocampus is reciprocally connected, either directly or indirectly, with all neocortical association areas. It receives upstream, through the parahippocampal, perirhinal, and entorhinal cortices, projections from unimodal and polymodal neocortical association areas and projects downstream, through the fornix, to the hypothalamus, the anterior thalamus, the anterior cingulate gyrus, and the orbitofrontal cortex. Lesions to the fornix result in amnesia without confabulation, whereas confabulation has been described for lesions involving all the previously mentioned neural structures, but sparing the hippocampus.

If the hippocampus is the neural correlate of TC, then its function is to *temporalize* information. This does not mean that the hippocampus has a subjective intentional life, like monitoring theories assume for the (anthropomorphized) frontal lobe (see Dalla Barba, 2002, for a discussion of the *homunculus* fallacy and the anthropomorphization of the brain), but that information, normal or distorted, assumes a personal, temporal dimension when processed by the hippocampus. In normal conditions, the hippocampus accomplishes its function very well, being able to capture the uniqueness of events—that specific walk I had yesterday afternoon along the Bastille Canal and not the walk I take each day there, or the conference I will give tomorrow at 5 p.m. and not a general talk I will be giving in the future. It is now well known that the hippocampus is crucial for rapid, single-trial learning of flexibly integrated what-where-when information (Henke, 2010). Consistently, it has been shown that long-term potentiation following a single train of high-frequency tetanic stimulation can be induced in the hippocampus (Trepel & Racine, 1998). Thus, the normal function of the hippocampus is to temporalize unique information. This function is the result of the interaction between the hippocampus and less stable patterns of modification in neocortical unimodal and polymodal association areas.

As we have seen, in keeping with some aspects of the MCTT, these patterns of modification of the neocortical association areas are atemporal and aspecific. They are atemporal because they do not represent the past, the present, or the future. They are aspecific because they do not contain any information specifying that they are representing episodes, meanings, rules, procedures, or algorithms. These patterns of modifications of the neocortical areas are made temporal and specific by the hippocampus, which processes them as temporal and specific. The patterns of modification that events produce in the neocortical areas can be expressed in behavior atemporally and aspecifically, as, for example, in priming effects. In priming effect, single, unique past events influence current performance and behavior without being temporalized (i.e., they are not expressed as elements of a personal past, present, or future). This is known since the pioneering clinical reports by Korsakoff, Claparède, and others.

So, at present, the hippocampus is the best candidate as the neural correlate of TC, although it is involved in other functions, like reaction to novelty, single trial learning, and "unconscious" episodic memory (Henke 2010).

## 6. HIPPOCAMPUS CONFABULATION AND AMNESIA

In its classical form, confabulation is observed for lesions in the mammillary bodies and the dorsomedial thalamic nucleus. It is well known that confabulation is frequently observed following lesions in the orbitofrontal cortex and basal forebrain. These uncontroversial observations lead researchers to consider lesions to these neural structures as crucial for confabulation to occur. However, as mentioned earlier, confabulations are observed for lesions in more than 20 anterior and posterior cortical and subcortical areas, which are all directly or indirectly connected to the medial temporal lobe and to the hippocampus. The orbitofrontal cortex is one of these structures to which the hippocampus projects through the fornix, mammillary bodies, and thalamus. Lesions to not only the mammillary bodies and the thalamus, but also the basal forebrain, produce confabulations (Schnider, 2008). In short, with the exception of lesions involving the fornix, damage at any point of the pathways running downstream of the hippocampus produce confabulation, provided that the hippocampus is, at least partially, preserved. If the hippocampus is severely damaged bilaterally, the result is deep amnesia without confabulation (Dalla Barba & La Corte, 2013).

Brain damage involving areas projecting from upstream to a preserved hippocampus are also known to produce confabulation (Attali et al., 2009; Dalla Barba, 1993a, b; De Anna et al., 2008). As stated elsewhere (Dalla Barba & La Corte, 2013), lesions to temporoparietal association areas, or to their projections to the hippocampus, may produce confabulated memories and plans, which may differ in content from confabulations observed from lesions involving the orbitofrontal cortex, or structures and pathways downstream of the hippocampus. Lesions downstream of the hippocampus produce semantically appropriate confabulations, either provoked or spontaneous. Lesions upstream of the hippocampal circuit produce more implausible and semantically anomalous confabulations. Therefore, the hippocampus is likely to be the core temporal device that temporalizes information provided either directly by temporoparietal association areas, or, indirectly, through the Papez's circuit, by diencephalic, basal forebrain, and orbitofrontal structures. Lesions upstream or downstream sparing the hippocampus may all produce confabulation. This model is presented in Figure 6.1.

Figure 6.1A depicts the normal functioning of the circuit. Figure 6.1B shows a complete bilateral damage to the medial temporal lobe and the hippocampus with the consequent loss of TC, resulting in deep amnesia. Figure 6.1C describes lesions to temporoparietal areas or their disconnection to the hippocampus, resulting in semantically anomalous confabulations. Figure 6.1D shows lesions downstream of the hippocampal circuit producing the most common form of

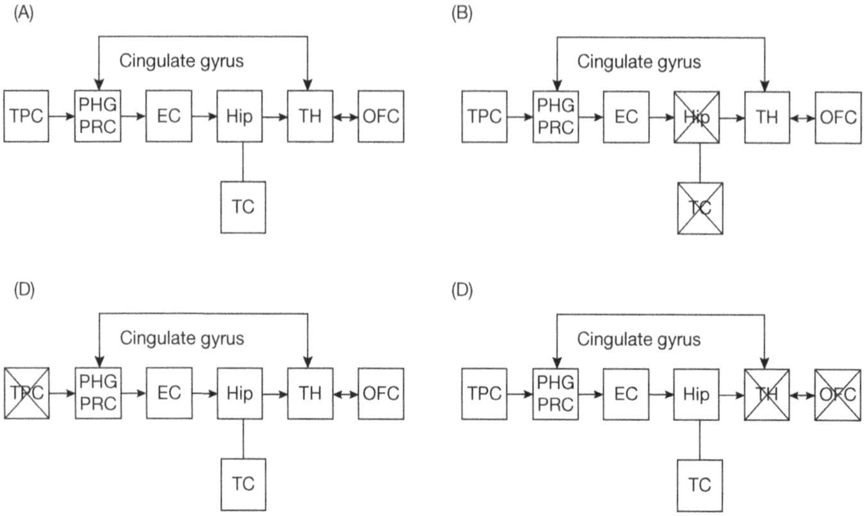


**Figure 6.1.** A schematic cognitive and neuroanatomical model of normal and pathological functioning of memory and temporal consciousness (TC). (A) Normal functioning of memory and TC: the hippocampus, which is the neural correlate of TC, temporalizes information received directly from the TPC, or indirectly from the OFC and the TH, through the cingulate gyrus, allowing individuals to remember their personal past, to be oriented in their present, and to predict their personal future. (B) Amnesia: complete, bilateral lesions to the hippocampus abolish TC, preventing individuals from accessing their personal temporality (i.e., their past, present, and future). (C) Implausible or semantically anomalous confabulation: lesions to the TPC provide the hippocampus with distorted semantic information, inducing the hippocampus and TC to make semantically anomalous confabulatory errors concerning the individual's personal temporality. (D) Plausible, or semantically appropriate confabulation: lesions to the TH and the OFC provide the hippocampus with plausible but erroneous information, inducing the hippocampus and TC to make plausible confabulatory errors concerning the individuals' personal temporality.
Abbreviations: TPC, temporoparietal cortex; PHG, parahippocampal gyrus; PRC, perirhinal cortex; EC, entorhinal cortex; Hip, hippocampus; TH, thalamus; OFC, orbitofrontal cortex; TC: temporal consciousness.
Adapted with permission from Dalla Barba & La Corte (2013).

confabulation, which are mainly plausible, semantically coherent, and indistinguishable from true memories, unless one is aware of the patient's past, present, and future situation.

The anatomical basis of confabulation has been a puzzling issue. Confabulation was originally described in alcoholic patients (Korsakoff, 1889), who were later shown to have diencephalic lesions (Victor, Adams, & Collins, 1971), but were then observed in patients with chronic infections, traumatic brain lesions, subarachnoid hemorrhage, brain tumors, and other pathologies (see Schnider 2008 for a review). Overall, more than 20 anterior and posterior brain lesion loci have been associated with confabulation (Dalla Barba & Boissé, 2010; Gilboa, Alain, Stuss, Melo, & Moscovitch, 2006; Gilboa et al., 2006; Dalla Barba & Boissé 2010). Confabulations are also a common finding in diffuse brain pathologies such as Alzheimer's disease and frontotemporal dementia. Therefore, it seems uncontroversial that confabulation lacks a specific neurobiological correlate, either in terms of pathology, or in terms of the location of the lesion. Here we propose that the neural correlate of confabulation is the (at least partial) integrity of the hippocampus in association with lesions in brain areas that project, directly or indirectly, to the hippocampus. Lesions upstream or downstream of the hippocampus may produce different types of confabulation through the disruption of different cognitive processes, but at least a partial integrity of the hippocampus is the necessary condition for confabulation to occur.

As far as the involvement in confabulation of specific regions within the hippocampus is concerned, at present no reliable data are available and, consequently, no conclusion is possible. However, it is reasonable to think that CA3 and the posterior hippocampus may be crucial for the normal functioning of TC. Neuroimaging data in normal subjects as well as animal studies show that various long-axis specializations arise out of differences between the anterior and posterior hippocampus (Poppenk, Evensmoen, Moscovitch, & Nadel, 2013). The anterior hippocampus is involved in coarse, global representations, whereas the posterior hippocampus is involved in fine-grained local representations. CA3 is known to be involved in pattern separation and pattern completion (Deuker, Doeller, Fell, & Axmacher, 2014; Rolls, 2013). Furthermore, CA3, compared to entorhinal cortex, subiculum, CA1–CA2, is relatively preserved in early Alzheimer's disease (Mueller et al., 2007; Mueller et al., 2010), a condition in which confabulations are present (Dalla Barba et al., 1999; De Anna, et al., 2008; Attali, et al., 2009). Fine-grained local representations, pattern separation, and pattern completion are processes possibly involved in TC's recognition of uniqueness (see earlier discussion). If these processes are disrupted, then multiplicity may be mistaken for uniqueness, because interfering, distorted information from other damaged brain areas or other hippocampal subfields prevents the normal functioning of the posterior hippocampus and CA3, and, consequently, of TC. Further research evaluating the role of specific hippocampal subfields in confabulation will provide possible support to what, at present, is mere speculation.

## 7. CONCLUSIONS

So far we have, briefly, discussed ontology, phenomenology, and theoretical and experimental issues converging on memory and temporality, using confabulation as a tool to demonstrate some characteristics, cognitive and neural, of TC. Where does this discussion take us, and how is it relevant to future-oriented MTT? MTT and FMTT are now largely used and accepted concepts. At the beginning of this work we provocatively stated that MTT does not exist. Indeed it does exist, but only if we accept the centrality of the present. Individuals can remember their past and project their future without moving an inch from the present they are anchored to. There is no "reviviscence" of past episodes, or "pre-living" of future events. We do not need memory traces to remember our past or to project our future. We need TC, and its neural correlate, the hippocampus, to travel in time, a travel that may take a wrong track, as in confabulation.

## REFERENCES


Addis, D., Cheng, T., Roberts, R., & Schacter, D. (2011). Hippocampal contribution to the episodic simulation of specific and future events. *Hippocampus, 21*(10), 1045–1052.

Attali, E., De Anna, F., Dubois, B., & Dalla Barba, G. (2009). Confabulation in Alzheimer's disease: poor encoding and retrieval of over-mearned information. *Brain, 132*, 204–212.

Benson, D. F., Djenderedjian, A., Miller, B. L., Pachana, N. A., Chang, L., Itty, L., & Mena, I. (1996). Neural basis of confabulation. *Neurology, 46*, 1239–1243.

Bonhoeffer, K. (1904). Der Korsakowsche Symptomenkoplex in seinen Beziehungen zu den verschiedenen Krankheitsformen. *Allgemeine Zeitung Psychiatrie, 61*, 744–752.

Botzung, A., Denkova, E., & Manning, L. (2008). Experiencing past and future personal events: functional neuroimaging evidence on the neural bases of mental time travel. *Brain and Cognition, 66*, 202–212.

Brentano, F. (1874). *Psychology from an empirical standpoint*. London: Routledge and Kegan Paul.

Dalla Barba, G. (1993a). Confabulation: knowledge and recollective experience. *Cognitive Neuropsychology, 10*(1), 1–20.

Dalla Barba, G. (1993b). Different patterns of confabulation. *Cortex, 29*, 567–581.

Dalla Barba, G. (2002). *Memory, consciousness and temporality*. Boston: Kluver Academic.

Dalla Barba, G., & Boissé, M.-F. (2010). Temporal consciousness and confabulation: is the medial temporal lobe "temporal"? *Cognitive Neuropsychiatry, 15*(1/2/3), 95–117.

Dalla Barba, G., Cappelletti, Y. J., Signorini, M., & Denes, G. (1997). Confabulation: remembering "another" past, planning "another" future. *Neurocase, 3*, 425–436.

Dalla Barba, G., Cipolotti, L., & Denes, G. (1990). Autobiographical memory loss and confabulation in Korsakoff's syndrome: a case report. *Cortex, 26*, 525–534.

Dalla Barba, G., Mantovan, M. C., Ferruzza, E., & Denes, G. (1997). Remembering and knowing the past: a case study of isolated retrograde amnesia. *Cortex, 33*(1), 143–154.

Dalla Barba, G., Nedjam, Z., & Dubois, B. (1999). Confabulation, executive functions and source memory in Alzheimer's disease. *Cognitive Neuropsychology, 16*(3/4/5), 385–398.

Dalla Barba, G., & Decaix, C. (2009). "Do you remeber what you did on March 13 1985?" A case study of confabulatory hypermnesia. *Cortex*, 45(5), 566–574.

Dalla Barba, G., & La Corte, V. (2013). The hippocampus, a time machine that makes errors. *Trends in Cognitive Sciences, 17*(3), 102–104. http://doi.org/10.1016/j.tics.2013.01.005.

De Anna, F., Attali, E., Freynet, L., Foubert, L., Laurent, A., Dubois, B., & Dalla Barba, G. (2008). Intrusions in story recall: when over-learned information interferes with episodic memory recall. Evidence from Alzheimer's disease. *Cortex, 44*, 305–311.

Deuker, L., Doeller, C. F., Fell, J., & Axmacher, N. (2014). Human neuroimaging studies on the hippocampal CA3 region: integrating evidence for pattern separation and completion. *Frontiers in Cellular Neuroscience, 8*, 64. http://doi.org/10.3389/fncel.2014.00064.

Gilboa, A., Alain, C., Stuss, D. T., Melo, B., & Moscovitch, M. (2006). Mechanism of spontaneous confabulations: a strategic retrieval account. *Brain, 129*, 1319–1414.

Hassabis, D., Kumaran, D., Vann, S., & Maguire, E. (2007). Patients with hippocampal amnesia cannot imagine new experiences. *Proceedings of the National Academy of Science*, *104*(5), 1726–1731.

Henke, K. (2010). A model of memory systems based on processing modes rather than consciuosness. *Nature Reviews Neuroscience*, *11*, 523–532.

Hodges, J. R., & McCarthy, R. A. (1993). Autobiographical amnesia resulting from bilateral paramedian thalamic infarction. *Brain*, *116*, 921–940.

Husserl, E. (1950). *Ideen zur einen reinen Phänomenologie*. Hen Haag: Martinus Nijhoff.

Ingvar, D. H. (1985). "Memory of the future": an essay on the temporal organization of conscious awareness. *Human Neurobiology*, *4*, 127–136.

Kartsounis, L. D., & de Silva, R. (2011). Unusual amnesia in a patient with VGKC-Ab limbic encephalitis: a case study. *Cortex: a Journal Devoted to the Study of the Nervous System and Behavior*, *47*(4), 451–459. http://doi.org/10.1016/j.cortex.2010.01.009.

Kikuchi, A., Chida, K., Misu, T., Okita, N., Takase, S., Nagata, T., . . . Itoyama, Y. (1999). [A case of limbic encephalitis associated with breast cancer developed in an HTLV-1 carrier]. *Rinshō Shinkeigaku = Clinical Neurology*, *39*(5), 555–559.

Klein, S. B., Loftus, J., & Kihlstrom, J. (2002). Memory and temporal experience: the effects of episodic memory loss on an amnesic patient's ability to remember the past and imagine the future. *Social Cognition*, *20*, 353–379.

Klein, Stanley B. (2015). Autonoetic consciousness: reconsidering the role of episodic memory in future-oriented self-projection. *Quarterly Journal of Experimental Psychology* (2006), 1–21. http://doi.org/10.1080/17470218.2015.1007150.

Kopelman, M. D. (1987). Two types of confabulation. *Journal of Neurology, Neurosurgery and Psychiatry*, *50*, 1482–1487.

Korsakoff, S. S. (1889). Etude médico-psychologique sur une forme des maladies de la mémoire. *Revue Philosophique*, *28*, 501–530.

Kwan, D., Carson, N., Addis, D., & Rosenbaum, R. (2010). Deficit in past remembering extend to future imagining in a case of developmental amnesia. *Neuropsychologia*, *48*, 3179–3186.

La Corte, V., Nathalie, Pradat, P., & Dalla Barba, G. (2011). Distorted Teporal Consciousness and preserved Knowing Consciousness in confabulation: a case study. *Behavioral Neurology*, *24*, 307–315.

La Corte, V., Serra, M., Boissé, M.-F., & Dalla Barba, G. (2010). Confabulation in Alzheimer's disease and amnesia: a qualitative account and a new taxonomy. *Journal of the International Neuropsychological Society*, *16*, 967–974.

Mercer, B., Wapner, W., Gardner, H., & Benson, F. (1977). A study of confabulation. *Archives of Neurology*, *34*, 429–33.

Merleau-Ponty, M. (1945). *Phénoménologie de la perception*. Paris: Librairie Gallimard.

Moscovitch, M. (1995). Confabulation. In Schacter DL, editor, *Memory distortion* (pp. 226–251). Cambridge, MA: Harvard University Press.

Mueller, S. G., Stables, L., Du, A. T., Schuff, N., Truran, D., Cashdollar, N., & Weiner, M. W. (2007). Measurement of hippocampal subfields and age-related changes with high resolution MRI at 4T. *Neurobiology of Aging*, *28*(5), 719–726. http://doi.org/10.1016/j.neurobiolaging.2006.03.007.

Mueller, S. G., Schuff, N., Yaffe, K., Madison, C., Miller, B., & Weiner, M. W. (2010). Hippocampal atrophy patterns in mild cognitive impairment and Alzheimer's disease. *Human Brain Mapping*, *31*(9), 1339–1347. http://doi.org/10.1002/hbm.20934.

Nahum, L., Ptak, R., Leemann, B., Lalive, P., & Schnider, A. (2010). Behaviorally spontaneous confabulation in limbic encephalitis: the roles of reality filtering and strategic monitoring. *Journal of the International Neuropsychological Society: JINS, 16*(6), 995–1005. http://doi.org/10.1017/S1355617710000780.

Nyberg, L., Kim, A., Habib, R., & Tulving, E. (2010). Consciousness of subjective time in the brain. *Proceedings of the National Academy of Science, 107*(51), 22356–22359.

Poppenk, J., Evensmoen, H. R., Moscovitch, M., & Nadel, L. (2013). Long-axis specialization of the human hippocampus. *Trends in Cognitive Sciences, 17*(5), 230–240. http://doi.org/10.1016/j.tics.2013.03.005.

Rolls, E. T. (2013). The mechanisms for pattern completion and pattern separation in the hippocampus. *Frontiers in Systems Neuroscience, 7*, 74. http://doi.org/10.3389/fnsys.2013.00074.

Rosenbaum, R. S., Wincour, G., Grady, C., Ziegler, M., & Moscovitch, M. (2007). Memory for familiar environments learned in the remote past: fMRI studies of healthy people and amnesic person with extensive bilateral hippocampal lesions. *Hippocampus, 17*, 1241–1251.

Sartre, J.-P. (1943). *L'être et le néant*. Paris: Gallimard.

Schacter, D. L., & Addis, D. R. (2007). The cognitive neuroscience of constructive memory: remembering the past and imaging the future. *Philosofical Transactions of the Royal Society. B, 362*, 773–786.

Schacter, D. L., Addis, D. R., & Buckner, R. (2007). Remembering the past to imagine the future. *Nature Review of Neuroscience, 8*, 657–660.

Schnider, A. (2008). *The confabulating mind: how the brain creates reality*. New York: Oxford University Press.

Schnider, A., von Däniken, C., & Gutbrod, K. (1996). The mechanisms of spontaneous and provoked confabulations. *Brain, 119*, 1365–1375.

Serra, M., La Corte, V., Migliaccio, R., Brazzarola, M., Zannoni, I., Pradat-Diehl, P., & Dalla Barba, G. (2014). Confabulators mistake multiplicity for uniqueness. *Cortex; a Journal Devoted to the Study of the Nervous System and Behavior, 58*, 239–247. http://doi.org/10.1016/j.cortex.2014.06.011.

Suddendorf, T., & Corballis, M. C. (2007). The evolution of foresight: what is mental time travel, and is it unique tu humans? *Behavioral and Brain Sciences, 30*, 299–351.

Szpunar, K. K., Addis, D. R., McLelland, V. C., & Schacter, D. L. (2013). Memories of the future: new insights into the adaptive value of episodic memory. *Frontiers in Behavioral Neuroscience, 7*, 47. http://doi.org/10.3389/fnbeh.2013.00047.

Szpunar, K. K., Spreng, R. N., & Schacter, D. L. (2014). A taxonomy of prospection: introducing an organizational framework for future-oriented cognition. *Proceedings of the National Academy of Sciences U S A, 111*(52), 18414–18421. http://doi.org/10.1073/pnas.1417144111.

Talland, G. A. (1961). Confabulation in the Wernicke-Korsakoff sindrome. *Journal of Nervous and Mental Disease, 132*, 361–381.

Trepel, C., & Racine, R. J. (1998). Long-term potentiation in the neocortex of the adult, freely moving rat. *Cerebral Cortex (New York, N.Y.: 1991)*, *8*(8), 719–729.

Tulving, E. (2002). Chronestesia: awareness of subjective time. In *Principles of frontal lobe functions* (pp. 311–325). New York: Oxford University Press.

Victor, M., Adams, R. D., & Collins, G. H. (1971). *The Wernicke-Korsakoff syndrome*. Phildelphia: Davis.

Wyke, M., & Warrington, E. (1960). An experimental analysis of confabulation in a case of Korsakoff's syndrome using a tachistoscopic method. *Journal of Neurology, Neurosurgery and Psychiatry, 23*, 327–333.

7

# The Role of Subjective Temporality in Future-Oriented Mental Time Travel

STANLEY B. KLEIN AND CHLOE STEINDAM

Future-oriented mental time travel (FMTT) consists in a set of neurocognitive functions that enable one to break free of the grasp of the perpetual "now" and consider how things might be in the "next." The adaptive functionality of such a mental competence is immense: to be capable of anticipating, imagining, and responding to contingencies that cannot be known with certainty—but whose consequences have fundamental significance for survival—provides an enormous selective advantage. While organisms vary greatly in the complexity and temporal reach of their ability to consider the future (e.g., Cheke & Clayton, 2010; Klein, 2013a; Suddendorf & Corballis, 1997), all living things possessed of motility and reasonably sophisticated neural function must, of adaptive necessity, be able to orient toward and react to personally relevant possibilities (Klein, 2013a). Indeed, several investigators have proposed that an organism's ability to anticipate and plan for the future is a major driving force of cortical evolution (e.g., Bar, 2011; Klein, Cosmides, Tooby, & Chance, 2002; Pezzulo, 2008; Tulving, 2005).

Psychologists have long held an interest in various aspects of anticipatory behavior. It was not until the first stirrings of the "cognitive revolution" in the 1950s, however, that serious and sustained attention to FMTT became a regular feature of research agendas. This now has developed into a thriving industry, with several hundred papers devoted to FMTT appearing in just the past 15 years.

While tremendous progress in our understanding of FMTT has been attained, these gains primarily have been in the explication of the behavioral characteristics, neuroanatomical correlates, and phylogenetic precursors of the human ability to orient toward the future. Considerably less attention has been devoted to the processes that enable a person to mentally navigate a nonexistent world of possibility. What, specifically, is future-oriented subjective temporality?

Although subjective temporality has long been a focus of philosophical and psychological investigation (for reviews, see Arstila & Lloyd, 2014; Roeckelein, 2000), it seldom has been addressed in the context of FMTT (there are, of course, exceptions; e.g., Dalla Barba, 2002; Klein, Loftus, & Kihlstrom, 2002; Tulving, 2005). We conjecture that this lack of attention is, in part, a consequence of the assumption that FMTT is an act of memory and imagination—and, since psychologists have trained their investigative skills on these faculties for much of the discipline's history, wherein lies the motivation to reinvent the "wheels" of memory and imagination?

## 1. INTRODUCTION

In this chapter we focus on the human ability to subjectively orient toward the future. First, however, a few words are required about the role of imagination in mental time travel (memory's role will be treated in section 3). FMTT experience typically entails some form of imagination (e.g., counterfactual, scenario construction, simulation, etc.). But to experience FMTT, imagination must be acted on by processes that enable temporal awareness—and it is those processes, not the imaginative act per se, that allow us to mentally project occurrent imagination from the "now" to the "next" (e.g., one can imagine a camping trip without situating those imaginings in *any* temporal context; Klein, Robertson, & Delton, 2010). In short, our capacity to move the mind through landscapes yet to be actualized is not, and cannot, be achieved solely via an act of imagination: absent additional processes, no imaginative movement through time is possible. These processes—which we refer to as subjective temporality—are the focus of this chapter.

For much of its existence as a topic of investigative effort, FMTT has been accorded the status of a reasonably well-circumscribed set of neurocognitive abilities. Despite surface variation due to the task employed, the mechanisms underlying FMTT are assumed to share a strong family resemblance (based largely on their supposed grounding in episodic memory; see section 2 and Klein, in press). However, when examined through an analytic lens, it becomes clear that the totality of phenomena labeled FMTT is more akin to a complex fractal set than a well-formed taxonomy. This, in turn, suggests that if we are going to understand FMTT we must be cognizant that it comes in different forms or types: one size does not fit all (for a recent review, see Klein, 2013a).

To anticipate our conclusions, we argue that although there is a marked tendency to treat FMTT as a unitary mental faculty that, despite task-driven outward variation, ultimately reduces to a common phenomenological state (underwritten primarily by episodic memory), FMTT is neither unitary nor is it necessarily beholden to episodic memory (on this latter point, see also Abraham & Bubic, 2015; Irish, Chapter 19 of this volume; Irish & Piguet, 2013; and Klein, 2103a). Rather, FMTT is diverse both in its experiential realizations and causal dependencies.[1] It is our contention that the phenomenological diversity found with FMTT is dependent not on the type of memory in play (e.g., episodic versus semantic), but on the type of subjective temporality associated with a particular memorial state (Klein, in press).

## 2. A BRIEF REVIEW OF THOUGHT ABOUT SUBJECTIVE TEMPORALITY AND MEMORY

Western interest in subjective temporality begins with ancient Greek philosophy.[2] Hesiod, in his *Theogony* (circa 700 B.C.), observes that a function of memory is to enable its possessor to subjectively transcend the stasis of the present. While he was not entirely clear whether temporal navigation was directionally biased (i.e., toward past or future), Aristotle (384 BC—322 B.C. was adamant: "The object of memory is the past" (cited in Sorabji, 1972, 13). The future, in contrast, is known not by memory but by acts of anticipation.

Aristotle's position dominated discourse on the relation between time and memory for almost two thousand years. Thus, Augustine (354–430 A.D.) contends, "The time of present things past is memory, the time of present things present is direct experience and the time of present things future is expectation" (*Confessions*, Book 11, chapter 20, heading 26). Although the concept of memory underwent a number of (often minor) changes (mostly in the hands of Scholastic authorities; e.g., Danziger, 2008; Yates, 1966), its relation to the past served as the stable ground for virtually all discourse.

It was not until the late nineteenth century that the idea that the subjective temporality enabled by memory is restricted to navigating one's past met serious challenge. Dissatisfaction with the Aristotelian view was stimulated by Darwin's work on natural selection. For example, Bradley (1887), adopting a functional perspective, argued that memory *must*, of adaptive necessity, be oriented toward the future. "Why," he asked,

> is our memory directed towards our incoming sensations and towards the [temporal] side from which change comes? . . . The answer, in a word, is practical necessity. . . . Life being a process of decay and of continual repair and struggle throughout against dangers, our thoughts, if we care to live, must mainly go the way of anticipation. We are concerned practically with what meets us and what we go to meet, and this practical concern has formed the main habit of our thought." (Bradley, 1887, 581–582; word in brackets added for textual clarification).

For much of the next century, treatment of the relation between memory and future-oriented temporal perspective remained largely in the hands of philosophers, while academic psychology continued to devote attention to the intuitively more appealing notion that memory is concerned with things past (thereby committing the logical error of assuming that since memory is *from* the past it must be *about* the past; e.g., Klein, 2013b, 2015).

### 2.1. The Emergence of FMTT and Its Relation to Memory

Psychology's interest in subjective temporality has been on display from the time the discipline first achieved academic independence (around the middle of the nineteenth century; for a review, see Fraisse, 1963). However, little of this work was devoted to explication of the role of memory in future orientation: to the extent that memory was partnered with subjective temporality, the questions asked were largely psychometric—for example, the role of memory in retrospective estimates of elapsed time, just noticeable differences between temporal intervals, and so on (e.g., Frankenhaeuser, 1959; Nichols, 1891; Ornstein, 1969; for a review, see Roeckelein, 2000).

Sustained interest in FMTT first appeared in the 1950s, as psychologists began to consider how thought about, or orientation toward, the future influences behavior (for reviews, see Cottle & Klineberg, 1974; Gorman & Wessman, 1977). These initial forays examined a diversity of topics—such as the part played by future orientation on life-span development (Klineberg, 1967; Wohlford, 1966), delinquent behavior (Davids, Kidder, & Reich, 1962), emotions (Melges & Fougerousse, 1968), impulse control (Klineberg, 1968; Rozek, Wessman, & Gorman, 1977; Siegman, 1961), motivation (Nuttin, 1985), academic achievement (Teahan, 1958), and mental illness (Krauss & Ruiz, 1967; Wallace, 1956). Most of this work treated future-oriented thought as a variable that people might use or be influenced by: the cognitive and neural underpinnings (e.g., memory) of FMTT seldom figured in discussion (this most likely was due to hard-line positivism's informal ban on "black-box" psychology).

Following a brief period of benign neglect, psychological investigation of subjective temporality was reinvigorated by three conceptual meditations—Tulving (1985), Ingvar (1985), and Suddendorf (1994). However, the questions addressed had changed: influenced by the theoretical commitments of the cognitive revolution, inquiry now was trained on the relation between future-oriented thought and its memorial underpinnings. The first few empirical treatments soon followed (e.g., Dalla Barba, Cappelletti, Signorini, & Denes, 1997; Klein, Loftus, & Kihlstrom, 2002; Williams, Ellis, Tyers, Healy, Rose & MacLeod, 1996), and in less than two decades the relation between memory and FMTT had grown into a thriving research industry. This interest shows no sign of abating—as evidenced by the increasing pace and broadening scope of the questions being asked, as well as special issues (e.g., *Quarterly Journal of Experimental Psychology*, in press), symposia (e.g., Society for Personality and Social Psychology [SPSP], 2015) and edited books (e.g., the present volume) devoted to the topic.

## 3. MEMORY, SUBJECTIVE TEMPORALITY, AND FMTT

Although Bradley's prophetic observations waited nearly 100 years before receiving serious attention, by the mid-1980s psychologists had begun to consider what (given the intellectual climate) seemed a radical possibility—that the evolved function of memory was to focus thought and behavior on the future, rather than on the past (for reviews, see Boyer, 2009; Klein, 2013b; Tulving, 2005). Within this framework, one particular type of long-term memory—episodic—was accorded special status as the neurocognitive basis of virtually all forms of FMTT (for recent reviews, see Addis & Schacter, 2012; Irish & Piguet, 2013; Klein, 2013a; Schacter, Addis, Hassabis, Martin, Spreng, & Szpunar, 2012; Suddendorf & Corballis, 2007; Szpunar, 2010).

The assumed bond between episodic memory and FMTT was widely accepted, and scholarly treatments of the pairing—impressive both in quantity and diversity—soon covered the academic landscape. Terms such as *episodic future thought* (Atance & O'Neill, 2005; Race, Keane, & Verfaellie, 2011; Schacter & Addis, 2007; Szpunar & McDermott, 2008) *episodic simulation/construction* (Addis, Cheng, Roberts, & Schacter, 2011; Hassabis, Kumaran, Vann, & Maguire, 2007; Hassabis & Maguire, 2007; Schacter, Addis, & Buckner, 2008), *episodic self-projection* (e.g., Buckner & Carroll, 2007), and *episodic foresight* (e.g., Atance & Sommerville, 2014; Suddendorf, 2010) became the "coin of the realm."

One possible reason for the tight focus on episodic memory is that—with two notable exceptions (Attance & O'Neill, 2001; Klein, Loftus, & Kihlstrom, 2002)—initial theory and research had examined *only* the effects of episodic memory on FMTT (Dalla Barba et al., 1997; Suddendorf & Corballis, 1997; Tulving, 1985; Wheeler, Stuss, & Tulving, 1997; Williams et al., 1996). However, both logical concerns and empirical demonstrations recently have called into question both the exclusivity and conceptual warrant of the assumed relation between episodic memory and FMTT (these are treated extensively in Klein, in press, and will not be elaborated here).

## 3.1. Broadening the Scope of Memory's Participation in FMTT

In what follows, we focus on the question of whether episodic memory is necessary for FMTT. As our brief review in the previous section suggests, this was, and continues to be, the "received view." However, in the first years of the new millennium, Klein, Loftus, and Kihlstrom (2002) demonstrated that semantic memory can support certain forms of FMTT in the absence of *any* functioning episodic memory. Specifically, as a result of an anoxic episode brought on by cardiac arrest, patient D. B. suffered a total loss of episodic memory. In contrast, his semantic knowledge largely was preserved (for case details, see Klein, Rozendal, & Cosmides, 2002). As anticipated from previous studies on the effects of episodic amnesia on FMTT (e.g., Dalla Barba et al., 1997), D. B. was unable to mentally project himself into the future. He could, however, imagine non-personal future scenarios.

Although this finding—that FMTT might be based on intact semantic knowledge—initially was ignored, investigators began to take notice as they encountered increasing difficulty fitting their FMTT findings to an exclusively episodic model (e.g., Addis & Schacter, 2012; Anderson, 2012; Cooper,Vargha-Khadem, Gadian, & Maguire, 2011; Irish, Chapter 19 of this volume; Maguire, Vargha-Khadem, & Hassabis, 2010). For example, patients with episodic amnesia due to hippocampal abnormalities sometimes retain the ability to imagine future experiences (e.g., Maguire et al., 2010; Mullally, Hassabis, & Maguire, 2012). Explanations adhering to the tenets of the episodic model suggested that these patients possibly possessed residual hippocampal function capable of underwriting future projection (e.g., Mullally et al., 2012). However, the findings also were consistent with the possibility that semantic knowledge, in the absence of episodic function, can sustain FMTT. It thus remained a live question whether semantic knowledge by itself, or in interaction with partially intact episodic memory (inferred from residual hippocampal activity), accounted for future-oriented imagination reported by these investigators.

Empirical support for the position that FMTT cannot be grounded solely in the offerings of episodic memory recently has been demonstrated by several labs. For example, Irish, Addis, Hodges, and Piguet (2012) showed that individuals suffering semantic dementia (with corresponding impairment of semantic memory) have difficulty constructing novel future scenarios despite largely intact episodic function. The authors concluded that semantic memory is a necessary component of future-oriented thought. Echoing these results, Duval, Desgranges, de La Sayette, Belliard, Eustache, and Piolino (2012) demonstrated that patients in early to moderate stages of semantic dementia experience significant impairment in future-oriented self-projection, regardless of whether the task performed was designed to access semantic or episodic memory. Taken together, these findings (and others like them; e.g., Costi, Navarro, Vallat-Azouvi, Brami, Azouvi, & Piolino, 2015) suggest that episodic memory is not sufficient for (at least certain forms of) FMTT.

It may not be necessary either (again, at least for certain forms of FMTT; e.g., Klein, in press). Consider the case of patient J. R. (Manning, Denkova, & Unterberger, 2012). Following surgical treatment for temporal lobe epilepsy, J. R. showed preserved episodic memory alongside selective loss of semantic function. Unlike the dementia patients studied by Duval et al. (2012), J. R. was able to recall his personal past and imagine a *personal* future. However, his ability to engage in mental time travel of an *impersonal* nature was seriously impaired in both temporal directions from the present. This finding, in combination with that of patient D. B., suggests a double dissociation between types of memory (episodic/semantic) and the ability to imagine types of future-oriented scenarios (personal/non-personal).

One particularly interesting case is that of patient R. B. (e.g., Klein & Nichols, 2012). Following a serious head injury, R. B. suffered a very unusual memory impairment: while he remained able to retrieve content typically associated with episodic memory (i.e., temporally, spatially, and self-referentially propertied), and was able to experience autonoetic consciousness (see section 3.3), he could not conjoin the two. Absent a linkage between retrieved content and autonoetic consciousness, no episodic memory is possible (e.g., Tulving, 1985, 1993; Wheeler, Stuss, & Tulving, 1997).

Yet despite his absence of episodic recollection he was able to form plans of a personal nature (surprisingly, he could form new episodic memories, i.e., postdating his injury; but this meager corpus—he was tested approximately 2–4 months post-injury—could not possibly provide a database sufficient for making plans, some of which were complex and temporally extensive). This case calls into question whether episodic memory is *necessary* for all forms of *personal* FMTT. Perhaps autonoetic consciousness, absent episodic recollection, is sufficient for self-referential FMTT (an explanation favored by Klein, in press)? Perhaps retrieved mental content, unaccompanied by recollective experience (e.g., Klein, 2015), can enable (at least certain kinds of) personal planning? Unfortunately, this is not the place for extended discussion of this fascinating case, which would take us far afield (as well as exceed our chapter's word limits).

Studies performed with mnemonically healthy individuals also have begun to question the sustainability of a purely episodic model of FMTT. For example, it has been shown that the ability to construct future-oriented simulations of novel events requires contributions from *both* episodic and semantic memory (Anderson, 2012; Irish, Chapter 19 of this volume). Along similar lines, work by D'Argembeau, Lardi, and Van der Linden (2012) and D'Argembeau and Mathy (2011) demonstrates that thinking about oneself in the future draws on resources from episodic and semantic memory.

In short, findings, mostly of recent vintage, pose a serious challenge to the explanatory adequacy of the episodic model of FMTT (for recent reviews, see Addis & Schacter, 2012; Irish, Chapter 19 of this volume; Irish & Piquet, 2013; Klein, 2013a, in press).

## 3.2. From Memory to Subjective Temporality

For present purposes, a central, but often overlooked, implication of the findings reported in section 3.1 is *not* that different types of memory are involved in FMTT. This no longer is in question. Rather, it is what these differential contributions tell us about the types of subjective temporality underwriting FMTT experience. The demonstration that FMTT is associated with semantic as well as episodic memory, taken in conjunction with the assumption that unique forms of subjectivity accompany different forms of memory (e.g., Tulving, 1985, 1993), led Klein and his colleagues (Klein, 2013a; Klein, Loftus, & Kihlstrom, 2002) to propose that FMTT cannot adequately be captured by a single type of temporal subjectivity: rather, the kind of subjective temporality enabling FMTT varies as a function of the type of memory called into play by task demands.

Moreover, as Klein (in press) has argued, it is the subjective temporality associated with memorial experience, not the memorial experience per se, that plays the foundational role in enabling mental time travel. Tasks that rely on episodic memory for their successful realization promote a form of subjective temporality that allows a person to project him- or herself into the future. Importantly, this entails the feeling that one is pre-experiencing the anticipated outcome. Klein and colleagues termed this "lived time." For example, if, when asked to imagine what you will watch on television tonight, you do so by recollecting information about what you have watched on this day of the week in the past, your temporal subjectivity will be the type that accompanies episodic memory (i.e., "lived time"). The phenomenology of "lived time" is pre-reflective—that is, past experience, when acted on by this form of temporal experience, results in a directly given, non-inferential feeling that the *self* is pre-living a future possibility (see Klein, in press). "Lived time" does not require additional considerations or deliberations to justify one's feeling that occurrent mental content points to future contingency.

In contrast, tasks that draw on semantic knowledge (e.g., when asked to anticipate environmental challenges likely to face the planet in the next decade, you formulate an answer by retrieving and reviewing what you know about ecological dangers looming on the horizon) are associated with *impersonal* forms of temporal experience. Klein and colleagues called this form of temporality "known time." Unlike "lived time," the phenomenology of "known time" is neither pre-reflective nor experienced as a pre-living. Rather, it is knowledge of impersonal possibilities that emerges from conceptual analyses performed on content available to consciousness. (Note that Klein, 2013a, argues that "known time" can, in principle, be self-referential. But the phenomenology is such that the self is treated as an object of conceptual analysis, rather than as first-person subjectivity.) For example, a person may infer from subsequent analysis that retrieved content refers, by logical implication, to the future (e.g., "I know town hall meetings are on Mondays—though I no longer recollect the circumstances in which I acquired this knowledge. Since today is Saturday, the meeting is in two days"). Thus, "known time" consists in the conceptual analysis of temporal features embedded in retrieved content, rather than the unmediated feeling (perhaps, but not necessarily, grounded in retrieved content; see section 3.5) that one is pre-experiencing a personal future.[3]

In sum, the difference between "lived" and "known" temporal experience resides in the phenomenology that each enables. In the former case, subjective temporality is directly given as part of one's occurrent mental state, whereas in the latter, subjective temporality is the product of inferential or interpretive acts (for more for extensive discussion, see Klein, 2013c, 2014a; in press). Further, in "lived time" temporal experience is felt as a pre-living; in contrast, the temporal experience enabled by "known time" is analytic, the outcome of inferential processing of chronologically relevant information. The take-away message is that if we hope to fully appreciate FMTT, we need to recognize that subjective temporality is not a single, undifferentiated phenomenon. Rather, there are different types of subjective temporality with different causal bases and phenomenological character.[4,5]

## 3.3. A Comparison of "Lived" and "Known" Time with Tulving's Distinction Between Autonoetic and Noetic Consciousness

"Lived time" maps reasonably well onto Tulving's (1985, 2005) proposal that episodic recollection entails a form of consciousness that he termed *autonoetic* (at least the future-oriented aspect of autonoesis; see note 5). According to Tulving, a person is autonoetically conscious when she "is capable of becoming aware of her own past as well as her own future; she is capable of mental time travel, roaming at will over what has happened as readily as over what might happen, independently of physical laws that govern the universe" (Tulving, 1985, 5). Autonoetic consciousness thus enables a person to relive earlier experiences as well as pre-live future possibilities (for related views, see Gardiner, 2001; Suddendorf & Corballis, 1997; Szpunar, 2010; Wheeler et al., 1997). Importantly, the temporal subjectivity made possible by autonoesis is directly given to awareness, rather than being based on inference or interpretation (e.g., Klein, 2013c; Markowitsch, 2003; Tulving, 2005). Hence, autonoeisis, like "lived time," is possible without, or apart from, conceptual analysis.

It is important to note that the adoption of episodic memory as the foundation of FMTT derives, at least in part, from an uncritical acceptance of a temporal symmetry between past- and future-oriented forms of autonoesis (e.g., Klein, 2013a; Merker, 2007; Perrin, Chapter 3 of this volume). Specifically, it is (often tacitly) assumed that past-oriented autonoetic consciousness is ontologically and phenomenologically isomorphic with future-oriented autonoesis, thereby sanctioning extension of the functional scope of the recollective act to "what is yet to be" as well as to "what once was" (for critique and discussion see Klein, 2013a; Perrin, Chapter 3 of this volume).

However, as a philosophers have argued, the phenomenological and causal natures of past and future (at least with respect to the everyday world of experiential reality; in the realms of special relativity [e.g., Hoffmann, 1998] and quantum reality [e.g., Einstein, Podolsky, & Rosen, 1935], questions of temporality become deeply puzzling) are notably asymmetric (e.g., Faye, 1989; Lockwood, 2005; McLure, 2005; Perrin, Chapter 3 of this volume). For example, while the accuracy of one's recollections can sensibly (though imperfectly; e.g., Von Leyden, 1961) be assessed by comparison with evidence (physical and mental) of a once existent past, there are no comparable criteria to guide assessment of the accuracy of future-oriented imagination. This follows directly from the definition of the future as "a mixture of coexisting latent possibilities as yet unresolved" (Merker, 2007, 326). All we can know is that "*all* possible futures, except *one*, will in fact *not* materialize, but not *which* one" (Merker, 2007, 326; emphases in original).

Of course, revisiting the past in anticipation of the future is not necessarily an act of imaginative futility. Past knowledge, conjoined with the experientially defensible assumption that natural laws ensure a level of uniformity in outcome, sanctions the reasonable expectation that one's pronouncements about uncertain, yet-to-exist eventualities will have some empirical warrant. But, unlike the past, knowledge of whether one's anticipatory beliefs are veridical remains evidentially opaque until "possibility" becomes "reality." The stubborn fact is that—as a consequence of both epistemological and ontological asymmetries—any simple acceptance of a past/future isomorphism is logically and phenomenologically problematic (for discussion in the context of mental time travel, see Klein, 2013a; Merker, 2007; Perrin, Chapter 3 of this volume). In this regard, Klein (2013a, in press) has called for the need to distinguish past- and future-oriented autonoetic consciousness in discussion of FMTT (see also Perrin, Chapter 3 of this volume; for a contrasting view, see Michaelian, Chapter 4 of this volume).

In contrast to "lived time," "known time" has no clear correlate in Tulving's taxonomy. Tulving maintained that the semantic memory is accompanied by noetic consciousness. Noesis "allows an organism to be aware of, and to cognitively operate on, objects and events, and relations among objects and events, in the absence of these objects and events" (Tulving, 1985, 3). However, unlike autonoesis, noetic experience takes place *absent* a sense of past or future (e.g., Tulving, 1993; Szpunar, 2010). Put simply, the objects of noetic consciousness are firmly situated in the "here and now."[6] Though "lived time" appears isomorphic with future-oriented auotonesis, "known time" has no obvious theoretical correlate in Tulving's (or Dalla Barba's; see note 6) system.

Thus, while the "lived" versus "known" conceptualization can account for both episodic- and semantic-based FMTT, the autonoetic versus noetic distinction (as currently configured) struggles to make sense of semantic forms of FMTT. In fact, Tulving and colleagues reject the possibility of *any* temporal experience accompanying semantic memory (e.g., Tulving, 1985; Szpunar, 2010). These definitional constraints on what constitutes temporal experience, however, can be loosened. By allowing that noetic consciousness can promote a form of temporal subjectivity based on conceptual analysis (i.e., "known time"), the construct can be modified to accommodate the type of subjective temporality associated with semantic-based FMTT.

## 3.4. "Lived" and Known" Time in Relation to Philosophical Conceptions of Time

The "multiple subjective temporality" postulate conforms nicely with a well-known distinction between types of temporality that traces to antiquity (e.g., Mooij, 2005; in the modern era this conceptualization most closely is associated with the work of McTaggart, 1908). The distinction trades on the difference between temporal experience conceived as (a) the constant flow from future to present to past, with temporal designators continually changing ontological status (i.e., that which once was future becomes present and that which is present will, with time's flow, recede further into the past), and (b) a fixed, "earlier-later" chronology in which temporal placement of events is invariant (e.g., October 9, 1952 is, and always will be, prior to October 9, 1953): time is frozen into an unchanging pattern of "before and after" (McTaggart, 1908; Papa-Grimadli, 1998; Rochelle, 1998). These two modes of temporal conceptualization—which McTaggart refers to as the A- and B-series, respectively—are not logically reducible, one to the other (e.g., Loizou, 1986; McTaggart, 1908; for an opposing view, see Cornish, 2011).

Like "lived time," the subjective temporality afforded by the A-series is pre-reflective. As McLure (2005) comments, apprehension of temporal flow is directly given to consciousness: "the 'eye' that espies the flow is within it, not behind it" (49). The B-series, in contrast, consists in a conceptual analysis of the events and happenings in terms of their chronological arrangement. It thus comports with the subjective temporality associated with "known time."

It is important to mention that persons deprived of their *feeling* of "lived time" (e.g., due to neural dysfunction) still can *know about* the temporal constituents of the A-series. For example, patients K. C. and D. B., both of whom suffered complete loss of episodic memory accompanied by an inability to experience temporal flow (a review of these cases can be found in Klein, 2014b; Klein & Gangi, 2010), remained fully able to construct grammatically sound propositions using the temporal designations of past, present, and future (that is, they conceptualized the A-series in terms of "known" rather than "lived" time).

Consider the following conversations with K. C. and D. B. (reported in Klein, 2014b). In response to the question "What is the future?" K. C. replies, "Events that haven't happened yet," while the question "What is the past?" is answered, "Events that have already happened." Asked "Can you change the past?" K. C. emphatically states, "No!" When queried "Can you change the future, and if so, how?" he observes, "Yes. By doing different things." To the question "Can something that happens in the future change what has happened in the past?" K. C. again responds with an unequivocal "No," while the query "If an event is in the future will it always stay in the future?" elicits the response "No. Because time moves on."

Patient D. B. also presents a nuanced understanding of temporality. In response to the question "What is the future?" he answers, "Things that haven't happened yet." He describes the past as "Things that happened before . . . but are not happening now." Asked, "Can you change the past, D. B. says, "Don't think so, unless you had a time machine or something. Don't think so . . . not really. Maybe in science fiction (laughs)." To the question "Can the past influence the present?" he replies, "Sure . . . that's the way things work."

In short, both patients know about the flow of time, despite being unable to experience temporal movement as part of their subjective reality. The future exists for K. C. and D. B., but not as the experienced flow of the now to the next. It is known but not felt.

## 3.5. Beyond Memory

Our focus has been on the type of subjective temporalities afforded by different kinds of memory. But new work suggests that we need to enlarge our conceptual net if we hope to capture the neurocognitive complexity of FMTT. For example, Zeithamova, Schlichting, and Preston (2012) have shown that other cognitive systems (e.g., perceptual) in addition to memory are involved in the ability to plan for future contingencies. Similar sentiments are voiced by de Vito, Gamboz, Brandimonte, Barone, Amboni, and Della Salla (2012), who argue that executive function plays a crucial role in Parkinson's patients' performance on future thinking tasks. Arzy, Collette, Ionata, Fornari, and Blanke (2009) have shown that brain areas associated with mental imagery must be taken into account for some forms of FMTT. These data are telling us that a full appreciation of FMTT requires that we recognize that neurocognitive mechanisms other than memory (and the subjective temporalities they entail) play an important role in our ability to orient toward the future (for discussion, see Klein, 2013a).

Recent research also suggests that certain forms of temporal self-projection may not rely on memory at all. For example, Kwan, Craver, Green, Myerson, Gao, Black, and Rosenbaum (2015; see also Kwan, Craver, Green Myerson, Boyer, & Rosenbaum, 2012) assert that memory impairment does not impact a patient's ability to make choices about future rewards (a task that seems to rely on decision rules rather than retrieved content). As another example, personal diachronicity (i.e., the sense that one is a temporal continuant) appears, at least in some of its manifestations, to be based on non-evidential feeling of continuity, rather than on factually or rationally based analyses (e.g., Klein, 2014b). Moreover, it seems perfectly reasonable to muse about one's future absent any grounding in considerations of the past (e.g., "I wonder what tomorrow will be like?").

While one might argue that these examples can be explained by positing that choice, personal diachronicity, and future-directed musings are mnemonically guided, this requires that one embrace all of these acts as species of memory. At this point, we maintain, the designation *memory* becomes so encompassing that it no longer is clear what the term picks out (for fuller discussion, see Klein, 2015). As Moyal-Sharrock (2009) observes, even allowing that most of our cognitive abilities involved memory at some stage in their development, and that this is taken as grounds for seeing them as the products of memory, "this is trivial ground . . . it is no more than to say that memory has played a role in their achievement, not that it continues to do so in their subsequent deployment [e.g., choice behavior or sense of self as a temporal continuant]" (226; text within brackets added for clarification).

In our view, the examples presented (a) highlight the multifaceted nature of the FMTT, and (b) suggest that we may need to enlarge our inventory of "forms of temporal subjectivity" (in this regard it is interesting to note that McTaggart discussed temporalities other than those captured by the A- and B-series). While some of the preceding examples *might* be accommodated by mnemonic processes (and their temporal accompaniments—though it is unclear how this would work in the case of felt diachronicity), to do so we have to expand the scope of the term *memory* to such a degree that it threatens to undermine its conceptual utility (for discussion, see Klein, 2015; Moyal-Sharrock, 2009).

## 4. CONCLUSIONS

Klein, Loftus, and Kihlstrom's (2002) proposed relation between types of memory and types of subjective temporality can be seen as a useful corrective to the (typically tacit) assumption that there is a single, unitary form of temporal experience. While it may be the case that the distinction they proposed ultimately will prove too limited to fully capture the richness of FMTT experience (see section 3.5), it is a step in the right—and, we believe, necessary—direction. What is not in doubt is that we need to be more analytically precise and nuanced about what it is we refer to when we use the label *FMTT* (e.g., Klein, 2013a, in press). Is it our ability to construct personal future scenarios, to construct public future scenarios, to plan for future contingencies, to anticipate events, to maintain our sense of personal continuity, to make future-oriented judgments, muse about future contingencies, to predict our future feelings, attitudes, and beliefs? Are these simply manifestations of superficial variations of a common underlying mental faculty—an inevitable, but theoretically uninteresting, consequence of using different tasks to probe a unitary ability to orient toward and imagine the future? Or do these variants rely—at least in part—on different neurocognitive mechanisms?

In this chapter we champion the latter possibility. Specifically, we argue that, in the absence of a more sophisticated understanding of the theoretical entailments of FMTT, the term runs a serious risk of becoming a "catch-all" phrase that diverts attention from different forms of subjective temporality—each with distinctive (as well as overlapping) neurocognitive and conceptual commitments—that enable FMTT in its various manifestations.

In summary, there are reasons (both empirical and conceptual) to suspect that no single form of future-oriented temporal subjectivity is sufficient to capture the complexity of FMTT experience. Nonetheless, we acknowledge that the idea that subjective temporality can be partitioned into subtypes remains, at present, an assertion in need of investigative and theoretical attention. For instance, while available evidence suggests that "lived time" is self-referential whereas "known time" is not (e.g., Klein, Loftus, & Kihlstrom, 2002), this assertion rests on the outcome of studies that did not orthogonally vary types of subjective temporality ("lived" versus "known") and self-referential processing (present versus absent). A necessary step toward understanding personal variants of mental time travel (into the past as well as the future) requires experimentally separating "lived" and "known" forms of temporality (perhaps by manipulating the type of memorial experience required for task performance) and examining their respective contributions to personal forms of FMTT.

## ACKNOWLEDGMENTS

We thank an anonymous reviewer for helpful comments.

## NOTES

1. This concern is not novel. For example, the first author has been voicing these concerns—to relatively little effect—since 2002. Others have as well (e.g., Addis & Schacter, 2012; Szpunar, 2010). However, discussion often (though not always; e.g., Suddendorf, 2010) has appeared piecemeal, surrounded by discussion of other issues. It is our hope that by placing our concerns in bold relief (i.e., providing them their own chapter), we might reduce the signal-to-noise ratio that inevitably attends embedding a topic within additional topics.
2. The study of subjective temporality found in Eastern wisdom traditions typically is less analytic than those in Western philosophical treatment. One reason for this is that discussion, where it exists, often is the byproduct of concern with other topics such as the question of past lives (Griffiths, 1992). That being said, a recent analysis of Buddhist texts reveals that certain writings (e.g., Vasubandhu's theory of mind) show striking similarities to Western thought concerning the role of the subjective time associated in acts of memory (Chadha, 2014).
3. There is no reason that conceptual analyses could not be performed on the offerings of episodic memory. And there likely are occasions on which they are. The point simply is that such analyses need not be undertaken in order for the individual to experience subjective temporality with this form of memory. In contrast, conceptual analyses are a necessary accompaniment for subjective temporality to emerge in the context of semantic knowledge.
4. It is, of course, possible to know about "lived time" without directly experiencing it. As will be shown in section 3.4, individuals can subject the idea of temporal flow to a purely conceptual analysis (i.e., "known time"). In fact, as reported in Klein (2014b), patients suffering complete failure of episodic memory retain preserved knowledge of the concepts of past and future. But, in virtue of their disability, they are incapable of experiencing temporal movement (i.e., "lived time").
5. In this chapter we restrict analysis of subjective temporality to its future orientation. While past-oriented temporality has many of the features described for its future-oriented brethren, there are profound logical and causal asymmetries between the two (see section 3.3). Accordingly, a simple extension of future-oriented temporal subjectivity to experience of the past is difficult to justify. We plan to discuss the nature of past-oriented temporal subjectivity at a later time (see also Perrin, Chapter 3 of this volume).
6. Work by Dalla Barba and colleagues (reviewed in Dalla Barba, 2002) on the relation between memory and temporal experience has a clear resemblance to Tulving's ideas on the topic. Like Tulving, Dalla Barba proposes that two modes of consciousness can accompany the retrieval of memory content—temporal consciousness (TC) and knowing consciousness (KC). TC, like autonoesis, is consciousness of time—it enables a person to become aware of something as part of his or her personal past, present, or future. KC, by

contrast, operates like noetic consciousness—it enables a person to become aware of something as an element of knowledge without that knowledge being situated in a temporal framework. Note: An updated version of the conceptual entailments of TC and KC can be found in Chapter 6 by Dalla Barba in the present volume.

## REFERENCES


Abrahamn, A., & Bubic, A. (2015). Semantic memory as the root of imagination. *Frontiers in Psychology, 6*, 325. doi: 10.3389/fpsyg.2015.00325.

Addis, D. R., Cheng, T., Roberts, R. P., & Schacter, D. L. (2011). Hippocampal contributions to the episodic simulation of specific and general future events. *Hippocampus, 21*, 1045–1052.

Addis, D. R., & Schacter, D. L. (2012). The hippocampus and imaging the future: where do we stand? *Frontiers in Human Neuroscience, 5, Article 173.*

Anderson, R. J. (2012). Imagining novel futures: the roles of event plausibility and familiarity. *Memory, 20*, 443–451.

Arstila, V., & Lloyd, D. (2014). *Subjective time: philosophy, psychology, and neuroscience of temporality.* Cambridge, MA: MIT Press.

Arzy, S., Collette, S., Ionata, S., Fornari, E., & Blanke, O. (2009). Subjective mental time travel: the functional architecture of projecting the self to the past and future. *European Journal of Neuroscience, 30*, 2009–2017.

Atance, C. M., & O'Neill, D. K. (2001). Episodic future thinking. *Trends in Cognitive* Sciences, *5*, 533–539.

Atance, C. M, & Sommerville, J. A. (2014). Assessing the role of memory in preschooler's episodic performance on episodic foresight tasks. *Memory, 22*, 118–128.

Augustine. (1997). *The confessions.* Translated by M. Boulding. Hyde Park, NY: New City Press.

Bar, M. (Ed.) (2011). *Predictions in the brain: using our past to generate a future.* New York: Oxford University Press.

Boyer, P. (2009). What are memories for? Functions of recall in cognition and culture. In P. Boyer & J. V. Wertsch (eds.), *Memory in mind and culture* (pp. 3–28). Cambridge, UK: Cambridge University Press.

Bradley, F. H. (1887). Why do we remember forwards and not backwards? *Mind, 12*, 579–582.

Buckner, R. L., & Carroll, D. C. (2007). Self-projection and the brain. *Trends in Cognitive Sciences, 11*, 49–57.

Chadha, M. (2014). A Buddhist explanation of episodic memory: from self to mind. *Asian Philosophy, 24*, 14–27.

Cheke, L. G., & Clayton, N. S. (2010). Mental time travel in animals. *WIREs Cogntive Science, 1*, 1–16.

Cooper, J. M., Vargha-Khadem, E., Gadian, D. G., & Maguire, E. A. (2011). The effect of hippocampal damage in children on recalling past and imagining new experiences. *Neuropsychologia, 49*, 1843–1850.

Cornish, D. (2011). Earlier and later if and only if past, present and future. *Philosophy, 86*, 41–58.

Costi, C., Navarro, B., Vallat-Azouri, C., Brami, M., Azouvi, P., & Piolino, P. (2015). Disruption of temporally extended self-memory system following traumatic brain injury. *Neuropsychologia*, *71*, 133–145.
Cottle, T. J., & Klineberg, S. L. (Eds.). (1974). *The present of things future: explorations in the human experience of time*. London: Collier Macmillan Publishers.
D'Argembeau, A., Lardi, C., & Van der Linden, M. (2012). Self-defining future projections: exploring the identity function of thinking about the future. *Memory*, *20*, 110–120.
D'Argembeau, A., & Mathy, A. (2011). Tracking the construction of episodic future thoughts. *Journal of experimental Psychology: General*, *140*, 258–271.
Dalla Barba, G. (2002). *Memory, consciousness and temporality*. Norwell, MA: Kluwer Academic Publishers.
Dalla Barba, G., Cappelletti, J. Y., Signorini, M., & Denes, G. (1997). Confabulation: remembering 'another' past, planning 'another' future. *Neurocase*, *3*, 425–436.
Danziger, K. (2008). *Marking the mind: a history of memory*. Cambridge, UK: Cambridge University Press.
Davids, A., Kidder, C., & Riech, M. (1962). Time orientation in male and female juvenile delinquents. *Journal of Abnormal and Social Psychology*, *64*, 239–240.
de Vito, S., Gamboz, N., Brandimonte, M. A., Barone, P., Amboni, M., & Della Salla, S. (2012). Future thinking in Parkinson's disease: an executive dysfunction? *Neuropsychologia*, *50*, 1494–1501.
Duval, C., Desgranges, B., de La Sayette, V., Belliard, S., Eustache, F., & Piolino, P. (2012). What happens to personal identity when semantic knowledge degrades? A study of the self and autobiographical memory in semantic dementia. *Neuropsychologia*, *50*, 254–265.
Einstein, A, Podolsky, B., & Rosen, N. (1935). Can quantum-mechanical description of physical reality be considered complete? *Physical Review*, *47*, 777–780.
Faye, J. (1989). *The reality of the future*. Odense: Odense University Press.
Fraisse, P. (1963). *The psychology of time*. London: Harper & Row Publishers.
Frankenhaeuser, M. (1959). *Estimation of time: An experimental study*. Stockholm: Almqvist & Wiskell.
Gardiner, J. M. (2001). Episodic memory and autonoetic consciousness: a first-person approach. *Philosophical Transactions of the Royal Society B*, *356*, 1351–1362.
Gorman, B. S., & Wessman, A. E. (eds.). (1977). *The personal experience of time*. New York: Plenum Press.
Griffiths, P. J. (1992). Memory in classical Indian Yogacara. In J. Gyatso (ed.), *In the mirror of memory: reflections on mindfulness and remembrance in Indian and Tibetan Buddhism* (pp. 109–131). Albany: State university of New York Press.
Hassabis, D., Kumaram, D., Vann, D. S., & Maguire, E. A. (2007). Patients with hippocampal amnesia cannot imagine new experience. *Proceedings of the National Academy of Sciences*, 104, 1726–1731.
Hassabis, D., & Maguire, E. A. (2007). Deconstructing episodic memory with construction. *Trends in Cognitive Sciences*, *11*, 299–306.
Hoffmann, B. (1998). *Relativity and its roots*. Mineola, NY: Dover Publications.
Ingvar, D. H. (1985). "Memory for the future": an essay on the temporal organization of conscious awareness. *Human Neurobiology*, *4*, 127–136.

Irish, M., Addis, D. R., Hodges, J. R., & Piguet, O. (2012). Considering the role of semantic memory in episodic future thinking: evidence from semantic dementia. *Brain, 135*, 2178–2191.

Irish, M., & Piguet, O. (2013). The pivotal role of semantic memory in remembering the past and imagining the future. *Frontiers in Behavioral Neuroscience, 7*, 27. doi: 10.3389/fnbeh.2013.00027.

Klein, S. B. (2013a). The complex act of projecting oneself into the future. *WIREs Cognitive Sciences, 4*, 63–79.

Klein, S. B. (2013b). The temporal orientation of memory: it's time for a change of direction. *Journal of Research in Applied Memory and Cognition, 2*, 222–234.

Klein, S. B. (2013c). Making the case that episodic recollection is attributable to operations occurring at retrieval rather than to content stored in a dedicated subsystem of long-term memory. *Frontiers in Behavioral Neuroscience, 7*, 3. doi: 103389/fnbeh.2013.00003.

Klein, S. B. (2014a). Autonoesis and belief in a personal past: an evolutionary theory of episodic memory indices. *Review of Philosophy and Psychology, 5*, 427–447.

Klein, S. B. (2014b). Sameness and the self: philosophical and psychological considerations. *Frontiers in Psychology: Perception Science, 5*, 29. doi: 10.3389/fpsyg.2014.00029

Klein, S. B. (2015). What memory is. *WIREs Cognitive Science, 6*, 1–38.

Klein, S. B. (in press). Autonoetic consciousness: reconsidering the role of episodic memory in future-oriented self projection. *The Quarterly Journal of Experimental Psychology.*

Klein, S. B., Cosmides, L., Tooby, J., & Chance, S. (2002). Decisions and the evolution of memory: multiple systems, multiple functions. *Psychological Review, 109*, 306–329.

Klein, S. B., & Gangi, C. (2010). The multiplicity of self: neuropsychological evidence and its implications for the self as a construct in psychological research. *Annals of the New York Academy of Sciences, 1191*, 1–15.

Klein, S. B., Loftus, J., & Kihlstrom, J. F. (2002). Memory and temporal experience: the effects of episodic memory loss on an amnesic patient's ability to remember the past and imagine the future. *Social Cognition, 20*, 353–379.

Klein, S. B., & Nichols, S. (2012). Memory and the sense of personal identity. *Mind, 121*, 677–702.

Klein, S. B., Robertson, T. E., & Delton, A. W. (2010). Facing the future: memory as an evolved system for planning future acts. *Memory & Cognition, 38*, 13–22.

Klein, S. B., Rozendal, K., & Cosmides, L. (2002). A social-cognitive neuroscience analysis of the self. *Social Cognition, 20*, 105–113.

Klineberg, S. L. (1967). Changes in outlook on the future between childhood and adolescence. *Journal of Personality and Social Psychology, 2*, 185–193.

Klineberg, S. L. (1968). Future time perspective and the preference of delayed reward. *Journal of Personality and Social Psychology, 8*, 253–257.

Krauss, H. H., & Ruiz, R. E. (1967). Anxiety and temporal perspective. *Journal of Clinical Psychology, 23*, 340–342.

Kwan, D., Craver, C. F., Green, L., Myerson, J., Boyer, P., & Rosenbaum, R. S. (2012). Future decision-making without episodic mental time travel. *Hippocampus, 22*, 1215–1219.

Kwan, D., Craver, C. F., Green, L., Myerson, J., Gao, F., Black, S. E., & Rosenbaum, R. S. (2015). Cueing the personal future to reduce discounting in intertemporal choice: Is episodic prospection necessary? *Hippocampus, 25*, 432–443.

Lockwood, M. (2005). *The labyrinth of time.* New York: Oxford University Press.

Loizou, A. (1986). *The reality of time*. Aldershot, UK: Gower Publishing.

Maguire, E. A., Vargha-Khadem, F., & Hassabis, D. (2010). Imagining fictitious future experiences: evidence from developmental amnesia. *Neuropsychologia, 48*, 3187–3192.

Manning, L., Denkova, E., & Unterberger, L. (2012). Autobiographical significance in past and future public semantic memory: a case study. *Cortex, 49*, 2007–2020.

Markowitsch, H. J. (2003). Autonoetic consciousness. In T. Kircher & A. David (eds.), *The self in neuroscience and psychiatry* (pp. 180–196). Cambridge, UK: Cambridge University Press.

McLure, R. (2005). *The philosophy of time*. New York: Routledge.

McTaggart, J. M. E. (1908). The unreality of time. *Mind, 68*, 457–484.

Melges, F. T., & Fougerousse, C. E. (1968). Time sense, emotions, and acute mental illness. *Journal of Psychiatric Research, 4*, 127–140.

Merker, B. (2007). Memory, imagination, and the asymmetry between past and future. *Behavioral and Brain Sciences, 30*, 325–326.

Mooij, J. J. A. (2005). *Time and mind*. Boston, MA: Brill.

Moyal-Sharrock, D. (2009). Wittgenstein and the memory debate. *New Ideas in Psychology, 27*, 213–227.

Mullally, S. L., Hassabis, D., & Maguire, E. A. (2012). Scene construction in amnesia: an fMRI study. *The Journal of Neuroscience, 32*, 5646–5663.

Nichols, H. (1891). *The psychology of time: historically and philosophically considered with extended experiments*. New York: Henry Holt.

Nuttin, J. (1985). *Future perspective and motivation*. Hillsdale, NJ: Lawrence Erlbaum.

Ornstein, R. E. (1969). *On the experience of time*. Middlesex, UK: Penguin Books.

Papa-Grimadli, A. (1998). *Time and reality*. Aldershot, UK: Ashgate.

Pezzulo, G. (2008). Coordinating with the future: the anticipatory nature of representation. *Minds Machines, 18*, 179–225.

Race, E., Keane, M. N., Verfaellie, M. (2011). Medial temporal lobe damage causes deficits in episodic memory and episodic future thinking not attributable to deficits in narrative construction. *Journal of Neuroscience, 31*, 10262–10269.

Rochelle, G. (1998). *Behind time: the incoherence of McTaggart's atemporal replacement*. Aldershot, UK: Ashgate.

Roeckelein, J. E. (2000). *The concept of time in psychology*. Westport, CT: Greenwood Press.

Rozek, F., Wessman, A. E., & Gorman, B. S. (1977). Temporal span and delay of gratification as a function of age and cognitive development. *Journal of Genetic Psychology, 131*, 37–40.

Schacter, D. L., & Addis, D. R. (2007). The cognitive neuroscience of constructive memory: remembering the past and imagining the future. *Philosophical Transactions of the Royal Society B, 362*, 773–786.

Schacter, D. L., Addis, D. R., & Buckner, R. L. (2008). Episodic simulation of future events: concepts, data, and applications. *Annals of the New York Academy of Science, 1124*, 39–60.

Schacter, D. L., Addis, D. R., Hassabis, D., Martin, V. C., Spreng, R. N., & Szpunar, K. K. (2012). The future of memory: remembering, imagining, and the brain. *Neuron, 76*, 677–694.

Siegman, A. W. (1961). The relation between future time perspective, time estimation, and impulse control in a group of young offenders and in a control group. *Journal of Counseling Psychology, 25*, 470–475.

Sorabji, R. (1972). *Aristotle on memory*. Providence, RI: Brown University Press.

Suddendorf, T. (1994). *The discovery of the fourth dimension: mental time travel and human evolution*. Master's thesis, University of Waikato, Hamilton, New Zealand.

Suddendorf, T. (2010). Episodic memory versus episodic foresight: similarities and differences. *WIREs Cognitive Science, 1*, 99–107.

Suddendorf, T., & Corballis, M. C. (1997). Mental time travel and the evolution of the human mind. *Genetic, Social, and General Psychology Monographs, 123*, 133–167.

Suddendorf, T., & Corballis, M.C. (2007). The evolution of foresight: what is mental time travel, and is it unique to humans? *Behavioral and Brain Sciences, 30*, 299–313.

Szpunar, K. K. (2010). Episodic future thought: an emerging concept. *Perspectives on Psychological Science, 5*, 142–162.

Szpunar. K. K., & McDermott, K. B. (2008). Episodic future thought and its relation to remembering: evidence from ratings of subjective experience. *Consciousness and Cognition, 17*, 330–334.

Teahan, J. E. (1958). Future time perspective, optimism, and academic achievement. *Journal of Abnormal and Social Psychology, 57*, 379–380.

Tulving, E. (1985). Memory and consciousness. *Canadian Psychology/Psychologie canadienne, 26*, 1–12.

Tulving, E. (1993). What is episodic memory? *Current Directions in Psychological Science, 2*, 67–70.

Tulving, E. (2005). Episodic memory and autonoesis: uniquely human? In H. S. Terrace and J. Metcalfe (eds.), *The missing link in cognition: origins of self-reflective consciousness* (pp. 3–56). Oxford: Oxford University Press.

Von Leyden, W. (1961). *Remembering: a philosophical problem*. New York: Philosophical Library.

Wallace, M. (1956). Future time perspective in schizophrenia. *Journal of Abnormal and Social Psychology, 52*, 240–245.

Wheeler, M. A., Stuss, D. T., & Tulving, E. (1997). Toward a theory of episodic memory: the frontal lobes and autonoetic consciousness. *Psychological Bulletin, 121*, 331–354.

Williams, J. M. C., Ellis, N. C., Tyers, C. Healy, H., Rose, G., & MacLeod, A. K. (1996). The specificity of autobiographical memory and imageability of the future. Memory *& Cognition, 24*, 116–125.

Wohlford, P. (1966). Extension of personal time, affective states, and expectation of personal death. *Journal of Personality and Social Psychology, 3*, 559–566.

Yates, F. A. (1966). *The art of memory*. Chicago: University of Chicago Press.

Zeithamova, D., Schlichting, M. L., & Preston, A. R. (2012). The hippocampus and inferential reasoning: building memories to navigate the future. *Frontiers in Human Neuroscience, 6*, 1–14.

# 8

# Time Is Not of the Essence

## *Understanding the Neural Correlates of Mental Time Travel*

FELIPE DE BRIGARD AND BRYCE S. GESSELL ■

The precise relationship between memory and imagination has been a matter of debate for centuries (e.g., Aristotle [Barnes, 1984]; Hobbes, 1668; Hume, 1739; Russell, 1921). But at no time has this debate seen a more remarkable development than in the last two decades, as numerous behavioral, neuropsychological, and neuroimaging studies have consistently shown that our capacity to remember episodes that occurred in the past and our ability to imagine possible events that may occur in the future are profoundly intertwined. Although somewhat cautiously, many of these results have been interpreted as providing evidence in favor of the view that episodic memory (i.e., our capacity to remember past personal events) and episodic future thinking (i.e., our ability to imagine possible future personal events) (Atance & O'Neill, 2001; Szpunar, 2010) should be seen as two operations of a single cognitive system that enables "mental time travel" (MTT) (Tulving, 1985): the allegedly uniquely human psychological capacity through which we "mentally project ourselves in time" by entertaining mental simulations of events whose contents are either about the past or about the future (Suddendorf & Corballis, 1997). According to this perspective, then, the psychological system for MTT operates upon representations with temporal or "tensed" contents. This view naturally leads to the hypothesis that it is the *temporal* nature of the content of the mental simulations involved in the aforementioned results that accounts for the common engagement of neural mechanisms during episodic memory and future thinking. Time, as it were, is thought to be of the essence.[1]

However, as we review in the present chapter, the story is not so simple. As it turns out, there is a large amount of evidence that is difficult to reconcile with the view that what accounts for the common engagement of the neural substrates of episodic memory and future thinking is the fact that the contents of such mental simulations are tensed. Indeed, we argue that much of this evidence actually suggests, somewhat paradoxically, that when it comes to the neural mechanisms of mental time travel, time is not of the essence—or, at least, not essential as part of the *content* of the simulation. Instead, we suggest that what accounts for this common engagement of brain regions is that the mental simulations they support *unfold* in time, regardless of whether or not they are *about* time. To use a distinction familiar to philosophers (Cussins, 1990; Dennett, 1978; Hurley, 1998): we argue that these results are better explained if time is not considered essential for the *content* of the simulation, but rather for its representational vehicle—or *structure*, as we prefer to call it, for reasons that will soon become clear.

This chapter is divided into five sections. In section 1 we review critical experimental results that have been thought to support the view that episodic memory and future thinking share a common neural substrate. We then explain how these results have been interpreted as providing evidence in favor of a cognitive system for MTT. In section 2 we review evidence suggesting that the temporal nature of the content of the mental simulation is not necessary for the engagement of the neural substrate thought to be involved in MTT. To account for these discrepant results, in section 3 we use a familiar philosophical distinction between representational contents and representational vehicles to explain the difference between the content and the structure of a temporal simulation. We show how these two dimensions—content and structure—are orthogonal to one another, and how focusing on the temporal structure rather than the tensed content of the mental simulation may help to accommodate both the evidence and the counterevidence reviewed in sections 1 and 2. Our discussion in this section will also deal with the issue of atemporal simulations—that is, simulations that are neither of the past nor the future. Next, in section 4, we compare our proposal to some prominent accounts of the brain activation during MTT, namely the episodic simulation hypothesis (Schacter & Addis, 2007) and the scene construction hypothesis (Hassabis & McGuire, 2007). We argue that our account is not only compatible with these hypotheses but also explains why they are not incompatible with each other. To anticipate, we suggest that scene construction is best understood as a hypothesis about the structure of the mental simulation, while the episodic simulation hypothesis is best understood as an account of its content. Finally, in section 5, we discuss three possible objections and some implications that may follow from our view.

## 1. THE TENSED-CONTENT VIEW

Although it is customary to cite Tulving (1985) when introducing the notion of mental time travel and its relation to episodic memory, the truth is that this idea predates him by centuries. Here is Saint Augustine, for instance, reflecting upon the contents he finds within the "vast court of his memory":

> There be all which I remember, either on my own experience, or other's credit. Out of the same store do I myself with the past continually combine fresh and fresh likenesses of things which I have experienced, or, from what I have experienced, have believed: and thence again infer future actions, events and hopes, and all these again I reflect on, as present. (Augustine, 1991)

Indeed, the idea that memories of past experiences both guide and constrain our imaginations of possible states of affairs found further reaffirmation in the work of

British Empiricists (Hobbes, 1668; Hume, 1739; Locke, 1689). What is unquestionable, though, is that Tulving provided the first piece of empirical evidence in support of the conjecture that our capacity to imagine possible future events—episodic future thinking (Szpunar, 2010)—critically depends on our capacity to remember episodes from our personal past (Tulving, 1985). In an oft-quoted exchange, Tulving (E. T.) asked amnesic patient K. C. (referred to here as N. N.) what he will be doing tomorrow (Tulving, 1985, 4):

> E. T. "Let's try the question again about the future. What will you be doing tomorrow?" (There is a 15-second pause.)
> N. N. smiles faintly, then says, "I don't know."
> E. T. "Do you remember the question?"
> N. N. "About what I'll be doing tomorrow?" E. T. "Yes. How would you describe your state of mind when you try to think about it?" (A 5-second pause.)
> N. N. "Blank, I guess."

Patient K. C. was 30 years old when he suffered a severe head injury in a motorcycle accident. A number of regions in K. C.'s brain were significantly compromised (Rosenbaum et al., 2005), including the medial temporal lobes (MTL), damage to which had been previously associated with impairments in episodic memory (Scoville & Milner, 1957). Indeed, although K. C. could still remember many facts about the world from before his accident—that is, his semantic memory was preserved—he was unable to remember many personally experienced events from his past (i.e., retrograde amnesia), and he was incapable of forming new episodic memories about events that he experienced after the accident (i.e., anterograde amnesia). This observation—that in addition to having his episodic memory compromised, K. C. was also incapable of conjuring up imaginations about possible personal future events—constituted a substantial piece of evidence for the claim that episodic memory and future thinking may share common neural mechanisms enabling MTT.

Further support for this claim came from another well-known case. In 2002, Klein, Loftus, and Kihlstrom (2002) described the case of D. B., a 78-year old man who suffered severe retrograde and anterograde amnesia as a result of anoxic encephalopathy. Although unfortunately there is no information about the integrity of D. B's brain after the accident, his neuropsychological evaluation clearly revealed that despite a profound deficit in episodic memory, his performance in other cognitive tasks, including semantic memory, scored within normal limits. Confirming Tulving's prior observation, Klein and colleagues demonstrated that D. B. was not only incapable of remembering past autobiographical events but was also unable to imagine possible personal future episodes. However, to test whether or not D. B.'s deficits in episodic future thinking extended to non-personal or "semantic" future events, Klein and colleagues asked him to remember issues in the public domain (e.g., medical, environmental, political) that were important in the past and to think of possible issues in such domains that might be important in the future. Since D. B.'s performance in this task did not differ from that of controls, Klein and colleagues suggested a dissociation between our capacity to remember past and imagine possible future personal events and our ability to remember past and imagine possible future non-personal events—a sort of "semantic" future thinking.

Around the same time, Atance and O'Neill (2001) published an influential paper providing further neuropsychological and developmental evidence to the effect that episodic memory and future thinking are not only profoundly interconnected but also critically dissociated from semantic memory and semantic future thinking. It only took two more years for cognitive neuroscience to produce neuroimaging evidence supporting these observations. A 2003 positron emission tomography study by Okuda et al. showed common engagement of brain regions, specifically MTL structures, during both episodic memory and future thinking, in contrast with a control task requiring semantic retrieval. Moreover, Okuda and colleagues found that many areas in the MTL showed equivalent or even greater levels of activation during thoughts about a possible personal future than during thoughts about an actually experienced past event.

Behavioral studies soon followed suit, as a number of results showing further parallels between episodic memory and future thinking started to emerge. For instance, using a variation of the memory characteristics questionnaire (Johnson et al., 1988), D'Argembeau and Van der Linden (2004) showed similar effects of valence and temporal distance for both episodic memory and future thinking. In a subsequent study, D'Argembeau and Van der Linden (2006) showed that certain dimensions known to affect memory for past events, such as emotion regulation and visual imagery, have parallel effects in episodic future thinking. Likewise, Spreng and Levine (2006) showed that the temporal distribution of both episodic past and future thoughts could be modeled on a logarithmic scale, as it was more common for subjects to generate events closer to the moment of the simulation than more remote events, suggesting common effects of temporal distance for both episodic memory and future thinking.

2007 was, however, the *annus mirabilis* for research in the neural correlates of mental time travel, as three different labs reported three independent studies yielding highly consistent results. In the first study, Szpunar, Watson, and McDermott (2007) asked subjects to either remember a past personal event, imagine a possible personal future, or imagine a fictitious non-personal event featuring Bill Clinton. Their analysis revealed that thinking about a personal past event and a personal future event, but not thinking about a non-temporal and non-personal event, commonly engages the medial prefrontal cortex (mPFC)—mainly BA 10—the posterior cingulate cortex (pCC), the parahippocampal cortex, and the superior occipital lobe toward the cuneus. In a second study, also employing fMRI, Addis, Wong, and Schacter (2007) used concrete words to cue participants, who were asked to either remember a past episode, imagine a possible future event, or create a sentence. Consistent with previous results, they found remarkable overlap between episodic memory and future thinking, as compared to sentence creation, in regions known to be associated with episodic recollection (Cabeza & St. Jacques, 2007; St. Jacques & De Brigard, 2015). Notably, this common engagement occurred in many of the same regions found by Szpunar and collaborators (2007): mPFC—including BA 10—MTL, pCC, and left precuneus. Unlike Szpunar et al. (2007), though, Addis and collaborators did find greater hippocampal activity during episodic memory and future thinking relative to the control condition, suggesting both that the hippocampus could have been involved in the non-personal simulations in Szpunar et al.'s study (see discussion later in this chapter), and that the hippocampus might be indispensable for MTT.

This last observation—that the hippocampus is critical for both episodic memory and future thinking—received even stronger support from the third study, in which Hassabis and colleagues (2007) asked 5 patients with hippocampal amnesia, and 10 healthy controls, to imagine possible new experiences in response to verbal cues (e.g., "Imagine you're lying on a white sandy beach in a beautiful tropical bay"). Participants' descriptions of their simulations were transcribed and coded to assess how rich, detailed, and spatially coherent they were. The results of this study clearly showed that amnesic patients' descriptions of their simulations were less rich, contained fewer details, and were less spatially coherent than the descriptions produced by controls. Indeed, a further study by Race and colleagues (2011) replicated this result, and found that the deficit in the patients' descriptions was independent of their narrative abilities (see also Verfaellie, Race, & Keane, 2012, and Bartsch et al., 2011, for further support).

In the last eight years, the number of studies reporting overlaps in the neural structures underlying episodic memory and future thinking has skyrocketed (Schacter et al., 2012), consolidating the view that thinking about the past and imagining the future engages a common core brain network (Schacter, Addis, & Buckner, 2007). This common core network substantially overlaps with the so-called *default network* (Spreng & Grady, 2010; Spreng, Mar, & Kim, 2009). These results are so robust that even studies purporting to report counterevidence, upon closer inspection, actually turn out to lend credence to the claim that episodic memory and future thinking share a common neural substrate. Consider a study conducted by Squire and collaborators (2010), in which six individuals with hippocampal damage and eight age-matched controls were asked to remember past events and imagine possible future events while being audio recorded. In line with previous studies, Squire et al. transcribed their participants' descriptions and tallied the number of internal (episodic) and external (semantic) details (Levine et al., 2002), as well as the degree of spatial coherence following Hassabis et al.'s (2007) method. Surprisingly, Squire and colleagues found no differences in the descriptions of patients and controls, and showed that patients had no trouble either remembering past or imaging future events. But this result, far from casting doubt on the relationship between episodic memory and future thinking, suggests exactly the opposite: unlike the Hassabis et al. patients, those in the Squire et al. study did not have retrograde amnesia, so their capacity to remember the past and imagine the future was preserved (Maguire et al., 2010). This finding simply indicates that not all hippocampal damage leads to impairments in mental time travel.[2]

What should we make of these reported cases of hippocampal damage where episodic memory and future thinking are preserved? One possibility is that a difference in the etiology of the hippocampal damage and/or in the progression of the disease may modulate the functional role played by the hippocampus during MTT. Maguire, Vargha-Khadem, and Hassabis (2010), for instance, reported that Jon, a well-described patient with developmental amnesia (Vargha-Khadem et al., 1997), was nonetheless able to imagine possible new experiences, suggesting that an early accident in an ontogenetically underdeveloped hippocampus, as opposed to the abrupt damage in a fully developed adult hippocampus, may influence the degree to which this brain structure is required for mental time travel. Another possibility, not incompatible with the first, is that the patients who were able to generate imaginations about possible future events still had some remaining hippocampal tissue that was recruited during their mental simulations. Although the extent to which Squire et al.'s patients have preserved hippocampal tissue is unclear, we do know that both Jon, the developmental amnesia patient just mentioned, as well as one of Hassabis et al.'s patients, P01, had about 50% of hippocampal tissue intact. Might it be possible that this reduced amount of hippocampal tissue was engaged during their successful episodic future simulations? To answer this question, Mullally, Hassabis, and Maguire (2012) asked patient P01 to engage in future simulation while undergoing fMRI and found that, in fact, his capacity to imagine fictitious experiences was associated with activation in the remaining hippocampal tissue.

Taken together, the experimental results surveyed here strongly suggest that episodic memory and future thinking commonly engage a consistent set of brain regions that includes the mPFC, pCC, MTL (especially the hippocampus), and lateral parietal and retrosplenial cortices, likely corresponding to the functionally well-characterized default network (Andrews-Hanna, Smallwood, & Spreng, 2014; Buckner, Andrews-Hanna, & Schacter, 2008). What accounts for the common engagement of these brain regions during episodic memory and future thinking? A natural hypothesis is readily available: since both episodic memories and future thoughts have temporal or *tensed* contents—that is, they operate upon mental representations that are either about the past or about the future—they tap into the same neural mechanisms and are actually distinct processes of a single brain system responsible for our capacity to entertain tensed mental contents: a single brain system for MTT (Suddendorf, Addis, & Corballis, 2009; Suddendorf & Corballis, 1997, 2007; Tulving, 2002). According to this *tensed-content view*, then, what accounts for the engagement of the same brain regions during episodic recollection and future thinking is a common feature of the (intentional) content of the mental simulations entertained during each task: namely, that both are about time.

## 2. IS TIME OF THE ESSENCE?

According to the tensed-content view, entertaining a tensed mental simulation—that is, thinking about a past or a future event—is necessary to engage the brain structures associated with a system for mental time travel. However, in the last few years a number of experimental results have cast doubt on this claim. Indeed, some of the very same results that are sometimes heralded in support of the content-based view do not provide univocal evidence in its favor. Consider the study by Hassabis and colleagues (2007) on individuals with hippocampal amnesia. Their participants were asked to imagine *new* experiences, but they were not instructed to mentally place such experiences in a possible future. Presumably, to imagine a new experience is to entertain a mental simulation of a possible event that, even though it has not happened yet, could nonetheless occur in the future. But it need not be this way. To imagine a new experience may just mean to think of a possible hypothetical situation that may or may not occur in one's life, regardless of its precise temporal location. For all we know, the mental simulations entertained by the participants in the Hassabis et al. study may not have featured a temporal stamp in their contents—that is, they may not have been tensed at all (Hassabis & Maguire, 2009, explicitly acknowledge this possibility).

Other studies have yielded results that can be seen as providing evidence against the tensed-content view. For instance, Addis and colleagues (2009) found very similar patterns of brain activation within the default network when participants were instructed to simulate an event with familiar people, objects, and places, either in a possible future or in a possible past. This result speaks against the tensed-content view—and, indeed, against a strict interpretation of the traditional formulation of MTT—in an intriguing way. Consider Tulving's initial influential formulation:

> A normal healthy person who possesses autonoetic consciousness is capable of becoming aware of her own past as well as her own future: she is capable of mental time travel, roaming at will over what *has happened* as readily as over what *may happen*, independently of physical laws that govern the universe. (Tulving, 1985, 5; emphasis added)

If there were a brain system dedicated to mental time travel, as defined by Tulving, then it should be sensitive to mental simulations whose contents are about what *has happened* in one's past and what *may happen* in one's future, but not necessarily to contents depicting events that did *not* happen in one's past. After all, when we remember, we engage in a cognitive activity that aims to bring to mind contents representing previously experienced events, whereas this is not the case when we engage in simulations about events we know did not occur. Of course, although unlikely, it may be possible that the participants in Addis et al.'s (2009) study were simply unsure as to whether or not the imagined past events actually occurred in their past, so they may have tensed the imagined contents as if they had occurred, while granting them less certainty than they did to remembered events. However, further studies speak against this interpretation. For instance, De Brigard and colleagues (2014) asked participants to engage in episodic counterfactual thinking—that is, thoughts about alternative ways in which past personal events could have occurred but did not, or about events that did not occur at all, but might have (De Brigard & Giovanello, 2012)—while undergoing fMRI. Consistent with Addis et al.'s (2009) result, they found engagement of the default network not only when participants were remembering their past but also when they were engaged in episodic counterfactual thought: that is, when they were thinking about alternative ways in which past personal events, which they knew had occurred, could have been different (see also Van Hoeck et al., 2013).

Indeed, neuroimaging evidence suggesting that mentally simulated contents need not be temporally tensed to engage the default network is rapidly accumulating. For instance, Hassabis et al. (2007) reported default network activation during non-temporal simulations (see also Summerfield et al., 2010). Similarly, D'Argembeau and colleagues (2008) instructed participants to imagine general and explicitly tenseless simulations where they engaged in routine activities such as brushing their teeth. They found that these non-tensed simulations produced activation in the medial temporal lobe, and other default network regions, at the same level as activation for explicitly future simulations (see also D'Argembeau et al., 2010). Relatedly, transient episodes of mind-wandering, whereby our attention shifts from ongoing stimuli toward self-generated thoughts, also seems to recruit the default network, even when it is unclear whether or not the simulated self-generated thought is located at a particular time (Christoff et al., 2009). Default network activity has also been reported in non-temporal tasks involving spatial navigation (Spreng, Mar, & Kim, 2009), mentalizing (Hyatt, Calhoun, et al., 2015; Spreng & Andrews-Hanna, 2015), narrative comprehension (Mar, 2011), and counterfactual thoughts about other people (De Brigard et al., 2015). Although it is still possible that all these disparate tasks involve placing the simulated mental content at a particular time, there is no prima facie reason to believe that a temporal component is common to all of them (we return to the issue of atemporal simulations in section 3).

The picture against the tensed-content view is further strengthened by a number of behavioral studies suggesting similarities in performance between tasks involving mental time travel and other kinds of mental simulations. For instance, using a variation on the memory characteristics questionnaire (Johnson et al., 1988) and the autobiographical interview (Levine et al., 2002), De Brigard and Giovanello (2012) reported similar effects of outcome valence in the mental simulation of episodic future and counterfactual thinking as compared to episodic memories. Similar measures were used by de Vito et al. (2012) to investigate the effects of self-relevance and familiarity of setting in imagining episodic future versus atemporal events. De Vito et al. found that, when the familiarity of the context of the simulation as well as the level of self-relevance are kept fixed, imagining the event at a possible future time versus at no particular time at all makes no difference to the phenomenology of the mental simulation. Taken together, these results indicate that the temporal stamp in the content of the mental simulation does not seem to play a prominent role in their phenomenology, suggesting that a tensed content may not be essential for the engagement of the cognitive processes common to episodic memory and future thinking.

But perhaps the most powerful evidence against the tensed-content view comes from recent studies in individuals with hippocampal amnesia. These results show not only parallel difficulties in the simulation of non-temporal events—suggesting that a tensed content is not necessary for the engagement of the hippocampus—but also perfectly preserved capabilities to entertain certain kinds of temporal thoughts. This suggests that a tensed-content may not even be sufficient for hippocampal engagement. On the one hand, initial evidence in favor of the hippocampus being required for non-tensed simulations comes from an interesting study by Rosenbaum and collaborators (2009). Aware of K. C.'s difficulties in remembering the past and thinking about the future, Rosenbaum and colleagues asked K. C. to generate fictional events and to reconstruct well-known fairy tales and Bible stories. Surprisingly, his performance at both tasks was significantly worse than age-matched controls. Further evidence comes from a recent study by Romero and Moscovitch (2012) in which older adults and young adults with MTL damage were cued with sets of words and asked to generate fictional events that included the items referred to by the cuing words. Critically, the events were not supposed to be set at any particular time. Their results showed that both older adults and young individuals with MTL damage generated mental simulations that exhibited less coherence and were more unstructured than their matched controls.

On the other hand, evidence against the hippocampus being required to entertain tensed thoughts comes from a number of recent neuropsychological studies investigating decision-making processes in individuals with amnesia. In one such study, Kwan and collaborators (2013) asked K. C. to engage in a task that required the valuation of future rewards. Surprisingly, despite the fact that K. C. is profoundly impaired when it comes to constructing vivid and coherent simulations of future events, he systematically discounted the value of future rewards on par with controls. These results strongly suggest that thinking about the value of future rewards and thinking about a possible future event are two kinds of tensed thoughts, only one of which depends on the hippocampus. Further evidence for this claim comes from other studies conducted on K. C. For instance, Craver and colleagues (2014) asked K. C. a number of explicit questions about the future, many of which paralleled those used by Klein and colleagues (Klein, 2002), and they found that K. C. was able to produce perfectly normal responses. Consider this excerpt, from a conversation reported in Craver et al. (2014; S. R. = Shayna Rosenbaum, the interviewer):

S. R.: What is the future?
K. C.: Events that haven't happened yet.
S. R.: What is the past?
K. C.: Events that have already happened.
S. R.: Can you change the past?
K. C.: No.
S. R.: Can you change the future?
K. C.: Yes.
S. R.: How?
K. C.: By doing different things.
S. R.: Does what happened in the past influence what happens in the future?
K. C.: Yes.
S. R.: Does what you do now influence what happens in the future?
K. C.: I guess so.
S. R.: Does what you do now change what has happened in the past?
K. C.: No.

And the conversation goes on, with questions about what it means to do something that may affect the future, and what it would mean to be able to travel in time. To all these questions K. C. gave perfectly normal answers, demonstrating not only that he had an appropriate understanding of the concept of time in the absence of episodic memory (contra Hoerl, 1999), but also that he could entertain tensed thoughts—that is, thoughts about times other than the present—without a functioning hippocampus. Indeed, not only was K. C. able to talk about time and discount possible future rewards on par with controls, but he also exhibited regret anticipation and other normal temporal attitudes, as measured by Zimbardo's Temporal Perspective Inventory (Kwan et al., 2013).

Similar studies with other individuals with amnesia have also revealed that they are still able to entertain tensed thoughts. Stuss and Guzman (1988), for instance, reported the case of J. V., a 50-year-old man with profound retrograde amnesia but only mild anterograde amnesia. Despite having forgotten both personal and non-personal information about his past, he was able to relearn these facts, and was able to think about them again—although, as reported by Stuss and Guzman (1988), "for these past facts that have been relearned, J.V. has no feeling of personal warmth, intimacy, or belonging that the memory is his, or that he was somehow involved" (27).

Klein and Nichols (2012) made similar observations regarding patient R. B., who was able to remember specific situations from his personal past even though he lacked the feeling of warmth and ownership that usually accompanies our episodic memories. Taken together, these case studies—K. C., J. V., and R. B—although different in detail and etiology, strongly suggest that individuals with impairments in episodic memory and future thinking can nonetheless entertain all sorts of tensed thoughts. Consequently, we surmise that these, along with the behavioral and neuroimaging results reviewed earlier, constitute evidence against the tensed-content view, according to which what accounts for the engagement of the same brain regions during episodic recollection and future thinking is the fact that time features in the intentional content of the mental simulations entertained during each task. Given this counterevidence, time just does not seem to be of the essence.

The challenge, then, is to offer an account that can reconcile all the evidence reviewed in section 1—which suggests an intimate relation between the neural substrates underlying our capacities to remember past personal episodes and to simulate possible future personal events—with all the evidence reviewed in section 2, which speaks against using the temporal content of such thoughts as the unifying factor accounting for the engagement of this common neural substrate. To be fair, many authors have recently considered other views to try to account for this apparent discrepancy, with varying degrees of success. Buckner and Carroll (2007), for instance, suggested that this common brain network supports thoughts that essentially involve self-projection; Hassabis and Maguire (2009) suggested instead that it supports thoughts that involve the construction of a spatial scene; and more recently, Andrews-Hanna, Smallwood, and Spreng (2014) suggested that what these thoughts may have in common is that they are social, goal-directed, and self-generated. Despite the merits of these views (which we discuss later in the chapter), we want to suggest a different approach that may provide a novel way to understand the neural correlates of MTT. Specifically, we suggest that what may account for the common engagement of these brain mechanisms during MTT simulations is not something about the content of these thoughts, but rather something about their structure. But to do that, we must first clarify the distinction between the content and the structure of a simulation.

## 3. THE CONTENT-STRUCTURE DISTINCTION

A traditional and fundamental distinction in the philosophy of mind is that between *intentional objects* and *intentional contents*. Mental states are intentional: they are *about* something (Brentano, 1874). Intentional objects are those things that mental states are about. They are so-called to distinguish them from mere actual objects, since the objects of our mental states need not exist. Both the paper with which this page is made as well as Santa Claus can be intentional objects, insofar as one can entertain mental states about them; yet only one of them exists. But that which our mental states are about is different from the *content* of our mental states. After all, one can entertain two different thoughts about the exact same object. Lois Lane, for instance, could entertain two entirely different thoughts—even contradictory thoughts—about Superman and Clark Kent, even though both Superman and Clark Kent are the same object. She can do that because the individual referred to by these names can be present to Lois Lane's mind in different modes. The mode in which an intentional object is mentally presented to us is the *intentional content* of a mental state (Crane, 2009; Husserl, 1913).

A fundamental project in the philosophy of mind for the last century has been to understand how mental contents relate to their objects. One popular view—often known as *representationalism*—is to think of intentional contents as the information carried by a representation, which in turn is related to its intentional object in some particular way. The precise relationship between the content of the representation and its object is a matter of ongoing debate, as there are a number of prominent theories in the offing and no consensus in sight (e.g., Pitt, 2013). For our current purposes, though, what matters is that most cognitive psychology and neuroscience assumes some form of representationalism. Indeed, many cognitive psychologists and neuroscientists see their research projects as that of uncovering the nature of the representations involved in a particular cognitive task. The same, of course, goes for cognitive neuroscientists studying MTT.

In the philosophy of mind, however, representationalists often draw an important distinction that is seldom acknowledged in contemporary cognitive neuroscience, namely that between *intentional content* and *representational vehicle*. An intentional content, as we mentioned, represents an intentional object, but a certain intentional content is different from the actual particular in the world that does the representing—that is, the thing in the world that carries the information about the relevant intentional object. Think of Andy Warhol's famous mass-produced portraits of Marilyn Monroe. They all were about Marilyn Monroe—she was their intentional object—and they all portrayed her more or less in the same way, that is, carrying more or less the same intentional content (we say "more or less" for reasons that will become clearer later). But Warhol used a number of different materials to create these portraits: silkscreen, textiles, metallic papers, plastic, and so on. Indeed, many artists after him have used all sorts of things to recreate this exact same portrait, with materials as varied as Lego bricks and gummy bears. All these diverse materials constitute different representational vehicles with the same content.

The content/vehicle distinction highlights the fact that there is a difference between the properties of the content of a representation and the properties of the vehicle that carries that content (Dennett & Kinsbourne, 1992; Hurley, 1988). Warhol's portrait of Marilyn made of plastic and the one made of gummy bears may well both depict Marilyn Monroe as being glamorous, but only one of these representations is gooey. This is because being glamorous is a property of the content of the representation, while being gooey is a property of (one of) the representational vehicles carrying such content. The same goes for mental representations. Properties of the intentional contents of a mental representation should not be confused with properties of the representational vehicles carrying the intentional contents. Given that most contemporary representationalists assume that mental states are multiply realizable, most also accept that intentional contents can be carried by different representational vehicles. In fact, the content/vehicle distinction also holds for views that take brain mechanisms to be the representational vehicles of our mental representations. For instance, representationalists who follow the Fodorian tradition of the language of thought consider that the representational vehicles carrying the intentional content of our mental states must have certain physical characteristics (e.g., symbolic locality, sentential structure, etc.), whereas representationalists who follow the connectionist tradition consider that those very same intentional contents must be carried by representational vehicles with a different physical configuration (e.g., distributed, non-local, non-symbolic, etc.). Still, both agree that the representational vehicles must be some kind of brain mechanism.

Unfortunately, most philosophical discussions on representational vehicles have been confined to examples in which the vehicles only vary along *physical* dimensions, such as size, weight, locality, or viscosity (Clark, 2008; Elpidorou, 2013; Rowlands, 2010). Even philosophers who agree that representational vehicles are neural mechanisms tend to confine their disagreements about their possible differences to physical dimensions alone—for example, local versus distributed representations, symbolic versus non-symbolic, and so on. But restricting the range of variations in representational vehicles to physical dimensions overlooks the fact that mental contents tend to vary across a *temporal* dimension as well. This oversight is most likely due to the fact that the philosophical tradition tends to consider mental contents as discrete entities that one can pinpoint to a precise moment during which they perdure, unchanged. As a result, the analogy of a mental representation to a sentence or a picture is enticing, as the vehicles of these kinds of representations need not change with time. But to think that all our intentional contents are discrete and temporally stable mental sentences or pictures utterly simplifies the often continuous and changing nature of the intentional contents we are aware of.[3] For although it is true that we frequently bring to mind thoughts whose contents are relatively stable and short-lived—like pictures and sentences—we also habitually entertain complex mental representations whose contents are experienced as dynamic and unfolding in time. In these cases, information is carried not only by the physical configuration of the representational vehicle but also by its temporal structure.

As an analogy, consider the pictures in Figure 8.1. Panel A depicts a stable representation of a tennis player hitting an overhead ball. As in comic books, the information about the trajectory and force of both the racquet and the ball is

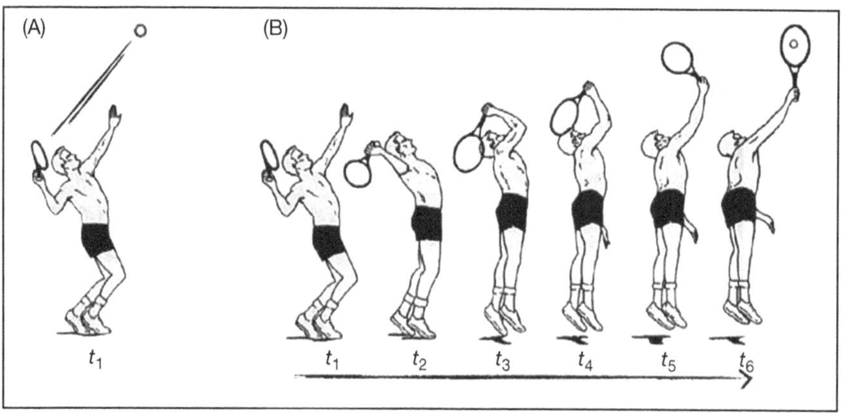


**Figure 8.1.** Schematic example of a static (A) versus dynamic (B) representation.

conveyed by motion lines. Despite the fact that the drawing conveys movement, the image is static. Now compare this image with a string of drawings displayed, one after another, as in panel B. Here, a single image by itself fails to convey information about movement and trajectory, as this information is encoded in the dynamical structure of the sequence of pictures, rather than by a single drawing alone.[4] Thus, although both representations are supposed to carry the same content,[5] they do so via two different vehicles. On the one hand, there is a representational vehicle whose structure is *static* and carries information about movement and trajectories with motion lines; on the other hand, there is a representational vehicle with a *dynamic* structure that carries information about movement and trajectories in the succession of the sequence of drawings (see Weiskopf, 2010, for a similar example).

Let us bring the analogy back to the realm of mental representation. Our suggestion is that some intentional contents are carried by representational vehicles whose structures can vary along a temporal dimension—from more static to more dynamic—even if they do not vary in their physical dimensions (i.e., even if both are constituted by distributed neuronal networks, say). Consequently, a mental representation with tensed content (i.e., a thought about the past or a possible future) can nonetheless fail to have a temporal or dynamic structure and have instead a rather non-temporal or static structure. In this sense, whether or not a mental representation has a temporal content (i.e., is tensed) is independent of whether or not it has a temporal structure (i.e., is dynamic). In other words, a temporal component in the content of a mental representation is orthogonal to the temporal dimension of its representational structure.

These considerations suggest a model in which time varies along two axes, content and structure, which in turn segregates temporal mental representations into four quadrants (Figure 8.2). First, there are simulations that involve mental representations with non-tensed contents carried by static representational vehicles. We suggest that paradigmatic semantic memories—such as the thought that triangles have three angles or that 2 is a prime number—belong to this quadrant. Second, there are thoughts that involve entertaining tensed contents carried by static representational

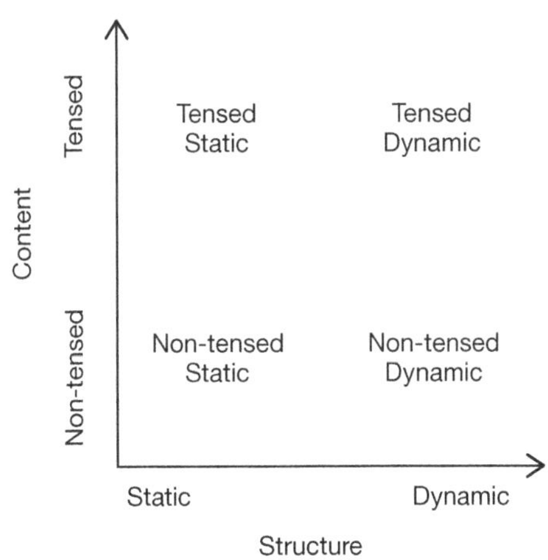


**Figure 8.2.** Two temporal dimensions of mental simulations.

vehicles. Mental representations involving semantic yet temporal information—such as the memory that one graduated from high school or the thought that one has a business trip next month—are presumably within this category. Third, there are mental simulations involving dynamic representational vehicles carrying non-tensed contents. Non-temporal imaginations and certain counterfactual and hypothetical simulations whose representational contents unfold over time would presumably fit this description. Finally, there are thoughts that involve mental simulations of tensed contents with a dynamic representational structure. Paradigmatic examples of episodic memories and episodic future thoughts, in which a mental simulation with a dynamic structure unfolds over time, arguably belong in this quadrant.

We will explore these distinctions further in the next section, where we bring them to bear on our discussion about the neural correlates of MTT. In brief, our suggestion is that the disagreements between the empirical results reviewed in sections 1 and 2 can be resolved if we relinquish the idea that time is essential to the content of the mental simulation and focus instead on its critical role for the structure of the simulation.

## 4. THE DYNAMIC-STRUCTURE VIEW

In section 1, we surveyed results from a number of studies suggesting a common core brain network underlying episodic memory and future thinking. To account for this common neural recruitment, a number of authors have endorsed, with varying degrees of commitment, what we called *the tensed-content view.* According to the tensed-content view, what explains the engagement of this core brain network during episodic memory and future thinking is the fact that the intentional content of the mental simulations entertained during each task is tensed. Since they operate upon representations with a temporal content, these brain structures likely support a system for mental time travel. However, in section 2, we reviewed a number of empirical results that speak against the tensed-content view, as they suggest that entertaining tensed contents is neither necessary nor sufficient for the engagement of the core brain network—or, at the very least, for the engagement of certain pivotal areas such as the hippocampus.

But the inconsistencies between these two lines of evidence dissipate, we suggest, if one adopts instead a *dynamic-structure view,* according to which what accounts for the common engagement of the core brain network is the fact that the structure of the representation is dynamical, that it unfolds over time, regardless of whether or not it is about time. Consider, for instance, both K. C.'s (Tulving, 1985) and D. B's (Klein et al, 2002) performance in their interviews. The report on K. C. suggests that he was unable to generate coherent thoughts when asked to engage in complex and dynamic simulations of both past and future events. The same occurred with D. B. However, when D. B. was asked to generate thoughts that did not involve the generation and maintenance of a dynamic representation, but rather of information carried by a somewhat more static representation (likened to semantic knowledge), D. B. showed no impairment. The same considerations apply to the more recent report by Craver and collaborators (2014), which clearly demonstrates that K. C. had the capacity to entertain tensed thoughts—that is, thoughts about time—as long as it did not require the generation and maintenance of dynamic simulations. His telegraphic answers seem to reflect the fleeting and non-dynamic nature of the mental representations they express.

These telegraphic answers are also characteristic of the amnesic patients from the Hassabis et al. (2007) study. For example, when asked to imagine "lying on a white sandy beach in a beautiful tropical bay," patient P03 reported that "all I can see is the colour of the blue sky and the white sand," and when probed further simply responded that he could not see anything else, that it felt "like I'm kind of floating." The lack of dynamicity of the mental representation that comes to mind is even clearer in their report from patient P05, who was asked to imagine "standing in the main hall of a museum containing many exhibits."

> Well, there's big doors. The openings would be high, so the doors would be very big with brass handles, the ceiling would be made of glass, so there's plenty of light coming through. Huge room, exit on either side of the room, there's a pathway and map through the centre and on either side there'd be the exhibits. [pause] I don't know what they are [pause]. . . . (1727)

At this point P05 reports that not much more comes to mind. Contrary to Hassabis et al.'s interpretation, this excerpt clearly suggests that P05 was able to imagine a clear spatial scene. It seems unlikely that someone who cannot entertain a mental representation of a spatial scene could generate such a detailed description of a museum room. What this description lacks, though, is dynamicity. It is as though P05 were describing a static picture of a space, rather than a dynamic scene whose components unfold over time. Indeed, there is evidence that individuals with MTL damage are capable of describing such scenes, as long as they are stable pictures (Gaesser et al., 2010; Race et al., 2011; Race et al., 2013). A similar interpretation is available for the results reported in the studies by Rosenbaum and collaborators (2009), as well as Romero and Moscovitch (2012), both of which asked individuals with MTL damage to imagine non-temporal fictional events. These patients' episodically impoverished answers as well as their telegraphic narratives strongly suggest that their simulations lacked dynamicity.

An interpretation consistent with the dynamic-structure view is also available for the neuroimaging results reviewed in sections 1 and 2. For instance, while Szpunar et al. (2007) found no difference in hippocampal engagement between episodic memory, future thinking, and mental simulations involving an imagined Bill Clinton, Addis et al. (2007) did find greater recruitment of the hippocampus for episodic memory and future thinking relative to a control condition is which participants had to construct a sentence. The dynamic-structure view can readily explain this difference, we think, for while it is likely that a mental simulation of a fictitious event involving Bill Clinton can easily have a dynamic structure, the construction of a sentence is paradigmatically a mental process that involves mostly static representational vehicles. The recruitment of the core brain network—and the MTL in particular—during non-MTT simulations as varied as fictitious events (Hassabis et al, 2007), non-temporal episodes (D'Argembeau et al., 2008), imagined past events (Addis et al., 2009), and counterfactual thoughts (De Brigard et al., 2013; van Hoeck et al., 2013) can be accounted for by the fact that all of these simulations involve the generation of dynamic representations. In contrast, tensed thoughts that do not require the deployment of dynamical representations—such as tasks involving future temporal discounting or regret anticipation (Benoit et al., 2011; Craver et al., 2014; Kwan et al., 2013)—neither seem to require the MTL nor engage the core brain network, at least to the same extent.

Another important virtue of the *dynamic-structure view* is that it is not only consistent with the scene construction (Hassabis & McGuire, 2007) and the episodic simulation (Schacter & Addis, 2007) hypotheses, allegedly the two most prominent accounts of the common brain activation during MTT simulations, but it also explains why these two accounts are perfectly compatible with each other. Scene construction, as defined by Hassabis and Maguire, consists in "the process of mentally generating and maintaining a complex and coherent scene or event" (299). The dynamic-structure view is entirely compatible with this perspective as long as the generated and maintained scene or event is thought to be dynamic. As we remarked, individuals with MTL damage—such as P05—can generate and maintain mental scenes as long as they are static. More precisely, their representations can be spatial insofar as their contents can feature spatial relations among components, as was clearly the case with the excerpt from P05's narration quoted earlier. But if the simulation requires the deployment of a dynamic representation of a scene, then it is likely that it would require the involvement of the core brain network, especially the MTL. As such, we suggest that it is best to understand the scene construction view as a hypothesis about the structure rather than the content of the simulation—an interpretation that, we believe, is congenial to the spirit of Hassabis and Maguire's approach.

A similar interpretation is available for a more recent study by Mullally and Maguire (2014) exploring counterfactual thinking in individuals with MTL damage. In this study, Mullally and Maguire asked 6 patients with bilateral hippocampal damage, and 10 controls, to engage in a counterfactual generation and a counterfactual inference task, both of which involved short vignettes. They found no difference in performance in either task between the two groups. However, they also recruited two independent cohorts of healthy participants and asked them to evaluate the degree to which the antecedents in counterfactual conditionals included in both tasks required the generation of mental representations of spatially coherent scenarios. Based on the ratings of these two independent cohorts, Mullally and Maguire split the counterfactual antecedents in two groups: strongly spatial and weakly spatial. Data from both MTL patients and controls were then reanalyzed, and it was found that, as compared to controls, the patients' performance was inferior in the counterfactual reasoning tasks that required the generation of strong—but not weak—spatial mental representations. Unfortunately, it is not clear that Mullally and Maguire's method to assess the degree of spatial strength in the required simulation provides a reliable measure across participants. In our opinion, given the low *N*s of the independent cohorts that rated the antecedents—19 in one case, 9 in the other—it is unlikely that their averaged scores represent a reliable scale to employ as an independent measure to split trials. Nonetheless, even if an appropriate sampling procedure were indeed to reveal a reliable difference in the degree of spatial coherence across counterfactual antecedents, it is still possible that those that are perceived as more strongly spatial are precisely the counterfactual antecedents that involve more dynamic structures. Perhaps a future experiment in which the dynamic structure involved in the mental simulation is kept constant across strong and weak spatial antecedents could help to disentangle this issue. Until then, the results from this study speak in favor of the scene-construction hypothesis as much as the dynamic-structure view.

It is worth noting that the scene-construction hypothesis has also been supported by a wealth of evidence indicating that there are pyramidal neurons with spatial receptive fields in the cornu ammonis (CA) of the hippocampus (O'Keefe & Nadel, 1978), as well as cell assemblies in the entorhinal cortex that map the surrounding space as a grid (Moser et al., 2008). These, along with a large number of related discoveries (Moser et al., 2008), have bolstered the view that the hippocampus and surrounding areas in the MTL (i.e., entorhinal, perirhinal, and parahippocampal cortices) are critical for spatial navigation and representation (Buckner, 2010). Given the engagement of these central regions of the core brain network during MTT simulations, it is natural to account for these common activations by the fact that the mental representations involved are spatial. However, there is also a large amount of evidence clearly indicating that place and grid cells are not only sensitive to space but also to sequential information about spatial navigation. Landmark studies by O'Keefe and Recce (1993) as well as Skaggs and collaborators (1996) showed that the moment of the firing of a place cell in relation to a navigational sequence has a precise timing relation with oscillations in the theta band. More recently, Foster and Wilson (2007) showed that place cells in CA1 in the hippocampus are "timed-locked" to theta oscillations, effectively demonstrating that, prior to performing a learned sequence, such place cells can "pre-play" the forthcoming action by way of firing in succession. These results, we think, indicate that the hippocampus and related areas may not only be coding for space but also for the dynamic relation among the components of a spatial representation. The fact that the representation is spatial may be as essential to the activation of the hippocampus as the fact that the relation among its components is dynamic. And this is precisely what the dynamic-structure view suggests.

The MTL has also been shown to be critical for the encoding and retrieval of all kinds of relations among perceptually distinct experiential components (Cohen & Eichenbaum, 1993; Eichenbaum & Cohen, 2001; Konkel & Cohen, 2009). This line of evidence has buttressed the long-standing view that the hippocampus may act as a "relational binder" whose computational role is not so much to store copies of past experiences but rather to store information about how the components of such experiences were related at encoding and how they should be reactivated together at retrieval (McClelland, 1995). This perspective, along with ample evidence pointing to the reconstructive character of memory (De Brigard, 2011, 2014; Michaelian, 2011; Schacter, Norman, & Koutstaal, 1998), has been marshaled in support of the constructive episodic simulation hypothesis (Schacter & Addis, 2007). According to this hypothesis, what accounts for the common engagement of the core brain network during MTT simulations is the fact that the same reconstructive processes that allow the reconstruction of episodic memories enable the recombination of episodic components during thoughts about possible future events. As a result, given that both episodic memories and future thoughts are cognitive processes that involve the construction of a mental representation from previously experienced episodic details, the fact that they both engage a similar underlying mechanism should come as no surprise.

Thus understood, however, the constructive episodic simulation hypothesis speaks to the constitution of the intentional contents of the representations carried by the identified brain mechanisms, rather than to the structure of the representations. In other words: what the episodic simulation hypothesis suggests is that there is a commonality in the representational processes during episodic memory and future thinking because both representational contents are similarly constituted. Whether or not the structures of such representations are dynamic is something to which the constructive episodic simulation hypothesis remains open. Insofar as the constructive episodic simulation hypothesis is about the content of the representations involved in MTT simulations, it is not incompatible with the scene-construction proposal, which—as we argued earlier—is a view about the structure of their representations.

The difference between an intentional content and a representational structure not only helps to dissipate the apparent inconsistencies in the results reviewed in sections 1 and 2, but it also helps to clarify what the scene construction and the constructive episodic simulation hypotheses are about, as well as to understand why they are perfectly compatible with each other. Moving forward, the challenge is to investigate whether there are other kinds of imaginative contents constituted by episodic details, to what degree they engage the same neural mechanisms as those involved in certain MTT simulations, and to what extent they require the deployment of representations with a dynamic structure.

## 5. CONCLUSIONS

In this chapter we put forth two main theses. First, we argued that what accounts for the commonalities in brain activation between certain kinds of simulations is not so much that their contents are tensed but the fact that their representations are dynamic. As such, we advocated for a dynamic-structure view of the neural correlates of mental time travel rather than a tensed-content view. Second, we suggested that while the scene-construction view should be seen as a hypothesis about the representational structure of the simulations involved in mental time travel, the constructive episodic simulation hypothesis is better understood as a proposal about the constitution of the intentional contents carried by such representations. As such, not only are they not incompatible, they are probably complementary.

There are, however, at least three possible objections that may put pressure against our theses. Here is the first. In section 3, when we introduced the philosophical distinction between intentional content and representational vehicle, we remarked that the same intentional content could be carried by two different representational vehicles. Next, we suggested that a representational vehicle with a static structure is as capable of carrying the same tensed content as a representational vehicle with a dynamic structure. But a case could be made to the effect that representations with dynamic structures carry more information than representations with static structures—akin to the claim that analog representations carry more information that their digital equivalents (Dretske, 1981). If this is the case, then the content of a dynamic representation will always differ, however slightly, from the content of a correspondent static representation. And if so, it is still possible that it is something about the content of a representation typically deployed in a dynamic structure, rather than something about the structure per se, that accounts for the common activation of the core brain network during the mental simulations reviewed earlier. Perhaps—to give an example close to the current literature—it is only self-referential, goal-oriented, and subjectively plausible hypothetical simulations that tend to be deployed in dynamic structures and, as such, only these recruit the core brain network.

Whether or not there is one unique kind of content that recruits the core brain network is, of course, an empirical question. Indeed, it seems to be the question leading a lot of the research on the cognitive neuroscience of mental simulation. For instance, noticing the similarities in brain activation between past and future thinking, on the one hand, and theory of mind, navigation, and moral cognition, on the other, Buckner and Carroll (2007) suggested that since all these processes had in common an element of self-projection, then the deployment of mental representations with a self-projection content may account for the common activation of the core brain network. But there are a number of results that are difficult to fit with this hypothesis, as they suggest not only that critical regions of the core brain network may not be necessary for certain tasks that involve self-projection (e.g., temporal discounting or regret anticipation; Craver et al., 2014) but also that they may not be sufficient (e.g., counterfactual thoughts involving others and not oneself; De Brigard et al., 2015). So the strategy has been to refine the hypothesis in order to exclude those contents that may not recruit the core brain network, and to try to come up with a characterization of a particular kind of content that can effectively capture those and only those representations that do recruit it. In the long run this strategy may work, but we worry that it could involve resorting to such a gerrymandered content description that its explanatory and predictive values might diminish. In this sense, exploring the simpler hypothesis that what unifies these common activations is something about the structure of the simulation, rather than their content, may be more beneficial both practically and theoretically.

The second objection is related to the first. As mentioned, the distinction between content and vehicle demands the possibility of sameness of content with only a difference in vehicle. However, if content is understood as the information carried by the representation, and there is a difference in the information carried by a dynamic versus a static representation, then there is a difference in content and not merely a difference in vehicle. But, if so, it is perfectly possible that the dynamicity of a mental representation may *not* be a property of the representational vehicle but rather of the intentional content, in which case the orthogonality between these two dimensions (see Figure 8.2) would be questionable. This is an important objection, but it may ultimately be terminological. For one, if we restrict the notion of representational vehicle to the *physical* substrate of the representation, then by fiat we would be ruling out the possibility of vehicles differing in dynamic, as opposed to physical, properties. This may force us to have to capture the distinction between dynamic and non-dynamic representations at the level of intentional content, which—again—is often understood as the information carried by a mental representation. However, such a notion of content is still somewhat vague, as it comprises both informational message and informational format. Our sense is that a broader definition of representational content in terms of carried information may still allow us to draw a distinction between contents that carry either temporal or non-temporal information in more or less dynamic formats. If so, then, the dynamic/non-dynamic distinction we draw here will re-emerge, albeit as a feature of the content of the representation, rather than the vehicle. Indeed, one of the motivations for us to talk about dynamic *structures* rather than dynamic *vehicles* is that we want to allow for the possibility that the two-way contrast between static/dynamic versus temporal/non-temporal representations may eventually be better captured by an adequate theory of the contents of mental simulations. We take this objection, then, as an invitation for further philosophical research on the notion of intentional content as it applies to MTT representations.

A third objection challenges the claim that there is such a thing as a static representation. After all, every mental content presumably takes some time to be consciously experienced by a subject, however fleetingly. Even entertaining a brief sentence in one's mind, such as "Washington, D.C., is the capital of the United States," or a quick image of a single object, like a pair of scissors, is a process that presumably requires at least a few milliseconds. As such, if even the prototypical examples of allegedly static representations take some time to unfold, then they are actually dynamic, in which case the dynamic/static distinction loses its grip. This is a fair point, and we certainly agree (perhaps *contra* the philosophical tradition in analytic philosophy) that even the shortest-lived of our consciously experienced mental contents may require some time to unfold. Nonetheless, this does not invalidate the static/dynamic distinction drawn here, as we do not think of it as a dichotomous distinction that exhaustively splits every mental state into either static or dynamic. Instead, our view is that every mental content falls somewhere in a continuum of structural dynamicity, with some contents close to being fully static, and some being extremely dynamic—even if there are not specific instances of mental representations that are fully static or fully dynamic. What these extremes may be is indeed a puzzling empirical question, the answer of which may involve understanding the informational limits of internal attention and working memory (De Brigard, 2011), as well as the differences in entertaining simulations in different modalities. Indeed, an attractive question that arises from the dynamic-structure view is whether or not simulations in diverse modalities differentially engage the default network. Would entertaining a dynamic mental narrative in a purely linguistic format, devoid of any visual or auditory elements, be sufficient to engage the default network? Future experiments comparing dynamic representation with linguistic versus non-linguistic formats may be able to tell whether or not the structural dynamicity of the representation is sufficient to engage the default network.

Which brings us to the last point: Are we implying that the function of the default network is to generate and maintain representations with a dynamic structure? After all, some of the most recent proposals about the function of the default network seem to try to define its function by way of characterizing some kind of representational content that is uniquely associated with its engagement, such as socially relevant hypothetical thinking (De Brigard et al., 2014) or self-generated goal-oriented thoughts (Andrews-Hanna et al., 2014). Our view is that the evidence is insufficient, at this point, to be able to adjudicate between competing views, and more research is needed to understand not only what is the best way to characterize the nature of the mental representations supported by the brain's default network, but also what the function of this brain system may be.

What we do believe, however, is that bringing in a hypothesis about the structure of the representation, rather than its content, may contribute to move this research project forward.

## ACKNOWLEDGMENTS

This paper was presented at the University of Michigan in Ann Arbor, in August 2015, at the meeting of the Moral Psychology Research Group. We thank the members of the audience for their comments. Special thanks to John Doris, Edouard Machery, Chandra Sripada, and Walter Sinnott-Armstrong for their insightful challenges and suggestions.

## NOTES

1. Not everyone who endorses the claim that the common engagement of the neural substrates of episodic memory and future thinking evidences a single system for MTT *also* endorses explicitly the claim that what accounts for this common engagement is the fact that the contents of the mental simulations are tensed. But to many, this assumption looms large, and only now is it starting to be scrutinized (Schacter et al., 2012). Our intention is to bring this assumption to the forefront and discuss it.
2. Another study that sometimes is referenced in relation to lack of hippocampal activity during MTT tasks was conducted by Nyberg and colleagues (2010). In this study, participants were asked either to imagine walking on a familiar path in a possible past, a possible future, or the present, or to remember a specific instance in which they actually walked that path. In an attempt to isolate brain regions differentially involved in MTT, the authors contrasted the temporal conditions against the present condition, and found no hippocampal activity. Nyberg and colleagues interpret this null finding as evidence that the hippocampus may be involved more in the episodic information rather than the temporal aspect of the simulation. In our view, this result is difficult to interpret for methodological reasons. For one, this study is severely underpowered. They recruited only five participants, and reported that not all of them were successful at training. The final number of participants scanned is not reported. In addition, there are only 20 observations per participant for five different conditions, the fMRI data of which was then fit to an analysis of variance model. As a result, it is likely that these null results are simply due to this study being severely underpowered. Given our methodological qualms with this particular study, we mention it only in a footnote rather than in the main text. Further studies may be able to confirm these observations.
3. This observation isn't new, but is worth repeating. William James's famous allegory of the "stream of consciousness" was primarily a criticism against Hume's "train of thoughts" for very similar reasons (James, 1890). The tradition in phenomenology, starting from Husserl (1913), also rejects the view of experiential contents as being discrete and discontinuous with the rest of our conscious experience.
4. Paraphrasing Wittgenstein (1953, 54), if we think of a man climbing a mountain, we don't find the entertained mental representation to be ambiguous with that of

a man descending the same mountain backward. The dynamical nature of these kinds of representations eliminates potential referential ambiguities.

5. The example here suggests that there may be a difference in content as well, and a case could be made to the effect that the representation in panel B carries information that the representation in A does not. This is indeed an important observation with relevant consequences for the forthcoming discussion. As such, it will be considered in section 5.

## REFERENCES


Addis, D. R., Pan, L., Vu, M. A., Laiser, N. & Schacter, D. L. (2009). Constructive episodic simulation of the future and the past: distinct subsystems of a core brain network mediate imagining and remembering. *Neuropsychologia*, *47*(11), 2222–2238.

Addis, D. R., Wong, A. T., & Schacter, D. L. (2007). Remembering the past and imagining the future: common and distinct neural substrates during event construction and elaboration. *Neuropsychologia 45*(7), 1363–1377.

Andrews-Hanna, J. R., Smallwood, J. & Spreng, N. R. (2014). The default network and self-generated thought: component processes, dynamic control, and clinical relevance. *Annals of the New York Academy of Sciences, 1316*, 29–52.

Atance, C. M., & O'Neill, D. K. (2001). Episodic future thinking. *Trends in Cognitive Sciences, 5*(12), 533–539.

Augustine, Saint. (1991). *Confessions*. Translated by H. Chadwick. Oxford: Oxford University Press.

Barnes, J. (1984). *The complete works of Aristotle*. Oxford: Oxford University Press.

Bartsch, T., Döhring, J., Rohr, A., Jansen, O., & Deuschl, G. (2011). CA1 neurons in the human hippocampus are critical for autobiographical memory, mental time travel, and autonoetic consciousness. *Proceedings of the National Academy of Sciences U S A, 108*(42), 17562–17568.

Benoit, R. G., Gilbert, S. J., & Burgess, P. W. (2011). A neural mechanism mediating the impact of episodic prospection on farsighted decisions. *Journal of Neuroscience, 31*(18), 6771–6779.

Brentano, F. [1874] (1973). *Psychology from an empirical standpoint*. London: Routledge and Kegan Paul.

Buckner, R. L. (2010). The role of the hippocampus in prediction and imagination. *Annual Reviews Psychology, 61*, 27–48.

Buckner, R. L., Andrews-Hanna, J. R., & Schacter, D. L. (2008). The brain's default network: anatomy, function, and relevance to disease. *Annals of the New York Academy of Sciences, 1124*, 1–38.

Buckner, R. L., & Carroll, D. C. (2007). Self-projection and the brain. *Trends in Cognitive Sciences, 11*(2), 49–57.

Cabeza, R., & St. Jacques, P. (2007). Functional neuroimaging of autobiographical memory. *Trends in Cognitive Sciences 11*(5), 219–227.

Christoff, K., Gordon, A. M., Smallwood, J., Smith, R., & Schooler, J. W. (2009). Experience sampling during fMRI reveals default network and executive system contributions to mind wandering. *Proceedings of the National Academy of Sciences U S A, 106*(21), 8719–8724.

Clark, A. (2008). *Supersizing the mind: embodiment, action, and cognitive extension.* Oxford: Oxford University Press.

Cohen, N. J., & Eichenbaum, H. (1993). *Memory, amnesia, and the hippocampal system.* Cambridge, MA: MIT Press.

Crane, T. (2009). Is perception a propositional attitude? *Philosophical Quarterly, 59,* 452–469.

Craver, C. F., Kwan, D., Steindam, C., & Rosenbaum, R. S. (2014). Individuals with episodic amnesia are not stuck in time. *Neuropsychologia, 57,* 191–195.

Cussins, A. (1990). The connectionist construction of concepts. In M. A. Boden (ed.), *The philosophy of AI* (pp. 368–440). Oxford: Oxford University Press.

D'Argembeau, A., Stawarczyk, D., Majerus, S., Collette, F., Van der Linden, M., Feyers, D., Maquet, P., & Salmon, E. (2010). The neural basis of personal goal processing when envisioning future events. *Journal of Cognitive Neuroscience, 22*(8), 1701–1713.

D'Argembeau, A., & Van der Linden, M. (2004). Phenomenal characteristics associated with projecting oneself back into the past and forward into the future: influence of valence and temporal distance. *Consciousness and Cognition, 13*(4), 844–858.

D'Argembeau, A., & Van der Linden, M. (2006). Individual differences in the phenomenology of mental time travel: the effect of vivid visual imagery and emotion regulation strategies. *Consciousness and Cognition, 15*(2), 342–350.

D'Argembeau, A., Xue, G., Lu, Z. L., Van der Linden, M., & Bechara, A. (2008). Neural correlates of envisioning emotional events in the near and far future. *Neuroimage, 40,* 398–407.

De Brigard, F. (2011). The role of attention in conscious recollection. *Frontiers in Psychology,* 3, 29.

De Brigard, F., Addis, D., Ford, J. H., Schacter, D. L., & Giovanello, K. S. (2013). Remembering what could have happened: neural correlates of episodic counterfactual thinking. *Neuropsychologia,* 51(12), 2401–2414.

De Brigard, F., et al. (2014). Is memory for remembering? Recollection as a form episodic hypothetical thinking. *Synthese,* 191(2), 155–185.

De Brigard, F., & Giovanello, K. S. (2012). Influence of outcome valence in the subjective experience of episodic past, future, and counterfactual thinking. *Consciousness and Cognition, 21*(3), 1085–1096.

De Brigard, F., Spreng, R. N., Mitchell, J. P., & Schacter, D. L. (2015). Neural activity associated with self, other, and object-based counterfactual thinking. *NeuroImage,* 109, 12–26.

Dennett, D. (1978). *Brainstorms.* Cambridge, MA: MIT Press.

Dennett, D., & Kinsbourne, M. (1992). Time and the observer: the where and when of consciousness in the brain. *Behavioral and Brain Sciences, 15*(2), 183–201.

De Vito, S., Gamboz, N., Brandimonte, M. A., Barone, P., Amboni, M., & Della Salla, S. (2012). Future thinking in Parkinson's disease: an executive function? *Neuropsychologia, 50*(7), 1494–1501.

Dretske, F. (1981). *Knowledge and the flow of information.* Cambridge, MA: MIT Press.

Eichenbaum, H., & Cohen, N. J. (2001). *From conditioning to conscious recollection: memory systems of the brain.* Oxford: Oxford University Press.

Elpidorou, A. (2013). Having it both ways: consciousness unique, but not otherwordly. *Philosophia, 41,* 1181–1203.

Foster, D. J., & Wilson, M. A. (2007). Hippocampal theta sequences. *Hippocampus, 17*(11), 1093–1099.

Gaesser, D.,Sacchetti, D. C., Addis, D. R., & Schacter, D. L. (2010). Characterizing age-related changes in remembering the past and imagining the future. *Psychology and Aging, 26*, 80–84.

Hassabis, D., & Maguire, E. A. (2009). The construction system of the brain. *Philosophical Transactions of the Royal Society B Biological Sciences, 364*, 1263–1271.

Hassabis, D., & Maguire, E. A. (2007). Deconstructing episodic memory with construction. *Trends in Cognitive Sciences, 11*(7), 299–306.

Hassabis, D., Kumaran D, Vann, S. D., & Maguire, E. A. (2007). Patients with hippocampal amnesia cannot imagine new experiences. *Proceedings of the National Academy of Sciences U S A, 104*(5), 1726–1731.

Hobbes, T. (1668) *Leviathan* (Latin edition), in E. Curley (ed.), *Leviathan, with selected variants from the Latin edition of 1668*, Indianapolis: Hackett, 1994.

Hoerl, C. (1999). Memory, amnesia, and the past. *Mind and Language, 14*(2), 227–251.

Hume, D. (1739). *A treatise of human nature.* edited by L. A. Selby-Bigge, 2nd ed. revised by P. H. Nidditch, Oxford: Clarendon Press, 1975.

Hurley, S. L. (1998). Vehicles, contents, conceptual structure and externalism. *Analysis, 58*(1), 1–6.

Husserl, E. (1913). *Ideas Pertaining to a Pure Phenomenology and to a Phenomenological Philosophy—First Book: General Introduction to a Pure Phenomenology*, trans. F. Kersten. The Hague: Nijhoff 1982.

Hyatt, C. J., Calhoun, V. D., Pearlson, G. D., & Assaf, M. (2015). Specific default mode subnetworks support mentalizing as revealed through opposing network recruitment by social and semantic FMRI tasks. *Human Brain Mapping*, 36(8), 3047–3063.

James, William. (1890). *Principles of Psychology.* N.Y.: Henry Holt and Company.

Johnson, M. K., et al. (1988). Phenomenal characteristics of memories for perceived and imagined autobiographical events. *Journal of Experimental Psychology: General, 117*(4), 371–376.

Klein, S., & Nichols, S. (2012). Memory and the sense of personal identity. *Mind, 483*, 677–702.

Klein, S. B., Loftus, J., & Kihlstrom, J. F. (2002). Memory and temporal experience: the effects of episodic memory loss on an amnesic patient's ability to remember the past and imagine the future. *Social Cognition, 20*, 353–379.

Konkel, A., & Cohen, N. J. (2009). Relational memory and the hippocampus: representations and methods. *Frontiers in Neuroscience, 3*(2), 166–174.

Kwan, D., et al. (2013). Dissociations in future thinking following hippocampal damage: evidence from dis-counting and time perspective in episodic amnesia. *Journal of Experimental Psychology: General, 142*, 1355–1369.

Levine, B. et al. (2002). Aging and autobiographical memory: dissociating episodic from semantic retrieval. *Psychology and Aging, 17*(4), 677–689.

Locke, J. (1689). *An essay concerning human understanding.* Nidditch, P. H. (Ed.) Oxford: Oxford University Press, 1975.

Maguire, E. A., Vargha-Khadem, & Hassabis, D. (2010). Imagining fictitious and future experiences: evidence from developmental amnesia. *Neuropsychologia, 48*(11), 3187–3192.

Mar, R. (2011). The neural bases of social cognition and story comprehension. *Annual Review of Psychology, 62*, 103–134.

McClelland, J. L., McNaughton, B. L., & O'Reilly, R. C. (1995). Why there are complementary learning systems in the hippocampus and neocortex: insights from the successes and failures of connectionist models of learning and memory. *Psychological Review, 102*, 419–457.

Michaelian, K. (2011). Generative memory. *Philosophical Psychology, 24*(3), 323–342.

Moser, E. I., Kropff, E., & Moser, M. (2008). Place cells, grid cells, and the brain's spatial representation system. *Annual Reviews Neuroscience, 31*, 69–89.

Mullally, S., Hassabis, D., & Maguire, E. A. (2012). Scene construction in amnesia: an fMRI study. *Journal of Neuroscience, 32*(16), 5646–5653.

Mullally, S., & Maguire, E. A. (2014). Counterfactual thinking in patients with amnesia. *Hippocampus, 11*, 1261–1266.

Nyberg, L., et al. (2010). Consciousness of subjective time in the brain. *Proceedings of the National Academy of Sciences U S A, 107*(51), 22356–22359.

O'Keefe, J., & Nadel, L. (1978). *The hippocampus as a cognitive map.* Oxford: Oxford University Press.

O'Keefe, J., & Recce, M. L. (1993). Phase relationship between hippocampal place units and the EEG theta rhythm. *Hippocampus, 3*(3), 317–330.

Okuda, J., et al. (2003). Thinking of the future and past: the roles of the frontal pole and the medial temporal lobes. *NeuroImage, 19*, 1369–1380.

Pitt, D. (2013). Mental representation. *Stanford Encyclopedia of Philosophy.* (Fall 2013 Edition), Edward N. Zalta (ed.), URL = <http://plato.stanford.edu/archives/fall2013/entries/mental-representation/>

Race, E., Keane, M. M., & Verfaellie, M. (2011). Medial temporal lobe damage causes deficits in episodic memory and episodic future thinking not attributable to deficits in narrative construction. *Journal of Neuroscience, 31*(28), 10262–10269.

Race, E., Keane, M. M., & Verfaellie, M. (2013). Living in the moment: patients with MTL amnesia can richly describe the present despite deficits in past and future thought. *Cortex, 49*(6), 1764–1766.

Romero, K., & Moscovitch, M. (2012) Episodic memory and event construction in aging and amnesia. *Journal of Memory and Language, 67*, 270–284.

Rosenbaum, R. S., Köhler, S., Schacter, D. L., Moscovitch, M., Westmacott, R., Black, S. E., Gao, F., & Tulving, E. (2005). The case of K.C.: Contributions of a memory-impaired person to memory theory. *Neuropsychologia, 43*, 989–1021.

Rosenbaum, R. S., et al. (2009). Amnesia as an impairment of detail generation and binding: evidence from personal, fictional, and semantic narratives in K. C. *Neuropsychologia, 47*, 2181–2187.

Rowlands, M. (2010). *The new science of the mind: from extended mind to embodied phenomenology.* Cambridge, MA: MIT Press.

Russell, B. (1921). *The analysis of mind.* London: George Allen & Unwin.

Schacter, D. L., & Addis, D. R. (2007). The cognitive neuroscience of constructive memory: remembering the past and imagining the future. *Philosophical Transactions of the Royal Society London B Biological Sciences, 362*, 773–786.

Schacter, D. L., Addis, D. R., & Buckner, R. L. (2007). Remembering the past to imagine the future: the prospective brain. *Nature Reviews Neuroscience, 8*, 657–661.

Schacter, D. L., et al. (2012). The future of memory: remembering, imagining, and the brain. *Neuron*, 76(4), 677–694.

Schacter, D. L., Norman, K. A., & Koutstaal, W. (1998) The cognitive neuroscience of constructive memory. *Annual Reviews Psychology, 49*, 289–318.

Scoville, W. B., & Milner, B. (1957). Loss or recent memory after bilateral hippocampal lesions. *Journal of Neurology, Neurosurgery, and Psychiatry, 20*(11), 11–21.

Skaggs, W. E., et al. (1996). Theta phase precession in hippocampal neuronal populations and the compression of temporal sequences. *Hippocampus, 6*, 149–172.

Spreng, R. N., & Andrews-Hanna, J. R. (2015). The default network and social cognition. In A.W. Toga (Ed.), *Brain Mapping: An Encyclopedic Reference*. Academic Press: Elsevier, pp. 165–169.

Spreng, R. N. & Grady, C. L. (2010). Patterns of brain activity supporting autobiographical memory, prospection, and theory of mind, and their relationship to the default mode network. *Journal of Cognitive Neuroscience, 22*(6), 1112–1123.

Spreng, N. R., & Levine, B. (2006). The temporal distribution of past and future autobiographical events across the lifespan. *Memory and Cognition*, 34(8), 1644–1651.

Spreng, R. N., Mar, R. A., & Kim, A. S. (2009). The common neural basis of autobiographical memory, prospection, navigation, theory of mind, and the default mode: a quantitative meta-analysis. *Journal of Cognitive Neuroscience, 21*(3), 489–510.

Squire, L. R., et al. (2010). Role of the hippocampus in remembering the past and imagining the future. *Proceedings of the National Academy of Sciences U S A, 107*, 19044–19048.

St. Jacques, P., & De Brigard, F. (2015). In D. R. Addis, M. Barense, and A. Duarte (eds.), *The Wiley handbook on the cognitive neuroscience of memory*. New York: John Wiley and Sons, 2015.

Stuss, D. T., & Guzman, D. A. (1988). Severe remote memory loss with minimal anterograde amnesia. *Brain and Cognition, 8*, 21–30.

Suddendorf, T., Addis, D. R., & Corballis, M. C. (2009). Mental time travel and the shaping of the human mind. *Philosophical Transactions of the Royal Society B Biological Sciences, 364*, 1317–1324.

Suddendorf, T., & Corballis, M. C. (1997). Mental time travel and the evolution of the human mind. *Genetic, Social, and General Psychology Monographs, 123*(2), 133–167.

Suddendorf, T., & Corballis, M. C. (2007). The evolution of foresight: what is mental time travel, and is it unique to humans? *Behavioral and Brain Sciences, 30*(3), 299–313.

Summerfield, J. J., Hassabis, D., & Maguire, E. A. (2010). Differential engagement of brain regions within a core network during scene construction. *Neuropsychologia, 48*, 1501–1509.

Szpunar, K. (2010). Episodic future thought: an emerging concept. *Perspectives on Psychological Science, 5*, 142–162.

Szpunar, K., Watson, J. M., & McDermott, K. B. (2007). Neural substrates of envisioning the future. *Proceedings of the National Academy of Sciences U S A, 104*(2), 642–647.

Tulving, E. (1985). Memory and consciousness. *Canadian Psychology, 26*(1), 1–12.

Tulving, E. (2002). Episodic memory: from mind to brain. *Annual Review of Psychology, 53*, 1–25.

Van Hoeck, N., et al. (2013). Counterfactual thinking: an fMRI study on changing the past for a better future. *Social Cognitive and Affective Neuroscience, 8*, 556–564.

Vargha-Khadem, F., et al. (1997). Differential effects of early hippocampal pathology on episodic and semantic memory. *Science, 277*, 376–380.

Verfaellie, M., Race, E., & Keane, M. M. (2012). Medial temporal lobe contributions to future thinking: evidence from neuroimaging and amnesia. *Psychological Belgica*, *52*(2–3), 77–94.

Weiskopf, D. (2010). Embodied cognition and linguistic comprehension. *Studies in History and Philosophy of Science*, *41*, 294–304.

Wittgenstein, L. (1953). *Philosophical Investigations*. G.E.M. Anscombe and R. Rhees (eds.), G.E.M. Anscombe (trans.), Oxford: Blackwell.

# PART IV

# The Self in Future-Oriented Mental Time Travel

9

# Future Mental Time Travel and the Me-Self

LILIANN MANNING

One of the most frequently cited contributions from James's *Principles of Psychology* (1890) to the current study of the self is the division of the self into the "I-self," the self as a knower, and the "Me-self," the self as known.[1] In general terms, James's I-self and Me-self distinction has been elaborated on and updated to produce present-day theoretical models, but independently of the existing expanded versions, the foundations remain in the groundwork that James laid out. A recent example of updated work based on the division of I-self and Me-self is the sense of self and memory model put forward by Prebble et al. (2013). Within the concept of the Me-self (i.e., the content of self), some of whose characteristics are tackled in the present chapter in relation to episodic future thought, Prebble et al.'s (2013) model includes two functions according to the temporal dimension: the present moment or across time. The *self-concept* based on Conway's theory (Conway, 2005; Conway & Pleydell-Pearce, 2000; Conway et al., 2004b) is the content of the self in the present moment. Its normal functioning allows us to experience a sense of unity in our mental representations of who we are. The content of self across time, on the other hand, allows us to experience *semantic continuity*. Regarding the relationships between the sense of self and the memory systems, in contrast with the I-self (see section 1) that is exclusively based on episodic autobiographical memory, the Me-self is associated with both semantiticized and episodic autobiographical memory.

It is acknowledged that the experienced unity of present and temporally extended disparate aspects of sense of self denote a paradox of selfhood. The sense of self has been called "synchronic unity" for the different selves in the present moment and, in the case of the concept of cross-temporal self, it is the notion of sameness of the person, despite important changes across time, that constitutes the "diachronic unity." This notion has been considered by Klein (2014) in relation to the epistemological, that is, the known-self (Me-self) and the ontological or knower-self (I-self).

Turning back to James's foundational notions, relatively less frequently mentioned in the literature is the further subdivision of the Me-self into the material, social, and spiritual selves. Referring to the "social me," James claimed that "*In each kind of self, material, social, and spiritual, men distinguish between the immediate and actual, and the remote and potential* [. . .]. Of all these wider more potential selves, the potential social self is the most interesting" (James, 1890, 315; italics in the original). The notion of potential selves became, among others, Markus and Nurius's (1968) starting point to hypothesize the "possible selves," which in turn, influenced the concept of self-defining future projections (D'Argembeau et al., 2012).

With regard to the "spiritual self," James's definition was as follows: "By the Spiritual Self, so far as it belongs to the Empirical Me, I mean a man's inner or subjective being, his psychic faculties or dispositions [. . .]. These psychic dispositions are the most enduring and intimate part of the self, that which we most verily seem to be" (James, 1890, 296). The contemporary version of what James so defined would be the ability of human beings to reflect on their own personal attributes, which is an important part of the self-concept (Conway et al., 2004a; and the conceptual self-knowledge in Prebble et al.'s, 2013 model). Moreover, it has also been considered as a central component of the future Me-self (e.g., Adamson et al., 2007; D'Argembeau et al., 2012; see further discussion later in this chapter).

Interactions between future mental time travel (FMTT) and the mental representation of who we are could be viewed as the influence of future projects on the unitary self, at any present moment. Complementarily, interactions between FMTT and the way we mentally represent ourselves across time would be associated with the capacity to integrate non–real event representations in a unitary self.

In summary, James's (1890) *I-self* and *Me-self* division entails a subjective versus objective dimension of the self. The *Me-self* encompasses the material (our body), social (how others see us), and spiritual (defined earlier) selves. On the other hand, Prebble et al. (2013) specified the I-self as the conscious, phenomenological experience of selfhood and the Me-self as the mental representations, comprising all the things that we perceive and know about ourselves. The authors highlighted the associations between memory system and sense of self (I-self with episodic memory and Me-self with both episodic and semanticized memories). They considered the I-self and the Me-self in two temporal dimensions, present and across time. The paragraphs in the Introduction focused on the Me-self, highlighting that Prebble et al.'s (2013) model operationalized the *Me-self in the present* in terms of the *self-concept*, and the *Me self across time* in terms of *semantic continuity*.

As mentioned earlier, the present chapter focuses on possible interactions between the Me-self and FMTT; however, in order to tackle the likely influence on the Me-self of the phenomenological I-self, the latter will be commented on in the perspective of FMTT, in the first section of the chapter. To this end, Tulving's (2005) conceptualization of the "traveler self" (i.e., the self that *makes* the mental time travel) is summarized, and new perspectives, on the bases of Prebble et al. (2013), are mentioned.

The second section confronts episodic processes as the *sine qua non* condition for FMTT, with semantic representations associated with FMTT. The outcome includes the Me-self and, consequently, a wider depiction of FMTT. One of the first authors, to our knowledge, to have suggested that our understanding of *episodic* future thought would be improved by taking into account the role played by the autobiographical semantic system has been Szpunar (2010). The proposal to study FMTT constructions as being the product of both episodic and semantic memory systems is the central notion of the second section and, more generally, the chapter.

The two last sections develop the interactions between the self-concept and FMTT, with section 3 focusing on the semantic continuity of the self-concept and commenting on Conway's (2005) model of the Self Memory System, and Prebble et al.'s (2013) theoretical suggestions associated with that concept. Finally, in section 4, the future selves are discussed on the bases of the contributions of both pioneer and current studies.

## 1. THE I-SELF IN FUTURE MENTAL TIME TRAVEL

The present section is an attempt to provide a general notion about the role that the I-self could play in FMTT, by raising the following points: (i) the definition (by different authors) of the I-self; (ii) the link between the I-self in the present moment with the I-self across time and the consequences of damage of the former on the latter, namely, on autonoetic consciousness; (iii) the central role of autonoetic consciousness in mental time travel, following Tulving's (1985, 2005) contribution, particularly on the "traveler self" concept.

The concept of the I-self as conceived by James has been expanded, among others, by Gallagher (2000), who suggested the minimal self as part of the division between the minimal self and the narrative self. Gallagher's phenomenological concept of minimal self is limited to that which is an immediate subject of experience. More precisely, "[e]ven if all of the unessential features of self are stripped away, we still have an intuition that there is a basic, immediate, or primitive 'something' that we are willing to call a self" (Gallagher, 2000, 15). The I-self in Prebble et al.'s (2013) model, in the present moment, includes the pre-reflective self-experience, described as an integral feature of our conscious experience of the world. These online processes of sense of self have also been defined as a goal attainment-modular window, open on our mind's reality, a slice of subjective reality (Conway et al., 2004). James's (1890), Prebble et al.'s (2013), Gallagher's (2000), or Conway et al.'s (2004) phenomenological conception of the I-self includes an embodied dimension, particularly elaborated by Gallagher (2000), who also drew the distinction between the sense of self-agency and the sense of self-ownership for actions.

Why should this basic sense of self be mentioned in a chapter dealing with FMTT? The answer is suggested by Prebble et al.'s (2013) autobiographical memory and sense of self model. The I-self in the present moment (i.e., the subjective sense of self) is a "basic" function thought to be necessary for more complex functions to be achieved. Consequently, it is directly associated in a unidirectional way (i) with phenomenological continuity and (ii) with the self-concept (the present moment Me-self; see earlier discussion). Moreover, different levels of subjective self-experience show also direct and unidirectional links in a hierarchical order. Thus, within the present moment I-self, pre-reflective self-experience, must be functional for self-awareness to be achieved. The latter is considered as "a vital precursor to autonoetic consciousness and episodic memory" (Prebble et al., 2013, 823), which is part of the function of the I-self across time. From a clinical perspective, the presence of delusions, auditory hallucinations, and thought insertion in schizophrenic patients, having been linked with breakdown in sense of agency and ownership of experience (Gallagher, 2000; Klein et al., 2013b; Sass & Parnas, 2003), is thought to illustrate the present pre-reflective self-experience impairment (Prebble et al., 2013). Going one step further, it can be speculated that episodic memory deficits observed in schizophrenia, with generation of specific future events being particularly impaired (e.g., D'Argembeau et al., 2008b), may also reflect dysfunctions of the minimal, pre-reflective self, through impaired sense of agency and sense of ownership.

As mentioned in the preceding paragraph, in Prebble et al.'s (2013) model, the present moment processes in the I-self also comprises the notion of self-awareness, generally conceived as the capacity of meta-representational activity, which allows the individual to tag his or her different mental representations (imagining, remembering, knowing, thinking, etc.). Functional damage to self-awareness would result in a loss of autonoetic consciousness and, consequently, altered phenomenological continuity.

The concept of *autonoetic consciousness* was suggested by Tulving in 1985 and elaborated by Wheeler et al. in 1997. More recently, Tulving updated his views on the similarities and differences in the nature of consciousness (Tulving, 2005). In Tulving's model, the level of consciousness (anoetic, noetic, and autonoetic) determines the different memory systems (Tulving, 1972, 2001, 2005). In the case of autonoetic consciousness, it entirely embodies episodic memory, and autonoetic episodic memory differs from other kinds of memory in that its operations require a self—more precisely, a traveler self since autonoetic consciousness, as a cognitive capacity, is inherently related to mental time travel. Tulving's key concepts are explained in the following definition: ". . . The mental time travel allows the 'owner' of episodic memory ('self'), through the medium of autonoetic awareness, to remember one's own previous 'thought-about' experiences, as well as to 'think about' one's own possible future experiences" (Tulving, 2005, 9). Additionally, Tulving argued that the expression of mental time travel is possible in the "self's autonoetic consciousness" and, owing to its inherent relation to episodic memory, to subjectively apprehended time. The latter is described as resulting from the interaction between physical time and autonoetic episodic memory. Subjective time that extends into the future, FMTT, represents a key feature of autonoesis that pervades an individual's whole life (Tulving, 2001, 2002, 2005). Expanding on Tulving's concept of subjective time, Szpunar (2011) reminds us that the notion of autonoetic consciousness highlights awareness of one's self existing in subjective time (while chronesthesia underscores awareness of the subjective time in which one's self exists).

With regard to the self's autonoetic consciousness studied in his renowned patient, K. C. (whose cognitive pattern of performance inspired Tulving to distinguish noesis and autonoesis), Tulving's assumptions were that the patient's self-awareness was abnormal and that he had no functioning autonoetic, traveler self: "Because he is perfectly well aware of his timeless self—self in the present—it seems reasonable to attribute his difficulties with personal past and personal future to deficient, perhaps largely lacking, autonoesis" (Tulving, 2005, 27). It is interesting to note that K. C. had learned his new post-brain injury self. It seems that K. C.'s ability to learn his new self, despite massive autobiographical memory impairment, illustrates what Klein and Lax (2010) call "[t]he remarkable resilience of knowledge of one's traits in the face of neural and cognitive impairment" (939). Going back to Tulving's fascinating report on his patient, he informs that K. C. compared the request of imagining what he was going to do in the immediate future, namely, after his interview with Tulving, with the request to look for a chair in an empty room. A possible question that remains unanswered is whether K. C. could have used his semantic memory (i.e., his updated personality traits) to imagine a personal future scene or event. More generally, it is not known what form of continuity might be possible to attain on the Me-self alone.

Summarizing, Tulving's theoretical model posits the relation self/memory exclusively with the episodic memory system. The semantic system is limited in his model, to a previous, necessary phase for the episodic memory to evolve. And it is exclusively the episodic memory system, imbued with autonoetic consciousness, that allows a given event to be encoded and later retrieved as a personally experienced event. The traveler self is manifested across mental time, since some quality of the self at the encoding phase must also be "felt" when traveling in subjective time to retrieve the event (Tulving, 2001).

## 2. SEMANTIC REPRESENTATIONS PLAY A PART IN FUTURE MENTAL TIME TRAVEL

In this section, the aim is to show that the semantic knowledge-based Me-self plays a key role in FMTT. To that end, two points of view are tackled in the present section. The first is held by Klein (2013, 2015; Klein et al., 2002). His observations are against the notion of FMTT as a unitary concept and the exclusive association between episodic memory system and FMTT. The second view, initiated by Szpunar (2010), deals with the role that the semantic memory system plays in the production of episodic future thought. Recent and crucial clinical data corroborating Szpunar's standpoint are discussed.

Klein's (2015) starting point is that the right question has not been asked with regard to FMTT, owing, among others, to the emphasis given to methodological issues at the expense of conceptual thinking. Psychologists seem to have put, Klein argues, the cart before the horse: empirical results give rise to theoretical constructions. On these bases, he suggests that semantic and episodic memory systems cannot be distinguished in terms of memory of the content of an experience, but in terms of how the content was experienced. Klein (2015) highlights the fact that some clinical observations show that in some patients, autonoetic awareness is damaged, not memory processes. The author suggests that there is no episodic content per se, since the content does not belong in a memory system. It is only when a given content is intrinsically associated with autonoetic awareness (a non-inferential feeling of a past personal event) that the content can be qualified as being an episodic content.

Based on the study of patient D. B., Klein et al. (2002) suggested that semantic memory can support certain forms of FMTT in the absence of any functioning episodic memory. Consequently, Klein's (2013) contention is that FMTT is not a unitary concept beholden to episodic memory, but it is dependent on the type of subjective temporality activated with a given recollection. Klein (2013) adds that, in principle, semantic memory must support semantic aspects of the self, but we do not know yet how to probe it.

Klein et al. (2002) proposed the introduction of "live time" and "known time" to account for personal and impersonal FMTT, respectively. Klein et al.'s (2002) patient D. B., whose lived time was densely impaired together with his preserved known time, and Manning et al.'s (2013) patient J. R., who presented the mirror image of D. B.'s cognitive profile, illustrate, argues Klein (2013), the double dissociation between types of memory and the capacity to simulate future personal and impersonal projections. To conclude, Klein and Steindam (Chapter 7 of this volume) propose that the notion of lived time is isomorphic with Tulving's autonoesis, while known time does not seem to have a place in the distinction between autonoetic and noetic consciousness.

Turning now to Szpunar's (2010) view of the relationships between FMTT and the semantic memory system, it is worth noting that he put into the question the notion that episodic autobiographical memory, characterized by autonoetic consciousness, virtually fully supports future episodic thought. As noted in the previous section, the raison d'être of the episodic system's evolution was hypothesized in terms of providing experienced events as the basic fabrics from which to weave future scenarios (Suddendorf & Corballis, 1997, 2007; Corballis, 2014). As rich and useful as this perspective has been for the last three decades, it left limited space to study the potential role of other memory systems in the ability to imagine specific personal episodes that may occur in the future.

A few years ago, when Szpunar (2010) asked whether episodic future thought relies exclusively on episodic memory, the answer seemed to be "no." Simulating personal future episodes makes use of other sources of information than exclusively episodic memory and, consequently, his question has been a turning point in the investigation of personal future thought. Szpunar (2010) put forward the possibility that episodic memory, as the unique source of information for generating personal future episodes, could be modulated by considering the precise context and circumstances in which the generation of personal future episodes actually takes place. The rationale was that episodic future thinking (and very likely, potential beneficial effects on future behavior) could be grounded, in general terms, in whatever memory system would be readily accessible (i.e., a system containing relevant memory representations to simulate a given personal future scenario). As the author concludes, "these claims may require a reevaluation of the prevailing conceptual framework of constructive episodic simulation" (Szpunar, 2010, 157).

The possibility of modulating episodic future thought, by alluding to the degree of availability of other systems' representations, implies, in the first place, considering the participation of the semantic memory system. More precisely, the *episodic* future instance that the individual imagines is likely to be drawn, *also*, from semantic knowledge through accessing abstracted or semanticized memories.

An indirect confirmation of Szpunar's (2010) suggestion of the participation of the semantic system in the construction of future episodic events comes from the observation that healthy subjects construct more vividly imagined future events when they involve more familiar elements, particularly events with more familiar people (besides being more detailed and more plausible) than imagined events with low ratings on such characteristics (Szpunar et al., 2013). A direct corroboration of Szpunar's proposal is suggested by the conclusions drawn by some clinical studies. Irish et al. (2012) explored the contribution of semantic memory to episodic future thinking in two neurological conditions that differed in the degree of medial temporal lobe damage, semantic dementia, and Alzheimer's disease. The authors reported that the Alzheimer's disease cohort showed similar impairment across past and future condition simulations. On the other hand, notwithstanding a relatively intact episodic retrieval for recent past events, the semantic dementia patients had significantly impaired scores in episodic future thinking. Importantly, neuroimaging structural analysis (VBM) showed that the left inferior temporal gyrus and bilateral temporal poles, known to support semantic memory, presented atrophy and correlated significantly with deficits in episodic future thinking in the semantic dementia cohort. Irish et al. (2012) claimed that simulation of future episodes "probably relies on general conceptual knowledge that can be harnessed irrespective of task demands, providing scaffolding into which specific episodic details can be integrated" (10). A further clinical study carried out by De Vito et al. (2012) in a group of Parkinson's disease patients demonstrated concomitantly impaired personal episodic future simulation and preserved autobiographical memory and non-temporal, fictitious constructive simulations. Although the authors' favored explanation for this pattern of performance strongly relied on altered executive functioning, they also considered an alternative account. Comparing personal episodic future imaging with non-temporal fictitious future simulation tasks, De Vito et al. (2012) considered a problem in processing self-awareness in subjective time affecting, obviously, only the personal future task. The authors dismissed the suggestion, arguing that their patients presented with neither time disorientation nor confabulations. However, Prebble et al.'s (2013) hypothesized relations between levels of subjective self-experience and episodic memory may account for a dissociation between past autonoetic recollections and future autonoetic imagining. By the same token, impaired personal future thought would not be invariably observed with time disorientation and/or confabulations. With regard to the association with the Me-self, as mentioned earlier, the level of functioning of self-awareness (in the phenomenological I-self) influences both phenomenological continuity and the self-concept in the Me-self. Importantly, the normal functioning of the self-concept is deemed a necessary condition for semantic continuity (Prebble et al., 2013).

In summary, by unraveling the role of semantic memory in episodic future thinking, researchers have demonstrated that the interactions between semantic and episodic memory are more complex and richer than previously thought, and they have contributed to study the role of the semantic knowledge-based Me-self in imagining temporally determined, plausible, and personal future events.

More to the point, on theoretical and clinical bases, a new perspective has been adopted in the study of the relationship between FMTT and memory systems. This perspective is grounded in the notion that FMTT is supported by the semantic system's relevant memory representations. The impact of the Szpunar's (2010) clinically corroborated proposal should be the systematic consideration of the knowledge-based Me-self in the investigation of episodic future thinking.

## 3. THE ME-SELF IN FUTURE MENTAL TIME TRAVEL

The conceptual self, as part of the Me-self, is the capacity of human beings to reflect on their own personal attributes, which constitutes the basis of the formation of a complex knowledge structure stored in long-term memory. Mental representations of this knowledge, in the form of summaries of an individual's concept on his or her own self (Kihlstrom & Klein, 1997), can be accessed without the participation of episodic memory (see discussion later in this chapter). The conceptual self was defined by Markus and Wurf (1987) as a dynamic aspect of the self, not only reflecting ongoing behavior, but also facilitating (future) behavior. More recently, Conway (2005; Conway et al., 2004a; Conway and Pleydell-Pierce, 2000) integrated the conceptual self, defined as the collection of traits, beliefs, values, physical and material characteristics, social position, and different roles that we attribute to ourselves, within the Self Memory System model (SMS). A general view of the SMS brings to light its conception as hierarchically higher and phylogenetically more recent than the episodic memory system, comprising two central components, the working self and the autobiographical memory knowledge base. The former is defined as a complex set of active goals and associated self-images, while the latter organizes autobiographical information at different levels of abstraction. This autobiographical memory knowledge base was introduced by Conway et al. (2004b) under the designation of "Long-term Self" to encompass two aspects of non-immediate knowledge of the self, the autobiographical knowledge base and the conceptual self. As stated by the authors, the introduction of the autobiographical memory knowledge base was thought to characterize the knowledge required by the working self to organize and instantiate active goal processes. In 2005, Conway modified the previous model and proposed that the autobiographical memory knowledge base contains two types of representations, autobiographical knowledge and episodic memories.

Complementarily to Tulving's major contribution about the preeminence of the episodic memory system (e.g., Tulving, 1972, 1985, 2001, 2005), Conway (e.g., 2005; 2009) viewed the function of episodic memories as that of keeping a record of progress with short-term goals. In this perspective, most episodic memories are deemed to be lost soon after their formation, with only a few of them becoming the basis of autobiographical memories.

The conceptual self is conceived as the units of abstracted knowledge "that exist independently of specific temporally defined incidents, but are connected to autobiographical knowledge and the episodic memory system to activate specific instances that exemplify, contextualize, and ground their underlying themes or concepts" (Conway et al., 2004b, 500). Both goals and conceptual self-knowledge act as control processes in the everyday regulation of memory. Importantly, the neuroanatomical underpinnings of the conceptual self have been "detected" in different studies. Among others, Kelley et al.'s (2002) investigation explored potential neural substrates of self-referential processing using event-related fMRI. They asked their participants to make judgments about trait adjectives associated either with themselves or with other people. The authors observed that a separate region of the medial prefrontal cortex (MPFC) was selectively engaged during self-referential processing. Consequently, they suggested that this processing is functionally dissociable from other forms of semantic processing, such as judging trait adjectives about others. D'Argembeau et al. (2008a) found the same MPFC region to be significantly activated in an fMRI study contrasting not only present self-relevant trait adjectives with present close friend–relevant trait adjectives, but also present self- with past self-relevant trait adjectives, in a group of college students. The authors suggested that "young adults are still in a phase of formation and consolidation of identity during the college years, and the ability to differentiate between present and past selves probably plays an important role in the creation of a stable identity" (250). Along these lines and more to the point, D'Argembeau et al. (2010) also investigated the neural bases that are specifically related to self-referential processes of goal-directed future-event simulation. In an fMRI study, their participants were asked to imagine future events that were related to their personal goals (personal future events, e.g., getting married next summer) and future events that were plausible but unrelated to their personal goals (non-personal future events, e.g., learning some new skill), and a control task (everyday activities). The authors were able to document an increased activation of the MPFC and the posterior cingulate cortex (PCC) when the participants imagined personal goal processing relative to non-personal imagined events. Moreover, the authors also documented common brain regions, namely, the MPFC and the PCC, supporting self-referential processing in the semantic (judgments about their own traits) and episodic domains (simulating personal vs. non-personal future events).

On a more theoretical perspective, the grounding aspects of the conceptual self seem to be marked in memory by highly accessible specific memories that come first to mind through self-trait processing (Conway, 2005; see also Kelley et al., 2002, or D'Argembeau et al., 2008a). Self-traits have frequently been studied by means of the "I am Test" (Kuhn & McPartland, 1954). This task is based on the notion that the self is also (and importantly) the result of our interactions with other people (our family, friends, and more generally, our social and cultural group). In that context, if a person is asked to state who he or she is, the responses could highlight different aspects of the self (physical, emotional, social, etc.), but in every case they would reflect the knowledge-based Me-self in relation to the social group.

It is worth highlighting that the conceptual self corresponds to the concept that Klein (Klein & Loftus, 1993; Klein & Gangi, 2010; Klein & Lax 2010) named "self-traits semantic knowledge summaries" or personality trait semantic representations. Importantly, these are represented separately and therefore are accessed independently from episodic memory. The high level of accessibility that characterizes these non-temporally specified self-structures is also suggested by the notion that developmentally, it is from sets of episodic memories that early non-verbal conceptual knowledge is abstracted (Conway, 2009).

Autobiographical memory, in Conway's theory (Conway, 2005; Conway & Pleydell-Pierce, 2000; Conway et al., 2004a), emerges from the tension between the self-coherence and an adaptive correspondence with reality. More precisely, the central principle of the SMS is that memory and crucial aspects of the self constitute a system characterized by its coherence. Importantly, self-concept knowledge is supported by memories of experiences that are coherent with the self and as such are more accessible than dissonant memories, which may jeopardize the beliefs about the self (Conway, 2005).

The characteristics attributed to the conceptual self that could be seen as interacting with episodic future thought are based on the fact that the conceptual self's contents are stored in the semantic system. It has been suggested (Szpunar,

2010) and demonstrated (Irish et al., 2012) that in semantic dementia, the atrophy of the anterior temporal lobe (strongly associated with the semantic system) prevents the construction of a framework to support the integration of detailed future simulations. Moreover, the conceptual self is highly accessible (Conway, 2005; Klein & Lax, 2010; Klein & Gangi, 2010). Accessibility to information stored in long-term memory containing memory representations relevant for the actual construction of a future event is a decisive factor for successful future simulation (Szpunar, 2010). Finally, in Conway's SMS model (Conway & Pleydell-Pierce, 2000; Conway, 2005), coherence is maintained between memories, conceptual knowledge, goals, and the conceptual self by controlling accessibility to autobiographical knowledge and episodic memories. By so doing, the working self "selects" goal-relevant self-defining experiences that provide constraints on what the self has been and can currently be, and more importantly, what it might become in the future.

## 4. POSSIBLE SELVES

The notion of the self-concept extending both backward and forward through time appears in James's (1890, [1892] 2012), work with the use of the term "potential social Me" (see the Introduction). Recently, several authors have suggested that the potential/possible selves likely play a more important role in guiding and motivating behavior and episodic future thought than episodic past recollections (e.g., Adamson et al., 2007; Bertsen & Bohn, 2010; D'Argembeau & Van der Linden, 2004; Markus & Nurius, 1986). But "[h]ow can something, like future, that does not exist in physical reality exert any influence on something that does, like human action? The solution is this: It is not the non-existent physical future that does the influencing but rather the fully existent mental future of the individual" (Tulving & Szpunar, 2012, 259).

Adamson et al. (2007) carried out quantitative and qualitative analyses in a group of late adolescents, aiming at the detection of the consistency or otherwise that they manifested in relation to their self-concept. Self-concept consistency was defined as the ability to be able to describe oneself in a non-contradictory manner, while self-concept inconsistency involved diffused and/or conflicted self-descriptions. Adamson et al. (2007) investigated the link between self-consistency/inconsistency and future orientation. They defined the latter as the extent to which people can imagine their futures in terms of goals, hopes, expectations, and plans. Interestingly, the content of the participant's self-reflections, such as work or further education, when envisaging the future, was generally similar independently of the consistency/inconsistency of self-concept dimension. However, the authors' evaluations went beyond the content of the individual's future simulations and could demonstrate that the quality of the adolescents' futures did vary in association with the self-concept consistency/inconsistency results. Briefly, the young people with inconsistent self-concepts described significantly more pessimism and apprehension about their future, and more confusion about who they were in comparison with subjects with a consistent self-concept. More recently, D'Argembeau et al. (2012) reported that those individuals, in their sample of young adults, whose self-esteem was higher than the rest of the group generated future events that were more positive. It is likely that higher self-esteem is more readily observed in a consistent self-concept, and in that sense D'Argembeau et al.'s finding could be viewed as corroborating Adamson et al.'s study, in individuals belonging in the next age range.

Future orientation in Adamson et al. (2007) is most closely related to the self-concept labeled "possible selves" by Markus and Nurius (1986). These latter authors viewed the self-concept as "a system of affective-cognitive structures about the self that lends structure and coherence to the individual's self-relevant experiences" (955). The possible selves are the ideal selves that we would very much like to become, also the selves we could become, and the selves we are afraid of becoming. The possible selves provide a conceptual link between cognition and motivation since their cognitive contents (hopes, fears, goals, and threats) are considered to be meaningful organizing elements capable of giving the self a direction. Through this dynamic, possible selves provide not only the bases to evaluate the present self, but also the incentives for future behavior.

With regard to their grounds, possible selves emerge from an individual's past experiences and reflect personal concerns and investment. Their influence on how information about the self is processed is ubiquitous since they determine which stimuli are selected for attention, which stimuli are remembered, and what type of inference is drawn. Important as the past experiences (and most particularly self-defining past experiences) are, the influence of future selves on who one is as a person has been deemed more important for the grounding of the individual's self than that of past experiences, as mentioned earlier. Possible selves are thought to form the self in a particularly dynamic manner because they are ideas about the self that are not as well anchored in experienced reality. This implies that possible selves comprise the self-knowledge that is the most vulnerable and responsive to changes in the environment. It follows that self-knowledge of what is possible to achieve is motivational since it frames behavior and guides its course. Since possible selves carry also associated affective states, they serve to select among future behaviors on affective bases, which are known to be paramount in decision-making (e.g., Damasio, 1999; Eslinger & Damasio, 1985).

After the publication of Markus and Nurius's (1986) study, the number of papers on possible selves increased rapidly. However, a general dissatisfaction was voiced, particularly from a methodological standpoint (e.g., Packard & Conway, 2006).

In their pioneering study, Markus and Nurius's (1986) possible selves included different mental semantic representations: personality (e.g., "I will be happy"), lifestyle (e.g., "I will travel widely"), physical (e.g., "I will be in good shape"), among others. These mental representations did not require the participants to imagine particular events relying on the episodic memory system. Recently, D'Argembeau et al. (2012) sought to investigate not only future episodic thought by asking their participants to imagine detailed events, but they also selected topics more readily meaningful for the participants. Their rationale was that, being meaningful, those mental representations would contribute more efficiently to the person's sense of self than more abstract semantic representations of possible selves. The authors introduced the construct of "self-defining future projection" (SDFPs), regarded as the future counterpart of the self-defining memories (SDM; Blagov & Singer, 2004; Lardi et al., 2010; Singer & Moffitt, 1991; Sutin & Robins, 2005). More precisely, SDFPs were conceptualized as the mental representations that may ground an individual's conception of him- or herself in the future. Relying on Conway's SMS model (Conway, 2005; Conway et al., 2004), D'Argembeau et al. (2012) suggested that SDFPs and self-conceptions may interact in a reciprocal way. To test that suggestion, they explored the potential correlations between SDMs and SDFPs in the two variables found in SDM to show variations from one individual to the other (i.e., specificity and integrative meaning). The suggestion was based on the fact that there is an individualized reflective process that manifests itself in a particular style of constructing self-defining information independently of the temporal direction. The results confirmed the authors' hypothesis since they revealed that the narrative of past and future events showed the same internal structure, particularly integrative meaning. Interestingly, the majority of SDFPs did not refer to specific events, which is interpreted by the authors as an indication of flexibility in drawing information from episodic and also semantic autobiographical systems to construct SDFP events.

Any form of possible selves could be necessary for self-continuity (Singer, 1995), in terms of the construction of a temporal chronology of our conceptual self-knowledge (Prebble et al., 2013). However, in order to have an overview of the separation of I-self and Me-self in association with possible selves, it could be useful to underscore that semanticized memories that are not only essential for semantic continuity, but actually used in future imaginings, were precisely semanticized from episodic memories grounded in phenomenological continuity and therefore based on the I-self. In other words, possible selves, most particularly episodic future thinking, require a functional I-self at the beginning of the process (see section 1).

In summary, Markus and Nurius's (1968) "possible selves" were considered to be meaningful organizing elements capable of giving the self a direction. Along these lines, Adamson et al. (2007) demonstrated an association between the quality of the future their youngsters imagined for themselves, and some characteristics of the self-concept. Moreover, D'Argembeau et al. (2012) refined the general concept of possible selves by investigating future episodic thinking and contributed with the notion of SDFP. The concept of future, possible, or potential selves has been associated with the semantic-based knowledge Me-self and more specifically to semantic continuity (see Introduction). However, a brief consideration in the present section serves as a reminder that semantic continuity is constructed on an intact I-self.

## 5. CONCLUSIONS

Among other skills, humans have acquired through evolution the ability to mentally travel in time (Suddendorf & Corballis, 1997) and the ability to reflect about themselves (D'Argembeau & Salmon, 2012). The latter was already considered by William James in 1890 as part of the Me-self, the self as known. The relationships between these human capacities have not been readily apparent. The indisputable importance of the episodic memory system in mental time travel, as well as the richness of the relationship between the phenomenological "traveler self" and mental time travel left little space to interrogate other memory systems. This began, on the one hand, with some investigations on semanticized memories (e.g., Conway, 2005) and, most particularly, self-trait semantic-knowledge summaries (e.g., Klein & Gangi, 2010; Rathbone et al., 2009), which showed that the individual's self-concept was preserved in the context of massive episodic memory impairment. On the other hand, during the last years, FMTT has been shown to possess some behavioral (e.g., D'Argembeau et al., 2010) and neuroanatomical (e.g., Addis & Schacter, 2008) distinctive characteristics, explored in the healthy individual. Moreover, some clinical work corroborated the differences between past and future simulations (e.g., De Vito et al., 2012; Irish et al., 2012). Complementarily, a few theoretical works (e.g., Klein, 2015; Conway, 2005; Szpunar, 2010) expressed the necessity of nuancing the preeminence of the episodic memory system in order to better apprehend FMTT. The ground was prepared, so to speak, to revive James's potential selves and to expand on Markus and Nurius's (1986) possible selves. Presently, it appears that the Me-self–FMTT interactions may contribute efficiently to improving our understanding of the nature of self-projections in time.

## NOTE

1. Historically, the first philosopher to reflect on a division of the self seems to be Plotinus (204/5–270 A.D.). His separation comprised the corporeal and the rational selves.

## REFERENCES


Adamson, L., Ferre-Wreder, L., & Kerpelman, J. (2007). Self-concept consistency and future orientation during the transition to adulthood. *Young*, *15*, 91–112.

Addis, D., & Schacter, D. (2008). Constructing episodic simulation: temporal distance and detail of past and future events modulate hippocampal engagement. *Hippocampus*, *18*, 227–237.

Bertsen, D., & Bohn, A. (2010). Remembering and forecasting: the relation between autobiographical memory and episodic future thinking. *Memory & Cognition*, *38*, 265–278.

Blagov, P. S., & Singer, J. (2004). Four dimensions of self-defining memories, specific meaning, content and affect, and their relationships to self-restraint, distress, and repressive defensiveness. *Journal of Personality*, *72*, 481–511.

Conway, M. (2005). Memory and the self. *Journal of Memory and Language*, *53*, 594–628.

Conway, M. (2009). Episodic memories. *Neuropsychologia*, *47*, 2305–2313.

Conway, M., Meares, K., & Standart, S. (2004a). Images and goals. *Memory*, *12*, 525–531.

Conway, M., & Pleydell-Pearce, C. (2000). The construction of autobiographical memories in the self-memory system. *Psychological Review*, *107*, 261–288.

Conway, M., Singer, J., & Tagini, A. (2004b). The self and autobiographical memory: correspondence and coherence. *Social Cognition*, *22*, 491–529.

Corballis, M. (2014). Mental time travel: how the mind escapes from the present. *Cosmology*, *18*, 139–145.

Damasio, A. *The feeling of what happens*. Reading, UK: Vintage, 1999.

D'Argembeau, A., Feyers, D., Majerus, S., Collettel, F., Van der Linden, M., Maquet, P., & Salmon, E. (2008a). Self-reflection across time: cortical midline structures differentiate between present and past selves. SCAN, 3, 244–252.

D'Argembeau, A., Lardi, C., & Van der Linden, M. (2012) Self-defining future projections: exploring the identity function of thinking about the future. *Memory, 20*, 110–120.

D'Argembeau, A., Ortoleva, C., Jumentier, S., & Van der Linden, M. (2010). Component processes underlying future thinking. *Memory & Cognition, 38*, 809–819.

D'Argembeau, A., Raffard, S., & Van der Linden, M. (2008b). Remembering the past and imagining the future in schizophrenia. *Journal of Abnormal Psychology, 117*, 247–251.

D'Argembeau, A., & Salmon, E. (2012). The neural basis of semantic and episodic forms of self-knowledge: insights from functional neuroimaging. In C. Lopez-Larrea (ed.), *Sensing in nature*. New York: Landes Bioscience and Springer Science+Business Media.

D'Argembeau, A., Stawarczyk, D., Majerus, S., Collette, F., Van der Linden, M., Feyers, D., Maquet, P., & Salmon, E. (2010). The neural basis of personal goal processing when envisioning future events. *Journal of Cognitive Neuroscience, 22*, 1701–1713. doi: 10.1162/jocn.2009.21314.

D'Argembeau, A., & Van der Linden, M. (2004). Phenomenal characteristics associated with projecting oneself back into the past and forward into the future: influence of valence and temporal distance. *Consciousness and Cognition, 13*, 844–858.

De Vito, S., Gamboz, N., Brandimonte, A., Barone, P., Amboni, M., & Della Sala, S. (2012). Future thinking in Parkinson's disease: an executive function? *Neuropsicologia, 50*, 494–1501.

Eslinger, P., & Damasio, A. (1985). Severe disturbance of higher cognition after bilateral frontal lobe ablation. *Neurology, 35*, 1731–1741. doi: 10.1212/WNL.35.12.1731.

Gallagher, S. (2000). Philosophical conceptions of the self: implications for cognitive sciences. *Trends in Cognitive Science, 4*, 14–21.

Irish, M., Addis, D., Hodges, J. & Piguet, O. (2012). Considering the role o semantic memory in episodic future thinking: evidence from semantic dementia. *Brain, 135*, 2178–2191.

James, W. (1890). *The principles of psychology*, Vol. 1. New York: Holt. doi: 10.1037/11059-000.

James, W. (1892). *Psychology: the briefer course*. New York: Holt (reprinted and edited by Gordon Allport in 1961/2012).

Kelley, M., Macrae, C. M., Wyland, C. L., Caglar, S., Inati, S., & Heatherton, T. (2002). Finding the self? an event-related fMRI study. *Journal of Cognitive Neuroscience, 14*, 785–794. doi: 10.1162/08989290260138672.

Kihlstrom, J. F., & Klein, S. (1997). Self-knowledge and self-awareness. In J. G. Snodgrass & R. L. Thompson, (eds.), *The self across psychology: self-awareness, self-recognition, and the self-concept* (Vol. 818, pp. 5–17). New York: Annals of the New York Academy of Sciences.

Klein, S. (2013). The complex act of projecting oneself into the future. *WIREs Cognitive Sciences, 4*, 63–79.

Klein, S. (2014). Sameness and the self: philosophical and psychological considerations. *Frontiers in Psychology, 5*, 29. http://dx.doi.org/10.3389/fpsyg.2014.00029.

Klein, S. (2015). Autonoetic consciousness: reconsidering the role of episodic memory in future-oriented self-projections. *The Quarterly Journal of Experimental Psychology*. Doi:10.1080/17470218.2015.1007150

Klein, S., Altinyazar, V., & Metzl, M. (2013b). Facets of self in schizophrenia: the reliability and accuracy of trait self-knowledge. *Clinical Psychological Science*,

Klein, S., & Gangi, C. E. (2010). The multiplicity of self: neuropsychological evidence and its implications for the self as a construct in psychological research. *Annals of the New York Academy of Sciences, 1191*, 1–15. doi: 10.1111/j.1749-6632.2010.05441.x.

Klein, S., & Lax, M. (2010). The unanticipated resilience of trait self-knowledge in the face of neural damage. *Memory, 18*, 918–948.

Klein, S., & Loftus, J. (1993). Behavioral experience and trait judgments about the self. *Personality and Social Psychology Bulletin, 19*, 740–745. doi: 10.1177/0146167293196009.

Klein, S., Loftus, J., & Kihlstrom, J. F. (2002). Memory and temporal experience: the effects of episodic memory loss on an amnesic patient's ability to remember the past and imagine the future. *Social Cognition, 20*, 353–379.

Kuhn, M., & McPartland, T. (1954). An empirical investigation of self-attitudes. *American Sociological Review*, 19, 68.

Lardi, C., D'Argembeau, A., Chanal, J., Ghisletta, P., & Van der Linden, M. (2010). Further characterization of self-defining memories in young adults: a study of a Swiss sample. *Memory, 18*, 293–309.

Manning, L., Denkova, E., & Unterberger, L. (2013). Autobiographical significance in past and future public semantic memory: a case study. *Cortex, 49*, 2007–2020.

Markus, H., & Nurius, P. (1986). Possible selves. *American Psychologist, 41*, 954–969.

Markus, H., & Wurf, E. (1987). The dynamic self-concept: a social psychological perspective. *Annual Review of Psychology, 38*(1), 299–337. doi: 10.1146/annurev.ps.38.020187.001503

Packard, B., & Conway, M. (2006). Methodological choice and its consequences for possible selves research. *Identity: An International Journal of Theory and Research, 6*, 251–271.

Prebble, S., Addis, R., & Tippett, L. (2013). Autobiographical memory and sense of self. *Psychological Bulletin*, 139, 815–840.

Rathbone, C., Moulin, C., & Conway, M. (2009). Autobiographical memory and amnesia: using conceptual knowledge to ground the self. *Neurocase, 15*, 405–418.

Sass, L., & Parnas, J. (2003). Schizophrenia, consciousness and the self. *Schizophrenia Bulletin, 29*, 427–444.

Singer, J. A. (1995). Seeing one's self: locating narrative memory in a framework of personality. *Journal of Personality, 63*(3), 429–457.

Singer, J. A., & Moffitt, K. (1991–1992). An experimental investigation of generality and specificity in memory narratives. *Imagination, Cognition, and Personality, 10*, 235–258.

Suddendorf, T., & Corballis, M. (1997). Mental time travel and the evolution of the human mind. *Genetic Social and General Psychology Mongraphs, 123*, 133–167.

Suddendorf, T., & Corballis, M. (2007). The evolution of foresight: what is mental time travel, and is it unique to humans? *Behavioral and Brain Sciences, 30*, 299–351.

Sutin, A. R., & Robins, R. W. (2005). Continuity and correlates of emotion and motives in self-defining memories. *Journal of Personality, 73*, 793–824.

Szpunar, K. (2010). Episodic future thought: an emerging concept. *Perspectives on Psychological Science, 5*(2), 142–162.

Szpunar, K. (2011). On subjective time. *Cortex*, 47, 209–211.

Szpunar, K., Addis, D., McLelland, V. (2013). Memories of the future: new insights into the adaptive value of episodic memory. *Frontiers in Behavioral Neuroscience.* doi: 10.3389/fnbeh.2013.00047.

Tulving, E. (1972). Episodic and semantic memory. In E. Tulving & W. Donaldson (eds.), *Organisation of memory* (pp. 381–403). New York: Academic Press.

Tulving, E. (1983). *Elements of episodic memory.* Oxford: Clarendon.

Tulving E. (1985). Memory and consciousness. *Canadian Psychologist*, 26, 1–12.

Tulving, E. (2001). Episodic memory and common sense: how far apart? *Philosophial Transactions of the Royal Society London B*, 356, 1505–1515.

Tulving, E. (2002). Episodic memory: from mind to brain. *American Review of Psychology*, 53, 1–25.

Tulving, E. (2005). Episodic memory and autonoesis: uniquely human? In H. Terrace & J. Metcalfe (eds.), *The missing link in cognition: self-knowing consciousness in man an animal* (pp. 3–56). New York: Oxford University Press.

Tulving, E., & Szpunar, K. (2012). Does the future exist? In B. Levine & F. Craik (eds.), *Mind and the frontal lobes: cognition, behavior and brain imaging* (pp. 248–263). New York: Oxford University Press.

Wheeler, M., Stuss, D., & Tulving, E. (1997). Toward a theory of episodic memory: the frontal lobes and autonoetic consciousness. *Psychological Bulletin*, *121*, 331–354.

10

# The Role of Personal Goals in Future-Oriented Mental Time Travel

ARNAUD D'ARGEMBEAU ■

All we ever know directly is the immediate present, and yet our mental life is filled with psychological time. Indeed, all sorts of memories, thoughts, and mental images that do not reflect current sensory input populate our minds, many of which are subjectively felt as referring to events that happened in our personal past, or that might occur in our personal future. Important progress has recently been made in understanding how such event representations are constructed and elaborated (Schacter et al., 2012), and research in fact suggests that much of the same representational systems and mental processes are involved in imagining different sorts of events or scenes, irrespective of whether or not the represented situation is located in time (Hassabis & Maguire, 2007). If this is indeed the case, then what gives us the subjective feeling that an event currently represented in our minds refers to our personal past or future? This is a perplexing question that has been much discussed and investigated, especially with respect to the recollection of past events (e.g., Johnson, 2006; Klein, 2014; Tulving, 2002).

The aim of this chapter is to address this same question as it relates to our sense of the future: What gives us the subjective feeling that a currently imagined event refers to something that might happen in our personal future, rather than, say, a mere fantasy? My main contention is that, in order for an event representation to be felt as truly belonging to one's future, it has to be meaningfully linked to autobiographical knowledge—to personal goals and general expectations that one has about oneself and one's life. This idea is developed in the first section of the chapter, in which it is argued that imagining personal future events not only involves scene or event construction, but also autobiographical knowledge that contextualizes events within the individual's life story. A prominent form of such contextualizing knowledge is personal goals, and an important function of future-oriented mental time travel may indeed be to allow the detailed representation or simulation of future possibilities in the service of effective goal pursuit. Support for this view is presented in the ensuing sections, with recent findings showing that personal goals influence the construction and organization of future-oriented thoughts, and contribute to the subjective sense of pre-experiencing one's personal future.

## 1. CONTRIBUTION OF AUTOBIOGRAPHICAL KNOWLEDGE TO EPISODIC FUTURE THINKING

Prospection, or future-oriented thinking, encompasses a variety of different forms of cognitions about what might happen in the future, which can relate to the self, others, and/or the world (Szpunar, Spreng, & Schacter, 2014). Here I focus on future-oriented cognitions that are autobiographical in nature, that is, thoughts and mental images that are dealing with an individual's own personal future. One particular type of such autobiographical prospection—often referred to as *episodic future thought*—has attracted much attention in the past few years and refers to the mental simulation of unique (one-time) events that might happen in one's personal future (Atance & O'Neill, 2001; Schacter, Addis, & Buckner, 2008; Suddendorf & Corballis, 2007; Szpunar, 2010). Although not without flaws (Gilbert & Wilson, 2007), episodic future thinking allows us to mentally "pre-experience" possible events before they happen: we can imagine what it would be like to be in a particular future situation, for example, by picturing its setting, characters, and action in our mind's eye (D'Argembeau & Van der Linden, 2004).

In the past few years, substantial evidence has accumulated showing that episodic future thinking involves the retrieval and integration of relevant informational components (e.g., representations of locations, objects, people, actions, and emotions) to construct mental scenes or event simulations (Schacter et al., 2012). These pieces of information composing event representations likely come from various sources, with some of them being extracted from single past experiences stored in episodic memory (Schacter & Addis, 2007; Suddendorf & Corballis, 2007), whereas others are based on semantic knowledge or schemas that have been formed over multiple past experiences (Anderson, 2012; Klein, 2013b; Rubin, 2014; Szpunar, 2010). The role of both episodic and semantic memory in the construction of episodic future thoughts is indeed supported by extensive empirical data (for reviews, see Irish & Piguet, 2013; Schacter et al., 2012).

The main argument developed in this chapter is that this process of scene construction or event simulation is not sufficient in order for an imagined event to be felt as future oriented. An additional ingredient is that the imagined event has to be owned by the individual, such that the represented content is subjectively perceived as referring to something that might happen in his or her personal future (see Klein, 2013a, for discussion of the role of personal ownership in the recollection of past events). Here I suggest that one important (though by no means the only) factor that contributes to this sense of ownership is the link between the imagined event and general knowledge about the self and one's life. On this view, the connection with autobiographical knowledge is part of what makes someone feel that an imagined event truly belongs to his or her future life.

What is the nature of this autobiographical knowledge that would contribute to the owning of imagined events? It likely consists of a variety of beliefs, expectations, intentions, hopes, and fears concerning our future selves and our future lives.

In the same way that knowledge about past selves is integral to autobiographical memories (Conway, 2001, 2005; Fivush, 2011), beliefs and expectations about our personal future may influence and constrain what particular events we envision. Such future-oriented autobiographical knowledge may consist of abstract representations of possible selves (e.g., personality traits, preferences, social roles, professional occupation, and lifestyle; Markus & Nurius, 1986) and future life periods (e.g., "when I'll have children" or "when I'll own my own business"), as well as representations of future events that are not specific in nature (often referred to as "general events"), including summary representations of extended events (e.g., going on vacation in Italy next summer) and categories of events (e.g., family parties) that could happen in our lives (Anderson & Dewhurst, 2009; D'Argembeau & Mathy, 2011). There is indeed evidence that people frequently access such general representations of their personal future in their daily lives (D'Argembeau, Renaud, & Van der Linden, 2011). This future-oriented autobiographical knowledge may in part consist of cultural life scripts (i.e., shared knowledge about a series of events that represents a prototypical life course in a given culture, such as graduation, marriage, and children; Berntsen & Bohn, 2010) or may be more idiosyncratic (i.e., based on personal experiences or interest).

My contention is that an event will be felt as belonging to one's personal future (and thus will be associated with a sense of mental time travel) only when it is linked with such general autobiographical knowledge. On this view, a mental scene such as "walking in a sunny garden next year" (cf. de Vito, Gamboz, & Brandimonte, 2012) will be subjectively experienced as a "real" future event only if it can be meaningfully related to general knowledge and expectations about one's future life—for example, imagining walking in the gardens of the palace of Versailles that one intends to visit during a holiday in France planned next summer. Absent this connection with autobiographical knowledge, an imagined event cannot be meaningfully experienced as a future happening.

A further proposal is that knowledge about personal goals is particularly important for this contextualization of episodic future thoughts (see also Conway, 2005, for further discussion of the role of goals in memory for past events). Goals are cognitive representations of desired states or outcomes (Austin & Vancouver, 1996), and personal goals may be defined as personally important objectives that individuals pursue in their daily lives (Emmons, 1986; Klinger, 2013; Little, 1983; McAdams, 2013). These goals focus on either a positive or negative possibility, and serve a motivational function in guiding the individual toward potential positive outcomes or away from potential negative ones (Elliot & Friedman, 2007). One important dimension along which personal goals vary is their level of abstractness or generality (Austin & Vancouver, 1996): people pursue some goals that are very concrete and specific (e.g., striving to arrive at a particular meeting on time), whereas other goals are abstract and general (e.g., being a good or successful person). These different goal representations are thought to be organized in a hierarchy in which higher-order goals (e.g., becoming an artist) determine the content of lower-order goals (e.g., taking painting lessons). Within this framework, episodic future thinking may be conceived as allowing the detailed representation of specific events, plans, and outcomes that are related to higher-order personal goals. In other words, an important function of episodic future thinking may be to provide a detailed (quasi-experiential) representation or simulation of what it would be like to be in a desired end-state, and to mentally try out various steps and envisage potential obstacles in achieving this state.[1] According to this view, personal goals drive and constrain future event representations, and recent findings indeed suggest that goal-related knowledge plays an important role in the construction and organization of episodic future thoughts.[2]

## 2. ROLE OF PERSONAL GOALS IN THE CONSTRUCTION OF EPISODIC FUTURE THOUGHTS

According to the view presented in the preceding section, the construction of future event representations is not random, but is instead guided by more general autobiographical knowledge, including knowledge about personal goals. If this hypothesis is correct, one would expect that, when attempting to generate episodic future thoughts, people would frequently activate general knowledge about their personal future in order to imagine specific events. D'Argembeau and Mathy (2011) tested this hypothesis using a think-aloud procedure (Fox, Ericsson, & Best, 2011) in which participants verbalized the content of their thought flow (i.e., reported whatever came to mind) while they attempted to imagine a series of specific events that might happen in their personal future. The results showed that in 84% of trials, participants reported abstract autobiographical representations (i.e., personal semantic information and/or general events) before describing a specific event (see Table 10.1 for an example of protocol obtained in that study), thus providing initial support for the idea that the activation of general autobiographical knowledge plays an important role in the construction of episodic future thoughts (D'Argembeau & Mathy, 2011, Study 1).

Next, D'Argembeau and Mathy (2011) sought to further investigate the nature of this general knowledge and, more specifically, to test the hypothesis that knowledge about personal goals plays an important role in envisioning possible future events. If representations of future events are indeed driven (at least in part) by personal goals, then possible future events should be more easily generated in response to cues referring to personal goals compared with other classes of personal semantic information (e.g., familiar persons and locations). This hypothesis was tested using a future-event fluency task (MacLeod & Byrne, 1996). Participants were asked to generate as many future events as possible for 60 seconds in relation to a series of personal goals, familiar persons, or familiar locations, which were previously obtained using a questionnaire (see Table 10.1 for examples of reported personal goals). It was found that the number of produced future events was higher in the personal goal condition than in the person and location conditions (D'Argembeau & Mathy, 2011, Study 2).

While this latter finding shows that future events are more easily generated in response to personal goals, participants were not explicitly instructed to imagine specific episodes (i.e., discrete events that take place in a specific place at a specific time), and they probably seldom did so because they had to generate as many events as possible under time constraints. Therefore, another study was conducted to further test the hypothesis that knowledge about personal goals facilitates the construction of specific event representations. According to this hypothesis, when attempting to envision specific events that might happen in their future, people

*Table 10.1.* Examples of Responses Produced in Studies Showing the Role of Personal Goals in Episodic Future Thinking

**Example of think-aloud protocol produced by a participant in D'Argembeau and Mathy (2011, Study 1).**

**Task:** The participant had to imagine a specific future event in response to the cue-word "friend" and was instructed to say aloud everything that came to mind during the imagination process.

**Protocol:** *"Friend . . . um, well that makes me think about some of my friends, and Gilles in particular . . . um, Gilles went to Canada and would like to go there again and I plan to go with him, to visit North America. I would really like to go, it could happen in three or four years, I need to save some money first. We would start on the east coast of Canada and then would go to the east coast of the US, we would rent a car and we have friends there to visit . . . um, it remains to be planned in detail but as I said we have friends there, my friend Danielle lives there on the east coast, near New York, and I'd really like to visit her one day during our trip. I see myself arriving in front of her place, I imagine hugging her after all this time we haven't seen each other."*

In this example, the participant starts accessing general knowledge about an extended event (a trip to North America) and then imagines a specific situation that might happen during this extended event (visiting a friend).

**Examples of personal goals provided by participants in D'Argembeau and Mathy (2011, Studies 2 & 3) and Lehner and D'Argembeau (2015)**

Having a child; moving in with Mary; working in a consulting company; graduating from college; going in vacation in Egypt; publishing my first article; buying new clothes; getting my driving license; creating a music band; going to grad school; losing weight.

**Examples of clustered and non-clustered events in D'Argembeau and Demblon (2012)**

*Clustered events*:

Event 1: a reception planned with all my friends
Event 2: getting my hair done just before the reception

*Non-clustered events*:

Event 1: the party planned for a friend's birthday
Event 2: going to a club for another friend's birthday

In the first example, the two events are part of the same event sequence (i.e., going to a reception), whereas in the second example the two events involve the same type of activity (i.e., going to a party) but do not belong to a same event sequence.

**Examples of goal-related and goal-unrelated events used in D'Argembeau et al. (2010b)**

*Goal-related event*: buying a present for Clara in three months
*Goal-unrelated event*: buying a clock in a secondhand market in three months

In that study, the descriptions of both kinds of events were personalized (i.e., based on an individualized pre-scan interview) and were matched for temporal distance.

would first access abstract knowledge about their personal goals, and then progressively specify goal features (e.g., consider the most likely place and time, the persons who would be present, the actions that might be possible, and so on) in order to construct a specific event representation. For example, a person might first consider the fact that she plans to go to India next summer (a personal goal) and then construct a specific experience that might happen in relation to that goal (e.g., she might imagine a specific place she would like to visit during her trip). If this hypothesis is correct, then increasing the accessibility of goal-related knowledge should facilitate the construction of specific future events.

To test this hypothesis, D'Argembeau and Mathy (2011, Study 3) asked participants to provide information about their personal goals (see Table 10.1 for examples of goals) and other personal features (i.e., personally known individuals and familiar locations), which was used to create cues for a future-event construction task that took place one week later. On each trial of this task, participants were asked to imagine a specific future event in relation to one of the goals, persons, and locations they previously reported, and they were also instructed to verbalize the content of their thought flow during the construction process. In line with the previously mentioned findings, the results showed that, on most trials, participants accessed general personal knowledge before producing a specific event. Interestingly, however, it was found that participants directly produced a specific event more frequently when they were cued with their personal goals (35% of trials) compared to other classes of personal information (27% and 18% of trials for person and location cues, respectively). Furthermore, even when participants recruited generative processes to construct specific future events (i.e., in the absence of direct access to a particular episode), they generated specific events more easily when cued with personal goals compared to familiar persons and locations. These findings suggest that activating knowledge about personal goals may be a common first step in episodic future thinking, which is used to guide the construction of relevant events.

In all the previously mentioned studies, episodic future thoughts were intentionally generated by the participants (in response to task instructions), so it could be that the influence of personal goals on the construction of episodic future thoughts is restricted to voluntary cognition. This does not seem to be the case, however. Indeed, a recent study has shown that involuntary future-oriented thoughts (i.e., thoughts that occur with no preceding attempt to generate them) are frequently related to personal goals (Cole & Berntsen, 2016). Furthermore, there is evidence that mind-wandering episodes (i.e., the spontaneous occurrence of thoughts that are unrelated to one's immediate circumstances; Smallwood & Schooler, 2006) often involve goal-related future thoughts (e.g., Baird, Smallwood, & Schooler, 2011; Stawarczyk, Cassol, & D'Argembeau, 2013). Indeed, it has been suggested for a long time that the thematic content of mind wandering and daydreams is determined by the individual's goals (see Klinger, 2013).

Once possible future events have been constructed (voluntarily or involuntarily), they may be retained as "memories of the future"—that is, memories of previous future event simulations (Ingvar, 1985; Szpunar, Addis, McLelland, & Schacter, 2013). Recent observations suggest that such memories of the future are often rapidly and directly accessed (i.e., with no search effort) in response to relevant cues. In a series of studies, Jeunehomme and D'Argembeau (2016) asked participants to produce episodic future thoughts in response to cue words. For each trial, response time was recorded and participants were asked to determine whether the future event directly came to their mind in response to the cue, or whether some active search was necessary for producing the event (see also Uzer, Lee, & Brown, 2012). It was found that many episodic future thoughts (around 60%, on average) were rapidly and directly produced. Interestingly, most of these thoughts had already been contemplated on at least one previous occasion and thus consisted of memories of the future, rather than newly imagined future events. Furthermore, the probability of direct access increased with the personal significance assigned to the future thought. This suggests that previously formed future thoughts may be particularly accessible when they are goal-relevant (e.g., anticipated events, plans, and outcomes that are conducive or obstructive to reaching personal goals). The rapid access to such representations may help guide decisions and behavior more efficiently and effectively, thus contributing to successful goal pursuit.

## 3. ROLE OF PERSONAL GOALS IN THE ORGANIZATION OF EPISODIC FUTURE THOUGHTS

Other studies suggest that many future events are not represented in isolation from one another and that autobiographical knowledge (in particular, knowledge about personal goals) contributes to linking and organizing episodic future thoughts in higher-order themes and sequences. To investigate the organizational structure of episodic future thought, D'Argembeau and Demblon (2012) used an adaptation of an event-cueing paradigm originally devised by Brown and Schopflocher (1998). Participants first imagined a series of future events, and each of these events (referred to as the cueing events) was used to cue the imagination of another related future event (referred to as the cued event). Then, participants looked back at each pair of events to answer questions about various ways in which the cueing and cued events could be related to each other (i.e., the two events could involve the same persons, the same location, and/or the same activity; one event could cause the other; one event could be included in the other; both events could be part of a single broader event). The results of two studies showed that the cueing and cued events were frequently embedded in an *event cluster*, meaning that they were causally and/or thematically related to each other (see Table 10.1 for examples of clustered versus non-clustered events). This finding suggests that episodic future thinking involves higher-order knowledge structures that link and organize imagined events in broader themes and causal sequences.

These studies further suggest that goal-related knowledge plays an important role in linking imagined events to each other and in organizing them in overarching themes and causal sequences. Indeed, it was found that the frequency of event clusters in episodic future thought increased with the personal importance attributed to the cueing events (D'Argembeau & Demblon, 2012). If one assumes that important events are mostly events that are relevant to personal goals, this suggests that goal-related knowledge helps structure imagined events in broader event sequences. More direct evidence for the organizational role of personal goals comes from an unpublished study in which we manipulated the kind of cues provided to initiate the imagination of future events: we found that future events that were produced in response to goal-related cues were more often part of an event cluster than future events that were elicited by cues referring to other kinds of personal information (i.e., known individuals and familiar locations). Finally, other findings suggest that personal goals might be particularly important for organizing events that are expected to happen in the distant future, whereas other principles (such as chronological order) are also at play for organizing near future events (Demblon & D'Argembeau, 2014).

## 4. NEURAL CORRELATES OF PERSONAL GOAL PROCESSING IN EPISODIC FUTURE THOUGHT

The evidence reviewed in the previous sections suggests that personal goal processing is often an integral part of episodic future thinking. Neuroimaging studies provide further support for this idea. Recent studies have shown that episodic future thinking is underpinned by a core network of brain regions that comprises medial prefrontal, posterior cingulate, and restrosplenial cortices, medial and lateral temporal regions, posterior inferior parietal lobules, and parts of the lateral prefrontal cortex (for a review, see Schacter et al., 2012). The extensiveness of this network fits with the view that projecting oneself into future scenarios is a complex mental ability that depends on an ensemble of various cognitive processes (D'Argembeau, Ortoleva, Jumentier, & Van der Linden, 2010a; Klein, 2013b; Schacter et al., 2012). To the extent that personal goal processing is an important component of episodic future thinking, it should be associated with specific activations within this core brain network.

To investigate brain regions supporting goal processing during episodic future thinking, D'Argembeau et al. (2010b) asked participants to imagine future events that were related to their personal goals and future events that were plausible but unrelated to their goals (see Table 10.1 for examples of goal-related versus goal-unrelated future events). When contrasting these two conditions, it was found that midline brain regions corresponding to key nodes of the core network (i.e., medial prefrontal and posterior cingulate cortices) were more activated when participants were thinking about goal-related than goal-unrelated future events. This suggests that cortical midline structures may support the processing of personal goals during episodic future thinking, an interpretation that fits with other studies of personal goal processing (e.g., Johnson et al., 2006). Spreng et al. (2010) have further shown that imagining taking various steps and actions to achieve personal goals depends on the functional coupling of the core network with the frontoparietal control network, a system supporting cognitive control (Vincent, Kahn, Snyder, Raichle, & Buckner, 2008). Thus, the frontoparietal control network may also contribute to personal goal processing, perhaps by monitoring and integrating future-oriented thoughts in coherent sequences to achieve imagined end-states. A recent study has further clarified the role of distinct brain areas in personal goal processing by showing that imagining what it would be like to achieve a given goal mainly involves the medial prefrontal cortex, whereas constructing a detailed plan of how to reach this goal engages regions from both the core network and the frontoparietal control network (Gerlach, Spreng, Madore, & Schacter, 2014).

With the exception of the few studies described in the preceding paragraph, most neuroimaging studies of episodic future thinking did not explicitly investigate the contribution of goal-related processes. Yet, if goal processing is an integral part of episodic future thinking, it is likely that goal-related representations were often spontaneously activated, and some of the detected activations may thus reflect such goal-related processes. We recently tested this hypothesis by performing meta-analyses of neuroimaging studies of episodic future thinking, on the one hand, and of personal goal processing, on the other hand, to determine the commonalities and differences in the brain regions associated with these two areas of research (Stawarczyk & D'Argembeau, 2015). If personal goals are a key component of episodic future thinking, then substantial overlap should be observed between these two domains in terms of associated brain activations. A conjunction analysis revealed that episodic future thinking and personal goal processing were indeed associated with overlapping activation in several regions of the core network and, most notably, in the medial prefrontal cortex (mPFC). This finding fits well with our previous study in which we explicitly manipulated goal processing (D'Argembeau et al., 2010b) and provides further support for the view that the mPFC plays a key role in the processing of personal goals and that such processing is an important component of episodic future thinking. An intriguing possibility is that the mPFC might mediate the integration of imagined experiences with higher-order autobiographical knowledge structures (including personal goals), thus contextualizing events within the individual's life story. This proposal remains speculative, however, and awaits further investigation.

## 5. PERSONAL GOALS AND AUTONOETIC CONSCIOUSNESS IN EPISODIC FUTURE THOUGHT

Having reviewed the evidence that personal goal processing is often an integral part of episodic future thinking, I now turn to the central question addressed in this chapter: the idea that to be owned by the individual, and thus subjectively felt as belonging to his or her future life, an imagined event has to be meaningfully linked to autobiographical knowledge and, in particular, to personal goals.

The subjective sense of re-experiencing past events is a hallmark of autobiographical remembering that has been much discussed in the literature (e.g., Brewer, 1996; Conway, 2001; Klein, 2014; Rubin, Schrauf, & Greenberg, 2003; Tulving, 2002). Tulving (1985, 2002; Wheeler, Stuss, & Tulving, 1997) put forth the idea that recollective experience is a manifestation of autonoetic consciousness, which allows one to mentally represent and become aware of subjective experiences in the past, present, and future (referred to as mental time travel). This sense of recollection may not be an intrinsic property of memory content but instead the manner in which that content is experienced at the time of retrieval—through temporal subjectivity (Klein, 2013a). In line with this view, there is evidence suggesting that memory contents and autonoetic consciousness are indeed dissociable (Klein, 2013a; Moulin, Conway, Thompson, James, & Jones, 2005).

While the notion of autonoetic consciousness has been much less discussed and investigated in relation to future-oriented thinking (but see Dalla Barba & Boissé, 2010; Klein, 2016; Suddendorf & Corballis, 2007; Szpunar, 2011; Tulving, 2005), phenomenological ratings suggest that people can indeed have the subjective feeling of traveling into their personal future to "pre-experience" possible happenings (Arnold, McDermott, & Szpunar, 2011; D'Argembeau & Van der Linden, 2004, 2006; Szpunar & McDermott, 2008). What shapes this sense of pre-experiencing is not yet fully understood, but it is likely multidetermined. Recently, D'Argembeau and Van der Linden (2012) investigated the specific contribution of various event properties by assessing the degree to which naturally occurring variations in autonoetic consciousness are predicted by different event features and sensory-perceptual qualities of episodic future thoughts. Participants imagined a series of future events and assessed the phenomenological characteristics of their future thoughts using rating scales. These scales measured the sensory-perceptual qualities of future thoughts (i.e., overall vividness, amount of visual details, amount of other sensory details, clarity of imagined location, clarity of imagined persons and objects, and visual perspective), autonoetic consciousness (i.e., the feeling of pre-experiencing the event, and the sense of mentally traveling to the time when the event would happen), and various event properties (i.e., temporal distance, familiarity of the imagined location, familiarity of the imagined persons and objects, importance of the event with regard to personal goals, affective valence, and amount of previous thought about the event). Multilevel regression analyses were then used to investigate what variables predicted variation in ratings of autonoetic consciousness among episodic future thoughts.

It was found that autonoetic consciousness was significantly related to the sensory-perceptual qualities of episodic future thoughts, as well as multiple event properties, such as the familiarity of constitutive elements, the event's importance with regard to personal goals, and frequency of previous thought. To examine the independent contribution of these variables, D'Argembeau and Van der Linden (2012) introduced vividness and visual perspective first in a multilevel regression model (considering that autonoetic consciousness can be conceived as a meta-cognitive judgment based on sensory-perceptual qualities; Rubin et al., 2003), and then investigated whether other event properties were significant predictors of autonoetic consciousness after sensory-perceptual qualities had been taken into account. The simultaneous introduction of vividness and visual perspective into the model indicated that they independently contributed to autonoetic consciousness: ratings of autonoetic consciousness increased with increasing vividness of episodic future thoughts and decreased when participants adopted an observer perspective while imagining the event. These two variables accounted for 29% of the within-participants variance in autonoetic consciousness. Adding location familiarity and familiarity of persons/objects to the model did not result in a significantly better fit. On the other hand, adding ratings of the event's importance with regard to personal goals after vividness and visual perspective had already been included in the model resulted in a significantly better fit, showing that autonoetic consciousness increased when the imagined future events were more related to personal goals. Adding temporal distance and frequency of previous thought to the model did not result in a significantly better fit. Overall, then, the best and most parsimonious model was to use vividness, visual perspective, and goal relevance to predict autonoetic consciousness. This model accounted for 42% of the within-participants variance (i.e., variation among future thoughts) in autonoetic consciousness.

In another recent study, we experimentally manipulated the goal relevance of events to further investigate its contribution to autonoetic consciousness, over and above the process of constructing a complex mental scene (Lehner & D'Argembeau, 2015). Some observations suggest that certain aspects of the subjective experience associated with episodic simulation are not affected by whether or not the imagined events are located in time. In particular, de Vito et al. (2012) found that asking people to construct an "atemporal scene" (e.g., imaging an event such as "walking in a sunny garden" without locating it in time) and imagining the same scene as happening in the future (e.g., "imagine walking in a sunny garden next year") did not lead to any difference in ratings of sensorial details and feeling of experiencing the event. However, some of the future events constructed in that study may not be truly autobiographical because participants may not necessarily contextualize the proposed events within their own lives (e.g., someone might construct a scene of "walking in a sunny garden" without linking it to general expectations of what would happen in her life in a year from now). Thus, we aimed to replicate de Vito et al. (2012)'s findings while adding another condition in which the proposed future events would be more related to the individual's personal goals. If the central argument of the present chapter is correct, autonoetic consciousness should be enhanced for this latter kind of future events.

To test this hypothesis, we asked participants to imagine three types of events: atemporal scenes provided by the experimenter (e.g., "imagine walking in a sunny garden"), future events provided by the experimenter (e.g., "imagine walking in a sunny garden next year"), and future events related to personal goals. For this latter condition, participants were asked, one week before the experiment, to list a series of personal goals (see Table 10.1 for examples of reported goals) and their responses were used to construct cues for the episodic future-thinking task. Participants had to imagine two events of each kind, in as much detail as possible, and for each event they rated their subjective experience on various scales assessing the amount of sensorial and contextual information, as well as autonoetic consciousness.

The results replicated de Vito et al. (2012)'s findings that atemporal scenes and future events did not differ in terms of associated amount of sensorial details and feelings of experiencing the situation, when these events were generated in response to cues provided by the experimenter. More important, however, when participants imagined future events that were related to their personal goals, they felt they were experiencing the situation and were visiting their personal future to a greater extent than when they imagined future events that were cued by the experimenter. On the other hand, these two kinds of events did not differ in terms of associated amount of sensorial and contextual details. These results thus show that autonoetic consciousness depends not only on the sensory-perceptual qualities of episodic future thoughts, but also on the importance of envisioned events with regard to personal goals.

Taken together, these two studies indicate that scene construction and personal significance both contribute to the subjective experience of pre-experiencing one's personal future. This does not necessarily mean that autonoetic consciousness is determined or caused by these factors (see Klein, 2013a, for further discussion of the idea that autonoetic consciousness does not depend on memory content) but, at the very least, these data demonstrate that the probability or degree to which autonoetic consciousness is applied to a representational content varies according to some event features, notably its relevance to personal goals.

## 6. CONCLUSIONS

The subjective sense of mentally visiting one's personal future remains an elusive phenomenon that is probably determined by multiple factors. In this chapter, I have proposed that this feeling requires not only event simulation processes, but also the contextualization of imagined situations within higher-order knowledge about the self and one's life. Knowledge about personal goals might play a particularly important role in this respect, one of the main functions of episodic future thoughts being to allow the representation of specific events that incarnate personal goals and the construction of scenarios that implement efficient ways to achieve these goals. In line with this view, recent behavioral and neuroimaging research suggests that personal goal processing is an integral part of episodic future thinking that contributes to the construction and organization of imagined events. Furthermore, the subjective sense of pre-experiencing one's personal future is enhanced when envisioning events that are related to personal goals. Forging meaningful links between imagined scenarios and personal goals may be part of what gives us the subjective feeling that an event currently represented in our minds belongs to our personal future.

## ACKNOWLEDGMENTS

Arnaud D'Argembeau is supported by the Fonds de la Recherche Scientifique-FNRS, Belgium.

## NOTES

1. Although it is beyond the scope of this chapter to develop this point, it is worth noting that this simulation of possible futures might actually increase the likelihood that goal-related actions are carried out (see, e.g., Altgassen et al., 2015; Gollwitzer, 1999).
2. The main argument defended in this chapter is that higher-order goals play an important role in driving and contextualizing episodic future thoughts, but it should be noted that the relationship between goals and episodic future thinking is likely bi-directional; indeed, goals may in part be shaped and refined by the contemplation of specific future possibilities.

## REFERENCES

Altgassen, M., Rendell, P. G., Bernhard, A., Henry, J. D., Bailey, P. E., Phillips, L. H., & Kliegel, M. (2015). Future thinking improves prospective memory performance and plan enactment in older adults. *The Quarterly Journal of Experimental Psychology, 68*(1), 192–204.

Anderson, R. J. (2012). Imagining novel futures: the roles of event plausibility and familiarity. *Memory, 20*(5), 443–451.

Anderson, R. J., & Dewhurst, S. A. (2009). Remembering the past and imagining the future: differences in event specificity of spontaneously generated thought. *Memory, 17*(4), 367–373.

Arnold, K. M., McDermott, K. B., & Szpunar, K. K. (2011). Imagining the near and far future: the role of location familiarity. *Memory & Cognition, 39*(6), 954–967.

Atance, C. M., & O'Neill, D. K. (2001). Episodic future thinking. *Trends in Cognitive Sciences, 5*(12), 533–539.

Austin, J. T., & Vancouver, J. B. (1996). Goal constructs in psychology: structure, process, and content. *Psychological Bulletin, 120*, 338–375.

Baird, B., Smallwood, J., & Schooler, J. W. (2011). Back to the future: autobiographical planning and the functionality of mind-wandering. *Consciousness and Cognition, 20*(4), 1604–1611.

Brewer, W. F. (1996). What is recollective memory? In D. C. Rubin (ed.), *Remembering our past: studies in autobiographical memory* (pp. 19–66). Cambridge, UK: Cambridge University Press.

Brown, N. R., & Schopflocher, D. (1998). Event clusters: an organization of personal events in autobiographical memory. *Psychological Science, 9*(6), 470–475.

Cole, S. N., & Berntsen, D. (2016). Do future thoughts reflect personal goals? Current concerns and mental time travel into the past and future. *The Quarterly Journal of Experimental Psychology, 69*(2), 273–284.

Conway, M. A. (2001). Sensory-perceptual episodic memory and its context: autobiographical memory. *Philosophical Transactions of the Royal Society of London B Biological Sciences, 356*(1413), 1375–1384.

Conway, M. A. (2005). Memory and the self. *Journal of Memory and Language, 53*, 594–628.

Dalla Barba, G., & Boissé, M.-F. (2010). Temporal consciousness and confabulation: is the medial temporal lobe "temporal"? *Cognitive Neuropsychiatry, 15*, 95–117.

D'Argembeau, A., & Demblon, J. (2012). On the representational systems underlying prospection: evidence from the event-cueing paradigm. *Cognition, 125*(2), 160–167.

D'Argembeau, A., & Mathy, A. (2011). Tracking the construction of episodic future thoughts. *Journal of Experimental Psychology: General, 140*, 258–271.

D'Argembeau, A., Ortoleva, C., Jumentier, S., & Van der Linden, M. (2010a). Component processes underlying future thinking. *Memory & Cognition, 38*, 809–819.

D'Argembeau, A., Renaud, O., & Van der Linden, M. (2011). Frequency, characteristics, and functions of future-oriented thoughts in daily life. *Applied Cognitive Psychology, 25*, 96–103.

D'Argembeau, A., Stawarczyk, D., Majerus, S., Collette, F., Van der Linden, M., Feyers, D., . . . Salmon, E. (2010b). The neural basis of personal goal processing when envisioning future events. *Journal of Cognitive Neuroscience, 22*(8), 1701–13.

D'Argembeau, A., & Van der Linden, M. (2004). Phenomenal characteristics associated with projecting oneself back into the past and forward into the future: influence of valence and temporal distance. *Consciousness and Cognition, 13*(4), 844–858.

D'Argembeau, A., & Van der Linden, M. (2006). Individual differences in the phenomenology of mental time travel: the effect of vivid visual imagery and emotion regulation strategies. *Consciousness and Cognition, 15*(2), 342–350.

D'Argembeau, A., & Van der Linden, M. (2012). Predicting the phenomenology of episodic future thoughts. *Consciousness and Cognition*, *21*(3), 1198–206.

Demblon, J., & D'Argembeau, A. (2014). The organization of prospective thinking: evidence of event clusters in freely generated future thoughts. *Consciousness and Cognition*, *24*, 75–83.

De Vito, S., Gamboz, N., & Brandimonte, M. A. (2012). What differentiates episodic future thinking from complex scene imagery? *Consciousness and Cognition*, *21*, 813–823.

Elliot, A. J., & Friedman, R. (2007). Approach-avoidance: a central characteristic of personal goals. In B. R. Little, K. Salmela-Aro, & S. D. Phillips (eds.), *Personal project pursuit: goals, action and human flourishing*. Mahwah, NJ: Lawrence Erlbaum.

Emmons, R. A. (1986). Personal strivings: an approach to personality and subjective well-being. *Journal of Personality and Social Psychology*, *51*, 1058–1068.

Fivush, R. (2011). The development of autobiographical memory. *Annual Review of Psychology*, *62*, 559–82.

Fox, M. C., Ericsson, K. A., & Best, R. (2011). Do procedures for verbal reporting of thinking have to be reactive? A meta-analysis and recommendations for best reporting methods. *Psychological Bulletin*, *137*(2), 316–44.

Gerlach, K. D., Spreng, R. N., Madore, K. P., & Schacter, D. L. (2014). Future planning: default network activity couples with frontoparietal control network and reward-processing regions during process and outcome simulations. *Social Cognitive and Affective Neuroscience*, *9*(12), 1942–51.

Gilbert, D. T., & Wilson, T. D. (2007). Prospection: experiencing the future. *Science*, *317*(5843), 1351–1354.

Gollwitzer, P. M. (1999). Implementation intentions: strong effects of simple plans. *American Psychologist*, *54*, 493–503.

Hassabis, D., & Maguire, E. A. (2007). Deconstructing episodic memory with construction. *Trends in Cognitive Sciences*, *11*(7), 299–306.

Ingvar, D. H. (1985). "Memory of the future": an essay on the temporal organization of conscious awareness. *Human Neurobiology*, *4*, 127–136.

Irish, M., & Piguet, O. (2013). The pivotal role of semantic memory in remembering the past and imagining the future. *Frontiers in Behavioral Neuroscience*, *7*, 27.

Jeunehomme, O., & D'Argembeau, A. (2016). Prevalence and determinants of direct and generative modes of production of episodic future thoughts in the word cueing paradigm. *The Quarterly Journal of Experimental Psychology*, *69*(2), 254–272.

Johnson, M. K. (2006). Memory and reality. *American Psychologist*, *61*, 760–771.

Johnson, M. K., Raye, C. L., Mitchell, K. J., Touryan, S. R., Greene, E. J., & Nolen-Hoeksema, S. (2006). Dissociating medial frontal and posterior cingulate activity during self-reflection. *Social Cognitive and Affective Neuroscience*, *1*(1), 56–64.

Klein, S. B. (2016). Autonoetic consciousness: re-considering the role of episodic memory in future-oriented self-projection. *Quarterly Journal of Experimental Psychology*, *69*(2), 381–401.

Klein, S. B. (2013a). Making the case that episodic recollection is attributable to operations occurring at retrieval rather than to content stored in a dedicated subsystem of long-term memory. *Frontiers in Behavioral Neuroscience*, *7*, 3.

Klein, S. B. (2013b). The complex act of projecting oneself into the future. *Wiley Interdisciplinary Reviews: Cognitive Science*, *4*, 63–79.

Klein, S. B. (2014). Autonoesis and belief in a personal past: an evolutionary theory of episodic memory indices. *Review of Philosophy and Psychology, 5*, 427–447.

Klinger, E. (2013). Goal commitments and the content of thoughts and dreams: basic principles. *Frontiers in Psychology, 4*, 415.

Lehner, E., & D'Argembeau, A. (2015). The role of personal goals in the feeling of pre-experiencing imagined future events. Manuscript submitted for publication.

Little, B. R. (1983). Personal projects: a rationale and method for investigation. *Environment and Behavior, 15*(3), 273–309.

MacLeod, A. K., & Byrne, A. (1996). Anxiety, depression, and the anticipation of future positive and negative experiences. *Journal of Abnormal Psychology, 105*, 286–289.

Markus, H., & Nurius, P. (1986). Possible selves. *American Psychologist, 41*, 954–969.

McAdams, D. P. (2013). The psychological self as actor, agent, and author. *Perspectives on Psychological Science, 8*, 272–295.

Moulin, C. J., Conway, M. A., Thompson, R. G., James, N., & Jones, R. W. (2005). Disordered memory awareness: recollective confabulation in two cases of persistent deja vecu. *Neuropsychologia, 43*(9), 1362–78.

Rubin, D. C. (2014). Schema-driven construction of future autobiographical traumatic events: the future is much more troubling than the past. *Journal of Experimental Psychology General, 143*(2), 612–30.

Rubin, D. C., Schrauf, R. W., & Greenberg, D. L. (2003). Belief and recollection of autobiographical memories. *Memory and Cognition, 31*(6), 887–901.

Schacter, D. L., & Addis, D. R. (2007). The cognitive neuroscience of constructive memory: remembering the past and imagining the future. *Philosophical Transactions of the Royal Society of London B Biological Sciences, 362*(1481), 773–786.

Schacter, D. L., Addis, D. R., & Buckner, R. L. (2008). Episodic simulation of future events: concepts, data, and applications. *Annals of the New York Academy of Science, 1124*, 39–60.

Schacter, D. L., Addis, D. R., Hassabis, D., Martin, V. C., Spreng, R. N., & Szpunar, K. K. (2012). The future of memory: remembering, imagining, and the brain. *Neuron, 76*(4), 677–94.

Smallwood, J., & Schooler, J. W. (2006). The restless mind. *Psychological Bulletin, 132*(6), 946–958.

Spreng, R. N., Stevens, W. D., Chamberlain, J. P., Gilmore, A. W., & Schacter, D. L. (2010). Default network activity, coupled with the frontoparietal control network, supports goal-directed cognition. *Neuroimage, 53*(1), 303–17.

Stawarczyk, D., Cassol, H., & D'Argembeau, A. (2013). Phenomenology of future-oriented mind-wandering episodes. *Frontiers in Psychology, 4*, 425.

Stawarczyk, D., & D'Argembeau, A. (2015). Neural correlates of personal goal processing during episodic future thinking and mind-wandering: an ALE meta-analysis. *Human Brain Mapping, 36*, 2928–2947.

Suddendorf, T., & Corballis, M. C. (2007). The evolution of foresight: what is mental time travel and is it unique to humans? *Behavioral and Brain Sciences, 30*, 299–351.

Szpunar, K. K. (2010). Episodic future thought: an emerging concept. *Perspectives on Psychological Science, 5*, 142–162.

Szpunar, K. K. (2011). On subjective time. *Cortex, 47*(3), 409–11.

Szpunar, K. K., Addis, D. R., McLelland, V. C., & Schacter, D. L. (2013). Memories of the future: new insights into the adaptive value of episodic memory. *Frontiers in Behavioral Neuroscience, 7*, 47.

Szpunar, K. K., & McDermott, K. B. (2008). Episodic future thought and its relation to remembering: evidence from ratings of subjective experience. *Consciousness and Cognition, 17*, 330–334.

Szpunar, K. K., Spreng, R. N., & Schacter, D. L. (2014). A taxonomy of prospection: introducing an organizational framework for future-oriented cognition. *Proceedings of the National Academy of Science U S A, 111*(52), 18414–18421.

Tulving, E. (1985). Memory and consciousness. *Canadian Psychologist, 26*, 1–12.

Tulving, E. (2002). Episodic memory: from mind to brain. *Annual Review of Psychology, 53*, 1–25.

Tulving, E. (2005). Episodic memory and autonoesis: uniquely human? In H. S. Terrace & J. Metcalfe (Eds.), *The missing link in cognition: origins of self-reflective consciousness* (pp. 3–56). Oxford: Oxford University Press.

Uzer, T., Lee, P. J., & Brown, N. R. (2012). On the prevalence of directly retrieved autobiographical memories. *Journal of Experimental Psychology: Learning Memory and Cognition, 38*(5), 1296–308.

Vincent, J. L., Kahn, I., Snyder, A. Z., Raichle, M. E., & Buckner, R. L. (2008). Evidence for a frontoparietal control system revealed by intrinsic functional connectivity. *Journal of Neurophysiology, 100*(6), 3328–3342.

Wheeler, M. A., Stuss, D. T., & Tulving, E. (1997). Toward a theory of episodic memory: the frontal lobes and autonoetic consciousness. *Psychological Bulletin, 121*(3), 331–354.

# PART V

# Functional Perspectives

11

# Temporal Perspectives In Imagination

## On the Nature and Value of Imagining the Future

DOROTHEA DEBUS

### 1. "FUTURE-DIRECTED MENTAL TIME TRAVEL"

Contemporary experimental psychologists and neuroscientists have recently set out to organize some of their work around a new experimental paradigm, the paradigm of "mental time travel." The central idea of this new and important work is the insight that human beings can be aware of, and can direct their attention toward, both the past and the future—in memory and in foresight, respectively—and that there might be important similarities between both those ways of being aware of, and directing our attention toward, events, processes, states of affairs, and objects that are not present at the time of the relevant mental occurrence, but that instead lie in the past or the future.

In a target paper on the topic, Suddendorf and Corballis, prominent proponents of the new paradigm, sketch the core idea by suggesting that we human beings have a

> general faculty of mental time travel that allows us not only to go back in time [in memory], but also to forsee, plan, and shape virtually any specific future event [in imagination].[1]

Empirical research on mental time travel highlights the fact that we need a philosophical account of our ability to "direct our mind" toward the past, as well as an account of our ability to "direct our mind" toward the future. Philosophers have given some attention to our ability to "direct our mind" toward the past by remembering the past, but our ability to "direct our mind" toward the future has so far not received much attention by philosophers at all. In an attempt to change this, the present chapter sets out to consider, from a philosophical perspective, cases of "*future-directed* mental time travel," that is, cases in which we "direct our mind" toward the future. More specifically, I here focus on one particularly important kind of future-directed mental time travel, namely our ability to *sensorily imagine* future events.

## 2. SENSORY IMAGINATIONS OF FUTURE EVENTS

Sensory imaginations (or "S-imaginations") are imaginations that have experiential characteristics. They are those cases of imagination that characteristically "correspond to our use of the distinct senses."[2] S-imaginations of future events are particularly *vivid* forms of being mentally related to the future, and they are therefore bound to play an especially important role in our engagement with the future.

And indeed, all mature healthy human subjects can, and often do, sensorily imagine future events. For example, you might engage in a little self-experiment and try to imagine what you are going to do this coming Sunday. If you engage in this task carefully, it is likely that you will experience some S-imaginations of relevant future events.

Alternatively, consider the following case:

> Anna is a scientist working in cancer research. She is presently thinking about what kind of experiments she will conduct during the next couple of years; then she starts daydreaming a bit and considers the possibility of her finding a cure for cancer; she thinks about how she will secure funding for a big study that will lead to the momentous discovery, and then she has vivid visual and auditory experiences as of checking the crucial experiment's results over and over again, then jumping up and down ecstatically, screaming for joy and running around the lab hugging her colleagues.

Cases of the present kind, that is, cases in which subjects S-imagine future events, stand at the heart of this chapter. More precisely, the present chapter addresses two main questions:

> **First Main Question**: How can we account for the *nature* of cases in which subjects S-imagine future events?
>
> **Second Main Question**: What is the *value* of our ability to S-imagine future events?

Before we set out to address these two questions, we should be clear about the concepts in play here, and a clarificatory stipulation might be in order: in the present context, in order for a subject to be said to S-imagine a future, past, or present event, it is necessary that the imagined event be presented *to the subject herself* as "temporally located" in the relevant way—that is, it is necessary that the imagined event be presented *to the subject herself* as future, past, or present. It is important to make this usage quite explicit, because it is possible to use the expression "to S-imagine a future (or past, or present) event" in a different way: consider a case in which someone S-imagines an event *without* imagining the event as temporally located *at all*. However, as it happens, an event just like the one that is S-imagined does eventually occur. One *might* describe such a case by saying that at the time of imagining, the subject did in fact S-imagine a future event, without herself being aware of it. However, in the following, this kind of usage is *excluded*.

Rather, in order for a subject to be said to S-imagine a future (or past, or present) event in the present context, it is necessary that the imagined event be presented *to the subject herself* as "temporally located" in the relevant way. Similarly, when I say, in the following, that an experience of some event is "temporally directed," or that the relevant experienced event is "temporally located" in some way, or that the experience has a certain "temporal orientation," I mean to imply that the relevant event is presented *to the S-imagining subject herself* in the relevant way—that is, that the event is presented *to the S-imagining subject herself* as temporally directed or temporally located in the relevant way, or as having a certain temporal orientation.

## 3. TWO CHARACTERISTIC FEATURES

Apart from being able to S-imagine *future* events, all mature healthy human subjects also can, and often do, S-imagine *present* events and *past* events. It also seems possible to S-imagine events with no particular temporal location at all.[3]

More specifically, subjects can, and do, S-imagine *present events* that actually do *not* occur at present—for example, you can S-imagine yourself presently working as a doctor in a hospital; subjects can, and do, S-imagine *present events* that actually *do* occur at present—for example, you can S-imagine your friend, who is a doctor in a hospital, presently working there; subjects can, and do, S-imagine *past events* that actually did *not* occur in the past—for example, you can S-imagine how, as a child, you once went on a holiday to the moon; subjects can, and do, S-imagine *past events* that actually *did* occur in the past—for example, you can S-imagine some event which took place in your life when you were very young, which your parents have described to you very carefully, but which you yourself cannot now remember; last not least, subjects also can, and do, S-imagine events with *no particular temporal location at all*—for example, you can S-imagine the earth and the sun moving relative to each other.

Thus, we can S-imagine actual as well as non-actual, merely possible, or counterfactual events with various temporal locations. But then, what exactly are the *differences* between cases in which a subject S-imagines *future* events, and cases in which a subject S-imagines *past* or *present* events, or events with *no* temporal location at all?

We might attempt to answer this question as follows: when a subject S-imagines a past, present, or future event, the subject herself is *aware* of the relevant events as "temporally located" in different ways—as having occurred in the past, as occurring in the present, or as to occur in the future. Thus, S-imaginations of future events differ from S-imaginations of events without *any* temporal location insofar as S-imaginations of future events *do* have a temporal orientation; and they differ from S-imaginations of *past* and *present* events insofar as S-imaginations of future events have a *different* temporal orientation than S-imaginations of past and present events. We therefore have reason to hold that

> **(Specific Temporal Orientation Claim)** S-imaginations of future events have a specific temporal orientation; when a subject S-imagines a future event, the imagined event is imagined as "temporally located" in the future.

Second, it also seems plausible to accept that in some sense the imagined future is "open":

> **(Openness Claim)** From the subject's own point of view at the time of S-imagining a *future event*, the future is *open* in some important way, and this is a constitutive feature of the S-imagination of the future event.

By contrast, from the subject's own point of view at the time of S-imagining a *past or a present event*, the past and the present are *not open* in the same way (i.e., in the way in which the future is open), and this (the past or present's *not* being open in that way) seems a constitutive feature of S-imaginations of past and present events.

What exactly it might mean to say that when a subject S-imagines a future event the future is "open" in a characteristic way is something we will have to explore later. For the time being, the Opennness Claim tries to capture a rough intuition that many people seem to have about relevant cases, but it does not yet offer any clear account of how exactly the intuition might be spelled out, or why we should accept it as true.

The two claims sketched here—the "Specific Temporal Location Claim" as well as the "Openness Claim"—each try to capture a characteristic feature of S-imaginations of future events. In an attempt to get clearer about the nature of S-imaginations of future events, it should be helpful to consider features that *are* characteristic of S-imaginations of future events and that might *distinguish* those mental occurrences from other S-imaginations. Thus, in an attempt to further our philosophical understanding of S-imaginations of future events, I suggest that we consider and explore both the "Specific Temporal Orientation Claim" and the "Openness Claim" in turn.[4]

## 4. SPECIFIC TEMPORAL ORIENTATION

What does it take for an experienced event to be "temporally located" in any particular way, and what does it take for an experienced event to be temporally located in the future?[5]

In response to this question, someone might first suggest that we should consider "temporal feelings." Indeed, so they might say, when a subject experiences a certain event and the experienced event is temporally located in a specific way, the experience does (at least usually) have the relevant temporal orientation in virtue of the experience's being conjoined with a certain specific "temporal feeling": when a subject experientially remembers, or S-imagines, a past event, so they might say, the experience is conjoined with a feeling of pastness; when a subject perceives, or S-imagines, a present event, the experience is conjoined with a feeling of presentness; and when a subject S-imagines a future event, the experience is conjoined with a feeling of futureness. Indeed, Russell once made some such suggestion with respect to memory, talking about a "feeling of pastness,"[6] and some philosophers have recently come to give this suggestion renewed attention.[7]

However, an account that relies on such "feelings" in an attempt to account for the various possible temporal orientations of S-imaginations in general, and that relies on a "feeling of futureness" to account for the future-directedness of S-imaginations of future events more particularly, will encounter some considerable problems. First, a defender of such a "feeling theory" will have to explain how we might be able to *individuate* relevant "feelings," and *which* mental occurrences precisely are referred to by the terms "feeling of futureness," "feeling of pastness," and "feeling of presentness." Next, the defender of a "feeling theory" will have to develop an answer to the question of why we should accept that there are any such feelings, that is, what reason we have to accept that any such feelings ever occur. Third, a critic might ask whether, assuming that relevant "temporal feelings" sometimes do occur, it is plausible to hold that a relevant feeling occurs in all, or at least in most, cases in which a subject S-imagines a future event; if the answer is that relevant feelings do not occur in all, and maybe not even in most, relevant cases (which seems a plausible answer), a defender of a "feeling theory" of the temporal orientation of S-imaginations of future events will have to account for the temporal orientation of those S-imaginations that do have a relevant temporal orientation but which are not accompanied by any relevant feelings in a different way, which in turn might provoke the question of why we need the original theory at all.

It might well be possible for a defender of a "feeling theory" to offer convincing answers to all these questions, and to solve the problems that the present set of questions raises. However, the problems highlighted here seem serious enough to consider possible alternative accounts of the temporal orientation of S-imaginations in general, and of S-imaginations of future events more specifically, and one such possible alternative account actually does seem rather promising.

The alternative account I have in mind here departs from the observation that the mental events under consideration—namely, S-imaginations of future events—usually occur in a rich context of other mental events and states of affairs. This context, so runs the present suggestion, the context in which relevant experiences occur, usually grounds the temporal orientation of the experiences in question. Thus, when a subject experiences a certain event and the experienced event is temporally located in a specific way, the experience does, at least usually, have the relevant temporal orientation in virtue of its occurrence in a context of temporally indexed beliefs (and other mental states and events) whose content is relevantly related to the content of the S-imagination. More specifically, when a subject S-imagines a certain event and the S-imagined event is temporally located in the future, the S-imagination does, at least usually, have its future-directed orientation in virtue of its occurrence in a context of future-directed beliefs (as well as future-directed desires and intentions) whose content is relevantly related to the content of the S-imagination.

## 5. CONTEXT MATTERS

More precisely, we might formulate the present suggestion as follows:

**(Context Claim)** In order for an S-imagined event to be temporally located in the future, that is, in order for a subject to be said to S-imagine a future event, it is usually *sufficient* that the relevant S-imagination occur in a context of future-directed thoughts, or beliefs (or future-directed desires, or intentions, or other future-directed mental states and events) whose content is relevantly related to the content of the S-imagination.

This, then, is the suggestion. But why should we *accept* the Context Claim? *How* might the context in which an S-imagination of a future event occurs ground its specific (namely, its future-directed) temporal orientation? And why should we accept that a context of future-directed thoughts or beliefs (or future-directed desires, or intentions, or other future-directed mental states and events) is *usually sufficient* for an S-imagination that occurs in such a context to be future-directed itself, that is, why should we accept that a relevant context is usually sufficient for the relevant S-imagined event to be temporally located in the future?

There are various ways in which the first of these questions (namely, the question as to *how* a relevant context might ground an S-imagination's temporal orientation) might be answered, because there are various kinds of relevant contexts that might play the role outlined by the Context Claim. I will here explore two kinds of relevant contexts: contexts that are provided by *a question* which has been posed; and contexts that are provided by a set of beliefs and thoughts on the basis of which the subject would be able to tell a relevant *story*.

First, then, it seems plausible to hold that sometimes a *question* can provide a context that grounds the temporal orientation of an S-imagination of a future event. Indeed, something like this probably happened earlier when I asked you to imagine what your plans for this coming Sunday were, a question that was intended to prompt you to S-imagine future events and activities that you plan to engage in this coming Sunday.

Everyday examples like the present one suggest that it is, more generally, plausible to accept that if an S-imagination occurs "in response" to a question about the future, the S-imagination is future-directed simply because the relevant question has directed the subject's attention toward the future. Thus, more generally, it seems we have good reason to hold that

**(Question Claim)** the context in which an S-imagination of a future event occurs can ground the S-imagination's specific (namely, future-directed) temporal orientation if the relevant context is one in which the subject explicitly asks herself a *question about the future*, and the S-imagination occurs "in response" to that question.

However, S-imaginations of future events obviously do not always occur in response to any particular question, and there might be various other ways apart from the Question Claim in which we might spell the Context Claim out. One important such further way of explicating the Context Claim is offered by the Narrative Claim, which says that

**(Narrative Claim)** the context in which an S-imagination of a future event occurs can ground the S-imagination's specific (namely, future-directed) temporal orientation if the relevant context provides the subject with a set of thoughts or beliefs about the future on the basis of which the subject could *tell a story about the future* which includes a reference to the content of the relevant imaginary experience.

For example, when Anna imagines finding a cure for cancer, Anna has various *thoughts* about what kinds of experiments she will conduct during the next couple of years, then she thinks about how to secure funding for a big study, and *then* she has vivid visual and auditory experiences as of checking the crucial experiment's results over and over again, and as of jumping for joy. It seems plausible to accept that the occurrence of the S-imagination in the context of relevant thoughts about the future does ground the S-imagination's temporal orientation, just as the Narrative Claim suggests. Anna takes the relevant S-imagined event to be an S-imagined future event simply because the relevant S-imagined event fits nicely into the story that she presently tells herself about the future.

Thus, exemplary cases like the present one seem to support the Narrative Claim; more generally, we can also develop an argument in support of the Narrative Claim, as follows:

*Argument in Support of the Narrative Claim*

(i) In order for some description to count as a story, it is necessary that the relevant description places the things it describes in a spatiotemporal framework.

(ii) If a subject is able to tell a story about the future that includes a reference to the S-imagined event, this usually implies that the relevant event is conceived of as occurring within the temporal framework that the story describes, namely during a period of time in the future.

(C) Thus, just as the Narrative Claim says, the context in which an S-imagination of a future event occurs can ground the S-imagination's specific (namely, future-directed) temporal orientation if the relevant context provides the subject with a set of thoughts or beliefs about the future on the basis of which the subject could tell a story about the future that includes a reference to the content of the relevant imaginary experience.

However, an opponent might question premise (ii) of this argument. Indeed, so the opponent might say, premise (ii) of the present argument must be false, because we can conceive of various cases in which a subject experiences an S-imagination, and is able to tell a story about the future that includes a reference to the S-imagined event, but nevertheless does *not* conceive of the S-imagined event as occurring within the temporal framework that the story describes, namely the future.

In order to support this claim, our opponent might suggest that we engage in a brief self-experiment. Do take a moment to think about your likely movements during the day tomorrow, so the opponent might suggest. Now take any particular point in space which you just thought about passing through during your day tomorrow; try to S-imagine a good friend walking toward you in the relevant location, but try to S-imagine this event as *not temporally located at all.*

It seems possible to do all this, and in doing so you have shown that it is possible to S-imagine an event, to be in a position in which one is able to tell a story about the future that could include a reference to the S-imagined event, while the

S-imagined event itself is nevertheless quite explicitly *not* conceived of as occurring within the temporal framework that the story describes, namely the future. The opponent concludes that this in turn shows that premise (ii) of the above Argument in Support of the Narrative Claim must be false.

In response, we should point out that premise (ii) of the Argument in Support of the Narrative Claim quite explicitly relates to *usual, everyday* cases. It states that "if a subject is able to tell a story about the future that includes a reference to the S-imagined event, this *usually* implies that the relevant event is conceived of as occurring within the temporal framework that the story describes, namely during a period of time in the future." Thus, premise (ii) means to leave room for the possibility of defeating circumstances; it grants that there might well be some unusual circumstances in which the role that a relevant narrative context would normally play in grounding the temporal orientation of a certain S-imagination is defeated.

And indeed, the particular circumstances which the opponent created for us in suggesting the self-experiment that we engaged in a moment ago should count as unusual and defeating circumstances. For after we had set out to think about the future (i.e., our day tomorrow), the opponent then instructed us quite explicitly to S-imagine an event with a *spatial* location that we had just thought about passing through tomorrow, but *without* any *temporal* location. And while we certainly *can* explicitly set out to S-imagine an event in a certain spatial location as having no particular temporal location whatsoever, quite simply by forming an additional, explicit belief to the effect that the S-imagined event is to be thought of as not temporally located in any specific way at all, when experiencing S-imaginations we usually do *not* form any such additional, explicit beliefs.

What premise (ii) is concerned with are *ordinary*, everyday cases such as Anna's imagining finding a cure for cancer. In those ordinary cases, subjects do *not* intentionally manipulate and actively determine the temporal orientation (or presence or absence of any such temporal orientation) of their S-imaginations by forming additional, explicit beliefs about the temporal orientation of relevant S-imagined events, and it seems plausible to accept that in those ordinary, everyday contexts, a narrative context *can* play the role described by premise (ii).

Thus, we can successfully defend premise (ii) against the opponent's objection. However, the opponent's objection has helped us to understand the Narrative Claim quite generally, and premise (ii) of the argument in its support more particularly, much better. We find that we have reason to hold that, if a subject is able to tell a story about the future that includes a reference to the S-imagined event, this implies that the relevant event is conceived of as occurring within the temporal framework that the story describes, namely during a period of time in the future, *unless* the subject forms some additional, explicit beliefs about the temporal location of the S-imagined event that contradict this default assumption. Thus, the explicit formation of a belief about the temporal orientation of an S-imagination can "overrule" the role that a narrative context might otherwise play in grounding the relevant S-imagination's temporal orientation, and such an additional, explicit belief about the temporal orientation of a certain S-imagination can therefore defeat the temporal orientation of an S-imagination that would otherwise be provided by the narrative context. But as premise (ii) quite explicitly makes room for such defeating circumstances, we have good reason to continue to endorse that premise. This in turn also means that we have good reason to continue to take it that the Argument in Support of the Narrative Claim is sound, which in turn means that we have good reason to endorse its conclusion, the Narrative Claim.

The Question Claim and the Narrative Claim offer us two ways in which the Context Claim can be developed. As indicated earlier, the Context Claim might also be spelled out in various other ways—for example, in some cases in which a subject S-imagines a future event, the context that grounds the S-imagination's future-directedness might be provided by a context of future-directed desires and future-directed intentions. There is no time to explore this suggestion here, but it is important to note that while the Question Claim and the Narrative Claim describe some important ways in which context might ground an S-imagination's temporal orientation, it is plausible to assume that other features of an S-imagination's context might ground its temporal orientation, which in turn means that there are bound to be other ways in which the Context Claim might be developed.[8]

Last but not least, it seems important to note that in order for a subject to think the thoughts, endorse the beliefs, and ask the questions that form the contexts of S-imaginations of future events as spelled out by the Narrative Claim and the Question Claim, the subject has to possess and employ the concept of the future. We thus find that in those cases in which the subject's ability to S-imagine a future event depends on its occurrence in a context as described by the Narrative Claim or the Question Claim, the subject's ability to S-imagine the relevant future event depends on the subject's possession of the *concept of the future*.

In at least many cases, a subject's ability to *S-imagine* a future event therefore depends on her ability to *think* about the future. This does not seem problematic—the ability to S-imagine a future event seems a rather sophisticated ability, and so the fact that a subject might need to rely on her understanding of the concept of the future in order to S-imagine future events does seem perfectly acceptable—but it seems worth noting that our ability to S-imagine future events frequently does depend on our ability to think about the future.[9]

The Context Claim in turn helps us to explain the first characteristic feature of S-imaginations of future events, namely their specific temporal orientation. As we said earlier, we have reason to hold that

> **(Specific Temporal Orientation Claim)** S-imaginations of future events have a specific temporal orientation; when a subject S-imagines a future event, the imagined event is imagined as "temporally located" in the future.

As we have now seen, we can account for the specific temporal orientation of S-imaginations of future events with the help of the Context Claim:

> **(Context Claim)** In order for an S-imagined event to be temporally located in the future, that is, in order for a subject to be said to S-imagine a future event, it is usually *sufficient* that the relevant S-imagination occur in a context of future-directed thoughts, or beliefs (or future-directed desires, or intentions, or other future-directed mental states and events) whose content is relevantly related to the content of the S-imagination.

## 6. THE OPENNESS OF THE FUTURE FROM THE SUBJECT'S OWN POINT OF VIEW

Next, we should consider the second characteristic feature of S-imaginations of future events, which we earlier sketched as follows:

> **(Openness Claim)** From the subject's own point of view at the time of S-imagining a *future event*, the future is *open* in some important way, and this is a constitutive feature of her S-imagination of the future event.

So then, what exactly might this mean? How might we spell out the intuition that the Openness Claim tries to capture?[10]

In a first attempt at answering this question, someone might suggest that the openness at hand is some form of metaphysical openness. Indeed, so one might say, from the subject's own point of view at the time of S-imagining a *past* event, there is a clear fact of the matter as to whether the event that the subject S-imagines did or did not occur in the past, even if the subject herself does not know whether any *specific* event that she S-imagined did or did not occur in the past. Similarly, from the subject's own point of view at the time of S-imagining a *present event*, there is a clear fact of the matter as to whether the event that the subject S-imagines is or is not presently occurring, even if the subject herself does not know whether any *specific* event that she S-imagines does or does not presently occur. By contrast, so one might suggest,

> **(Metaphysical Openness Claim)** from the subject's own point of view, at the time of S-imagining a *future event*, there is *no fact of the matter* as to whether the event that the subject S-imagines will or will not occur.

However, the Metaphysical Openness Claim will encounter some formidable opposition. Determinists, for example, will reject the Metaphysical Openness Claim on the following grounds:

> *Argument Against the Metaphysical Openness Claim*
>
> (i) According to determinism, whatever happens next is fully determined by the present state of affairs, together with the laws of nature.
>
> (ii) But then, determinism is true.
>
> (iii) So whatever happens anytime in the future is fully determined by how things are now, together with the laws of nature.
>
> (C) So at least as far as *rational* human beings are concerned, the Metaphysical Openness Claim should be false.

One might try to resist the present argument by rejecting determinism. However, under the assumption that determinism is true, we have to conclude that, at least for any *rational* human being, from the subject's own point of view at the time of S-imagining a future event, the future is *not metaphysically open*.

Instead, it might be suggested that we can offer an epistemological reading of the Openness Claim. For when a subject S-imagines a *present* or a *past* event, the subject herself can (although she sometimes does not) *know* whether the relevant S-imagined event is actual or not. By contrast, so it might be suggested,

> **(Epistemological Openness Claim)** when a subject S-imagines a *future* event, the subject herself *could not possibly* know whether the relevant S-imagined event is actual or not.

Indeed, given our epistemological limitations at present, no one is presently able to know very much about the future at all, but we do know quite a few things about the past and the present. This might speak in favor of the Epistemological Openness Claim. However, the determinist will say that once we have gained full knowledge of the laws of nature, and once we have learned to describe the overall state of affairs at any particular moment in time in the appropriate detail, we will be able to know about the future, just as we can know about the past and the present. Thus, according to the determinist, the fact that we presently are not in a position to know about the future is a contingent matter, due to our present epistemological limitations; but one day soon, so the determinist holds, it will be the case that when a subject S-imagines a *future* event, the subject herself *is* able to know whether the relevant S-imagined event is actual or not. Under the assumption that determinism *is* true, we therefore have to conclude that from the subject's own point of view at the time of S-imagining a future event, the future is *not* epistemologically open in any interesting sense either.

## 7. AGENTIAL OPENNESS

However, I think there is another way in which we can more successfully spell out the intuition captured by the Openness Claim. This alternative departs from the observation that when a subject S-imagines a *present* or a *past* event, the subject is aware of the fact that she herself, or others, could not possibly have any influence whatsoever on bringing about, or preventing, the actual onset of the S-imagined past or present event: The past is gone, so whatever a subject might S-imagine about the past now has already happened or has not happened in the past, but in either case it cannot now be made to happen any more; and present events are occurring at present, so while one might still act in ways that might change a relevant event's further development, it is impossible to act now to bring about, or prevent, the actual onset and occurrence of a presently occurring event.[11] By contrast, so runs the present suggestion,

> **(Agential Openness Claim)** when a subject S-imagines a *future* event, the subject is aware of the fact that she herself, or others, *might act* in ways that could bring about, or prevent, the actual future onset and occurrence of the very event that she S-imagines now. The subject's awareness of this fact is a constitutive feature of an S-imagination of a future event.

Indeed, I think the Agential Openness Claim successfully captures what the Openness Claim tries to get at. For example, according to the Agential Openness Claim, we should hold that when you S-imagine what you are going to do next Sunday, you are aware of the fact that you yourself, or others, might act in ways that could bring about, or prevent, the actual future onset and occurrence of the very event that you S-imagine now, and your awareness of this fact is a constitutive feature of your S-imagination of the future event.

Furthermore, while our earlier attempts at spelling out the Openness Claim with the help of the Metaphysical Openness Claim and the Epistemological Openness Claim turned out not to be viable due to concerns raised by the determinist, the Agential Openness Claim is perfectly compatible with determinism. According to the determinist, all our actions are *fully determined* by how things were before, together with the laws of nature; but the determinist does accept that we *do act*, and that our actions in turn do bring about, or prevent, the onset of other events in the future. Furthermore, the determinist has no reason to deny that we can *be aware* of relevant actions being related to other events in relevant ways. Thus, the determinist can accept the Agential Openness Claim.[12]

However, an opponent might worry that the Agential Openness Claim "over-intellectualizes" things. Indeed, so the opponent might say,

> *First Objection to the Agential Openness Claim*
>
> (i) When subjects S-imagine future events they do not usually *think* about the fact that they themselves or others might act in ways that could bring about, or prevent, the actual future onset and occurrence of the very event that they S-imagine now.
>
> (ii) But then, the Agential Openness Claim entails that when subjects S-imagine future events they *do* think about this fact.
>
> (C) Thus, the Agential Openness Claim must be false.

In response to the present objection, we should reconsider (and ultimately deny) the objection's premise (ii). According to the Agential Openness Claim, when a subject S-imagines a future event, the subject *is aware* of the fact that she herself, or others, might act in ways that might bring about, or prevent, the actual future onset and occurrence of the very event that she S-imagines now. This does *not* entail that when a subject S-imagines a future event, the subject *thinks* about the fact that she herself, or others, might act in ways that might bring about, or prevent, the actual future onset and occurrence of the event that she S-imagines now, as the opponent assumes. Rather, a subject who S-imagines a future event can be aware of the fact that she herself, or others, might act in ways that could bring about, or prevent, the actual future onset and occurrence of the very event that she S-imagines now in a more *primitive way*. We all have a *primitive understanding* of the fact that human beings, ourselves included, can, and do, engage in a wide range of different actions, and that those actions in turn affect what happens next. All it takes for a subject who S-imagines a future event to be aware of the fact that she herself, or others, might act in ways that might bring about, or prevent, the actual future onset and occurrence of the very event that she S-imagines now is such a *primitive understanding* of the fact that human beings, ourselves included, engage in a wide range of different actions, and that those actions in turn affect what happens next.

However, an opponent might insist that the Agential Openness Claim turns out to be problematic, and indeed false, once we realize that it is possible to S-imagine future events that are so extremely unlikely that they are more or less impossible (such as, for example, your being crowned the monarch of France in your old age), and future events that are beyond all human influence and are thus in some sense inevitable (such as, for example, the eventual demise of planet Earth). These are events with respect to which a rational S-imagining subject should, so the opponent holds, be fully aware that she herself, and others, will *not* be able to act in ways that could bring about, or prevent, the actual future onset and occurrence of the relevant event, which in turn means that the Agential Openness Claim must be false. More generally, so the opponent will say,

> *Second Objection to the Agential Openness Claim*
>
> (i) According to the Agential Openness Claim, when a subject S-imagines a *future* event, the subject is aware of the fact that she herself, or others, *might act* in ways that could bring about, or prevent, the actual future onset and occurrence of the very event that she S-imagines now, and the subject's awareness of this fact is a constitutive feature of an S-imagination of a future event.
>
> (ii) However, a subject can S-imagine some future events with respect to which the subject herself is fully aware that she herself, and others, will *not* be able to act in ways that could bring about, or prevent, the actual future onset and occurrence of the relevant events.
>
> (C) The Agential Openness Claim must therefore be false.[13]

In response to the present objection, we should consider the objection's premise (ii) more carefully. The future events that the opponent has in mind when formulating premise (ii) are events that seem more or less impossible (such as your being crowned the monarch of France in your old age) and events that seem inevitable (such as the eventual demise of planet Earth). Everybody will agree with the opponent that it is possible to S-imagine such future events, but I think we should reject the opponent's claim that a subject who S-imagines such events "is fully aware that she herself, and others, will *not* be able to act in ways that could bring about, or prevent, the actual future onset and occurrence of the relevant event."

Quite on the contrary, when, for the sake of the argument, I now try to S-imagine being crowned queen of France in my old age—an event which I (gladly) realize is most extremely unlikely to occur—I am well aware of the fact that I myself, and others, *might be* able to act in ways that could bring the relevant event about (or, indeed, in ways that could prevent it from occurring); for example, as far as actions that would promote the occurrence of the relevant event are concerned, the royalist cause in France might suddenly receive a lot of renewed support from the French people, I might set out to support that cause, I might do some research into the history of my family and might, who knows, find some evidence that shows that I have a claim on the French throne—and so I might eventually be crowned the queen of France in my old age. Of course, as I said at the outset, luckily all this is extremely unlikely to happen, but when I S-imagine being crowned queen of France in my old age, it certainly does *not* seem to be the case that I am "fully aware that I myself, and others, will *not* be able to act in ways that could bring about, or prevent, the actual future onset and occurrence of the relevant event"—quite on the contrary, I might (as I have just done) start thinking about, and thus be aware of, a range of (potentially outlandish, or daring, or radically life-changing) things I myself, or others, might do in order to bring about the (extremely unlikely) future event that I S-imagine. Thus, with respect to cases where subjects S-imagine future events that are highly unlikely to occur and therefore seem more or less impossible, premise (ii) of the opponent's Second Objection to the Agential Openness Claim is false.[14]

The second kind of case that the opponent has in mind when formulating premise (ii) of her second objection is cases in which a subject S-imagines future events that seem inevitable, for example the eventual demise of our planet, or the death of any particular living creature. When S-imagining any of those cases, so the opponent might say, a reasonable subject will be fully aware that she herself, and others, will *not* be able to act in ways that bring about, or prevent, the actual future onset and occurrence of the relevant events, which in turn means that the objection's premise (ii) is true.

However, we have reason to resist the opponent's present line of thought. Of course, so we might grant the opponent, it is inevitable that every particular living creature will eventually die, and it is also inevitable that our planet will eventually cease to exist, at least in the way we know it, and any reasonable subject will be aware of this fact. But it seems not at all obvious—and indeed, it turns out to be false—that when I S-imagine some such future event as the demise of our planet, I am fully aware that I myself, and others, will not be able to act in ways that will bring about, or prevent, the actual future onset and occurrence of the relevant S-imagined event.

Indeed, it seems impossible to *S-imagine* the *general fact* that our planet will eventually cease to exist. Thus, when a subject *S-imagines* the future demise of our planet, she will have to S-imagine a somewhat more specific scenario. For example, she might S-imagine the demise of our planet as we know it as the result of some political escalation leading to a disastrous act of atomic warfare; in this case, it seems more than likely that the subject will quite explicitly be aware of the fact that she herself, and others, will be able to act in ways that will bring about, or prevent, the actual future onset and occurrence of the S-imagined event. (The S-imagining subject might become politically active in an attempt to reduce the risk of atomic warfare, and politicians in relevant positions can press relevant buttons or refrain from pressing them.) Alternatively, a subject might S-imagine the demise of our planet as a result of some explosion, or a major planetary collision; once more, however, it seems likely that she will conceive of any such future event as one that scientists might well be able to prevent—very powerful tools might be developed in the future with which our planet might be saved from the particular S-imagined explosion, or the particular S-imagined collision.

Thus, even in cases in which a subject knows that it is inevitable that something (such as a living creature, or a planet) will perish eventually, there seems reason to accept that whenever a subject S-imagines a particular future event of the relevant entity's perishing, she is aware that she herself, or others, might be able to act in ways that could bring about, or prevent, the actual future onset and occurrence of that particular S-imagined event of destruction.

The opponent is therefore once more unsuccessful in supporting premise (ii) of her Second Objection to the Agential Openness Claim. We can conclude that at least for the time being, we have no good reason to accept premise (ii), and that the Second

Objection to the Agential Openness Claim is therefore unsuccessful. Accordingly, we can continue to endorse the Agential Openness Claim and hold that

> **(Agential Openness Claim)** when a subject S-imagines a *future* event, the subject is aware of the fact that she herself, or others, *might act* in ways that could bring about, or prevent, the actual future onset and occurrence of the very event that she S-imagines now. The subject's awareness of this fact is a constitutive feature of an S-imagination of a future event.

The Agential Openness Claim in turn helps us to spell out the intuitive idea that in some sense, the imagined future is open, while the imagined past and the imagined present are not open in the same way. Thus, in helping us to spell out the intuitive idea that the imagined future is open in some way, the Agential Openness Claim in turn helps us to capture one important characteristic feature of S-imaginations of future events.

Having reached this point, it might well be useful to take stock of what we have done so far in this chapter. As we said at the outset, the present chapter aims to address two questions: first, the question of how we can account for the *nature* of cases in which subjects S-imagine future events, and second, the question of what the *value* of our ability to S-imagine future events might be. In an attempt to address the first question, we have meanwhile discussed two characteristic features of S-imaginations of future events in some detail. First, as the Specific Temporal Orientation Claim says, S-imaginations of future events do have a specific temporal orientation; when a subject S-imagines a future event, the imagined event is imagined as "temporally located" in the future. We have seen how the Specific Temporal Orientation Claim can be developed with the help of the Context Claim, which in turn can be developed further with the help of, among others, the Question Claim and the Narrative Claim. Second, according to the Openness Claim, from the subject's own point of view at the time of S-imagining a future event, the future is open in some important way, and this is a constitutive feature of her S-imagination of the future event. We have seen why neither the Metaphysical Openness Claim nor the Epistemological Openness Claim provide us with a satisfactory understanding of the Openness Claim, but we then also saw that the Openness Claim can be successfully developed with the help of the Agential Openness Claim.

In discussing the two characteristic features of S-imaginations of future events that are formulated by the Specific Temporal Orientation Claim and the Openness Claim, respectively, I do hope to have made a useful contribution to an attempt at answering the present chapter's First Main Question. Next, I suggest we now turn our attention to the Second Main Question.

## 8. AXIOLOGICAL ISSUES

The Second Main Question of the present chapter is the question of what the *value* of our ability to S-imagine future events might be. Indeed, at first sight it might seem as if S-imaginations of future events are "axiologically neutral," that S-imaginations of future events are just idle, "dangling" elements of our mental lives, parts of daydreams with no value whatsoever. This first impression, however, turns out to be false.

Quite on the contrary, as I hope to show in the following, S-imaginations of future events can, and sometimes do, have value. The value that S-imaginations of future events might have is grounded in the fact that

> **(Effect Claim)** a present S-imagination of a future event can have an *effect* on how things develop *with respect to the S-imagined future event*, because the experience of an S-imagination of a future event can, and often does, make the S-imagining subject engage in certain actions with respect to the S-imagined future event that she would not otherwise have engaged in, and these actions will in turn have an effect on whether or not the relevant event will occur, and on the precise way in which the event will occur if it does.

We can explain this further with the help of the observation that

> **(Reason and Motivation Claim)** an S-imagination of a future event can make a subject engage in certain actions with respect to the S-imagined future event by giving the subject a *reason* to act in the relevant way. In giving the subject a reason to act in certain ways with respect to the S-imagined future event, an S-imagination of a future event might also *motivate* the subject to act in relevant ways.

Ordinary everyday examples might be offered to illustrate that and why both the Effect Claim and the Reason and Motivation Claim are true. For example, consider the case of Lisa:

> Lisa will be moving house next week. Lisa presently S-imagines the day of the move, and while she S-imagines the sofa being carried into the house she realizes that the sofa could not possibly be carried through the narrow door of the new house in one piece. Because of her S-imagination of the relevant future event, she disassembles the sofa before the day of the move, and once the moving day has arrived, the sofa is moved into the new house in parts.

In the story as told, Lisa's S-imagination of next week's move does have an effect on how things develop with respect to the S-imagined event, namely the move, just as the Effect Claim has it, and it also seems plausible to accept that the S-imagination of the future event, namely the move, does have an effect on the relevant event because it makes the subject engage in certain actions with respect to the S-imagined future event that she would not otherwise have engaged in (that is, she dismantles the sofa, which she would not have done had she not S-imagined the future event), and this in turn eventually does have an effect on the S-imagined event, because the S-imagined event of getting stuck with the sofa in the door actually does not occur. It also seems plausible to accept that, just as the Reason and Motivation Claim would have us expect, the subject's S-imagination of the future event of the move makes her engage in certain actions with respect to the S-imagined future event by giving her a *reason* to act in the relevant way—the S-imagination of the future event of getting stuck in the door gives the subject a *reason* to disassemble the sofa, because she wants to avoid a situation in which the move gets held up by the sofa getting stuck—and the S-imagination of the future event also clearly does motivate her to act, that is, to disassemble the sofa.

Similarly,

> when asked what he is going to do this coming Sunday, Paul S-imagines going on a bike ride around the nearby countryside. So far, Paul hadn't thought much about what to do this coming Sunday, and might well have ended up just staying at home. However, his S-imagination of the relevant future event now spurs him on to engage in various preparatory activities, and once it is Sunday, he does go on the bike ride.

Again, it seems clear that the present example instantiates both the Effect Claim as well as the Reason and Motivation Claim. Paul's S-imagination of a bike ride this coming Sunday does have an effect on how things develop with respect to the S-imagined future event—the S-imagination makes Paul engage in actions (namely, relevant preparatory activities) with respect to the S-imagined future event that he would not otherwise have engaged in, and these actions do in turn have an effect on whether or not the relevant event will occur, for it is only due to the relevant preparatory activities that the bike ride does take place eventually. We can explain this further with the help of the observation that Paul's S-imagination of the future bike ride gives him *reason* to engage in relevant preparatory activities. Once he S-imagines the future event, and S-imagines the beauty of the familiar countryside, he realizes how much fun it should be to get out on his bike on Sunday, and this in turn gives him a good reason to act in ways that will bring about the S-imagined future event in due course. The S-imagination gives him a reason to act in relevant ways with respect to the S-imagined future event, and the S-imagination of the future event also *motivates* him to act in relevant ways, just as the Reason and Motivation Claim has it. More generally, we therefore find that we have reason to accept both the Effect Claim and the Reason and Motivation Claim.

However, an opponent might say that the Effect Claim and the Reason and Motivation Claim are rather trivial claims. Indeed, so our opponent might point out, we can formulate analogous claims for various other mental events, for example for memories of past events. It seems clear that a subject's remembering a certain event might have an effect on future events because the relevant memory can, and often does, make the remembering subject engage in certain actions with respect to future events which, had it not been for relevant memories, she would not have engaged in, and these actions will in turn have an effect on whether or not the relevant event will occur, and on the precise way in which the event will occur if it does. For example, assume that Paul vividly remembers that his friend Julie had said that she would be free on Sunday, so he calls her and arranges to go on a bike ride with her. The vivid memory of his conversation with Julie has an effect on the relevant future event—had Paul not remembered that Julie said she would be free on Sunday, he would not have asked her to come along, and then he would have gone on a bike ride without Julie, probably on his own. Thus, his memory of Julie's saying that she would be free on Sunday does have an effect on the future event, and it does have that effect because it gives Paul a reason to call Julie and make relevant arrangements.

The fact that cases like the present one are possible prompts the opponent to hold that the two claims that we have just formulated with respect to S-imaginations of future events, namely the Effect Claim and the Reason and Motivation Claim, are not specific to the phenomenon under consideration here, namely our ability to S-imagine future events; rather, so the opponent concludes, the two claims summarize some rather commonplace, general facts about our mental lives and are therefore not particularly interesting with respect to the specific phenomenon of S-imaginations of future events.

However, I think the Effect Claim and the Reason and Motivation Claim are of greater interest to our present inquiry than the opponent is willing to admit. Of course the opponent is right in pointing out that all sorts of mental (and physical) events can have an effect on future events, and that all sorts of mental (and physical) events can give a subject reason, and motivate a subject, to act in certain ways with respect to future events. However, the Effect Claim and the Reason and Motivation Claim go beyond this very general observation. As those two claims indicate, what is important and interesting about the role that *S-imaginations of future events* play with respect to the occurrence (or non-occurrence) of relevant future events is that a present *S-imagination* of a relevant future event might have an effect *on the very event that is S-imagined.* This seems a specific characteristic of S-imaginations of future events that is *not* shared by most other mental events which might have an effect on future events. For example, memories are of past events, so while a memory of a past event might well have an effect on the occurrence (or non-occurrence) of a certain future event, it could not possibly be the case that the event that is remembered is also the event that is then affected by the relevant mental occurrence. By contrast, human beings can represent future events by S-imagining future events, and a subject's S-imagination of a future event can then in turn have a direct effect on that very S-imagined future event; this, or so I think, is not a trivial observation, but is actually quite a remarkable feature of our ability to S-imagine future events.

One kind of mental phenomenon that will have characteristics that are similar to those described by the Effect Claim and the Reason and Motivation Claim is our ability to *think* about future events. But as S-imaginations of future events are more vivid than mere thoughts about relevant future events, *S-imaginations* of future events will often have greater motivational power than mere *thoughts* about relevant future events.

## 9. PUTTING S-IMAGINATIONS OF FUTURE EVENTS TO (GOOD) USE

Sometimes a present S-imagination of a future event is of value because it has a *good effect* on how things develop with respect to the S-imagined future event. For example, Paul's going on a bike ride this coming Sunday will be good for his health and is therefore of value to him. As Paul's S-imagination of a relevant future event plays a part in bringing the relevant event about, his S-imagination of the relevant future event is of instrumental value.

However, at other times a present S-imagination of a future event is harmful, because it has a *bad effect* on how things develop with respect to the S-imagined future event. For example,

> Emma considers running for political office in her hometown. However, when S-imagining relevant future events, Emma is overwhelmed by the very many problems she would face, and she therefore decides not to run for office. Had she not S-imagined relevant future events, she would have run for office, would have been elected, and would have changed things for the better.

Thus, Emma's running for office, and then doing the relevant work, would have been of value overall, so her S-imagining relevant problematic future events, which puts her off, is harmful. More generally, we therefore find that, in virtue of the positive or negative value of the effects that S-imaginations of future events have on the occurrence (or non-occurrence) of the relevant S-imagined future events, S-imaginations of future events sometimes are of instrumental value, and sometimes they have the potential to cause harm.

Furthermore, everyday (self-)observation also suggests that mature healthy subjects usually do have at least a *primitive understanding* of the *ways* in which certain S-imaginations of future events might make them act (or not act) with respect to the relevant future events, and thus which effects relevant S-imaginations of future events might have on the actual occurrence (or non-occurrence) of relevant S-imagined future events. But then, on the basis of their understanding of the ways in which S-imaginations of future events might make them act (or not act) with respect to the relevant S-imagined future events, subjects also can, and sometimes do, *regulate* their own mental lives by *making* themselves *have* certain S-imaginations of future events in an attempt thereby to make themselves engage in future actions that are in turn likely to have desirable effects on the actual occurrence (or non-occurrence) of the relevant S-imagined future events. Thus,

> **(Shaping the Future Claim)** subjects sometimes set out to (and sometimes successfully do) shape the future by means of making themselves have certain S-imaginations of future events.

For example,

> Peter has a job interview next week. In preparation, he S-imagines the job interview; as he wants to perform well in the interview, he sets out to S-imagine the interview in a way that, as he understands, will have the effect of his performing well—he tries *not* to dwell too much on potential problems, but tries to S-imagine the future event in a clear and realistic way, which, as he understands, is conducive to his responding in a calm and competent manner in the actual interview.

Thus, Peter quite intentionally S-imagines the future event in a way that is conducive to his offering responses during the interview that will promote his chances of getting the job.

Similarly, Anna (who is a scientist working in cancer research) knows that finding a cure for cancer will, if it is going to happen at all, be the result of a lot of hard work. So in an attempt to motivate herself to work hard and thereby make it more likely that the relevant event will eventually occur, Anna does, every now and then, make herself S-imagine the future event of making the momentous discovery, and relevant events that would follow it.

Thus we have good reason to accept that subjects sometimes do set out to (and sometimes successfully do) shape the future by means of making themselves have certain S-imaginations of future events, just as the Shaping the Future Claim says. This observation in turn brings us back to the quote we encountered at the beginning of the present chapter. In introducing the phenomenon of "mental time travel," Suddendorf and Corballis claim that we human beings have a

> general faculty of mental time travel that allows us not only to go back in time [in memory], but also to forsee, plan, and shape virtually any specific future event [in imagination].[15]

In this chapter we have considered one particular kind of future-directed mental time travel, namely our ability to S-imagine future events, and we have now seen that, and how, subjects *can* shape the future with the help of this one particular kind of future-directed mental time travel, namely their ability to S-imagine future events.

Furthermore, the fact that S-imaginations of future events can play this role suggests that perhaps the greatest value that S-imaginations of future events have lies in the fact that subjects can, and sometimes do, put their ability to S-imagine future events to good use and thereby shape the future for the better. And quite analogously, maybe the greatest harmful potential that S-imaginations of future events have lies in the fact that subjects can, and sometimes do, put their ability to S-imagine future events to bad use and thereby shape the future for the worse.

Accordingly it might, in conclusion, also be plausible to hold that given their potentially substantial effects, we have good reason to *take care* of our S-imaginations of future events; that is, we have good reason to take care of *which* future events we set out to S-imagine, and we also have good reason to take care of *how* we set out to S-imagine those future events.

## ACKNOWLEDGMENTS

I first presented the material developed here in February 2015 during an inspiring workshop at the Institut Jean Nicod in Paris, and I am grateful to the workshop participants for their generous feedback. My thanks also go to the members of the Department of Philosophy at the University of York, who later provided further helpful comments in the context of a staff work-in-progress meeting, as well as to a group of excellent graduate students who participated in a Summer School on Moral Phenomenology at the Central European University in Budapest in the summer of 2015. I am especially grateful to Stan Klein and Denis Perrin, who acted as referees for this chapter and offered many helpful comments on an earlier draft. Thank you!

# NOTES

1. Suddendorf and Corballis (2007), 299. Published in the same year, Schacter, Addis, and Buckner (2007) offer another very useful review of research on "mental time travel" up to 2007.
2. Martin (2002), 403.
3. On the following set of issues see also Klein and Steindam (Chapter 7 of this volume).
4. Of course, besides the features described by the "Specific Temporal Orientation Claim" and the "Openness Claim," there are *other* features that are characteristic of S-imaginations of future events, and which will thus distinguish S-imaginations of future events from other S-imaginations. However, in the present context I suggest that we focus on the two characteristic features described by the "Specific Temporal Orientation Claim" and the "Openness Claim," respectively, because these two features seem important in an attempt to characterize the distinctive *phenomenology* of S-imaginations of future events—they are features that are accessible from a first-person perspective, that is, they are accessible to the subject who experiences a relevant S-imagination of a future event *herself*. (Thanks to Denis Perrin for prompting the present clarification.)
5. Note that the present question is also addressed by Klein and Steindam (Chapter 7 of this volume) from an empirical perspective, and further empirical work that is relevant to the present question can be found in Klein, Robertson, and Delton (2010).
6. See Russell (1989), Lecture IX ("Memory"), especially pp. 176–187.
7. See, e.g., Dokic's (2014) "two-tiered account" of episodic memory.
8. However, so someone might say, apart from a description of such potential other kinds of contexts (which it might not be necessary to offer here), something more general is still missing from our development of the Context Claim. So far, we have developed the Question Claim and the Narrative Claim in an attempt to explain *how* a context might ground an S-imagination's temporal orientation, and various other types of contexts might be described. But, so an interlocutor might say, describing particular kinds of contexts is not enough in an attempt to answer the question of how a context might ground an S-imagination's temporal orientation; rather, so our interlocutor might say, in order to develop the Context Claim fully, we still have to answer the more general question as to how *any* particular context (be it a question, a narrative, or any other possible context) might ground the temporal orientation of an S-imagination. How, so our interlocutor might ask, are relevant S-imaginations "bound up" with their relevant context so that the temporal orientation of the context grounds the temporal orientation of the S-imagination itself? In a first attempt at answering this question, a defender of the Context Claim might suggest that in ordinary cases, when an S-imagination occurs in a relevant context, the S-imagination does, from the experiencing subject's own point of view, *fit in* with the relevant context in which it occurs, and its fitting in with the relevant context from the subject's own point of view is all it takes for the S-imagination to be "bound up" with its context in such a way that the relevant context *can* ground the S-imagination's temporal orientation, from the subject's own point of view. However, our interlocutor might respond that this does not really answer her question. Rather, so the interlocutor might say, the question

was meant to highlight the fact that our account of the temporal orientation of S-imaginations is still in need of an account of the (sub-personal) *mechanisms* which make it the case that the temporal orientation of S-imaginations is grounded in their relevant context. This, in turn, we should simply agree with. The interlocutor is right in saying that we have not offered an account of any relevant sub-personal mechanisms and processes here, and such empirical research does lie beyond the remit of the present chapter. However, empirical scientists have recently started to explore relevant questions, so anyone who is interested in relevant sub-personal mechanisms should be encouraged to consult relevant recent empirical work (see, e.g., Klein [2013] and Klein [2015]). (Thanks to Stan Klein for prompting the present set of clarifications.)

9. Indeed, it seems worth noting this, especially as an opponent might object that the dependency just described, namely the dependency of a subject's ability to S-imagine a future event on her ability to think about the future, *is* problematic. The opponent might argue as follows: (i) A subject's ability to think about the future does in some way depend on a subject's ability to S-imagine future events; but then, (ii) if we also hold that a subject's ability to S-imagine future events depends on her ability to think about the future, we have run into a vicious circle; and (iii) the best way to avoid such a vicious circle is to give up on any account (such as the account which has been developed in the main text) that entails that a subject's ability to S-imagine future events does depend on her ability to think about the future. We therefore find that (iv) as this is entailed by the Context Claim, we should give up on the Context Claim. In response, and in defense of the Context Claim, we might reconsider and ultimately reject some of, or indeed all of, the three premises of the opponent's argument. First, we might question premise (i); indeed, it is not *obvious* that and why we should accept that a subject's ability to think about the future does in some way depend on a subject's ability to S-imagine future events, as premise (i) has it. But even if we grant the opponent this premise, we might next question their assumption, as set out in their premise (ii), that a relevant circle would have to be a *vicious* circle—quite on the contrary, so we might say, there might well be relations of mutual interdependence between various of our mental abilities, and at least some of those relations of mutual interdependence might be "virtuous"; there certainly is no obvious reason to take it that all such relations of mutual interdependence have to be "vicious," as the opponent seems to assume. (For an argument that aims to show that relevant interdepence relations need not be "vicious," see Debus [2013]). Lastly, we might also challenge premise (iii) of the opponent's objection. For even if we grant the opponent her premise (ii), we might still hold that the *best* way to avoid a relevant vicious circle would be to give up on the opponent's premise (i), rather than to give up on the claim that a subject's ability to S-imagine a future event depends on her ability to think about the future, as the opponent's premise (iii) suggests we should do. Thus, although the opponent's objection might seem very compelling at first sight, there are various ways in which the opponent's objection can be shown to be unsuccessful, and we can therefore continue to endorse the Context Claim. (Thanks to Marilee Coetsee and Denis Perrin, who both independently suggested that the present set of issues should be addressed here; I say much more on the theme of the present note in Debus [2013].)

10. The following two sections are indebted in some important respects to Ismael (2012).

11. Hoerl (1999) develops a similar point for the case of memory.
12. But then, so a critical reader might ask, how could one possibly hold that the Agential Openness Claim spells out a way in which the future is "open" that is compatible with the assumption that determinism is true? Shouldn't we just accept that if determinism is true, the future is *not* open in any way whatsoever? We might respond as follows: If determinism is true, whatever happens next is fully determined by how things were in the past, together with the laws of nature. So, if determinism is true, the future is *not* open in the way described by the *Metaphysical* Openness Claim. Nevertheless, even if determinism *is* true, there is an important *difference* between past and present events, on the one hand, and future events, on the other, insofar as a subject's *actions* with respect to the relevant events are concerned. For quite independently of the truth or otherwise of determinism, with respect to *past and present* events, it is *incompatible with the laws of nature* for someone to act (now or in the future) in such a way as to bring about, or prevent, the actual onset and occurrence of a relevant event (i.e., a past or present event), for we take it that backward causation goes against the laws of nature and thus is not possible in any context that is governed by the laws of nature. However, even on the assumption that determinism *is* true, *future* events are events of a kind such that, in *some* instances of events of the relevant kind, the subject herself, or others, *can and will, now or in the future*, act in ways that will in turn bring about, or prevent, the actual onset and occurrence of the relevant future event, and this is perfectly *compatible* with the laws of nature. Thus, a determinist might hold that the future is "open" for a subject who S-imagines a future event insofar as it is compatible with the laws of nature for present or future actions of the subject herself, or of others, to be a part of the causal chain that will eventually lead to the occurrence (or prevention) of the S-imagined future event; this is not possible (because it is not compatible with the laws of nature) as far as past and present events are concerned, which in turn explains why the past and the present are not "open" in the relevant sense. Subjects themselves in turn can be *aware* of this difference between past and present events, on the one hand, and future events, on the other, just as the Agential Openness Claim demands. The Agential Openness Claim therefore does offer a view according to which the future *is* open in a distinctive way, even if determinism is true. (Thanks to Denis Perrin for prompting the present set of clarificatory comments.)
13. Thanks to Tom Stoneham and Myrto Mylopoulos for suggesting (independently of each other) that the present objection should be considered here.
14. However, our opponent might want to pursue the present line of thought further and suggest instead that it is possible for someone, call her Mary, to S-imagine the future event of her being crowned queen of France in her old age, while Mary herself is, at the same time, fully and wholeheartedly convinced that the occurrence of such an event is totally and completely impossible. Although the event seems outlandish, it seems possible for it to occur, so Mary is *wrong* in endorsing the belief that it is totally and utterly impossible for her to be crowned queen of France in her old age. But Mary might of course nevertheless still *be* of that conviction. Assume this is so. Then, so the opponent will suggest, Mary's case as just sketched presents us with someone who is able to S-imagine a future event while at the same time being strongly convinced that it is impossible for the relevant future event to occur, and thus also being convinced that

she herself, and others, will *not* be able to act in any way that could bring about, or prevent, the actual future onset and occurrence of the relevant event. Thus, so the opponent will conclude, we have now described a case that successfully supports premise (ii) of the Second Objection to the Agential Openness Claim. In reply, we will have to point out that the opponent's present scenario does not support premise (ii) of the Second Objection as the opponent thinks it does. Premise (ii) of the Second Objection to the Agential Openness Claim says that "a subject can S-imagine some future events with respect to which the subject herself *is fully aware* that she herself, and others, will not be able to act in ways that could bring about, or prevent, the actual future onset and occurrence of the relevant events." But then, in order for a subject to be said to *be aware* of a certain state of affairs, the relevant state of affairs has to *obtain*. In the scenario just sketched, Mary is of the strong *conviction* that she herself, and others, will not be able to act in ways that could bring about, or prevent, the actual future onset and occurrence of the relevant events. However, as we said, this conviction is *false*. Accordingly, in the scenario as just sketched, Mary could not possibly be said to be *aware* that she herself, and others, will not be able to act in ways that could bring about, or prevent, the actual future onset and occurrence of the relevant events, and hence, the scenario just sketched cannot be offered in support of the Second Objection's premise (ii) either. (Thanks to Denis Perrin for prompting me to develop the material offered in the present note.)

15. Suddendorf and Corballis (2007), 299.

## REFERENCES

Debus, D. (2013). Thinking about the past and experiencing the past. *Mind and Language*, *28*, 20–54.

Dokic, J. (2014). Feeling the past: a two-tiered account of episodic memory. *Review of Philosophy and Psychology*, *5*, 413–426.

Hoerl, C. (1999). Memory, amnesia and the past. *Mind and Language*, *14*, 227–251.

Ismael, J. (2012). Decision and the open future. In A. Bardon (ed.), *The future of the philosophy of time* (pp. 149–168). New York: Routledge.

Klein, S. (2013). The complex act of projecting oneself into the future. *WIREs Cognitive Science*, *4*, 63–79.

Klein, S. (2015). What memory is. *WIREs Cognitive Science*, *6*, 1–38.

Klein, S., Robertson, T. E., & Delton, A. W. (2010). Facing the future: memory as an evolved system for planning future acts. *Memory and Cognition*, *38*, 13–22.

Martin, M. G. F. (2002). The transparency of experience. *Mind and Language*, *17*, 376–425.

Russell, B. (1989). *The analysis of mind*. London: Routledge.

Suddendorf, T., & Corballis, M. C. (2007). The evolution of foresight: what is mental time travel, and is it unique to humans? *Behavioral and Brain Sciences*, *30*, 299–351 (with Open Peer Commentary).

12

# Making Decisions about the Future

## *Regret and the Cognitive Function of Episodic Memory*

CHRISTOPH HOERL AND TERESA MCCORMACK ■

Episodic memory is the capacity to consciously recollect particular events from one's own past (Hoerl, 2007; Tulving, 1985). Empirical work in psychology suggests that it is a fairly sophisticated memory capacity that may develop later than other memory and learning abilities (McCormack & Hoerl, 1999) and that can be selectively impaired in amnesia (Squire & Zola, 1998). It is also often held to be unique to humans (Hoerl, 2008; Suddendorf & Corballis, 2007; Tulving, 2002)—echoing the long-standing philosophical trope that animals' lives are in some sense confined to the present (Aristotle, 1930; Bennett, 1964; Bergson, 1991; Nietzsche, 1983; Schopenhauer, 1999)—although there is considerable debate about this in comparative psychology. Existing discussions of episodic memory in philosophy are typically concerned with the idea that episodic recollection involves being in a distinct kind of epistemic state (Martin, 2001; Soteriou, 2008) and a state with a distinct kind of phenomenology (Hoerl, 2001; Hopkins, forthcoming). In both disciplines, the question as to what episodic memory is *for*—what distinctive evolutionary function it might serve—has, until relatively recently, rarely been asked (Baddeley, 1988), but is now becoming a very lively area of debate (Boyer, 2008; Klein, 2013; Suddendorf & Corballis, 2007). In fact, this question has arguably emerged as one of the most significant challenges for memory research. While learning about the general characteristics of the world (and retaining such information for future use) is clearly beneficial, of what benefit is it to the individual to be able to remember particular past events as such, given that those particular events will never come again?

This chapter will consider an approach to this issue that has not yet been explored, by examining how the set of cognitive abilities employed in episodic memory might be necessary for the emotion of regret. Regret, at least as we will understand this notion in the context of the present chapter, is by definition an emotion directed toward the past: one regrets a choice one has made, typically believing that if one had chosen differently a better outcome would have obtained. However, although when we experience regret it is directed toward the past, psychologists have argued that this emotion has an important effect on our future lives (Roese, 2005; Zeelenberg & Pieters, 2007). Having experienced regret over a choice, we may choose differently when faced with a similar choice again. Moreover, when making choices, we may anticipate and seek to avoid future regret.[1] What we will explore in this chapter is the role of episodic memory in regret, and hence future-oriented decision-making itself.

As we see it, a key advantage of exploring this issue is that it may make vivid how episodic memory, which itself is intrinsically backward-looking, may nevertheless play a vital role in shaping our decisions about the future. In the first section, we set up the puzzle of accounting for a distinctive function for episodic memory in future-oriented decision-making. Recent accounts of the function of episodic memory have sought to identify a role for it in such decision-making, but, as we show in the next section, it is not clear how the accounts that have been put forward so far include the backward-looking element that is an essential aspect of episodic memory. In the second section, we will consider psychological research that has demonstrated the different ways in which regret may inform decision-making, distinguishing between the effects of experiencing versus anticipating regret. Finally, we will then consider how experiencing and anticipating regret may be connected to mental time travel, and whether episodic memory can be seen to be essential in supporting the way of thinking about time that is involved in regret. Our argument will be that episodic memory, because of its backward-looking element, underpins the ability to experience regret, and that this an important way in which episodic memory impacts on and supports adaptive future-oriented decision making.

## 1. FINDING A DISTINCTIVE ROLE FOR EPISODIC MEMORY IN FUTURE-ORIENTED DECISION-MAKING: THE BASIC PUZZLE

Having originally been conceived of in relative isolation from other cognitive abilities, episodic memory has in recent years been re-conceptualized as one aspect of the more general capacity for engaging in what is sometimes referred to as "mental time travel" (Schacter & Addis, 2009; Suddendorf & Corballis, 1997, 2007). This has been partly in response to the growing recognition that there is also a distinctive, episodic way of thinking about future events (Atance & O'Neill, 2001), and in response to evidence from imaging studies and studies of cases of amnesia that such episodic future thinking relies on the same brain network as episodic memory (Schacter, Addis, & Buckner, 2007). However, part of the motivation behind it has also been the thought that the function of episodic memory must ultimately turn on its connection to the ability to engage in episodic future thinking, and the role that the latter plays in enabling certain forms of planning or decision-making (Klein, 2013). The basic line of thought here is that, although it stems from the past, episodic memory is primarily useful not because it is about the past, but because it allows us to prepare adaptively for what is to come (Klein, 2016).

One problem, though, with this general line of argument is that it threatens to do away with any distinctive function for specifically episodic memory (by contrast to other types of memory) in our mental lives. Indeed, a consideration of this line of argument can serve to demonstrate what the problem of accounting for the function of episodic memory comes to. To illustrate this, we will focus on one particularly persuasive and influential claim that has been made about the role of episodic memory, and episodic thinking more generally, in future-oriented decision-making: that of Boyer (2008). Recent accounts of episodic future thinking often emphasize that it is a type of mental simulation; the typical claim in the literature is that this simulation involves the flexible recombination of material from episodic memories of past experiences (Eacott & Easton, 2012; Suddendorf & Redshaw, 2013). Boyer's (2008) claim is that episodic memory is critical for future-oriented decision-making, insofar as it supports the mental simulation of future events that allows us to give due weight to the future rather than current benefits of certain choices.

The paradigmatic decision-making scenario envisaged in this context is one that involves delayed gratification. Boyer argues that episodic memory allows simulation of the future, and in virtue of this property it has a special link to the emotions: he states that the episodic system results in the generation of "emotions that bypass current goals as well as time discounting and, therefore, provide us with immediate counter-rewards against opportunistic motivation" (Boyer, 2008, 222). That is, insofar as the episodic system yields mental simulation of possible future events, it preferentially supports choices that may currently seem unattractive or have no immediate benefits but that are in the long term advantageous. This argument is on the face of it persuasive, because it focuses not just on the type of information about the world that the episodic system could deliver, but on the distinctive way in which this information is delivered, namely as input into a type of simulation.

Boyer focuses particularly on the idea that such simulation delivers a preview of a future emotional reward (see also Benoit et al., 2011). However, in virtue of the fact that it involves mentally "pre-experiencing" the future, it is plausible that episodic future thinking may also impact on other factors that are already known to affect temporal discounting, but which have not typically been explicitly linked to episodic thinking as such. Two examples of such factors are subjective temporal distance (Kim & Zauberman, 2009, 2013; Kim, Zauberman, & Bettman, 2012) and level of connectedness to one's future self (Bartels & Rips, 2010; Hershfield, 2011). Kim and Zauberman (2009) argue that the extent to which one discounts a future event is closely linked to how far away in the future it feels; a natural extension of this suggestion is that simulation of the future underpinned by the episodic memory system might reduce subjective temporal distance to the future and thus reduce temporal discounting. Similarly, Bartels and Rips (2010) argue that individuals feel to varying degrees psychologically connected to their future selves (e.g., oneself in the distant future may be represented as only partially overlapping with one's current self), and that this can help explain the typical temporal discounting function. Again, a plausible hypothesis is that episodic simulation may alter psychological connectedness to one's future self and may thus support future-oriented decision-making.

Not only are there good reasons to hypothesize that mentally simulating the future will lead to it being more valued, empirical studies have begun to emerge demonstrating that encouraging people to simulate the future impacts on the extent to which it is discounted (Benoit, Gilbert, & Burgess, 2011; Lin & Epstein, 2014; Peters & Buchel, 2010, 2011). The crucial question in the current context, though, is whether this can be used to support the idea that it is specifically *episodic* memory that does some crucial work in decision-making that involves simulation in this way. Mental simulation, of the type envisaged, clearly relies on past experience; even when we simulate a novel type of event, we can only do so to the extent that the elements out of which our simulation is composed correspond to experiences with which we are familiar (Lewis, 1990). In that sense, information that is in memory as a result of our own past experiences must play an important role in supporting simulation of the future. Yet, this does not mean that simulation requires that we can actually recall the particular past occasions on which we had those experiences (Hoerl, 2014). We may engage in such simulation, drawing on information from memory, without actually casting our minds back to our past at all. Indeed, from an adaptive point of view, engaging in recollection in the context of this type of decision-making can put us at a disadvantage (see also Schacter, Benoit, De Brigard, & Szpunar, 2015, for this point), given that drawing on memories of particular past events in predicting the future introduces systematic biases—when casting our mind back to the past, instead of remembering how things typically go, we in fact tend to remember occasions that were atypical (Morewedge, Gilbert, & Wilson, 2005).

What these considerations illustrate is that there is an obvious function that memory can play in decision-making in cases where the choice is between outcomes of a type one has already encountered. In such cases, memory can provide the necessary materials for simulating the future. But for memory to serve that function, the relevant memory does not need to include the capacity to recollect particular past events per se.[2] The basic problem here is that once the primary function of episodic memory is thought to be one of serving future-oriented decision-making, it becomes difficult to see why this type of memory, with its backward-looking element, should be seen to play a distinctive role in serving this function that cannot also be served by types of memory that do not include this element.

At the risk of laboring the point, the structure of the problem in accounting for a distinctive role for episodic memory could be put as follows. One approach to establishing a distinctive role for episodic memory in our mental lives would be to try to identify a type of information about past events that could only be yielded by episodic memory. For example, one might argue that episodic memory brings with it information about the spatiotemporal context in which remembered events occurred (see McCormack, 2001, for discussion). However, the immediate difficulty that arises is that any such information one might try to home in on that might become useful again in the future (including information about the context in which an event occurred) could in principle be supplied by other types of memory (Klein, 2013, 2016). Presumably for this reason, researchers trying to account for the function of episodic memory have instead typically turned to focus on the way in which episodic memory delivers information about the past, highlighting its distinctive phenomenology that involves "re-experiencing" events (Suddendorf & Corballis, 1997, 2007). They suggest that episodic future thinking has a parallel type of phenomenology: we simulate the future and as a result "pre-experience" it. The idea is then that episodic memory may play some special role in supporting such simulations (for example, the title of Schacter et al.'s, 2007, paper is "Remembering the Past to Imagine the Future"). This line of thought can also be seen to underpin Boyer's description of why episodic memory helps with future-oriented decision-making. But the problem with this suggestion is that it is not clear why the information used in such simulations need derive specifically from recollecting particular past events. Why should episodic memory, with its backward-looking element, be what is required for simulation of the future?

There are two immediately obvious responses to this puzzle, both of which have been put forward by Stanley Klein. One response here would be to bite the bullet and hold that there is indeed nothing distinctive about the function of episodic memory in this regard. Klein (2013, 2016) recognizes the puzzle when he argues that episodic memory is not necessary for future mental time travel, suggesting on empirical and theoretical grounds that purely semantic memory might be not only necessary but also sufficient even for the type of mental simulation that is typically thought to accompany future mental time travel. The other, not incompatible, response would be to deny that we should place any special weight on the backward-looking element of episodic memory. So, for example, Klein (2014; Klein, Robertson, & Delton, 2010, 2011) argues that even episodic memory should be viewed primarily as future-oriented, with its backward-looking element merely an evolutionary byproduct of a system that evolved to help us anticipate the future (though see Klein, 2016, for a somewhat different evolutionary explanation of the backward-looking element of episodic memory).

We agree with Klein's (2013, 2016) assessment that some sorts of future-oriented thinking and decision-making need not necessarily involve episodic memory, and that there is empirical evidence to support this assessment. For example, we have been considering Boyer's (2008) hypothesis that the episodic system may play a special role in the paradigmatic case of decision-making that involves temporal discounting. But, in experimental settings, amnesic patients who lack episodic memory for the past are not more inclined to discount the future than healthy controls (Kwan, Craver, Green, Myerson, & Rosenbaum, 2013). This suggests that if Boyer (2008) is correct that temporal discounting is affected by the ability to imagine the future, it is a type of imagination that does not rely on being able to recollect the past.[3]

However, we believe that, even if it is true that the adaptive function of episodic memory ultimately lies with the role it plays in future-oriented decision-making, there may be ways of fleshing out this thought that also recognize the importance of its backward-looking element. The challenge is to consider whether there are aspects of future-oriented decision-making that do seem to require a type of mental simulation that is itself intrinsically bound up with a capacity to recollect one's past (i.e., that have what we termed a backward-looking element). In the next section, we will argue that experiencing and anticipating the emotion of regret play a crucial role in future-oriented decision-making. One of the defining features of regret is its backward-looking nature: by necessity, one can only regret what is in the past (although, as will be discussed, one may anticipate future regrets). Nevertheless, there are good reasons to believe that regret shapes our future behavior. A consideration of regret provides an alternative way of thinking about how episodic memory might play an essential role in future-oriented decision-making, not despite its backward-looking element but because of it.

## 2. REGRET AND DECISION-MAKING

The idea that regret plays an important role in decision-making has gained much currency over the last two decades (for reviews, see Koch, 2014; Roese, 2005; Zeelenberg & Pieters, 2007). Regret is typically defined within the psychological literature as the negative emotion that one feels when one realizes or believes that an outcome could have been better if one had chosen differently. In psychology, the term *regret* is typically reserved for cases in which one had responsibility for the choice and hence the outcome (Weisberg & Beck, 2012; Zeelenberg, Van Dijk, & Manstead, 1998, 2000; though see Ordóñez & Connolly, 2000). It has been shown to be an extremely common emotion; for example, one study (Shimanoff, 1984; see also Saffrey, Summerville, & Roese, 2008) indicated that it was the most frequent negative emotion mentioned in everyday conversation—and one that has far-reaching effects on our behavior (Roese, 2005; Zeelenberg & Pieters, 2007).

One way to argue for a close relation between regret and future-oriented decision-making might be to point to its close relation with agency and personal responsibility: indeed, Zeelenberg and Pieters (2007) argue that regret is the only negative emotion that has a special relation with one's own choices, such that it is only experienced if one believes one has made a poor or suboptimal choice. However, for present purposes we want to focus on the temporal character of regret. Zeelenberg and Pieters (2007, 8) capture this character in their claim that "[r]egret bridges the past and the future in the present." What they mean by this is that although regret is experienced in the present, it is always past-directed; at the same time, although regret is directed toward our past, it has the capacity to exert a strong influence on our future.

Regret researchers typically distinguish between two broad ways in which regret may affect future decisions: through experienced regret or anticipated regret. In the former case, experiencing regret about a choice might have a subsequent effect on what one decides to do when faced with that decision again. In the case of anticipated regret, the basic idea is that, when faced with a decision, one anticipates the regret that might result as a consequence of one's different choice options, and this guides the choice that one makes. In both cases, regret is assumed to affect behavior primarily as a result of regret aversion (Zeelenberg & Beattie, 1997; Zeelenberg, Beattie, van der Pligt, & deVries, 1996), that is, because people attempt to avoid feeling regret and take steps to regulate the emotion when they do experience it (Zeelenberg & Pieters, 2007).

Turning first to the case of experienced regret, studies suggest that experiencing regret may impact on subsequent decision-making because, following a poor decision that one regrets, one may make a better choice the next time one is faced with the same type of decision (O'Connor, McCormack, & Feeney, 2014; Ratner & Herbst, 2005; Zeelenberg & Pieters, 2004). In this way, regret may have an impact on choices that are encountered repeatedly (e.g., whether to hit the snooze button when the alarm clock goes off, or whether to eat the large slice of chocolate cake). O'Connor et al. (2014) demonstrated the effect of experienced regret in a developmental study by showing that children who experienced regret about a simple choice (which of two boxes to pick in order to win a prize) were more likely to switch choices when faced with the same decision again than children who did not experience this emotion. In this way, regret can be seen as a functional or useful emotion that helps us to learn and adjust our behavior adaptively (Roese & Summerville, 2005). In O'Connor et al.'s (2014) terms, experiencing regret facilitates "adaptive switching."

However, there may be other circumstances in which the experience of regret has an impact, albeit more indirectly, on future decision-making, even if one is never faced with the same choice again. Beike, Markman, and Karadogan (2009) argue that the paradigmatic case of experienced regret is one of missed or lost opportunity. They showed that people report the most intense regrets about choices that cannot be revisited, because as a result of making the initial choice, certain opportunities were lost and were lost irreversibly. Loss of opportunity in this sense could occur for both minor (e.g., deciding whether to buy a souvenir T-shirt in Disneyland) as well as major life choices (e.g., deciding whether to propose marriage), but may be particularly clear in the case of certain major life choices. To give an example, if one decides not to have children, for women at least there will come a point in their lives in which that choice cannot be reversed. In such circumstances, it is not possible for one's regret to affect behavior through adaptive switching, because the option to switch is no longer available. Nevertheless, the fact that one experiences regret in the face of lost opportunities may have an impact on one's future choices. Beike et al. (2009, 395) argue that "regretting lost opportunities reminds people not to lose out on other current opportunities. Hence, regret motivates people to seize the day." If this is correct, experiencing regret about missed opportunities could have wide-ranging effects on subsequent choices, because when faced with another opportunity-providing choice, given one's previous experience, one anticipates feeling regret about not taking up this option, and this informs one's choice.

There are many other circumstances in which anticipated regret is likely to affect decision-making. Experimental studies on economic decision-making have demonstrated that the anticipation of regret is one factor that shapes choice patterns, with people avoiding choices that are likely to lead to regret (e.g., Camille et al., 2004; Coricelli et al., 2005; Coricelli, Dolan, & Sirigu, 2007; Mellers, Schwartz, & Ritov, 1999). The general idea is that people try to minimize future regret by assessing the level of regret that might result from each choice and considering how likely it is that choices might result in such regret. There are influential formal models of economic decision-making that are based on this type of principle (Bell, 1982; Loomes & Sugden, 1982). Studies of economic decision-making in brain-damaged patients who have lost the ability to experience or anticipate regret indicate that a lack of regret is associated with poorer choice (Camille et al., 2004; Coricelli et al., 2007). The powerful nature of anticipated regret has also been demonstrated in numerous studies in applied psychology, where levels of anticipated regret have been shown to be good predictors of subsequent health-related behavior and experimental manipulations to induce anticipated regret have been demonstrated to impact on choice (e.g., with regard to deciding whether to use contraception [Richard, van der Plight, & DeVries, 1996], get a vaccination [Chapman & Coups, 2006], or take up smoking [Conner, Sandberg, McMillan, & Higgins, 2006]; see Koch, 2014, for review).

Viewed this way, regret can be seen to be a functional or useful emotion that helps guide better decision-making (Roese, 2005; Zeelenberg & Pieters, 2007). Indeed, when questioned, people seem to value regret more than any other negative emotion because they believe it helps them to avoid repeating their mistakes and helps them to know how to act in the future (Saffrey et al., 2008). Although research indicates that regret has a positive function, avoiding future regret turns out to be very important for one's well-being, at least in the longer term. Research has also shown that high levels of lingering regret are associated with poorer mental and physical health (Torges, Stewart, & Miner-Rubino, 2005; Wrosch, Bauer, Miller, & Lupien, 2007; Wrosch, Bauer, & Scheier, 2005). For example, Wrosch et al. (2005) found that among elderly adults, intensity of regret predicted higher levels of depressive symptoms, poorer physical health, and lower levels of reported life satisfaction. These authors argue that when reflecting on how one's life has unfolded, having long-standing regrets about the path one has chosen can be detrimental in measurable ways to one's health (Wrosch et al., 2007).

## 3. REGRET AND EPISODIC THINKING

The overall aim of this chapter is to explore whether we might come to a better understanding of how the possession of episodic memory can be seen to play a distinctive role in future-oriented decision-making by considering the role that regret has in decision-making. In section 2, we reviewed research that has demonstrated how regret, although an intrinsically backward-directed emotion, can nevertheless play an important role in future-oriented decision-making. Having established the importance of regret in this regard, we now need to consider how this emotion may depend upon episodic memory abilities. The thought that regretting something (or at least paradigmatic cases of doing so) involves episodic recollection may already be intuitive enough. But we think it is worth exploring in more detail just how exactly regret and episodic memory might be seen to be connected, as it may allow us to uncover how decision-making, too, might be seen to trade crucially on episodic thinking. We will do this in two stages. First, we will characterize the way of thinking about time that seems to be involved in both experienced and anticipated regret. Second, given this characterization, we will then consider how episodic memory plays a distinctive role in regret and, as a consequence, future-oriented decision-making.

### 3.1. Ways of Thinking about Time and Regret

#### 3.1.1. Experiencing Regret

Regret is by definition a counterfactual emotion: it depends upon counterfactual thought insofar as it involves looking back at a past decision that has led to a certain outcome and thinking about how one would have felt had one made a different choice. We can identify two important aspects of the way of thinking about time that is recruited in this sort of counterfactual thought. First, it seems to require an ability to consider two different temporal perspectives on the same point in time, where the relevant point in time is the point at which the choice was made. One of these is looking backward from one's current location in time to the choice point. That choice is now in the past, and its consequences have become apparent. However, this is not all one is doing when one is experiencing regret: one is also thinking counterfactually about how things could have turned out differently. Such counterfactual thought seems to require being able, at least in principle, to think back to a point in time when one was considering a choice, but making the choice was still in the future. From that alternative, temporal perspective, there were different options available to the decision-maker and it was not yet settled which one she would take. Experiencing regret seems to hinge on being aware of the availability of these different options, which are now in the past but which were once possible ways the future could have unfolded, and a consideration of how they would have yielded alternative outcomes.

The second, related, aspect of this way of thinking about time that we want to emphasize is that it requires a capability to engage in what elsewhere we have described as "event-independent thinking about time" (McCormack, 2014, 2015; McCormack & Hoerl, 2008). What we are trying to capture by this term is the ability to think about particular times independently of the events that have occurred or that could occur at such times. A spatial analogy is helpful in clarifying this notion. When we think of a place, we can think about it independently of the objects that occupy it. For example, we can think of "that place over there where my desk is" as remaining "that place" even if it was occupied by a different object or no object at all. That is, we can think about space in an object-independent way. Similarly, at least as mature thinkers, we can think about points in the past or the future as being distinct from any specific events that may have happened or will happen at those points in time. That is, we can think *about particular times* per se, rather than just being able to think about events.

The sort of counterfactual thought about the past that one engages in when experiencing regret recruits event-independent thought about time. When one thinks back to the time before one's choice was settled, in considering possible alternative courses of action one could have taken, one is able to think about the choice point as a location in time at which different events could have happened. Counterfactual thought of this sort is intimately connected to grasping that there are multiple possible ways in which events can or could have unfolded (Beck, Robinson, Carroll, & Apperly, 2006). What we are arguing here is that such a grasp should also be seen as part of what it is to be able to think about time itself (for this argument in more detail, see McCormack, 2015).

Our purpose in unpacking the temporal abilities involved in experiencing regret is to demonstrate that there is a very robust sense in which regret might be said to involve mental time travel into the past. First, one needs to be able, at least in principle, to adopt a past temporal perspective at which having made one's choice was still in the future, and the choice itself was as yet undecided. Second, it is genuinely a particular past time that the subject revisits in her thinking, in so far as the subject thinks of that time as one in which she made a particular choice, but does not think of the time just *in terms of* the choice she made then, but as one at which she could have made a different choice.

#### 3.1.2. Anticipating Regret

In the previous section we mentioned the idea that decision-making can also turn on the anticipation of regret. That is to say, in some choices agents aim to minimize anticipated regret—they imagine looking back at their choice from a point of view later in time and thinking counterfactually about how they would have felt had they made a different choice. What they are concerned with is how they will look back on a decision that they have yet to take once the decision has been made and they have to live with its consequences. Note that this type of thinking, too, can be seen to involve mental time travel in a very robust sense, albeit now into the future. Indeed, it is arguable that, because of the more complex temporal structure of the simulation it involves, anticipating regret more clearly involves mental time travel than the type of mental simulation described in section 1. In that section, we considered existing claims (e.g., those of Boyer, 2008) about how mental simulation may impact on decision-making about the future. The type of mental simulation that is invoked in that context is typically characterized as mental time travel from one's present point in time to a future one at which the outcome of one's choice will obtain (e.g., imagining a future reward, which may reduce temporal discounting). In fact, it is far from obvious how much of an understanding or reasoning specifically about *time* that type of simulation actually requires on the part of the subject. Indeed, as we have suggested, there is a sense in which the purpose of such simulation is precisely to diminish the relevance of time, by allowing the subject to "pre-experience" the future outcome already, reducing subjective temporal distance to it, and forging a tighter connection between one's present self and one's future one.

Time itself, by contrast, plays a far more central role in the reasoning involved in anticipating regret, particularly with regard to the need to occupy a future perspective in time from which one is looking back at what will then be a potentially irrevocable decision, but is now an as yet undecided choice, and comparing the outcome of that decision with the counterfactual outcome that would have obtained had one's choice been different. It is again genuinely a different time the subject occupies in imagination in this type of reasoning, in virtue of the fact that this time is thought of in an event-independent way. That is to say, the imaginary future point of view occupied by the subject is conceived of as a time at which only one of the possible choice outcomes will obtain, but it is not thought of in terms of which such outcome will obtain, but as a time at which different outcomes may obtain, depending on the choice.

Thus, mental time travel abilities, both into the past and into the future, can be seen to play a crucial role in experiencing and anticipating regret. While anticipating regret involves future-directed mental time travel, though, and experiencing regret itself involves past-directed mental time travel, there is a backward-looking element that is an essential embedded ingredient in the future-directed thinking involved in anticipating regret. Because of this, we could say that anticipation of regret involves not just prospection (Buckner & Carroll, 2007) but prospective retrospection.

### 3.2. Episodic Memory and Regret

We have argued that there is a rich sense in which experiencing and anticipating regret involve mental time travel. However, what remains to be elaborated in more detail is how the ability to experience and anticipate regret depends essentially on memory that is specifically episodic, rather than some other type of memory or knowledge about the past gained in another way. As we mentioned earlier, it may seem intuitive that regret involves recollecting the past. Indeed, we think that it is highly likely that in many cases in which people experience regret, not only are they engaged in past-directed thought of the sort outlined earlier, but that episodically remembering the past is a central ingredient of such thought. That is, although the past-directed thought must have a counterfactual aspect, episodically remembering a past choice forms the backdrop for the associated counterfactual cognition. When we described the temporal structure of thought that seems to be involved in experiencing regret, we emphasized that it seemed to involve considering dual temporal perspectives: thinking of one's choice as being in the past but also having an awareness that there was a time at which this was not yet the case and alternative choices could have been made. What is intuitively compelling about the idea that regret typically involves episodic recollection is that in regretting a choice one seems to be not simply retrieving information about a past choice, but mentally revisiting a previously occupied point in time and dwelling on the possibilities that existed when one actually occupied that temporal perspective. And it is this sort of revisiting of the past that usually held to be the preserve of episodic memory. Indeed, as we will describe in the following, recent neuroscientific research strongly suggests that episodic memory is typically involved in the sort of counterfactual thought about the past that features in regret.

Note, though, that we want to make a different (and in fact stronger) claim about the link between episodic memory and regret than simply that this type of memory normally is a central ingredient of regret. For a start, insofar as we are seeking a role for episodic memory to play, via its connection with regret, in future-directed decision-making, we are also interested in ways in which episodic memory might be seen to be involved in underpinning the capacity to anticipate regret,[4] which we have argued plays an important role in such decision-making. Yet anticipating regret concerning a future decision cannot literally involve episodic recollection of that decision in the way that experiencing regret about a past decision can, and it is also not always a matter of recollecting past decisions of the same type, as we try to anticipate regret even when making entirely novel decisions. Moreover, even in the case of experienced regret, it is straightforward to think of instances in which one could regret a decision despite being unable to episodically remember making the decision itself. For example, you could switch on your computer in the morning to find out to your horror that in last night's drunken stupor, you sent an inadvisable email to your boss. You could still engage in past-directed mental time travel in such an instance, and deeply regret the incident, but under such circumstances your past-directed mental time travel would be based on reconstruction rather than recollection of the past. Rather than arguing that all experiences of regret must involve episodically remembering the past, we want to instead claim that episodic memory underpins the very ability to engage in the ways of thinking about the past and the future required for regret (both experienced and anticipated). As a result, regret based on reconstruction of, rather than episodic memory of, the past must be parasitic or dependent on being able to experience regret as a consequence of episodic memory.

A related line of thought can be found in Bertrand Russell, who seems to be describing what nowadays would be called episodic memory when he says that "[i]t is obvious that we often remember what we have seen or heard or had otherwise present to our senses, and that in such cases we are still immediately aware of what we remember, in spite of the fact that it appears as past and not as present" (Russell, 1912, 26). Russell then goes on to argue that "[t]his immediate knowledge by memory is the source of all our knowledge concerning the past: without it, there could be no knowledge of the past by inference, since we should never know that there was anything past to be inferred" (1912, 26). In short, his idea is that recognizing that there is a past in the first place, portions of which we can reconstruct from information obtained in the present, presupposes having knowledge of at least some portions of that past that is not thus reconstructed, through (what would now be called) episodic memory. Thus, insofar as regret, too, can sometimes be based on reconstructing the past from present information rather than actually episodically recollecting the past, this, too, is only possible against a backdrop of episodic memory.

One way of understanding Russell's argument here is in terms of an idea that Annette Baier (1976, 220) expresses by saying that "it is not an unimportant conceptual truth that memory is of times, while knowledge is of facts, and that times cannot be discontinuous, as known facts can be disjoint." As the context makes clear, the type of memory she has in mind here is, more specifically, episodic memory, which she explicitly contrasts with "the sort of human memory which is just information storage, just a capacity to regurgitate input" (1976, 219). The point she can be seen to be making is that an important dimension of the contrast between the two types of memory (which psychologists would call the contrast between episodic memory and semantic memory) is that, whereas the latter presupposes a separate understanding of whatever information is retained, and cannot explain such an understanding, the former can itself contribute to, or be a manifestation of, understanding—specifically an understanding of time itself as a domain in which the remembered events fall (see also Debus, 2013). It is this type of understanding, which we have also called event-independent thought about time, which also makes it possible for us to reconstructively populate the past with events we cannot actually remember and to think counterfactually about how, even though we did one thing at a certain point in the past, we could have done something else at the time.

Thus, the general claim about the connection between episodic memory and regret that we want to put forward is that, even though regret may sometimes occur in the absence of episodic memory for the particular past action that is its object, episodic memory has a key role to play in grounding the kind of understanding of time involved in regret (and also in anticipating regret). One way of understanding this claim is as a developmental one—that episodic memory is required for the acquisition of the relevant understanding of time. But a stronger, constitutive, reading of it is also possible—that is, that episodic memory abilities are also required to retain such an understanding. Whether it should only be understood in the former sense, or whether it can also be defended on the latter reading, is not our current concern (for a debate on this issue, see Hoerl, 1999, and Craver, 2012). Rather, we can ask whether, apart from the kinds of philosophical considerations mentioned earlier, there is also empirical support for the idea of a connection between episodic memory and regret of the general type we have suggested.[5] As it turns out, there are several recent findings in psychology that at least point in such a direction.

In line with the suggestion that episodic memory and regret are closely linked, recent neuroimaging studies have indicated that episodic memory/future thinking, and thinking counterfactually about a past event seem to depend, at least in part, on the same core network (often referred to as the Default Mode Network; see Schacter et al., 2012, for a review). This line of research has focused on what has recently been termed *episodic counterfactual thinking* (De Brigard, Addis, Ford,

Schacter, & Giovanello, 2013; De Brigard & Giovanello, 2012; De Brigard, Spreng, Mitchell, & Schacter, 2015), that is, counterfactual thinking that involves imagining how a specific episode in one's personal past could have been different, which is the sort of counterfactual thinking that is central to the experience of regret. Recent fMRI studies indicate that there is significant overlap between the brain regions involved when participants are asked to episodically remember an event from their past and when they are asked to engage in the sort of upward episodic counterfactual thinking associated with regret (De Brigard et al., 2013; Van Hoeck et al., 2013; Van Hoeck, Ma, Van Overwalle, & Vandekerckhove, 2010; Van Hoeck, Watson, & Barbey, 2015). This overlap is particularly notable when the counterfactual thinking involves imagining realistic alternatives to what actually happened to oneself (De Brigard et al., 2013) and is largely absent when the counterfactual thinking is impersonal (De Brigard et al., 2015). De Brigard et al. (2015) conclude from these studies that specifically episodic memory is typically heavily involved when thinking counterfactually about events in one's own past. Indeed, De Brigard (2014, 177) has argued that "the very cognitive system by means of which we operate with mental contents about personal events that did happen also allows us to process mental contents about events [that] we think could have plausibly happened in our lives."

Thus, recent neuroscientific evidence strongly supports the idea that there is a close link between episodic memory, counterfactual thinking, and regret, although of course evidence of this type is restricted to establishing correlations and leaves the precise nature of the relevant links between different cognitive abilities underdetermined. Interestingly, though, there is also some, albeit limited, empirical evidence suggestive of the more specific claim that episodic memory is necessary for the capacity to experience regret. This emerges from the only attempt we are aware of to examine regret in amnesia. Craver, Kwan, et al. (2014) interviewed K. C., whom they described as "the definitive case of episodic amnesia," about regret. What they showed was that K. C. had a good understanding of what regret was, in that he could give a generalized description of the nature of regret and the sort of things that might cause regret. Strikingly, though, K. C. was completely unable to describe any regrets that he had, and he was even unable to think of any regrets that anyone else (such as his mother) might have (see also Craver, Cova, et al., 2014). This is despite K. C. still possessing information about his past that remained intact in his semantic memory. Assuming that his responses to direct questioning provide reliable enough evidence, the case of K. C. thus suggests that, in the absence of episodic memory, individuals lack the mode of mental simulation that allows them to experience regret or even to imagine another's regrets.[6]

So far, we have primarily considered how experiencing regret may require episodic memory. However, as we pointed out in section 2, regret has an important impact on future-oriented behavior, not just as a result of experiencing it, but as a result of anticipating it. Is it plausible to argue that episodic memory also plays a role in the anticipation of regret? The answer to this question depends on what is meant by "plays a role." If what is meant is "plays a role developmentally," then making such a case is straightforward: if episodic memory is required for experiencing regret, then insofar as anticipating regret depends developmentally on being able to experience it (and the evidence suggests that it does; McCormack & Feeney, 2015), then episodic memory abilities are a developmental necessity for anticipated regret. Note, however, that in addition to arguing that episodic memory might be developmentally necessary for experiencing regret, at the start of this subsection we also suggested that it is intuitively very plausible that when actually experiencing regret, one typically is engaging in recollection and thus recruiting episodic memory. As we have already pointed out, it is less straightforward to see how episodic memory may play a similar role in the anticipation of regret.

Again, one answer that one might give here is that retrieval of information from episodic memory is not the critical ability that is involved in anticipation of regret (although there may indeed be many occasions on which such memory retrieval informs the future simulation). Rather, it is the mode of thinking about time that is instantiated in episodic memory that is critical in the anticipatory case. As we have described it, the anticipation of regret has not only a forward- but also a backward-looking element: one is imagining oneself looking back on a decision that will be in the past, but is currently in the future. And insofar as this backward-looking element of the simulation can be viewed as typically supplied by the episodic memory system, episodic memory processes can be seen to contribute to the anticipation of regret.

Finally, before leaving the issue of the connection between episodic memory and regret, it is worth considering further the different ways in which episodic memory might make its contribution depending on the nature of the regret. We pointed out in section 2 that regrets about missed opportunities may be particularly salient cases of regret (Beike et al., 2009). These are regrets that one cannot easily come to terms with because the previously presented opportunity no longer exists. What may be distinctive about such regrets is that on considering the missed opportunity, one often *regrets anew* one's choice, a process that re-involves episodic memory, at least in central cases. By contrast, experiencing regret regarding a repeatable choice can straightforwardly impact on decision-making without regretting anew. In the case of repeatable choices, it is plausible that, when faced with the same choice again, the most that memory need deliver is that one once came to regret this choice—information that could perhaps straightforwardly be supplied by semantic memory. It may even be the case that having experienced regret about a choice can affect one's subsequent decision-making without one having to remember that one once regretted that choice. That is, experiencing regret may have led one to think of that type of choice as a poor one, and to avoid it subsequently, even if one no longer has any memory of the experience of regret itself (and, arguably, even if one cannot remember ever even facing such a choice before).

However, this cannot be the means by which regret about lost opportunities can have an impact on future decisions. As described in section 2, the impact of these sorts of regrets must be more indirect. By definition, one cannot be faced with exactly the same type of choice again. In the case of these lingering regrets, it may be that one re-engages in the same sort of mental time travel that resulted in regret in the first place, which we have already argued typically involves episodic recollection. Then the question is how regretting a lost opportunity anew, which recruits episodic memory, may impact on decision-making. It was suggested in section 2 that, when faced with another opportunity-providing choice, one may anticipate regretting another sort of missed opportunity, and this may shape the choice that one makes. By this means, episodic memory can have a further, albeit more indirect, input into anticipated regret.

## 4. BACK TO THE QUESTION OF FUNCTION

We started this chapter by raising the question of the evolutionary function of episodic memory: Of what adaptive benefit is it to us to be able to remember particular past events as such, given that those particular events will never come again? The way in which we have approached this question is by looking at ways in which episodic memory might have a distinctive role in decision-making. In particular, we have looked at the role of regret in decision-making, and the sophisticated capacities for mental time travel recruited in experiencing regret and anticipating regret. And we have discussed reasons for thinking that the ability to recollect past events in episodic memory underpins these capacities for mental time travel. We want to conclude this chapter by considering how exactly the arguments we have put forward bear on the question of the evolutionary function of episodic memory itself.

It might perhaps be thought that showing how episodic memory underpins certain abilities recruited in particular forms of decision-making is sufficient to demonstrate how it can have an adaptive role. Yet, we have not quite closed the circle, as, in order to complete the answer about the evolutionary function of episodic memory, we also have to show that there are no other, considerably simpler, forms of decision-making processes that could essentially play the same role—in which case it is not clear why these specific ones have evolved—and also that the relevant forms of decision-making are in fact adaptive. These are the issues we want to address in what follows.

One type of decision-making that we have discussed turns on experienced regret. In section 2, we presented evidence that experienced regret has an influence on decision-making by facilitating adaptive switching. In a situation in which a subject has come to regret a particular type of choice they made in the past, this will lead them to avoid making the same choice again when encountering a choice situation of the same type in the future. There is a problem, though, with relying on this sort of case in framing claims about the evolutionary function of regret, and by extension of the episodic memory abilities that plausibly underpin the capacity to experience regret.

The problem here is somewhat similar to the problems we raised, in section 1, for attempts to frame claims about the evolutionary function of episodic memory more directly in terms of ways in which it might be involved in decision-making in situations in which the decision-maker has come across a choice situation of the relevant type before. In such choice situations, what is important is that past experience equips the subject with the means to know how to act next time she encounters a situation of the relevant type again. Of course, if the subject comes to regret having made the poor choice in the past, this can be one way for her to possess such knowledge. However, given the rather sophisticated cognitive capacities that we have argued are required to experience regret, and also given the emotional cost that feeling regret carries with it, even in situations in which it aids decision-making, it could be thought that this is a rather cost-intensive way of solving the problem of finding a way to learn from past experience how to make better decisions in the future. Insofar as the function of regret in these sorts of cases is simply to make salient what the right choice is to make next time, it is not obvious that the backward-looking aspect of regret plays an essential role in them, and it is possible to think of other, less cost-intensive ways of solving that problem.[7] It is therefore difficult to argue that the function for which regret has evolved is primarily to aid decision-making in this type of situation.

This suggests that a better context within which to frame claims about the evolutionary function of regret is in fact that of decision situations in which the subject faces choices of a type she has never made before. As we have argued, regret plays a role in such decision situations, too, because deliberation in such situations turns on minimizing anticipated regret. Yet here, too, it might in fact be thought that there is a particular kind of challenge that arises for the line of thought we have sketched, if it is to provide us with a potential insight into functions for which particular cognitive capacities have evolved.[8]

Our suggestion has been that decision-making in the context of major life choices, in particular in cases in which the subject is faced with a choice of a type she has never faced before, often involves reasoning about anticipated regret. That is to say, the subject's deliberations turn on whether, in looking back on her choice, she will come to feel regret when comparing it with the other choice(s) she could have made. Yet, it might be argued that it is hard to see why using such a decision-making procedure should be considered to be adaptive in the context of choice situations of a kind the subject has not encountered before—in particular, ones that involve potential choice options of a kind that the subject has never encountered before. This is because, in such a situation, the subject does not have enough to go on to figure out *which choice she will in fact come to regret* when comparing it with others, precisely because she has not been in the relevant situation before. Thus, if she comes to decide *that* she will regret certain choices more than others, there is an important sense in which that decision is not based on proper evidence, because she is not yet in possession of the relevant evidence that would allow her to make that decision (we will say more about this shortly). As such, there is no reason to think that using the decision procedure will reliably lead her to do what is in fact best for her—and it is therefore difficult to see how it could be adaptive. Indeed, it may be thought that using this decision procedure comes at a considerable cost, especially if the subject ends up pondering endlessly about whether or not she will come to regret her decision. Also, as we mentioned before, regret itself comes at a cost to one's welfare, especially in situations in which the decision that subject feels regret about cannot possibly be reversed; thus, the realization that one might unwittingly make a choice that one will come to regret this way is also likely to impact negatively on one's welfare to some extent.

One central ingredient in the line of thought just sketched has recently been discussed under the heading of *transformative choice*, where the claim has been made that choices of the kind just described, in which the subject faces the question as to whether she should make a decision of a type she has never made before, cannot be made rationally (Paul, 2014, 2015; Ullmann-Margalit, 2006). In particular, the argument has been that, whatever the decision-making processes involved in making such choices, there is an important sense in which these processes are unable to yield a normatively correct answer as to which is actually the better choice for the subject to make, because one or more of the choices open to the subject is potentially *transformative.*

L. A. Paul (2014) distinguishes in particular between two different ways in which choices can be transformative: epistemically transformative choices and personally transformative choices. In short, decisions can be epistemically transformative or personally transformative—and many major life decisions such as the decision to have a child are arguably typically both—because *experience* can be epistemically transformative and personally transformative.

Experience can be epistemically transformative, because it can provide the subject with a type of knowledge that she could not possess before undergoing the experience—namely, what it is like for her to undergo the experience itself. The only way to acquire this form of knowledge is by actually undergoing the relevant form of experience onself (Jackson, 1982). An epistemically transformative choice, thus, is a choice where what is at stake is whether or not one should undergo an experience of a type one is not yet familiar with.

Experience can also be personally transformative; an individual can change as a person in response to experience, insofar as experience can change what her interests and values are. As a result, choices, too, can be personally transformative insofar as they are choices between options that may change the subject in this way. And the existence of personally transformative experiences also affects decision-making in a broader way. Even if personal transformation is not the object of one's choice—that is, one's decision is not between whether or not one should undergo a (potentially) personally transformative experience—the existence of personally transformative experiences can make it the case that one faces choices about outcomes that one may come to view differently at the time because one's interests and values have changed (Pettigrew, 2015).[9]

A key theme of Paul's discussion of the nature of epistemically and personally transformative choices is that they imply that we are sometimes faced with choices in which at least one alternative, and sometimes each of them, constitutes, in an important respect, a step into the unknown. In both epistemically and personally transformative choice, we lack a crucial piece of information concerning the value we will attach to at least one of the outcomes we have to choose between: in epistemically transformative choice, we have not had the relevant experience before, and may be unable for that reason to determine how much we would value (or disvalue) having it; in personally transformative choice, we may not lack any relevant knowledge about the outcome itself, but we do not yet know how much our choice will affect the extent to which we will value it.

The key distinguishing features of transformative choices, then, are that the subject faces a genuine choice, but also knows that she lacks a crucial piece of knowledge about at least some of the outcomes she has to choose between, insofar as they involve situations of a type she has no experience of as yet, or insofar as her future self might attach different values to them than her present self does. In particular, as Paul argues, there is a distinctively first-personal aspect to the subject's epistemic lack: she lacks a specific type of epistemic basis for judging what value she herself will attach to the relevant outcomes, because she has no prior experience of being in the types of situations they involve or cannot anticipate to what extent her own values may change as a consequence of her decision.[10] As a result, she argues, a standard decision-theoretic model based on maximizing expected value cannot accommodate choices involving transformative experience—it provides no rational means for the subject to make her choice.

The existence of transformative choices, then, may be thought to raise the following problem for our overall argument. If it is right that certain options the subject must choose between in transformative choice situations are steps into the unknown, whose future value to her the subject has no rational way of assessing, it might seem that *whatever* decision-making processes the subject uses in such choice situations will be unable to reliably (or even probabilistically) lead to an outcome that is in fact better for her, when compared with the other possible outcomes. And this general argument might be thought to apply also to the decision-making process we have sketched, which turns on reasoning about anticipated regret, as much as to any other process the subject might use in transformative choice situations. Thus, no reason for thinking that this type of decision-making process is adaptive has yet been given.[11]

In fact, though, we believe that the issues we have discussed are somewhat orthogonal to those at stake in Paul's argument that, from a certain kind of normative perspective, transformative choices cannot be made rationally. It is important in this context to clarify that our claim is not that the decision-making processes that we have described, which turn on anticipating future regret, can somehow make up for the particular epistemic lack suffered by the subject facing transformative choices that Paul is concerned with. Within the context of the present chapter, one way of describing Paul's (2014) argument is in terms of the idea that the kind of decision-making based on simulation that we described in section 1 of this chapter is unavailable in the case of transformative choices, because the subject faces a choice of the relevant type for the first time. The difficulty, as she sees it, is that we are unable to judge what the subjective value of certain future outcomes may be to us, if they involve unfamiliar, transformative experiences. In section 3, we described a different, rather more sophisticated form of simulation centered on anticipating regret. In doing so, however, we did not mean to suggest that this form of simulation can somehow solve the problem Paul says cannot be solved using the simpler type of simulation when we are dealing with a case of transformative choice. In an important sense, that problem remains. What we are suggesting, in some sense, is that the more sophisticated form of simulation may be seen to help the subject come to a decision despite it.

But this still leaves the question as to how the ability to anticipate regret might be adaptive in the case of transformative choices, despite not being able to remove the epistemic lack that subjects face in those choices, that is, the fact that she cannot as yet know what value she will assign to at least some of the outcomes she has to choose between. We want to end with three brief and admittedly speculative comments on this issue.

First, while the ability to anticipate regret cannot make up for the epistemic lack involved in facing transformative choices, it plausibly plays a key role in one's being able to appreciate and take account of their potentially transformative character in the first place. Transformative choices carry the risk that one may end up with an outcome one is not happy with but which one cannot make undone, and regret about past such outcomes or about lost opportunities can alert one to the possibility of one's choice having such consequences again in the future. Appreciating the irrevocability of certain choices (Hoerl, 1999; Ullmann-Margalit, 2006), and of the effect they may have on oneself, can clearly be beneficial, for instance in the context of the thought that one is not ready to make the decision to have a child, because one's life situation is not yet settled enough to cope with some of the more stressful or negative experiences associated with having a child.

Second, awareness that a choice is potentially transformative, and that one may live to regret it, may also lead one to seek more third-person information about the extent to which others who have made the relevant choice have come to value their decision. While third-person information may not conclusively allow the subject to settle how she herself may feel upon making the relevant type of choice, it may at least increase her confidence as to whether or not that choice will be the right one for her to take as well (Dougherty, Horowitz, & Sliwa, 2015; Pettigrew, 2015). Note further that the decision-making process that we have described, which turns on anticipating regret, might also explain in this context how the subject's decision can be first-personal while being based on such third-person information. (Paul [2014] describes this as the question as to how decisions in the context of transformative choice can be *authentic*.) A concern with future regret, that is, might help explain how a subject can choose a certain future course of action *for herself*, even though doing so constitutes a step into the unknown for her, whose actual effect on her she is unable to predict, and all she has to go on in judging that it will be the right course of action is the testimony others have given her of their own cases.

Finally, a focus on how decision-making in the context of transformative choice is bound up with the ability to anticipate regret might also provide for a way of recognizing a benefit in transformative choice itself. Animals and young children presumably undergo experiences that are epistemically transformative for them, or that transform them as individuals. Yet, it is less plausible to think that this is something they make choices about as such (rather than it occurring as a mere consequence of their other choices). As adults, however, we do sometimes consciously seek out new experiences or personal transformation, even while knowing that doing so constitutes a step into the unknown. Moreover, there seems to be a particular kind of value that we attach to the very idea of having new experiences or doing something for the first time, rather than simply doing things we are already familiar with. As we mentioned earlier, for instance, the kinds of choices people seem to regret most take the form of lost opportunities. And while it is of course right that consciously seeking out new experiences and challenges may not always produce the best result for us (indeed, as we said two paragraphs ago, it carries with it specific risks), the fact that we can and value doing so is arguably part of what explains the vast gulf between the kinds of learning abilities that we find in animals and the distinctive way in which humans are able to take active control of their lives and their own acquisition of knowledge.

## ACKNOWLEDGMENTS

We thank Felipe De Brigard and an anonymous reviewer for very helpful comments on an earlier draft of this chapter. Part of the first author's work on this chapter was carried out within the context of the Spanish Government project no. FFI2012-35153, and the chapter was also supported by ESRC grant number ES/K000411/1 to the second author.

## NOTES

1. We will say more about this understanding of regret as an emotion directed at choices. Also, we will focus specifically on regret, as regret can be seen to play a particularly prominent role in decision-making. However, for other purposes it

might be useful to think of regret as just one example of a broader set of similarly structured emotions with both a backward-looking and a counterfactual ingredient, of which relief is also a member.

2. On similar grounds, Zentall, Clement, Bhatt, and Allen (2001) argue that tasks that require extensive prior training are unsuitable for demonstrating the existence of episodic memory in animals.
3. The alternative possibility is that the correct description of such patients is that they cannot properly imagine the future, and Boyer is wrong in claiming that mental time travel can moderate temporal discounting. Craver, Kwan, Steindam, and Rosenbaum (2014) argue for this interpretation.
4. We will say more about the relevance of this for our overall argument in section 4.
5. What we know about development is clearly consistent with at least the developmental reading of the claim mentioned earlier: children do not seem to be able to experience regret until around 5–6 years (O'Connor, McCormack, & Feeney, 2012; O'Connor et al., 2014; Weisberg & Beck, 2012), and cannot anticipate regret until around 8 years (McCormack & Feeney, 2015). Thus, these abilities emerge at a time when episodic memory abilities are already intact and robust. However, given the sophisticated temporal and counterfactual reasoning abilities involved in regret, it is unsurprising that it is a relatively late developing emotion, which makes it hard to argue for any more specific link with episodic memory on the basis of the developmental evidence.
6. Craver, Cova, et al. (2014) demonstrate that K. C. is susceptible to a particular type of effect in economic decision-making (the Allais paradox) that has previously been explained in terms of anticipated regret (Loomes & Sugden, 1982). One interpretation of this finding is that K. C. can in fact anticipate regret, even though he does not seem to experience it. Such an explanation is broadly compatible with the recent suggestion that the anticipation of regret may involve different brain regions from the experience of regret, and that the former can be intact even in the absence of the latter (Levens et al., 2014). However, Craver, Cova, et al. (2014) instead suggest that regret-based explanations of the paradox must be incorrect.
7. It is of course still true that regret, with its counterfactual element, provides for a *less* cost-intensive way of learning from past experience than even simpler ones, because it makes salient not only that one's choice was a poor one, but which alternative choices might have been better. In the absence of such counterfactual thought, learning may be slower, as it might be necessary to actually try out each of the alternatives one by one. This is one reason that counterfactual thought is often assumed to be particularly beneficial in correcting suboptimal behavior (see Epstude & Roese, 2008, for more detailed discussion of the adaptive benefits of counterfactual thinking).
8. See also Schacter et al. (2015, 20), who, immediately after stating that "[b]oth 'prefeeling' a possible future scenario and reminiscing about foregone past choices could [. . .] provide motivational incentives that foster more farsighted decisions" (in contrast to decisions driven by more immediate interests), go on to point out that "this mechanism need not always enhance the probability of making choices that are [in fact] beneficial in the long run."
9. This latter issue is less prominent in Paul's own discussion of transformative experience, and we will also largely set aside in what follows. There are in fact two different issues that decision-makers face in this situation: first, they may not be able tell to what extent their interests and values will change; but even to the extent

that they are able to do so, they face the question of which set of values they should choose to evaluate the desirability of the outcome.

10. Most of the existing critical responses to Paul's argument, in one way or another, target this element of her argument, and the implication that transformative choices pose problems separate from those involved in decision-making under other forms of uncertainty. The suggestion, in short, is that while testimony from others or observational knowledge of other's behavior who are in the kind of situation envisaged may not provide *conclusive* evidence about what things will be like in one's own case (because of what Paul refers to as the reference class problem), they can provide some evidence—and in this the choice situation is no different in kind from other choice situations in which we have only limited information about the likely outcomes of our actions. We will leave this issue to one side.
11. Note that the argument described in this paragraph is not one Paul herself makes. We are just describing one way in which one might think her argument bears on our position.

## REFERENCES


Aristotle. (1930). On memory and reminiscence (J. I. Beare, Trans.). In W. D. Ross (ed.), *The works of Aristotle* (vol. 3). Oxford: Oxford University Press.

Atance, C. M., & O'Neill, D. K. (2001). Episodic future thinking. *Trends in Cognitive Sciences*, *5*(12), 533–539.

Baddeley, A. (1988). But what the hell is it for? In M. M. Gruneberg, P. E. Morris, & R. N. Sykes (eds.), *Practical aspects of memory: current research and issues.* Vol 1: *Memory in everyday life* (pp. 3–18). Chichester, UK: John Wiley.

Baier, A. (1976). Mixing memory and desire. *American Philosophical Quarterly*, *13*(3), 213–220.

Bartels, D. M., & Rips, L. J. (2010). Psychological connectedness and intertemporal choice. *Journal of Experimental Psychology: General*, *139*(1), 49–69.

Beck, S. R., Robinson, E. J., Carroll, D. J., & Apperly, I. A. (2006). Children's thinking about counterfactuals and future hypotheticals as possibilities. *Child Development*, *77*(2), 413–426.

Beike, D. R., Markman, K. D., & Karadogan, F. (2009). What we regret most are lost opportunities: a theory of regret intensity. *Personality and Social Psychology Bulletin*, *35*(3), 385–397.

Bell, D. E. (1982). Regret in decision making under uncertainty. *Operations Research*, *30*(5), 961–981.

Bennett, J. (1964). *Rationality: an essay towards an analysis.* London: Routledge and Kegan Paul.

Benoit, R. G., Gilbert, S. J., & Burgess, P. W. (2011). A neural mechanism mediating the impact of episodic prospection on farsighted decisions. *Journal of Neuroscience*, *31*(18), 6771–6779.

Bergson, H. (1991). *Matter and memory* (N. M. Paul & W. S. Palmer, Trans.). Cambridge, MA: MIT Press.

Boyer, P. (2008). Evolutionary economics of mental time travel? *Trends in Cognitive Sciences*, *12*(6), 219–224.

Buckner, R. L., & Carroll, D. C. (2007). Self-projection and the brain. *Trends in Cognitive Sciences, 11*(2), 49–57.

Camille, N., Coricelli, G., Sallet, J., Pradat-Diehl, P., Duhamel, J. R., & Sirigu, A. (2004). The involvement of the orbitofrontal cortex in the experience of regret. *Science, 304*(5674), 1167–1170.

Chapman, G. B., & Coups, E. J. (2006). Emotions and preventive health behavior: worry, regret, and influenza vaccination. *Health Psychology, 25*(1), 82–90.

Conner, M., Sandberg, T., McMillan, B., & Higgins, A. (2006). Role of anticipated regret, intentions and intention stability in adolescent smoking initiation. *British Journal of Health Psychology, 11*, 85–101.

Coricelli, G., Critchley, H. D., Joffily, M., O'Doherty, J. P., Sirigu, A., & Dolan, R. J. (2005). Regret and its avoidance: a neuroimaging study of choice behavior. *Nature Neuroscience, 8*(9), 1255–1262.

Coricelli, G., Dolan, R. J., & Sirigu, A. (2007). Brain, emotion and decision making: the paradigmatic example of regret. *Trends in Cognitive Sciences, 11*(6), 258–265.

Craver, C. F. (2012). A preliminary case for amnesic selves: toward a clinical moral psychology. *Social Cognition, 30*(4), 449–473.

Craver, C. F., Cova, F., Green, L., Myerson, J., Rosenbaum, R. S., Kwan, D., & Bourgeois-Gironde, S. (2014). An Allais paradox without mental time travel. *Hippocampus, 24*(11), 1375–1380.

Craver, C. F., Kwan, D., Steindam, C., & Rosenbaum, R. S. (2014). Individuals with episodic amnesia are not stuck in time. *Neuropsychologia, 57*, 191–195.

De Brigard, F. (2014). Is memory for remembering? Recollection as a form of episodic hypothetical thinking. *Synthese 191*(2), 155–185.

De Brigard, F., Addis, D. R., Ford, J. H., Schacter, D. L., & Giovanello, K. S. (2013). Remembering what could have happened: neural correlates of episodic counterfactual thinking. *Neuropsychologia, 51*(12), 2401–2414.

De Brigard, F., & Giovanello, K. S. (2012). Influence of outcome valence in the subjective experience of episodic past, future, and counterfactual thinking. *Consciousness & Cognition, 21*(3), 1085–1096.

De Brigard, F., Spreng, R. N., Mitchell, J. P., & Schacter, D. L. (2015). Neural activity associated with self, other, and object-based counterfactual thinking. *Neuroimage, 109*, 12–26.

Debus, D. (2013). Thinking about the past and experiencing the past. *Mind and Language, 28*(1), 20–54.

Dougherty, T., Horowitz, S., & Sliwa, P. (2015). Expecting the unexpected. *Res Philosophica, 92*(2), 301–321.

Eacott, M. J., & Easton, A. (2012). Remembering the past and thinking about the future: is it really about time? *Learning and Motivation, 43*(4), 200–208.

Epstude, K., & Roese, N. J. (2008). The functional theory of counterfactual thinking. *Personality and Social Psychology Review, 12*(2), 168–192.

Hershfield, H. E. (2011). Future self-continuity: how conceptions of the future self transform intertemporal choice. *Annals of the New York Academy of Sciences, 1235*(1), 30–43.

Hoerl, C. (1999). Memory, amnesia, and the past. *Mind and Language, 14*(2), 227–251.

Hoerl, C. (2001). The phenomenology of episodic recall. In C. Hoerl & T. McCormark (eds.), *Time and memory: issues in philosophy and psychology* (pp. 315–338). Oxford: Oxford University Press.

Hoerl, C. (2007). Episodic memory, autobiographical memory, narrative: on three key notions in current approaches to memory development. *Philosophical Psychology, 20*(5), 621–640.

Hoerl, C. (2008). On being stuck in time. *Phenomenology and the Cognitive Sciences, 7*(4), 485–500.
Hoerl, C. (2014). Remembering events and remembering looks. *Review of Philosophy and Psychology, 5*(3), 351–372.
Hopkins, R. (forthcoming). Imagining the past: on the nature of episodic memory. In F. Dorsch & F. MacPherson (eds.), *Perceptual imagination and perceptual memory.* Oxford: Oxford University Press.
Jackson, F. (1982). Epiphenomenal qualia. *The Philosophical Quarterly, 32*(127), 127–136.
Kim, B. K., & Zauberman, G. (2009). Perception of anticipatory time in temporal discounting. *Journal of Neuroscience, Psychology, and Economics, 2*(2), 91.
Kim, B. K., & Zauberman, G. (2013). Can Victoria's Secret change the future? A subjective time perception account of sexual-cue effects on impatience. *Journal of Experimental Psychology: General, 142*(2), 328.
Kim, B. K., Zauberman, G., & Bettman, J. R. (2012). Space, time, and intertemporal preferences. *Journal of Consumer Research, 39*(4), 867–880.
Klein, S. (2013). The temporal orientation of memory: it's time for a change of direction. *Journal of Applied Research in Memory and Cognition, 2*, 222–234.
Klein, S. (2014). Autonoesis and belief in a personal past: an evolutionary theory of episodic memory indices. *Review of Philosophy and Psychology, 5*(3), 427–447.
Klein, S. (2016). Autonoetic consciousness: reconsidering the role of episodic memory in future-oriented self-projection. *Quarterly Journal of Experimental Psychology, 69*(2), 381–401.
Klein, S., Robertson, T. E., & Delton, A. W. (2010). Facing the future: memory as an evolved system for planning future acts. *Memory and Cognition, 38*(1), 13–22.
Klein, S., Robertson, T. E., & Delton, A. W. (2011). The future-orientation of memory: planning as a key component mediating the high levels of recall found with survival processing. *Memory, 19*(2), 121–139.
Koch, E. J. (2014). How does anticipated regret influence health and safety decisions? A literature review. *Basic and Applied Social Psychology, 36*(5), 397–412.
Kwan, D., Craver, C. F., Green, L., Myerson, J., & Rosenbaum, R. S. (2013). Dissociations in future thinking following hippocampal damage: evidence from discounting and time perspective in episodic amnesia. *Journal of Experimental Psychology: General, 142*(4), 1355–1369.
Levens, S. M., Larsen, J. T., Bruss, J., Tranel, D., Bechara, A., & Mellers, B. (2014). What might have been? The role of the ventromedial prefrontal cortex and lateral orbitofrontal cortex in counterfactual emotions and choice. *Neuropsychologia 54*, 77–86.
Lewis, D. (1990). What experience teaches. In W. G. Lycan (ed.), *Mind and cognition* (pp. 29–57). Oxford: Blackwell.
Lin, H., & Epstein, L. H. (2014). Living in the moment: effects of time perspective and emotional valence of episodic thinking on delay discounting. *Behavioral Neuroscience, 128*(1), 12–19.
Loomes, G., & Sugden, R. (1982). Regret theory: an alternative theory of rational choice under uncertainty. *Economic Journal, 92*(368), 805–824.
Martin, M. G. F. (2001). Out of the past: episodic recall as retained acquaintance. In C. Hoerl & T. McCormack (eds.), *Time and memory: issues in philosophy and psychology* (pp. 257–284). Oxford: Oxford University Press.
McCormack, T. (2001). Attributing episodic memory to animals and children. In C. Hoerl & T. McCormack (eds.), *Time and memory: issues in philosophy and psychology* (pp. 285–314). Oxford: Oxford University Press.

McCormack, T. (2014). Three types of temporal perspective: characterizing developmental changes in temporal thought. *Annals of the New York Academy of Sciences, 1326*, 82–89.

McCormack, T. (2015). The development of temporal cognition. In R. M. Lerner, L. S. Liben, & U. Mueller (eds.), *Handbook of child psychology and developmental science.* 7th edition, Volume 2: *Cognitive Processes* (pp. 624–670). Hoboken, NJ: Wiley-Blackwell.

McCormack, T., & Feeney, A. (2015). The development of the experience and anticipation of regret. *Cognition & Emotion, 29*(2), 266–280.

McCormack, T., & Hoerl, C. (1999). Memory and temporal perspective: the role of temporal frameworks in memory development. *Developmental Review, 19*(1), 154–182.

McCormack, T., & Hoerl, C. (2008). Temporal decentering and the development of temporal concepts. *Language Learning, 58*(Suppl 1), 89–113.

Mellers, B. A., Schwartz, A., & Ritov, I. (1999). Emotion-based choice. *Journal of Experimental Psychology: General, 128*(3), 332–345.

Morewedge, C. K., Gilbert, D. T., & Wilson, T. D. (2005). The least likely of times: how memory for past events biases the prediction of future events. *Psychological Science, 16*, 626–630.

Nietzsche, F. W. (1983). On the uses and disadvantages of history for life. *Untimely meditations.* Cambridge, UK: Cambridge University Press.

O'Connor, E., McCormack, T., & Feeney, A. (2012). The development of regret. *Journal of Experimental Child Psychology, 111*(1), 120–127.

O'Connor, E., McCormack, T., & Feeney, A. (2014). Do children who experience regret make better decisions? A developmental study of the behavioral consequences of regret. *Child Development, 85*(5), 1995–2010.

Ordonez, L. D., & Connolly, T. (2000). Regret and responsibility: a reply to Zeelenberg et al. (1998). *Organizational Behavior and Human Decision Processes, 81*(1), 132–142.

Paul, L. A. (2014). *Transformative experience.* Oxford: Oxford University Press.

Paul, L. A. (2015). What you can't expect when you're expecting. *Res Philosophica, 92*(2), 1–23.

Peters, J., & Buchel, C. (2010). Episodic future thinking reduces reward delay discounting through an enhancement of prefrontal-mediotemporal interactions. *Neuron, 66*(1), 138–148.

Peters, J., & Buchel, C. (2011). The neural mechanisms of inter-temporal decision-making: understanding variability. *Trends in Cognitive Sciences, 15*(5), 227–239.

Pettigrew, R. (2015). Transformative experience and decision theory. *Philosophy and Phenomenological Research, 91*(3), 766–774.

Ratner, R. K., & Herbst, K. C. (2005). When good decisions have bad outcomes: The impact of affect on switching behavior. *Organizational Behavior and Human Decision Processes, 96*(1), 23–37.

Richard, R., van der Pligt, J., & de Vries, N. (1996). Anticipated regret and time perspective: Changing sexual risk-taking behavior. *Journal of Behavioral Decision Making, 9*(3), 185–199.

Roese, N. J. (2005). *If only: how to turn regret into opportunity.* New York: Broadway Books.

Roese, N. J., & Summerville, A. (2005). What we regret most . . . and why. *Personality and Social Psychology Bulletin, 31*(9), 1273–1285.

Russell, B. (1912). *The problems of philosophy.* Oxford: Oxford University Press.

Saffrey, C., Summerville, A., & Roese, N. J. (2008). Praise for regret: people value regret above other negative emotions. *Motivation and Emotion, 32*(1), 46–54.

Schacter, D. L., & Addis, D. R. (2009). Remembering the past to imagine the future: a cognitive neuroscience perspective. *Military Psychology, 21*(Suppl 1), S108-S112.

Schacter, D. L., Addis, D. R., & Buckner, R. L. (2007). Remembering the past to imagine the future: the prospective brain. *Nature Reviews Neuroscience, 8*(9), 657–661.

Schacter, D. L., Addis, D. R., Hassabis, D., Martin, Victoria C., Spreng, R. N., & Szpunar, Karl K. (2012). The future of memory: remembering, imagining, and the brain. *Neuron, 76*(4), 677–694.

Schacter, D. L., Benoit, R. G., De Brigard, F., & Szpunar, K. K. (2015). Episodic future thinking and episodic counterfactual thinking: Intersections between memory and decisions. *Neurobiology of Learning and Memory, 117*, 14–21.

Schopenhauer, A. (1999). *Prize essay on the freedom of the will*. Cambridge, UK: Cambridge University Press.

Shimanoff, S. B. (1984). Commonly named emotions in everyday conversations. *Perceptual and Motor Skills, 58*(2), 514–514.

Soteriou, M. (2008). The epistemological role of episodic recollection. *Philosophy and Phenomenological Research, 77*(2), 472–492.

Squire, L. R., & Zola, S. M. (1998). Episodic memory, semantic memory, and amnesia. *Hippocampus, 8*(3), 205–211.

Suddendorf, T., & Corballis, M. C. (1997). Mental time travel and the evolution of the human mind. *Genetic, Social, and General Psychology Monographs, 123*(2), 133–167.

Suddendorf, T., & Corballis, M. C. (2007). The evolution of foresight: what is mental time travel, and is it unique to humans? *Behavioral and Brain Sciences, 30*(3), 299–313.

Suddendorf, T., & Redshaw, J. (2013). The development of mental scenario building and episodic foresight. *Annals of the New York Academy of Sciences, 1296*(1), 135–153.

Torges, C. M., Stewart, A. J., & Miner-Rubino, K. (2005). Personality after the prime of life: men and women coming to terms with regrets. *Journal of Research in Personality, 39*(1), 148–165.

Tulving, E. (1985). *Elements of episodic memory*. Oxford: Oxford University Press.

Tulving, E. (2002). Episodic memory: from mind to brain. *Annual Reviews of Psychology, 53*, 1–25.

Ullmann-Margalit, E. (2006). Big decisions: opting, converting, drifting. *Royal Institute of Philosophy Supplement, 81*(58), 157–172.

Van Hoeck, N., Ma, N., Ampe, L., Baetens, K., Vandekerckhove, M., & Van Overwalle, F. (2013). Counterfactual thinking: an fMRI study on changing the past for a better future. *Social Cognitive and Affective Neuroscience, 8*(5), 556–564.

Van Hoeck, N., Ma, N., Van Overwalle, F., & Vandekerckhove, M. (2010). Counterfactual thinking and the episodic system. *Behavioral Neurology, 23*(4), 225–227.

Van Hoeck, N., Watson, P. D., & Barbey, A. K. (2015). Cognitive neuroscience of human counterfactual reasoning. *Frontiers in Human Neuroscience, 9*, 420.

Weisberg, D. P., & Beck, S. R. (2012). The development of children's regret and relief. *Cognition & Emotion, 26*(5), 820–835.

Wrosch, C., Bauer, I., Miller, G. E., & Lupien, S. (2007). Regret intensity, diurnal cortisol secretion, and physical health in older individuals: evidence for directional effects and protective factors. *Psychology and Aging, 22*(2), 319–330.

Wrosch, C., Bauer, I., & Scheier, M. F. (2005). Regret and quality of life across the adult life span: the influence of disengagement and available future goals. *Psychology and Aging, 20*(4), 657–670.

Zeelenberg, M., & Beattie, J. (1997). Consequences of regret aversion 2: additional evidence for effects of feedback on decision making. *Organizational Behavior and Human Decision Processes*, *72*(1), 63–78.

Zeelenberg, M., Beattie, J., vanderPligt, J., & deVries, N. K. (1996). Consequences of regret aversion: effects of expected feedback on risky decision making. *Organizational Behavior and Human Decision Processes*, *65*(2), 148–158.

Zeelenberg, M., & Pieters, R. (2004). Beyond valence in customer dissatisfaction: a review and new findings on behavioral responses to regret and disappointment in failed services. *Journal of Business Research*, *57*(4), 445–455.

Zeelenberg, M., & Pieters, R. (2007). A theory of regret regulation 1.0. *Journal of Consumer Psychology*, *17*(1), 3–18.

Zeelenberg, M., van Dijk, W. W., & Manstead, A. S. R. (1998). Reconsidering the relation between regret and responsibility. *Organizational Behavior and Human Decision Processes*, *74*(3), 254–272.

Zeelenberg, M., van Dijk, W. W., Manstead, A. S. R., & van der Pligt, J. (2000). On bad decisions and disconfirmed expectancies: the psychology of regret and disappointment. *Cognition & Emotion*, *14*(4), 521–541.

Zentall, T. R., Clement, T. S., Bhatt, R. S., & Allen, J. (2001). Episodic-like memory in pigeons. *Psychonomics Bulletin and Review*, *8*(4), 685–690.

13

# The Mechanisms and Benefits of a Future-Oriented Brain

GIOVANNI PEZZULO ■

*Ho questa foto di pura gioia*
*È di un bambino con la sua pistola*
*Che spara dritto davanti a se*
*A quello che non c'è*[1]

From the song "Quello Che Non C'è," *Afterhours*, 2002

Over the past decade, researchers have increasingly acknowledged the fact that the brain not only is a stimulus-response device but also continuously generates predictions—from short-term predictions of the next sensory stimuli to more sophisticated forms of prospection. Predictive processing has been implied in virtually every form of cognitive processing in textbooks, such as perception, decision-making, emotional regulation, reasoning, and beyond, leading to the proposal that the brain is essentially a predictive device and is *functionally oriented toward the future* (Bar, 2007; Bubic et al., 2010; Friston, 2010; Klein, 2013a; Pezzulo & Castelfranchi, 2009).

Predictive abilities might have originated during evolution due to the demands of immediate action control, for example to overcome sensory noise and delays in the processing of increasingly larger brains, but successively they might have been *exapted*, and elaborated, paving the way for higher cognitive skills such as prospection and counterfactual thinking. Although there is considerable debate on this topic, one hypothesis is that the passage from sensorimotor control to cognitive abilities is one in which predictive and prospective abilities were reused and became partly internalized (for example, as "covert actions") to support cognitive functions (Buzsáki et al., 2015; Grush, 2004; Hesslow, 2002; Pezzulo & Castelfranchi, 2009). This hypothesis highlights the continuity between sensorimotor and higher cognitive skills, rather than describing them as separate domains in the brain (Cisek & Kalaska, 2010; Pezzulo et al., 2011).

## 1. DETACHMENT AND INTERNALLY GENERATED NEURAL INFORMATION PROCESSING

Whether or not this hypothesis holds, predictive processing and prospection—but also, as we will see, some forms of remembering—are associated to a specific "mode" of brain functioning: they require a certain *detachment* from the here and now and the current sensorimotor context. There can be various degrees of detachment, which link to the increasing demands of short- and long-term predictions, and of more sophisticated forms of prospection and imagination.

To better understand the link between prediction and detachment, let's consider first a case of non-predictive processing: a stimulus-response system. To work properly, a stimulus-response system requires a tight sensorimotor coupling, as each action is triggered by, and in turn produces, a stimulus that triggers the next action (Gibson, 1979; Gigliotta & Nolfi, 2008; Pfeifer et al., 2007). For this reason, the brain dynamics of a stimulus-response system need to be closely coupled with—in a sense, dominated by—sensory stimulations, feedback, and motor efferent information, such as corollary discharges of actions.

The situation is different if one instead assumes that perceptual processing and action control include predictive and top-down components, for example in (hierarchical) *predictive coding* and *active inference* architectures (Friston, 2010; Friston et al., 2010; 2015; Kiebel et al., 2008; Pezzulo et al., 2015). Predictive coding architectures for perceptual processing anticipate rather than being driven by stimuli. They use an internal *generative or forward model* (i.e., a model that captures the statistics of perceptual events) to generate top-down predictions and compare them with stimuli; what is propagated in a bottom-up way in the architecture is a prediction error (the difference between a prediction and a stimulus), not the stimulus itself, and this prediction error is used to "keep in register" or revise the higher level hypothesis. The dynamics of their bottom layer(s) remain quite sensitive to current stimulations, but those at their higher layer(s) have a higher degree of detachment from current stimulations and can unfold at timescales that are different from those of perception and action.

Predictive coding architectures and related methods that are today popular in machine learning, such as *deep belief networks* (Hinton, 2007a), do not necessarily need to be triggered by external stimuli but are *generative* architectures: they can be initialized to internally generate their expected inputs through their internal model, without being triggered by external events, thus in a certain sense "hallucinating" them (Hinton, 2007b). Normally, these architectures are limited in their future-orientedness and essentially use prediction in the service of immediate perceptual processing: they predict the current, not the future stimulus. However, they can be also used (with some modifications) in such a way that their generative processes are iterated over time, with or without feedback from the external stimuli, thus projecting perceptual predictions into the future. This process permits us to model, for example, how a visual percept unfolds over time (LeCun et al., 2015).

Let's now consider generative architectures in the context of action performance, not just perception. One example is *active inference* systems, in which the generation of an action (e.g., a hand movement) requires the prediction of the *next* proprioceptive state (e.g., of the next hand posture) as well as of the next sensory state (e.g., the visual consequences of the hand movement), but also in parallel the *attenuation* of the current sensations (e.g., the current position of the hand) (Friston et al., 2010). Active inference systems are essentially a]nticipatory and are guided by internal dynamics rather than stimulus-response rules, although they normally are kept in register by checking the sensory consequences of actions. Furthermore, active inference architectures can include *priors* encoded at higher levels that essentially represent distal objectives of goals (e.g., pressing four buttons in a row). These priors can be encoded, for example, in the form of dynamical systems that have itinerant dynamics and model transitions from one attractor to the other, where each attractor represents, say, a given sub-goal (e.g., the next button to be pressed). These priors can govern the system dynamics by unfolding top-down predictions about the various actions or sub-goals over time (Friston et al., 2012). As this example illustrates, active inference architectures are principally governed by internally generated predictions rather than external stimuli, and are functionally oriented toward the future—although, for the demands of immediate action control, this is usually a rather shortsighted future.

However, the active inference scheme introduced earlier, or related ones such as *planning-as-inference* (Toussaint, 2009), can be expanded to support various forms of long-term prediction, planning, and prospection. One way to do so is to iterate various cycles of predictive processing, through the internal model of the situation, thus essentially projecting the predictions into increasingly distal futures. For this, the system needs to simulate the action consequences over time; action can have exteroceptive, priopriaceptive, and/or interoceptive consequences, and thus this information has to be somewhat encoded in the generative model. This procedure permits an individual to plan action sequences, to generate imagined ("what if") scenarios, which can be then evaluated and compared with respect, say, to their ability to bring the agent closer to a desired goal. The same mechanism can, in principle, be used to simulate future-oriented plans that can be self-maintained and successively used for cognitive control (Botvinick & Toussaint, 2012; Friston et al., 2015; Kappen et al., 2012; Pezzulo, 2012).

As these examples illustrate, generative architectures offer one mechanistic hypothesis on the internal generation of stimuli, linking this process to predictive dynamics. Furthermore, generative architectures can support *prospective* processing and permit the construction and advance consideration of future and even imaginary scenarios.

## 2. INTERNALLY GENERATED SEQUENCES (IGS) IN THE RODENT HIPPOCAMPUS AS AN EXAMPLE OF INTERNALLY GENERATED BRAIN DYNAMICS

When engaged in forms of predictive/generative processing and prospection, the brain enjoys a significant *detachment* from the here-and-now of the current sensorimotor context (as well as the emotional/affective context, as I will discuss later). Importantly, these processes requires the brain to self-generate information rather than extract it from the sensory stimuli, or just keep the same stimulus information in working memory. Predictive systems such as those described earlier require mechanisms permitting the brain to internally generate and iterate predictions over time; for example, mechanisms for sampling them from an *internal model* of the task that describes how the relevant variables unfold in time (e.g., which are the sensory effects of actions).

One area where internally generated dynamics have been studied in great detail (not only in the context of prediction, but also of other functions) is the rodent hippocampus. It has been consistently reported that, during rest or sleep, the hippocampus "replays" spatial trajectories that the animal has experienced in the past: the same sequence of neurons that coded for spatial locations during the actual rodent navigation (place cells) can be reactivated endogenously (i.e., in the absence of triggering stimuli), in a forward or backward manner (Diba & Buzsáki, 2007; Dragoi & Tonegawa, 2011; Pastalkova et al., 2008; Pezzulo et al., 2014; Wikenheiser & Redish, 2013). Not only can these "internally generated sequences" (IGS) code for trajectories that have been already experienced by the animal, but they can also generalize to novel trajectories, such as combinations of already experienced trajectories (Gupta et al., 2010); they can be initiated at locations that are different from the current animal location (which implies that they do not need to be initiated by the animal's current sensory cues); they can code extended trajectories (e.g., up to ~10 m [Davidson et al., 2009]) and also novel paths that the animal will successively take (i.e., *preplays* [Dragoi and Tonegawa, 2011]). IGS have been initially linked to memory function such as consolidation. However, the fact that when the animals rest between goal-directed spatial navigation episodes the IGS are preferentially directed toward known goal sites and are predictive of future choices suggests that they might support a planning function rather than only the recall of previous experience (Pfeiffer & Foster, 2013). All these (p)replay events are detected in the hippocampus during the so-called sharp wave ripple (SWR) complexes: short (~50–120 ms) high-frequency oscillations (~140–200 Hz), which are typical when the animals are at rest or during sleep.

There is a second kind of hippocampal IGS. When rodents are engaged in a task, for example when they dwell in a maze, their hippocampal local field potential shows a different characteristic rhythmic oscillation, the *theta rhythm* (~6–10 Hz). Within the theta rhythm, short sequences of active place cells (*theta sequences*) have been long identified, which represent time-compressed ongoing behavioral trajectories (Foster & Wilson, 2007; Lisman & Redish, 2009). Theta sequences are formed very rapidly (Feng et al., 2015) and often act as a "moving window," coding for a (forward) sequence of spatial positions centered on the moving animal, just beyond, or just above it. However, recent studies have shown that when a rodent is engaged in a decision task, theta sequences can "sweep" far forward in space. For example, when a rat stops at a decision point in a T-maze, theta sequences can sweep far away from the animal, even if it does not move, and cover (parts of) one branch of the T-maze, and successively (parts of) the other branch of the T-maze (Johnson & Redish, 2007). These IGS have been associated to periods of "vicarious trial and error," when the rat stops and turns the head to the right and left, with the underlying hypothesis that they might support the serial deliberation between two choice options, or in other words, "what if" simulations (Pezzulo et al., 2014; Redish, 2013; van der Meer et al., 2012). To implement this form of deliberation, the hippocampus might work in concert with the ventral striatum: the hippocampus might sequentially generate possible action alternatives and trajectories in the form of IGS, and the ventral striatum might evaluate their quality and expected reward—an idea that is supported by the fact that at decision points, neurons in the ventral striatum show covert expectations of rewards (van der Meer & Redish, 2010). Furthermore, during goal-directed navigation, theta sequences tend to be preferentially directed toward goal sites and can be predictive of future choices (Wikenheiser & Redish, 2015). Overall, thus, hippocampal IGS and associated phenomena have been variously linked to goal-oriented spatial navigation, planning, and learning, and their neurobiology and dynamical aspects are increasingly studied (Buzsáki & Moser, 2013a; Lisman, 2015).

From a computational perspective, it has been proposed that the hippocampus forms generative models jointly with other brain areas, such as entorhinal cortex and ventral striatum; and in these architectures, hippocampal IGS support the prospective simulation of spatial trajectories and their value (Penny et al., 2013; Pezzulo et al., 2013). In these and other computational models, hippocampal "forward sweeps" and "replays" derive from an internal model (e.g., they might be series of samples from the model), and support a variety of operations such as decision-making, planning, and learning—for example, they can be used to plan spatial trajectories (Penny et al., 2013) or to train a behavioral controller off-line by feeding it with simulated experiences (Sutton, 1991).

## 3. PREDICTIVE PROCESSING AS AN INTEGRATIVE HYPOTHESIS FOR DETACHED BRAIN OPERATIONS?

To summarize the discussion so far, several cognitive operations, including forms of future-oriented cognition, require a form of *detachment*—or a certain independence from external stimulations—and a "brain mode" where it is guided by its internally generated dynamics, as in the case of IGS in the hippocampus (Pezzulo et al., 2014). Those internally generated dynamics might correspond to the functioning of an internal model, of the kind that is required by predictive processing in perceptual and motor domains. If one focuses on the ability of generative models to self-generate relevant information, one can even say that the brain is a *phantastic organon* (Friston et al., 2014)—where *phantastic* is used in the sense of the Greek word *phantastikos* (i.e., able to create mental images)—and is largely dominated by its imagination rather than by mere sensory stimulations.

Can a similar mechanism based on internally generated dynamics explain other forms of detached cognition? There are some indications that this might be the case. For example, the predictive system introduced earlier can be exploited in a "reverse way" so that it constructs a "what if" simulation about a past and not a future event; or, in other words, it re-samples and re-evaluates a past event. Linking back to the discussion of the hippocampus–ventral striatum system in rodent spatial navigation, a recent study has identified internally generated dynamics in situations when rodents retrospectively (or counterfactually) consider the value of "the path not chosen," suggesting that this mechanism entails a form of counterfactual thinking and the experience of *regret* when a wrong decision is made (Steiner & Redish, 2014).

This discussion raises the intriguing hypothesis that a mechanism based on internally generated dynamics might underlie other forms of detached and higher cognition, such as counterfactual thinking, and might even extend to the episodic memory, imagination, mind wandering, prospection, and "time travel" in the past and the future. This hypothesis is speculative at this stage and deserves future investigations. However, it might offer a much-needed integrative perspective explaining the similarities and differences between these abilities, potentially explaining why, and in which occasions, they—to some extent—recruit shared neuronal circuits (Hassabis & Maguire, 2009; Schacter et al., 2012, 2007). Furthermore, this hypothesis would explain the evolution of these advanced cognitive abilities as sophistications of predictive mechanisms and of internally generated dynamics, not requiring completely de novo brain systems (Hurley, 2008; Jeannerod, 2006; Pezzulo, 2011, 2008).

## 4. THE ADAPTIVE FUNCTIONS AND CORE MECHANISMS OF DETACHED COGNITION FROM THE ANGLE OF PREDICTIVE PROCESSING

The idea that detachment and internally generated neural information processing originated from the sake of predictive processing could help shed light on the adaptive benefits and the core mechanisms of detached cognition. Indeed, there are various situations, such as while mind wandering or fantasizing, when detaching from the current context and mentally generating imaginary situations seem deprived of adaptive benefits, and might run against task or goal achievement. Furthermore, a detached brain loses the adaptive benefits of a close sensorimotor coupling, and might even be at risk, as exemplified by the story narrated in Plato's *Theaetetus*, in which Thales of Miletus fell into a well while he was contemplating the stars. Finally, detached forms of cognition dominated by internally generated dynamics can go awry: because they are not "kept in register" by external stimulations, they can "drift" away from reality, and in certain occasions might generate severe deficits such as hallucinations or delusions (Friston et al., 2014; Montague et al., 2012). In the light of these risks, one might ask what are—if any—the adaptive functions of detached forms of cognition and how they emerged during evolution. Besides adaptive and evolutionary issues, one might ask whether different forms of detached cognition are based on different brain mechanisms, or whether they share "common core" principles and mechanisms.

These questions could be addressed by looking at detachment from the angle of predictive processing. The ability to endogenously generate and use predictions and representations of the future, rather than just classify and react to stimuli, gives the brain significant adaptive benefits, for example the ability to conceive and *coordinate with* future (predicted) states of affairs (Pezzulo, 2008). Furthermore, in some computational schemes such as *active inference* and *planning-as-inference*, simulating future situations is important for the formation of goals that can be successively used as "set points" to regulate behavior over long time horizons, as in cognitive (or self-) control (Pezzulo, 2012, 2008).

The adaptive benefits of predictive processing increase with the complexity of the situation that an animal needs to face. In simple choice situations, action selection is often considered to be largely guided by currently available cues, or by the past—for example, in the form of a learned history of reinforcements (Balleine & Dickinson, 1998; Rescorla & Wagner, 1972; Sutton & Barto, 1998). However, cue- and past-based selection methods might be more problematic in complex situations, and especially (but not only) in human social contexts. For example, in these contexts important cues (e.g., another's intentions) might not be perceptually available. Furthermore, most social situations are novel or unique, or at least not easily generalizable from previous situations that are apparently similar, and are highly context sensitive. One way to characterize these situations is in terms of "high outcome divergence": the same action can have truly different effects in two situations that have several cues in common. In these situations, basing action selection on a (weighted) average of previous situations might be ineffective. Rather, it can be more beneficial to base the decision on an analogy with a single, well-selected episode (or "case" or "rule") that resembles the current situation, and for which a solution is available; for example, using a form of *episodic control* (Lengyel & Dayan, 2008). Episodic control does not necessarily include predictive components. However, sometimes episodes cannot be simply remembered but need to be "constructed" de novo (say, by recombining elements of previous episodes) and/or projected into the future—all functions that have been associated with the hippocampus (Buckner, 2010; Hassabis & Maguire, 2009; Lisman & Redish, 2009).

Another hallmark of complex choice situations is that their contingencies, for example the value of states or actions, are not stationary but change from time to time; for this reason, actions need to be re-evaluated de novo during a choice. Computational modeling studies indicate that, to better deal with those complex choice situations, it is beneficial to use predictive systems that imagine future events, and use them to replace (or complement) unavailable cues or episodes. These systems are often called *model-based* because they require a model of the task, as opposed to stimulus-response or model-free methods that do not require it. Using their model, predictive systems can construct and evaluate situations de novo before a choice, rather than base it on a learned average of previous evaluations. In other words, these systems afford *prospective optimization* (Sejnowski et al., 2014). Developing efficient methods for prediction and model-based control, and assessing the biological plausibility of these methods, are active research areas (Daw & Dayan, 2014; Dolan & Dayan, 2013; Friston et al., 2010; Gelly et al., 2012; Pezzulo et al., 2013).

I have provided in the preceding text one example of the functioning of such predictive, model-based systems in the context of "what if" evaluations of spatial trajectories and the hippocampus–ventral striatum pathway. This is, however, a relatively simple case of prospective evaluation, as it regards the choice between immediate actions, such as turning left or right to reach a food location.

## 5. SIMULATING THE "EVALUATION CONTEXT": WHY DO WE NEED TO (RE)FEEL EXPERIENCES?

In more complex scenarios, for example in complex social contexts, actions can be more complex and have longer-terms effects. Not only is it difficult to learn or predict those long-term dependencies, but it is also difficult to consider the *future* value of action outcomes, that is, the value that the outcomes will have when they will be delivered, in the light of the fact that the situation can change in the meantime. Consider the case when, in a restaurant, I order a pie that I will only consume at the end of a dinner; it is possible that the (perceived) value of the pie is different when I order it, and when I will consume it, because at the end of the dinner I might be in a different physiological state, (e.g., full). Not only motivational but also external, contextual elements might change. This problem is present in many everyday cases, such as when we evaluate the benefits of an insurance plan, and is partially reflected in intertemporal choices, where subjects are asked to compare, say, $10 now versus $40 in 6 months.

In these situations, the "evaluation context" (a suitcase word that I use to indicate "internal" factors such as motivational, physiological, and affective elements, as well as "external," perceptual, and contextual elements) that I use to evaluate the predicted outcomes can be different from the one that will be in place when I will obtain the action outcomes, and disregarding this difference can be detrimental for the prospective optimization. One way to face this problem is to construct simulations that project into the future not only my predictions of the action outcomes, but also other contextual elements, as well as my own motivational and affective systems, and ultimately detach—to some extent—my evaluative system and the body itself (Pezzulo & Rigoli, 2011). In other words, in order to evaluate events in their correct future context (*will I appreciate the pie when I will consume it?*), it is necessary to internally generate a detailed simulation of the context, including some of its affective and bodily elements.

The projection into the future of the motivational and affective systems, with the alleged feelings and physiological conditions (e.g., higher heart rate in anticipation of a stressful event), can be characterized from a computational viewpoint in terms of a *motivational forward model* (Pezzulo & Rigoli, 2011), analogous to forward models used to predict action outcomes. The motivational forward model produces anticipations that allow one to evaluate action outcomes in light of the future (not the present) motivational, affective, and physiological state. The anticipations are used in the decision in two ways: "*to evaluate outcomes (e.g., the pleasure of eating a pie is evaluated differently at the beginning of a dinner, when one is hungry, and at the end of the dinner, when one is satiated), and as outcomes having a value themselves (e.g., the case of dread as a cost of waiting for punishment)*" and this procedure allows one to optimize decisions (Pezzulo & Rigoli, 2011, p. 1).

This study suggests that simulating the motivational, affective, and physiological conditions in which the outcomes will be obtained allows one to better select among them—or in other words, that the most appropriate context (or "frame of reference") to evaluate an outcome is not the one in which the decision is made, but the one in which the outcome will be delivered. In turn, this method requires constructing detailed simulations of the (often future) evaluation context, which—importantly—can include affective and motivational information. The involvement of affective and motivational systems might determine a "feeling" of the anticipated events, which aligns well with the proposal that adaptive decisions sometimes require one to "pre-feel" or to have "gut feelings" of future situations (Loewenstein, 1987). From this perspective, when it is necessary to internally generate (or recreate) the appropriate "evaluation context," *experiential* components might become essential for an accurate prospective evaluation, suggesting that in these occasions feeling or re-feeling experiences might be necessary or at least beneficial. Here, in a sense, the body itself can be considered as a proper part of the decision-making system, as it is a source of signals that are incorporated in the choice (Pezzulo, 2014). This perspective is coherent with the view that the body encodes relevant (e.g., affective) variables that are used for the choice (Mellers et al., 1999), and with "affect as information" theories (Clore and Bar-Anan, 2007).

Furthermore, it is likely that simulating the "evaluation context" benefits from the use of episodic and autobiographic information, in order to predict affective states (*how did I feel in a previous, similar situation?*) as well as detailed sensory and contextual elements (*who complained to me in a previous, similar situation?*). This raises the hypothesis that during the simulations, episodic and/or autobiographical content becomes integrated into an essentially model-based computation, giving rise to a constructive process that does not involve verbatim memories but, as is typical in model-based computations, (re)constructs them from a model. I will return to this point in the Conclusions.

Besides its importance for prospective optimization, the idea of simulating the (future) evaluation context points to a growing body of empirical literature (see Schacter et al., 2012, for a recent review). Various theories of prospective processing, mental time travel, and self-projection highlight the importance of affective, autobiographical, and even temporal elements, although the idea that these are core elements of all simulative processes is disputed (Buckner & Carroll, 2007; Hassabis & Maguire, 2009; Klein, 2013a, 2013b; Tulving, 1983). From the perspective presented here, detailed affective and autobiographical elements, sometimes deriving from episodic memories, should have a prominent role when the prospective forms of optimization discussed earlier are in play, such as when there is a need to detach the evaluation system and the body.

Along similar lines, it has been suggested that the adaptive value of mental time travel might lie in the imagination and emotional engagement with a non-present situation, such as a past or a future event: as affect is central to episodic memory, an episodic recall or simulation process might elicit situation-specific emotional rewards that act as a "choice calibration" process, limiting opportunistic decisions (Boyer, 2008). This idea has received some support by evidence indicating that episodic forms of prospection can promote farsighted decisions (Benoit et al., 2011; Peters & Büchel, 2010). This theory and others that emphasize the importance of affective components of past- and future-oriented cognition are compatible with the idea that *detaching* from the current situation might help evaluate events in their correct (affective and emotional) "frame of reference" (see Klein, 2013b; Michaelian, 2016).

The sophisticated predictive methods described here, which are based on anticipated motivational and affective states, are in principle powerful, but they are also prone to errors and can give rise to notorious forecasting biases (Gilbert & Wilson, 2007). One possible reason for these failures is that detaching the evaluation system and bodily/physiological states might be intrinsically problematic, because detachment might conflict with their important role in maintaining homeostasis (Pezzulo, 2014; Seth, 2013). In generative architectures, detachment links to the fact that self-generated predictions replace perceptual (exteroceptive, proprioceptive, and/or interoceptive) stimuli, as in the aforementioned examples of "hallucinatory" perceptual processes (Hinton, 2007b) and action execution in active inference (Friston et al., 2010). In turn, to be replaced by predictions, perceptual channels need to be temporarily attenuated or suppressed; in the language of predictive coding, they should be assigned a low precision, otherwise the process would produce undesired interferences between predictions and sensations (e.g., between the currently sensed hand position and its predicted or goal position). A hypothesis that remains to be investigated by future research is that these interference effects, which are certainly possible in every form of simulation or imagination, might be especially severe when interoceptive and bodily states are involved, given the risks (in terms of homeostasis and ultimately the survival of the organism) of lowering their precision.

## 6. CONCLUSIONS AND PROSPECTS FOR FUTURE RESEARCH

In the last decade we have gathered a great deal of knowledge on detached and future-oriented cognition; still we lack a comprehensive framework to fully understand them. Here I suggest that we can understand detached forms of cognition in terms of an internal generative model, which primarily (but not exclusively) supports predictive processing, and whose neural implementation is based on internally generated brain dynamics, as exemplified by IGS in the rodent hippocampus and related dynamical phenomena in other parts of the brain (Andrews-Hanna et al., 2014; Buzsáki, 2010; Buzsáki et al., 2015; Lisman, 2015; Pfeiffer & Foster, 2015; Rabinovich et al., 2015; Schuck et al., 2015; Voloh et al., 2015).

The focus on prediction does not mean that detached cognition cannot be retrospective—it can, as exemplified by the cases of episodic retrieval and regret discussed earlier. Furthermore, detachment can have several other roles besides those mentioned here; for example, Leslie (1987) has convincingly argued that *pretend play*, for example pretending that a banana is a telephone, requires a degree of detachment or decoupling from the object's original function.

It is also worth noting that the link between internally generated brain dynamics, predictive processing, and generative or forward models is not mandatory, as the self-generation and self-maintenance of information have been linked to other functions such as working memory, and to other computational mechanisms such as recurrent or attractor networks (Rolls & Treves, 1997). Here, the idea is that predictive processing requires a high degree of detachment, as it is not limited to the maintenance of stimulus information in recurrent networks, and it might have favored the rise of a generic, detached brain mode, which is based on internal models and the internal generation of information, and which might constitute a "core mechanism" across increasingly complex forms of detached and future-oriented cognition.

In this perspective, brain mechanisms for predictive control could have provided an initial form of detachment from current stimuli that served as a prerequisite to (successively) support imagination, future-oriented thinking, and other forms of higher cognition. If this hypothesis holds, higher cognition does not necessarily require a separated set of brain processes or a modularized brain, but can be realized as a *detached* form of sensorimotor processing. This continuity between sensorimotor and cognitive processing suggests intriguing parallels between the information processing of apparently disparate domains, for example that *"neuronal algorithms underlying navigation in real and mental space are fundamentally the same"* (Buzsáki & Moser, 2013b, p. 130), that *"visuomotor strategies and eye movements are reused for abstract thinking"* (Pezzulo et al., 2011, p. 7), that *"the off-line re-enactment of anticipatory mechanisms used for action control gives rise to (embodied) mental simulations"* (Pezzulo & Castelfranchi, 2009, p. 559), that *"situated simulations lie at the hearth of [...] the conceptual system"* (Barsalou, 2003, p. 513), that *"all imagery allows us to answer 'what if' questions by making explicit and accessible the likely consequences of being in a specific situation or performing a specific action"* (Moulton & Kosslyn, 2009, p. 1273), or that motor simulations that might underlie several forms of "motor cognition" such as action understanding and motor imagery involve *"the off-line rehearsal of neural networks involved in specific operations such as perceiving or acting"* (Jeannerod, 2006, p. 129).

This hypothesis, which remains to be fully tested, has several consequences for the empirical study of detachment and future-oriented cognition. One example is the debate on the nature and functioning of the brain *default network* (Buckner et al., 2008). This network was first identified in terms of a "baseline" activity or "resting state" (Raichle et al., 2001), suggesting that it might be antithetic to goal-directed processing or task achievement. However, recent studies suggest that under certain conditions it can flexibly couple with other brain networks and jointly support goal-directed tasks (Schacter et al., 2012). By thinking of the default network in terms of a brain mode dominated by internal dynamics, rather than in terms of a baseline brain activity, the link with goal-oriented task achievement becomes possible but not mandatory, and depends on task demands. For example, it is likely to be observed when the current task requires episodic content to be used in the service of prospection.

Another intriguing issue is assessing why the hippocampus and surrounding structures in the medial temporal lobe (MTL) have a key role in detached and future-oriented forms of processing. It has been proposed that these are important because they implement forms of memory formation, contextual processing, and scene construction (Hassabis & Maguire, 2009; Schacter et al., 2012). From the perspective of predictive coding architectures, it might also be worth noticing that some (but not all) of these areas lack a granular layer. According to biological schemes for predictive coding (Barrett & Simmons, 2015; Shipp et al., 2013), the lack of a granular layer might render these areas less permeable by bottom-up inputs and thus, in principle, simpler to detach.

These and several other aspects of the neurobiology of detached and future-oriented cognition remain to be clarified by future research. To this aim, a collaboration between empirical research and computational modeling can be beneficial (Pezzulo et al., 2011). Historically, computational theories have been influential in the field, such as the idea that the hippocampus and surrounding areas in the MTL might specialize in the rapid acquisition of memories that are successively transferred to the neocortex, with the two areas that closely interact for the formation and recall of events (Káli & Dayan, 2004; McClelland et al., 1995; Treves & Rolls, 1994). Another relevant stream of computational research suggests that the decision systems of (at least some) animals include both simpler control loops and more sophisticated ones, which increasingly involve the prefrontal cortex in loop with cortical and subcortical structures, and which increase the animals' proactivity and future orientedness (Koechlin, 2014; Redish, 2013; Sejnowski et al., 2014; Verschure et al., 2014). Finally, computational models of goal-directed navigation are helping to shed light on IGS and associated phenomena that I have discussed here (Erdem & Hasselmo, 2014; Penny et al., 2013; Pezzulo et al., 2013).

Although useful, these models need to be expanded to fully capture the complexity of future-oriented and detached forms of cognition. Facing this challenge might require the development of more detailed and biologically realistic models of predictive processing, and possibly the integration of principles of model-based (Dolan & Dayan, 2013) and episodic (Lengyel & Dayan, 2008) behavioral controllers, which are often considered separately. The former is often linked to semantic memory, and uses a (usually statistical) model, not "verbatim" memories, to generate simulations of the future, or even retrospective simulations. The latter, instead, uses specific recalled episodes and remembers for each situation the single action that led to the best outcome. Some forms of detached cognition, such as the episodic simulation of (future, atemporal, or imagined) events, seem to require elements of both.

One paradigmatic example is the constructive recombination of elements of past episodes to simulate a new (say, a future) episode. Consider, for example, the case of imagining a children's party. Here, one hypothesis is that semantic memories in the internal model include knowledge that "scaffold" the simulation (e.g., a child's party often includes a cake and takes place in the garden of the child's house), thus orienting the search of relevant episodes from memory (e.g., the time I visited this house and its garden; the time I ate a cake at another child's party). Furthermore, the generative dynamics of the internal model help project this episodic content in a new imagined situation, where it needs to blend coherently (e.g., imagining eating a cake in the garden of the child's house) (Glenberg, 1997). In this example, episodic content becomes part of an essentially model-based computation, constructing combined model-based and episodic simulations. At the same time, this constructive process can produce interference effects that are typical of memory recall and imagination (Hassabis & Maguire, 2009; Michaelian, 2011). Future research is needed to fully understand how the constructive process blends episodes and how it usually avoids—but sometimes produces—interferences, for example between incompatible episodes or between representations of past events and counterfactual or future events (see Michaelian, 2016).

Developing testable mechanistic models of the kind of processes described here might help shed light on complex cases of detached cognition, for which we still lack a detailed computational and empirical understanding.

## NOTE

1. I have this picture of pure joy; of a child with his gun; who shoots straight in front of him; to what is not there.

## REFERENCES

Andrews-Hanna, J. R., Smallwood, J., & Spreng, R. N. (2014). The default network and self-generated thought: component processes, dynamic control, and clinical relevance. *Annals of the New York Academy of Sciences, 1316*, 29–52. doi: 10.1111/nyas.12360.

Balleine, B. W., & Dickinson, A. (1998). Goal-directed instrumental action: contingency and incentive learning and their cortical substrates. *Neuropharmacology, 37*, 407–419.

Bar, M. (2007). The proactive brain: using analogies and associations to generate predictions. *Trends in Cognitive Sciences, 11*, 280–289.

Barrett, L. F., & Simmons, W. K. (2015). Interoceptive predictions in the brain. *Nature Reviews Neuroscience, 16*, 419–429. doi: 10.1038/nrn3950

Barsalou, L. W. (2003). Situated simulation in the human conceptual system. *Language and Cognitive Processes, 18*, 513–562.

Benoit, R. G., Gilbert, S. J., & Burgess, P. W. (2011). A neural mechanism mediating the impact of episodic prospection on farsighted decisions. *Journal of Neuroscience, 31*, 6771–6779. doi: 10.1523/JNEUROSCI.6559-10.2011.

Botvinick, M., & Toussaint, M. (2012). Planning as inference. *Trends in Cognitive Sciences, 16*, 485–488. doi: 10.1016/j.tics.2012.08.006

Boyer, P. (2008). Evolutionary economics of mental time travel? *Trends in Cognitive Sciences, 12*, 219–224. doi: 10.1016/j.tics.2008.03.003.

Bubic, A., von Cramon, D. Y., & Schubotz, R. I. (2010). Prediction, cognition and the brain. *Frontiers in Human Neuroscience, 4*, 1–15.

Buckner, R. L. (2010). The role of the hippocampus in prediction and imagination. *Annual Review of Psychology, 61*, 27–48, C1–8. doi: 10.1146/annurev.psych.60.110707.163508.

Buckner, R. L., Andrews-Hanna, J. R., & Schacter, D. L. (2008). The brain's default network: anatomy, function, and relevance to disease. *Annals of the New York Academy of Sciences, 1124*, 1–38. doi: 10.1196/annals.1440.011.

Buckner, R. L., & Carroll, D. C. (2007). Self-projection and the brain. *Trends in Cognitive Sciences, 11*, 49–57. doi: 10.1016/j.tics.2006.11.004.

Buzsáki, G. (2010). Neural syntax: cell assemblies, synapsembles, and readers. *Neuron, 68*, 362–385. doi: 10.1016/j.neuron.2010.09.023.

Buzsáki, G., & Moser, E. I. (2013a). Memory, navigation and theta rhythm in the hippocampal-entorhinal system. *Nature Neuroscience, 16*, 130–138. doi: 10.1038/nn.3304.

Buzsáki, G., & Moser, E. I. (2013b). Memory, navigation and theta rhythm in the hippocampal-entorhinal system. *Nature Neuroscience, 16*, 130–138. doi: 10.1038/nn.3304.

Buzsáki, G., Peyrache, A., & Kubie, J. (2015). Emergence of cognition from action. *Cold Spring Harbor Symposia on Quantitative Biology*, 024679. doi: 10.1101/sqb.2014.79.024679.

Cisek, P., & Kalaska, J. F. (2010). Neural mechanisms for interacting with a world full of action choices. *Annual Review of Neuroscience, 33*, 269–298. doi: 10.1146/annurev.neuro.051508.135409

Clore, G. L., & Bar-Anan, Y. (2007). Affect-As-Information. *Encyclopedia of Social Psychology*, http://www.sage-ereference.com/view/socialpsychology/n8.xml

Davidson, T. J., Kloosterman, F., & Wilson, M. A. (2009). Hippocampal replay of extended experience. *Neuron*, 63, 497–507.

Daw, N. D., & Dayan, P. (2014). The algorithmic anatomy of model-based evaluation. *Philosophical Transactions of the Royal Society of London B Biological Sciences, 369*, 20130478.

Diba, K., & Buzsáki, G. (2007). Forward and reverse hippocampal place-cell sequences during ripples. *Nature Neuroscience, 10*, 1241–1242. doi: 10.1038/nn1961.

Dolan, R. J., & Dayan, P. (2013). Goals and habits in the brain. *Neuron, 80,* 312–325. doi: 10.1016/j.neuron.2013.09.007.

Dragoi, G., & Tonegawa, S. (2011). Preplay of future place cell sequences by hippocampal cellular assemblies. *Nature,* 469, 397–401.

Erdem, U. M., & Hasselmo, M. E. (2014). A biologically inspired hierarchical goal directed navigation model. *Journal of Physiology Paris, 108,* 28–37. doi: 10.1016/j.jphysparis.2013.07.002.

Feng, T., Silva, D., & Foster, D. J. (2015). Dissociation between the experience-dependent development of hippocampal theta sequences and single-trial phase precession. *Journal of Neuroscience, 35,* 4890–4902.

Foster, D. J., & Wilson, M. A. (2007). Hippocampal theta sequences. *Hippocampus, 17,* 1093–1099. doi: 10.1002/hipo.20345.

Friston, K. (2010). The free-energy principle: a unified brain theory? *Nature Review Neuroscience, 11,* 127–138. doi: 10.1038/nrn2787.

Friston, K., Daunizeau, J., Kilner, J., & Kiebel, S. J. (2010). Action and behavior: a free-energy formulation. *Biological Cybernetics, 102,* 227–260. doi: 10.1007/s00422-010-0364-z.

Friston, K. J., Stephan, K. E., Montague, R., & Dolan, R. J. (2014). Computational psychiatry: the brain as a phantastic organ. *Lancet Psychiatry, 1,* 148–158. doi: 10.1016/S2215-0366(14)70275-5.

Friston, K., Rigoli, F., Ognibene, D., Mathys, C., FitzGerald, T., & Pezzulo, G. (2015). Active inference and epistemic value. *Cognitive Neuroscience, 6*(4), 187–214.

Friston, K., Shiner, T., FitzGerald, T., Galea, J. M., Adams, R., Brown, H., Dolan, R. J., Moran, R., Stephan, K. E., & Bestmann, S. (2012). Dopamine, affordance and active inference. *PLoS Computational Biology, 8,* e1002327. doi: 10.1371/journal.pcbi.1002327.

Gelly, S., Kocsis, L., Schoenauer, M., Sebag, M., Silver, D., Szepesvári, C., & Teytaud, O. (2012). The grand challenge of computer Go: Monte Carlo tree search and extensions. *Communications of the ACM, 55,* 106–113.

Gibson, J. J. (1979). *The ecological approach to visual perception.* Mahwah, NJ: Lawrence Erlbaum Associates.

Gigliotta, O., & Nolfi, S. (2008). On the coupling between agent internal and agent/environmental dynamics: development of spatial representations in evolving autonomous robots. *Adaptive Behavior - Animals, Animats, Software Agents, Robots, Adaptive Systems, 16,* 148–165. doi: http://dx.doi.org/10.1177/1059712308089184.

Gilbert, D. T., & Wilson, T. D. (2007). Prospection: experiencing the future. *Science, 317,* 1351–1354. doi: 10.1126/science.1144161.

Glenberg, A. (1997). What memory is for. *Behavioral and Brain Sciences, 20,* 1–55.

Grush, R. (2004). The emulation theory of representation: motor control, imagery, and perception. *Behavioral and Brain Sciences, 27,* 377–96.

Gupta, A. S., van der Meer, M. A. A., Touretzky, D. S., & Redish, A. D. (2010). Hippocampal replay is not a simple function of experience. *Neuron, 65,* 695–705. doi: 10.1016/j.neuron.2010.01.034.

Hassabis, D., & Maguire, E. A. (2009). The construction system of the brain. *Philosophical Transactions of the Royal Society of London B Biological Sciences, 364,* 1263–1271. doi: 10.1098/rstb.2008.0296.

Hesslow, G. (2002). Conscious thought as simulation of behaviour and perception. *Trends in Cognitive Sciences, 6,* 242–247.

Hinton, G. E. (2007a). Learning multiple layers of representation. *Trends in Cognitive Sciences, 11*, 428–434. doi: 10.1016/j.tics.2007.09.004.

Hinton, G. E. (2007b). To recognize shapes, first learn to generate images. *Progress in Brain Research, 165*, 535–547. doi: 10.1016/S0079-6123(06)65034-6.

Hurley, S. (2008). The shared circuits model (SCM): how control, mirroring, and simulation can enable imitation, deliberation, and mindreading. *Behavioral and Brain Sciences, 31*, 1–22.

Jeannerod, M. (2006). *Motor cognition*. Oxford: Oxford University Press.

Johnson, A., & Redish, A. D. (2007). Neural ensembles in CA3 transiently encode paths forward of the animal at a decision point. *Journal of Neuroscience, 27*, 12176–12189. doi: 10.1523/JNEUROSCI.3761-07.2007.

Káli, S., & Dayan, P. (2004). Off-line replay maintains declarative memories in a model of hippocampal-neocortical interactions. *Nature Neuroscience, 7*, 286–294. doi: 10.1038/nn1202.

Kappen, H. J., Gómez, V., & Opper, M. (2012). Optimal control as a graphical model inference problem. *Machine Learning, 87*, 159–182.

Kiebel, S. J., Daunizeau, J., & Friston, K. J. (2008). A hierarchy of time-scales and the brain. *PLoS Computational Biology, 4*, e1000209. doi: 10.1371/journal.pcbi.1000209.

Klein, S. B. (2013a). The temporal orientation of memory: it's time for a change of direction. *Journal of Applied Research in Memory and Cognition, 2*, 222–234. doi: 10.1016/j.jarmac.2013.08.001.

Klein, S. B. (2013b). The complex act of projecting oneself into the future. *Wiley Interdisciplinary Reviews Cognitive Sciences, 4*, 63–79. doi: 10.1002/wcs.1210.

Koechlin, E. (2014). An evolutionary computational theory of prefrontal executive function in decision-making. *Philosophical Transactions of the Royal Society of London B Biological Sciences, 369*, 20130474. doi: 10.1098/rstb.2013.0474.

LeCun, Y., Bengio, Y., & Hinton, G. (2015). Deep learning. *Nature, 521*, 436–444.

Lengyel, M., & Dayan, P. (2008). Hippocampal contributions to control: the third way. In J. C. Platt, Koller, D., Singer, Y., & Roweis, S. (eds.) *Advances in neural information processing systems 20* (pp. 889–896). Cambridge, MA: MIT Press.

Leslie, A. M. (1987). Pretense and representation: the origins of" theory of mind." *Psychological Review, 94*, 412.

Lisman, J. (2015). The challenge of understanding the brain: where we stand in 2015. *Neuron*, 86, 864–882. doi: 10.1016/j.neuron.2015.03.032.

Lisman, J., & Redish, A. D. (2009). Prediction, sequences and the hippocampus. *Philosophical Transactions of the Royal Society of London B Biological Sciences, 364*, 1193–1201. doi: 10.1098/rstb.2008.0316.

Loewenstein, G. (1987). Anticipation and the valuation of delayed consumption. *Economic Journal, 97*, 666–684.

McClelland, J. L., McNaughton, B. L., O'Reilly, R. C. (1995). Why there are complementary learning systems in the hippocampus and neocortex: insights from the successes and failures of connectionist models of learning and memory. *Psychological Review*, 102, 419–457.

Mellers, B., Schwartz, A., & Ritov, I. (1999). Emotion-based choice. *Journal of Experimental Psychology: General, 128*, 332–345. doi: 10.1037/0096-3445.128.3.332

Michaelian, K. (2016). *Mental time travel*. Cambridge, MA: MIT press.

Michaelian, K. (2011). Generative memory. *Philosophical Psychology, 24*, 323–342. doi: 10.1080/09515089.2011.559623.

Montague, P. R., Dolan, R. J., Friston, K. J., Dayan, P. (2012). Computational psychiatry. *Trends in Cognitive Sciences, 16*, 72–80. doi: 10.1016/j.tics.2011.11.018.

Moulton, S. T., & Kosslyn, S. M. (2009). Imagining predictions: mental imagery as mental emulation. *Philosophical Transactions of the Royal Society of London B Biological Sciences, 364*, 1273–1280. doi: 10.1098/rstb.2008.0314.

Pastalkova, E., Itskov, V., Amarasingham, A., & Buzsáki, G. (2008). Internally generated cell assembly sequences in the rat hippocampus. *Science, 321*, 1322–1327. doi: 10.1126/science.1159775.

Penny, W. D., Zeidman, P., & Burgess, N. (2013). Forward and backward inference in spatial cognition. *PLoS Computational Biology, 9*, e1003383. doi: 10.1371/journal.pcbi.1003383.

Peters, J., & Büchel, C. (2010). Episodic future thinking reduces reward delay discounting through an enhancement of prefrontal-mediotemporal interactions. *Neuron, 66*, 138–148. doi: 10.1016/j.neuron.2010.03.026.

Pezzulo, G. (2008). Coordinating with the future: the anticipatory nature of representation. *Minds and Machines, 18*, 179–225. doi: http://dx.doi.org/10.1007/s11023-008-9095-5.

Pezzulo, G. (2011). Grounding procedural and declarative knowledge in sensorimotor anticipation. *Mind & Language, 26*, 78–114.

Pezzulo, G. (2012). An active inference view of cognitive control. *Frontiers in Psychology, 478*. doi: 10.3389/fpsyg.2012.00478.

Pezzulo, G. (2014). Why do you fear the bogeyman? An embodied predictive coding model of perceptual inference. *Cognitive Affective and Behavioral Neuroscience, 14*, 902–911. doi: 10.3758/s13415-013-0227-x.

Pezzulo, G., & Castelfranchi, C. (2009). Thinking as the control of imagination: a conceptual framework for goal-directed systems. *Psychological Research, 73*, 559–577.

Pezzulo, G., Barsalou, L.W., Cangelosi, A., Fischer, M. H., McRae, K., & Spivey, M. (2011). The mechanics of embodiment: a dialogue on embodiment and computational modeling. *Frontiers in Psychology, 2*, 1–21.

Pezzulo, G., & Rigoli, F. (2011). The value of foresight: how prospection affects decision-making. *Frontiers in Neuroscience, 5*. http://journal.frontiersin.org/article/10.3389/fnins.2011.00079/full

Pezzulo, G., Rigoli, F., & Chersi, F. (2013). The Mixed Instrumental Controller: using Value of Information to combine habitual choice and mental simulation. *Frontiers in Psychology, 4*, 92. doi: 10.3389/fpsyg.2013.00092.

Pezzulo, G., Rigoli, F., & Friston, K. (2015). Active Inference, homeostatic regulation and adaptive behavioural control. *Progress in Neurobiology, 134*, 17–35.

Pezzulo, G., van der Meer, M. A. A., Lansink, C. S., & Pennartz, C. M. A. (2014). Internally generated sequences in learning and executing goal-directed behavior. *Trends in Cognitive Sciences, 18*, 647–657. doi: 10.1016/j.tics.2014.06.011.

Pfeifer, R., Lungarella, M., & Iida, F. (2007). Self-organization, embodiment, and biologically inspired robotics. *Science, 318*, 1088–1093.

Pfeiffer, B. E., & Foster, D. J. (2015). Autoassociative dynamics in the generation of sequences of hippocampal place cells. *Science, 349*, 180–183. doi: 10.1126/science.aaa9633.

Pfeiffer, B. E., & Foster, D. J. (2013). Hippocampal place-cell sequences depict future paths to remembered goals. *Nature, 497*, 74–79. doi: 10.1038/nature12112.

Rabinovich, M. I., Simmons, A. N., & Varona, P. (2015). Dynamical bridge between brain and mind. *Trends in Cognitive Sciences, 19*(8), 453–461.

Raichle, M. E., MacLeod, A. M., Snyder, A. Z., Powers, W. J., Gusnard, D. A., & Shulman, G. L. (2001). A default mode of brain function. *Proceedings of the National Academy of Sciences, 98*, 676–682. doi: 10.1073/pnas.98.2.676.

Redish, A. D. (2013). *The mind within the brain: how we make decisions and how those decisions go wrong.* Oxford: Oxford University Press.

Rescorla, R. A., & Wagner, A. R. (1972). A theory of Pavlovian conditioning: variations in the effectiveness of reinforcement and nonreinforcement. In A. H. Black & W. F. Prokasy (eds.), *Classical conditioning ii: current research and theory* (pp. 64–99). New York: Appleton-Century.

Rolls, E. T., & Treves, A. (1997). *Neural networks and brain function.* Oxford: Oxford University Press.

Schacter, D. L., Addis, D. R., & Buckner, R. L. (2007). Remembering the past to imagine the future: the prospective brain. *Nature Reviews Neuroscience, 8*, 657–661. doi: 10.1038/nrn2213.

Schacter, D. L., Addis, D. R., Hassabis, D., Martin, V. C., Spreng, R. N., & Szpunar, K. K. (2012). The future of memory: remembering, imagining, and the brain. *Neuron, 76*, 677–694. doi: 10.1016/j.neuron.2012.11.001.

Schuck, N. W., Gaschler, R., Wenke, D., Heinzle, J., Frensch, P. A., Haynes, J.-D., & Reverberi, C. (2015). Medial prefrontal cortex predicts internally driven strategy shifts. *Neuron, 86*, 331–340. doi: 10.1016/j.neuron.2015.03.015.

Sejnowski, T. J., Poizner, H., Lynch, G., Gepshtein, S., & Greenspan, R. J. (2014). Prospective optimization. *Proceedings of the IEEE, 102*, 799–811. doi: 10.1109/JPROC.2014.2314297

Seth, A. K. (2013). Interoceptive inference, emotion, and the embodied self. *Trends in Cognitive Sciences, 17*, 565–573. doi: 10.1016/j.tics.2013.09.007.

Shipp, S., Adams, R. A., & Friston, K. J. (2013). Reflections on agranular architecture: predictive coding in the motor cortex. *Trends in Neuroscience, 36*, 706–716. doi: 10.1016/j.tins.2013.09.004.

Steiner, A. P., & Redish, A. D. (2014). Behavioral and neurophysiological correlates of regret in rat decision-making on a neuroeconomic task. *Nature Neuroscience, 17*, 995–1002. doi: 10.1038/nn.3740.

Sutton, R. S. (1991). Integrated modeling and control based on reinforcement learning and dynamic programming. In D. S. Touretzky (ed.), *Advances in neural information processing systems* (pp. 471–478).

Sutton, R. S., & Barto, A. G. (1998). *Reinforcement learning: an introduction.* Cambridge, MA: MIT Press.

Toussaint, M. (2009). Probabilistic inference as a model of planned behavior. *Künstliche Intelligenz, 3*(9), 23–29.

Treves, A., & Rolls, E. T. (1994). Computational analysis of the role of the hippocampus in memory. *Hippocampus, 4*, 374–391. doi: 10.1002/hipo.450040319.

Tulving, E. (1983). *Elements of episodic memory.* Oxford: Clarendon Press.

van der Meer, M. A. A., & Redish, A. D. (2010). Expectancies in decision making, reinforcement learning, and ventral striatum. *Frontiers in Neuroscience, 4*, http://journal.frontiersin.org/article/10.3389/neuro.01.006.2010/abstract

van der Meer, M., Kurth-Nelson, Z., & Redish, A. D. (2012). Information processing in decision-making systems. *The Neuroscientist, 18*, 342–359.

Verschure, P., Pennartz, C. M. A., & Pezzulo, G. (2014). The why, what, where, when and how of goal-directed choice: neuronal and computational principles.

*Philosohical Transactions of the Royal Society of London B Biological Sciences, 369*, 20130483.

Voloh, B., Valiante, T. A., Everling, S., & Womelsdorf, T. (2015). Theta–gamma coordination between anterior cingulate and prefrontal cortex indexes correct attention shifts. *Proceedings of the National Academy of Sciences, 112*(27), 8457–8462. doi: 10.1073/pnas.1500438112.

Wikenheiser, A. M., & Redish, A. D. (2013). The balance of forward and backward hippocampal sequences shifts across behavioral states. *Hippocampus, 23*, 22–29. doi: 10.1002/hipo.22049.

Wikenheiser, A. M., & Redish, A. D. (2015). Hippocampal theta sequences reflect current goals. *Nature Neuroscience, 18*, 289–294. doi: 10.1038/nn.3909.

PART VI

# Evolutionary and Comparative Perspectives

14

# Evolutionary Perspectives on Prospective Cognition

JAMES M. THOM AND NICOLA S. CLAYTON

Niko Tinbergen (1963) described four complementary questions to be asked of any animal's behavior in order to understand it. Two of the questions seek proximate explanations for behavior: What are the material causes of behavior? And how does the behavior develop within the lifetime of an individual? These are questions of *mechanism* and *ontogeny*, and they are the primary focus of many psychologists and neuroscientists. While other chapters in this volume will explore at length the proximate causes of prospective cognition,[1] we will direct our attention to the other two questions, which concern the ultimate, evolutionary causes of prospection: function and phylogeny.

## 1. FUNCTION AND PHYLOGENY

Evolutionary explanations for an adaptation must enlighten on two key facts: first, the reproductive advantage that adaptation confers upon an individual animal (*function*); second, the constraints imposed on that adaptation by a species' evolutionary history (*phylogeny*). The first of these is easily understood. A large animal's size allows it to retain heat and to fight off rivals, but also imposes a metabolic cost, which reduces the resources that can be devoted to producing offspring and increases the risk of starvation. Natural selection will favor genes that contribute to a more optimal trade-off between cost and benefit, and future generations will cluster around this optimum. In fact, the trade-off between an adaptation's benefits and its metabolic cost is of central importance in the evolution of any new cognitive capacity. Cognitive processing requires a neural substrate, and big brains are expensive, making up 20%–25% of the human body's energy needs (Holliday, 1986; Mink et al., 1981). A *functional* explanation of prospective cognition must therefore (i) highlight the increased fitness associated with thinking about the future and (ii) assert that these benefits will at least compensate for the concomitant increase in metabolic requirements.

The second evolutionary question is that of *phylogeny*—the constraints imposed by an organism's evolutionary history. Natural selection typically acts upon a species gradually and continually, over many generations. A consequence of this is that an optimal adaptation is unreachable if it requires several generations of reduced fitness first. The canonical example is the vertebrate eye, which has a blind spot where the retinal inputs to the optic nerve converge. Because an eye with a blind spot is better than no eye at all, natural selection can only improve upon the existing eye, rather than start from scratch with a blind individual. Consequently, vertebrate eyes have been extensively adapted to a range of ecological demands while retaining the suboptimal blind spot (Land & Nilsson, 2002). Psychological adaptations like prospective mental time travel should then be evaluated as variations upon what came before. As an example, we have recently argued that continuity in hippocampal function across the animal kingdom may have constrained the independent evolution of prospective faculties in mammals and birds (Thom & Clayton, 2014).

To summarize, an attempt to understand the evolutionary origins of prospective mental time travel should elucidate the specific fitness advantages conferred by specific variations in those psychological capacities already present in a species. However, the ascription of an adaptive function to a cognitive faculty like mental time travel depends on our knowledge of that faculty, since we cannot infer a cause without understanding the effect. If we wish to take a comparative approach—relating species differences in cognition to variations in the requirements of the ecological niches these species occupy—we must first know the extent of these species differences in cognition. In the case of prospective mental time travel, the focus of most empirical work has been the extent of the difference between humans and the rest of the animal kingdom. Put simply, is prospective mental time travel unique to humans? Answering this proximate, mechanistic question is an essential first step in advancing any ultimate, evolutionary account of mental time travel. It is therefore this question, and how best to answer it, that is the topic of the remainder of this chapter.

## 2. THE NULL HYPOTHESIS

In setting out to identify non-human animals capable of prospective mental time travel we face a problem: How do we know if an observed behavior is driven by prospective cognition? Consider food-caching. An animal that stores food with the intention of retrieving it later would certainly seem to be exhibiting some form of prospective cognition. But it is also plausible that evolution has endowed the animal with a desire to cache food, oblivious to the future benefits of doing so. The task is to identify instances in which an animal acts for the future *with the future in mind.*

Lloyd Morgan's Canon, which has dominated interpretations of animal behavior in the comparative study of cognition, asserts that "[i]n no case is an animal activity to be interpreted in terms of higher psychological processes if it can be fairly interpreted in terms of processes which stand lower in the scale of psychological evolution and development" (Morgan, 1903, 59). In the case of food-caching, this means that we should assume that the second explanation (oblivious desire to cache) is true unless we have good reason to reject it. Ignoring Morgan's archaic use of "higher" and "lower," we can see this as an application of Occam's razor to the evolution of intelligence. It is, however, a flawed one. The correct locus of probability is the generation of mutations responsible for changes in psychological processing, not a subjective assertion of the complexity of the relevant process (though see Sober, 2005, and Fitzpatrick, 2008, for a more detailed discussion of Occam's razor in Morgan's Canon). An apparently simple behavioral rule might require a more improbable series of mutations than a cognitive capacity that we intuit to be more complex.

A second problem for Morgan's Canon comes from the particular aspects of cognition we choose to scour the animal kingdom for, and where we choose to look. Comparative psychology is not a comprehensive and dispassionate assay of all possible cognitive faculties in all possible species. Aside from the practical difficulties such an endeavor would prevent, researchers would still be limited by their own imaginations. We tend to look for those cognitive capacities that we see in ourselves, and that we intuitively believe to be most complex. We also typically look first to our closest relatives, the great apes. This preference is not without foundation; great apes look more like us than other animals, they have large brains like we do (Jerison, 1973), and exhibit some behaviors that we normally think of as very human, such as tool use (Van Lawick-Goodall, 1970). It is difficult, however, to see how Occam's razor can apply when we ask whether chimpanzees and bonobos share a particular intellectual faculty with humans. We already know the mutations responsible for the faculty in question have arisen because *we have that faculty*. The question is thus not "How many mutations were required?" but rather "When did these mutations occur?". We cannot apply a principle of parsimony to conclude that this faculty first emerged in a recent ancestor of modern humans and not a common ancestor of humans, chimpanzees, and bonobos.

It should be noted that the preceding criticism only applies in cases of *homology*, in which species share a trait because they both inherited it from a shared ancestor. We infer homology of any shared intelligence between humans and chimpanzees because it is more parsimonious to assume that this intelligence emerged once in a shared ancestor than twice in separate lineages. The latter possibility is one of *convergence*, and the criticism of Morgan's Canon from parsimony would not apply in this case. An example of convergence that we will discuss later in this chapter is that of intelligence, and particularly foresight, in the crow family. Because crows and humans are very distally related, our common ancestor is also the ancestor of all mammals and birds, namely a stem amniote. Unless we believe that all mammals and birds are capable of prospective cognition, we might conclude that a great many mutations were required to *lose* this faculty in other species. More likely, we would argue that this faculty has evolved independently in crows and humans, and that both have simply converged on the same solution to a problem. In this case, the mutations responsible for prospective capacities must have occurred at least twice if we believe that crows are capable of prospective mental time travel. In the absence of any evidence to the contrary, Morgan's Canon will lead us to the most parsimonious conclusion: crows cannot engage in prospective mental time travel.

It should be apparent from the preceding paragraphs that Lloyd Morgan's Canon is not a strong foundation for assumptions about intelligence in non-human animals, particularly with respect to our closest evolutionary cousins. It nonetheless reflects three beliefs on which to base our null hypothesis. The first is that cognitive explanations are likely to require a more improbable evolutionary history of mutations than a simpler explanation. This has some validity, though the caveats noted earlier should be kept in mind. The second belief is that natural selection will not tolerate the expense of an unnecessarily cognitive solution to a problem. This is justifiable insofar as we associate cognitive processing with energy requirements that impose opportunity costs. However, it should be noted that we have little definitive evidence concerning the metabolic costs of specific cognitive faculties. The third belief is intuitive: that there must be something that is qualitatively different about humans, and that something is likely to be an aspect of our intelligence. This third belief could prove to be incorrect, but is widely held and motivates a great deal of interest in animal cognition. These three beliefs supporting Morgan's Canon are not strongly justifiable, but neither are they obviously wrong. In the absence of a better alternative, we begin with a null hypothesis that no animal, excepting a human, is capable of prospective mental time travel.

The specific articulation of this principle most often tested is the Bischof-Köhler hypothesis, which states that "animals other than humans cannot anticipate future needs or drive states, and are therefore bound to a present that is defined by their current motivational state" (from Suddendorf & Corballis, 1997). Suddendorf and Corballis note in presenting this hypothesis that it retains a degree of ambiguity, relying on a common sense definition of "needs." The qualification of human uniqueness exemplifies the skepticism of Lloyd Morgan's Canon in limiting mechanistic explanations of non-human behavior to present drive states. Falsification of the Bischof-Köhler hypothesis is the current gold standard for claims of prospective cognition in non-human animals.

## 3. THE GREAT APES

Our closest relatives in the animal kingdom are the non-human great apes: orangutans, gorillas, chimpanzees, and bonobos. As we have posited earlier, the argument for the uniqueness of human intelligence may not rely as surely on phylogenetic parsimony in the case of the great apes. These animals are therefore the logical starting place for our search.

Suddendorf and Corballis (1997) summarize several anecdotal observations of great ape behavior that might indicate a capacity to plan for the future. These include a male chimpanzee following and sleeping near a female while she is in estrus (Goodall, 1986), and carrying straw from inside to make a warm nest on a cold night (De Waal, 1982). A more recent example comes from Furuvik zoo in Sweden, where a male chimpanzee had been collecting stones and pieces of concrete and later throwing them at visitors (Osvath, 2009). The collecting behavior had only been observed in the hours before the zoo opened (so in the absence of visitors to throw at), and the caches of projectiles were always discovered on the shoreline facing the visitors. These observations are intriguing, but their anecdotal nature limits their usefulness. In the absence of controlled experiments, it is difficult to draw any concrete conclusions (Suddendorf & Corballis, 2010; though see also Osvath & Karvonen, 2012, for a more systematic exploration of the Furuvik chimpanzee's stone throwing).

A recent study of apparent route planning in chimpanzees faces the same criticisms. Janmaat et al. (2014) observed the movements and nesting and feeding behaviors of five female chimpanzees in the wild for 275 days. The subjects traveled to a fruit tree each morning for breakfast, and departed earlier when competition for the fruit was likely to be greater (normally smaller fruit that could also be eaten by other, smaller animals). When the next day's breakfast was to be figs—which were typically depleted very quickly—the subjects also tended to position their nests en route to the breakfast tree. The chimpanzees therefore appeared to plan their routes in order to arrive at the breakfast tree before all the fruit had been eaten. Unlike the previous anecdotes, these observations were made systematically over a long period of time. However, when studying a wild population, it is impossible to control for confounds from previous life experience (Thom & Clayton, 2015), and controlled manipulations are required to make unambiguous predictions about an animal's behavior given differing present and future needs.

Two notable studies in the last decade attempted falsification of the Bischof-Köhler hypothesis in great apes in a more controlled empirical setting. Both studies focused on the selection and transport of tools needed to gain access to a food reward in the future. Mulcahy and Call (2006) tested bonobos and orangutans that had been trained to use a tool to access fruit juice in a testing room. The apes were then given the opportunity to select a suitable tool from an array of unsuitable alternatives, at a time when they were unable to access the apparatus containing the fruit juice. The subjects were then ushered out of the testing room, and were only given access after an hour. To successfully acquire the juice, therefore, the apes had to select the correct tool, carry it outside, keep it, and return to the testing room with it later. They did so on 40.9% of the trials, significantly more often than would be expected by chance. This finding was also seen in two subjects tested with a longer delay of 14 hours.

In order to establish that this behavior was carried out *with the future in mind*, Mulcahy and Call included a control experiment in which subjects were rewarded for bringing back the correct tool, but the apparatus was not present and the tool was not actually used. This condition therefore tested acquisition of an arbitrary contingency between bringing back the tool and getting the juice. As the authors predicted, the correct tool was returned reliably less often in this experiment than in the previous experiments when the tool had been used. However, two of the four subjects in this control experiment *never* returned the correct tool, and so had no opportunity to acquire the contingency being assessed. The other two performed similarly to those subjects in the other experiments (Suddendorf, 2006). It is therefore possible that the observed behavior can be attributed to long-delay instrumental conditioning (Cheke & Clayton, 2010), by which the tool could have become associated with the juice reinforcer, facilitated by spending the delay outside the testing room, reducing interference (Lett, 1975). Furthermore, we know nothing of the subjects' motivational states at the time of testing. This same criticism applies to a more recent attempt to replicate Mulcahy and Call's first experiment (Dufour & Sterck, 2008). The apes could plausibly have experienced a present desire for juice without entertaining any particular expectation of an encounter with the apparatus. In the absence of a clear cost to selecting and keeping the correct tool, we might expect a present desire for juice to drive the observed behavior.

Osvath and Osvath (2008) attempted to overcome this hurdle by introducing a self-control element, in which present and future needs are opposed by design. Two chimpanzees and an orangutan were given the opportunity to select a tool from a tray of alternative items. This tool could be used to access some highly valued fruit soup after a 70-minute delay. Initially, the alternative items were non-functional tools, and all three subjects reliably chose the functional tool. In the second experiment, however, one of the alternative items was a piece of the animal's favorite fruit. The subjects therefore faced a choice between an immediately available piece of fruit, and the prospect of future access to a more valuable fruit reward using the functional tool. All three subjects selected the tool on significantly more than the expected 0% of trials.

A third experiment controlled for the possibility that the functional tool had become a secondary reinforcer via its association with the fruit soup. The subjects were now offered two choices instead of one. The first was between a functional tool and several non-functional alternatives, and the second was between another functional tool, a piece of fruit, and some non-functional tools. If the tool had previously been preferred to the fruit because it had acquired its own reinforcing properties, then the apes would have been expected to choose the functional tool twice. Instead, each of the subjects selected the fruit in the second choice on every trial. The authors infer from their results that the apes were anticipating their future encounter with the fruit soup and their consequent need for a single functional item. In a final experiment, all subjects showed an above-chance preference for a novel functional tool over some equally novel non-functional alternatives. Osvath and Osvath argue that this final finding indicates that their subjects were pre-experiencing their interaction with the apparatus containing the fruit soup, to identify which tool they would need.

In response to Osvath and Osvath's (2008) paper, Suddendorf et al. (2009) issued a commentary urging caution. They note that tool selection in the second experiment was compared to 0%, not to chance. Osvath (2010) responds that chance is not the correct comparison, because the presence of an attractive alternative (the favorite fruit item) would lead a "future blind" ape to select the functional tool less often than chance. This argument is plausible, and it is unfortunate that it remains untested. A simple control could confirm it using a less attractive fruit alternative, which should result in more choices for the functional tool, and fewer for the fruit (and still none for the non-functional tools). Applying this control, the third experiment could also strengthen those findings. Since the value of the fruit is always greater than that of a second functional tool, the apes should behave identically and always choose the fruit.

The claims of Osvath and Osvath (2008), however, rest primarily on the validity of their third experiment, which purports to control for associative learning. Suddendorf et al. (2009) suggest that the apes may have held an expectation of an immediate encounter with the apparatus containing the fruit soup, which motivated their initial choice. This would seem surprising since the subjects were never given the opportunity to use the tool immediately after selection. However, one aspect of the apes' behavior does appear inconsistent with anticipation of future tool use. Suddendorf et al. note that the researchers were able to retrieve the tool after each testing session. If the apes expected to encounter the fruit soup within a trial based on past experience, why would they not expect to do so in future trials?

If they did, we might expect them to guard the functional tool for future use. An alternative to this is that the apes also understood that they would always be offered a new tool, based on past experience. One way to test this would be to provide a food option at *both* choices. Given foresight, the optimal behavior would be to pick the tool once, guard it between trials, and then benefit from both food rewards and the fruit soup on each subsequent trial.

Finally, Suddendorf and colleagues question whether the paradigm used by Osvath and Osvath is capable of testing the Bischof-Köhler hypothesis. Since it is not clear that desire for the fruit and desire for the fruit soup are "qualitatively" different motivational states, there may be nothing to dissociate from. This criticism betrays an ambiguity at the heart of the Bischof-Köhler hypothesis in its description of present and future "needs." The interpretation of needs as drive states is convenient for empirical work, but it is extremely limiting if applied universally. If the apes were indeed pre-experiencing a future encounter with the apparatus containing fruit soup, would anticipation of the need for the tool not be sufficient to be considered a type of episodic foresight? This problem is also apparent in interpreting other studies of tool transportation (e.g., Mulcahy & Call, 2006); it is not clear whether this behavior could ever satisfy the Bischof-Köhler hypothesis.

The evidence for planning in great ape tool transport is, while impressive, inconclusive. This is partly attributable to the considerable practical constraints involved when working with these animals. Beyond these constraints, however, we are left with a critical polysemy in the Bischof-Köhler hypothesis. The vague meaning of "needs" is clearly a problem for attempts at falsification. Despite the high degree of uncertainty in interpretation of the research presented so far in this chapter, it is difficult not to be impressed with the apes' behavior. When tested on a tool-transport paradigm similar to that of Mulcahy and Call (2006), long-tailed macaques only learned to select and transport tools following shaping with immediate rewards (Dekleva et al., 2012). Furthermore, this behavior mostly stopped with a longer delay of 20 minutes, while Mulcahy and Call's apes tolerated a 14-hour delay. If the great apes show a greater propensity for future-oriented behavior than some other primates, it is important to determine whether this is due to a greater capacity for foresight. We must therefore consider the evidence for prospective cognition in these other primates.

## 4. OTHER PRIMATES

Naqshbandi and Roberts (2006) compared the performance of squirrel monkeys and rats in a test of the Bischof-Köhler hypothesis. The animals were given a series of choices between smaller and larger quantities of thirst-inducing food. During the task, they had limited access to water, and choosing the smaller food reward resulted in earlier (but still delayed) access to the water. All the animals had a baseline preference for the larger food reward, and the authors predicted that an animal capable of foresight would switch its preference to the smaller food reward as the task progressed. The squirrel monkeys tested did so, but the rats did not, leading the authors to conclude that the squirrel monkeys were anticipating their future thirst and acting accordingly. In a follow-up control experiment with water availability matched for the two food rewards, the monkeys preferred the larger reward, as they had at baseline. This suggests that the dates had not simply become less attractive by association with thirst.

The findings of Naqshbandi and Roberts (2006) are promising, and the competition between hunger and thirst would seem to meet the requirement for present and future motivational states to be "qualitatively" different. Given the problems associated with small study samples, however, it should be noted that only two squirrel monkeys were tested. More recent work attempted to replicate this result in rhesus macaques (Paxton & Hampton, 2009). The macaques were tested using the same paradigm as the squirrel monkeys, but they continued to select the larger food reward despite the consequent increase in thirst. The authors made systematic variations to the procedure, and found that the macaques did switch to choosing the smaller food reward when the periods of water deprivation were decreased so that there was a 0-minute delay to the return of water when choosing the smaller amount. The macaques' failure in the original task might therefore be attributable to a failure to learn the contingency between choice and water availability over long delays.

The authors go further, noting that the Bischof-Köhler hypothesis does not state how distant "the future" needs to be. Since their macaques were not thirsty when given the choice involving the 0-minute delay (which took as long as needed for the experimenters to return access to water), they did seem to be anticipating thirst in the very near future and thus "... technically contradict[ing] the Bischof-Köhler hypothesis" (Paxton & Hampton, 2009, p. 243). The oddity of this should be striking. What is implied is either a capacity for foresight that cannot extend into the future, or a capacity to anticipate future events without the ability to learn to expect those future events. It is difficult to envisage the adaptive utility of either. Indeed, enabling an animal to learn about, and later anticipate, the long-term consequences of its action seems an important advantage of mental time travel. That this is not in any way captured by the Bischof-Köhler hypothesis gives good reason for caution in rejecting out of hand behavior that does not meet this particular criterion.

To summarize, the evidence for prospective mental time travel in a non-human primate is promising, but far from conclusive. There have been no undeniable and replicated falsifications of the Bischof-Köhler hypothesis. This is partly due to practical difficulties inherent in research, but also due to ambiguities in the hypothesis itself. In the next section, we consider attempts to challenge the Bischof-Köhler hypothesis in a completely different clade: the corvids.

## 5. THE CORVIDS

The corvids are a family of songbirds including crows, ravens, and jays. They share several important characteristics with primates that have made them attractive prospects for research into comparative cognition: they have large brains for their body sizes (Emery & Clayton, 2004; Jerison, 1973); some corvids are capable of using tools (Bird &d Emery, 2009; Taylor et al., 2007) and of manufacturing them (Bird & Emery, 2009; Weir & Kacelnik, 2006); and some species appear to demonstrate impressive understanding of social problems (Bugnyar & Heinrich, 2005; Emery & Clayton, 2001; Ostojic et al., 2013; Seed et al., 2008). Because corvids are very distant relatives to humans and other primates, their intelligence is believed to have evolved independently to and in parallel with our own (Emery & Clayton, 2004). Since this is a case of convergent evolution, Morgan's Canon can draw support from phylogenetic parsimony in a way it cannot not when we consider great ape intelligence. Nonetheless, our null hypothesis—the Bischof-Köhler hypothesis—is identical.

The focus of much research into corvid prospective cognition has been a single species: the Western scrub jay *(Aphelocoma californica)*. There are two reasons for this. The first is that scrub jays naturally engage in a future-oriented behavior suitable for exploration; they cache food for future consumption. The second reason is that scrub jays find and retrieve their caches using flexible memory recall that has been described as analogous to the retrospective component of human mental time travel—episodic memory (Clayton et al., 2003; Clayton & Dickinson, 1998).

Two studies of scrub jay caching behavior purport to demonstrate prospective mental time travel. Raby et al. (2007) describe an experiment in which scrub jays must "plan for breakfast" by distributing caches appropriately between two rooms in which they might wake the following morning. Correia et al. (2007) report a falsification of the Bischof-Köhler hypothesis, in which the birds were presented with opposing desires for present and future.

Raby et al. (2007) report two experiments, which took place in a set of three adjacent cages. In Experiment 1, the scrub jays were shut each morning into one of the end cages. One cage—the "breakfast room"—contained food; the other—the "no breakfast room"—did not. For the rest of the day, the birds had access to all rooms. Each evening, a caching tray was inserted into each of the two end rooms, but only non-cacheable powdered food was available. After three mornings spent in each of the end cages, the birds were given a caching test in which the evening's powdered food was replaced with whole (cacheable) pine nuts. The birds cached significantly more pine nuts in the "no breakfast room" than they did in the "breakfast room."

The second experiment was similar to the first, but the "breakfast" and "no breakfast" rooms were replaced by a "peanut room," in which peanuts were available for breakfast, and a "kibble room," in which kibble was available for breakfast. Therefore, specific satiety—consumption-dependent devaluation of a particular food type—would have resulted in satiety for peanuts in the peanut cage, and for kibble in the kibble room. At the caching test, the birds preferentially cached peanuts in the kibble room, and kibble in the peanut cage. In other words, they cached what they would want at recovery, and where they would want it.

It has been argued that these results contrast with the predictions of an associative account, according to which the animals would have preferentially cached the foods in the cages already associated with those foods. This argument is derived from the fact that rats (*Rattus norvegicus*) eat more in rooms associated with food (Petrovich et al., 2007), and that scrub jay caching behavior is partially controlled by the motivational system that governs eating (de Kort et al., 2007)—presumed to be relatively similar in birds and rats. However, the observed distribution of caches could also result from a preference to spread stores somewhat evenly. Such a preference would be entirely spatial, and would not entail foresight. We should be wary of post hoc heuristic accounts, since it is possible to end up with an overwhelming accumulation of "simple rules" that look decidedly less simple en masse. Nonetheless, a preference for spreading food stores is plausible because we have reason to think that scrub jays might already possess just such a preference. Scrub jays are scatter hoarders, unlike rats, which tend to store their caches in one or a few places (Vander Wall, 1990). This difference in storing preference must be underpinned by some psychological disparity. If this psychological disparity is a consequence of a capacity for foresight, then we should expect all scatter hoarders to be capable of foresight, which is a strong claim indeed. Conversely, if this psychological disparity between rats and scrub jays is not related to foresight, then so might be the difference in observed behavior in the "planning for breakfast" task. Because of this alternative explanation, the findings presented by Raby et al. (2007) cannot make a conclusive case for prospective mental time travel in scrub jays.

Correia et al. (2007) presented scrub jays with the opportunity to cache while their present desires were in conflict with the desires they would experience in the future, at retrieval. The birds' motivational states at caching and retrieval were manipulated by pre-feeding with either peanuts or kibble to induce specific satiety. Half of the birds were pre-fed the same food prior to both caching and recovery (the "Same" group), half were pre-fed different foods (the "Different" group). In the "Different" group this induced a conflict, because scrub jays typically prefer to cache what they want to eat right now (Clayton & Dickinson, 1999). On the first trial, when the birds had no knowledge of what their states would be at retrieval, both groups preferentially cached the non-pre-fed food. However, on trials 2 and 3, this pattern was reversed in the "Different" group but not in the "Same" group. The authors concluded that the birds in the "Different" group overcame their current motivational state in order to cache what they would want at retrieval, thus falsifying the Bischof-Köhler hypothesis.

A commentary on Correia et al.'s (2007) study produced two main criticisms (Suddendorf & Corballis, 2008). First, the birds in the "Same" group had been presented with more of the pre-fed food type across the course of the experiment than birds in the "Different" condition. This is true, but it is unclear why this should have affected the birds' motivational states. All birds in both groups had prior experience of the pre-fed food types from previous studies, and all birds had *ad libitum* access to a range of maintenance food types between trials. Furthermore, subsequent work with Eurasian jays (*Garrulus glandarius*)—another corvid species—replicated the findings of Correia et al. using a within-subject design that prevents this problem (Cheke & Clayton, 2012).

The second criticism focuses on the use of proportions of food types cached in the analysis, rather than absolute numbers. These proportions were criticized for reflecting a reduction in caching of the pre-fed food, rather than an increase in caching of the non-pre-fed food. In other words, the birds did not appear to anticipate a future need for the non-pre-fed food, but instead appeared to anticipate that they would *not* need the pre-fed food. However, this is precisely what one ought to expect given that they began on trial 1 by caching both food types (Clayton et al., 2008). If the birds had anticipated that they would want *more* of any food at retrieval, they would be wrong! The pre-feeding with one food type before retrieval reduced the value of that food type most, but also reduced the value of all foods through its effect on general satiety. Put in lay terms, knowing that you plan to have savory leftovers for dinner should stop you from buying yourself more savory food, not encourage you to buy extra dessert.

The evidence from scrub jay caching is suggestive of a capacity for representing a future motivational state that differs from that being presently experienced. The findings of Correia et al. (2007) and Cheke and Clayton (2012) particularly appear to contradict the Bischof-Köhler hypothesis. It is not, however, clear that we can conclude from this that scrub jays are able to engage in prospective mental time travel. It has been noted that many studies of scrub jay cognition focus on food-caching, perhaps implying a narrow focus for their abilities that contrasts with the broad range of future experiences that humans can envisage (Suddendorf & Corballis, 2010). However, it is entirely possible that the emphasis on caching is a matter of empirical convenience rather than a cognitive limitation. It could therefore be that scrub jays are indeed capable of foresight beyond a caching context, but we simply have not tested for it.

Recent work in Eurasian jays hints at an avenue for research into foresight in a non-caching context. Male jays share food with their partners in breeding season, and researchers asked whether they anticipate their partners' desires when doing so (Ostojic et al., 2013). They pre-fed the females on one of two food types in order to induce specific satiety, and then gave the males a choice of food types to share. The males preferentially selected and shared the non-pre-fed food, in accordance with their partners' motivational states. Importantly, this preference was only seen when the males could observe the pre-feeding, so the females were not just cueing the males' choices behaviorally. Subsequent work suggests that the males' food selection is partially driven by their own motivational state (Ostojic et al., 2014). The birds therefore needed to overcome their own desires in order to act to satiate their partner's desires. Given that the selection of food was necessarily made before sharing, and because the Bischof-Köhler hypothesis does not state how far into the future a "future need" should be (Paxton & Hampton, 2009), it could be argued that food-sharing does indeed provide evidence of foresight beyond caching.

Whether or not one finds the case of food-sharing convincing as a potential example of prospective cognition, the emphasis on caching in studies of scrub jay cognition ought not to be surprising, for two reasons. First, scrub jays devote much of their time, energy, and attention to caching, so that is the most obvious manifestation of their cognitive abilities. Consider the analogous example of tool use in corvids. New Caledonian crows make and use tools in the wild (Orenstein, 1972), and appear to exhibit understanding of novel physical problems (e.g. Weir & Kacelnik, 2006). Similar behavior was later reported from laboratory studies with rooks, which are not natural tool users (Bird & Emery, 2009). From observations of rooks in the wild, we would never have known that they could learn to manufacture tools, simply because they do not naturally experiment with objects as New Caledonian crows do. In the case of tool use, the rook studies were conducted because similar work had already been carried out with New Caledonian crows. In the absence of any unambiguous non-caching demonstrations of episodic foresight in a non-human animal, it is not obvious what form a non-caching paradigm should take. This leads us to a potentially more troubling concern. The Bischof-Köhler hypothesis is intended as a conservative test of non-human foresight, so there may be instances of behavior for which foresight might be a plausible explanation, but where competing explanations cannot be ruled out. It is possible that the narrow focus of the Bischof-Köhler hypothesis on anticipation of future needs so restricts the range of behaviors from which prospective cognition can be inferred that caching provides the only conclusive examples. If this is indeed the case, we must consider whether the Bischof-Köhler hypothesis continues to be a useful tool for research into future-oriented behavior.

## 6. THE BISCHOF-KÖHLER HYPOTHESIS RECONSIDERED

We have reviewed a range of studies in primates and corvids that purport to examine foresight. A consistent concern has been whether the behavior in these studies meets the standards of the Bischof-Köhler hypothesis, namely anticipation of future needs that differ from the animal's present needs. One problem has been the ambiguity of the word *needs*, and whether *motivational state* is an appropriate proxy. Even when behavior appears to contradict the Bischof-Köhler hypothesis, such as in the case of the Western scrub jay, we can still draw very few conclusions about the animal's prospective capacities when compared to our own.

It is instructive to consider application of the Bischof-Köhler hypothesis to our own species. Humans are certainly able to anticipate future needs. As an example, we must know that we will need to eat later in the week when going food shopping. If a holiday is approaching and family members are visiting for a large meal, we will buy more food than usual. This is clearly an example of acting for the future *with the future in mind*. Nonetheless, it is not necessarily the case that we dissociate completely from the context of our present motivational state when we engage in this behavior. Indeed, shopping for food is very much influenced by current hunger (Nisbett & Kanouse, 1969), even though that hunger is a temporary state, unlikely to affect our needs in the coming days. It is furthermore unclear whether the desire for food now and the anticipated desire for food later would be considered "qualitatively different" needs.

What we are left with, then, is a criterion that is neither sufficient nor necessary to indicate prospective mental time travel: one can imagine future events without dissociating from present desires, and act for future needs without pre-experiencing them. The Bischof-Köhler hypothesis has provided an important focus for comparative research into prospective cognition, and has encapsulated the principles laid out in Morgan's Canon. However, it has also proven highly restrictive as a null hypothesis. We believe that future research will benefit from the development of new criteria for future-oriented cognition.

## 7. CONTINUITY AND CRITERIA

The logical starting point for development of a behavioral criterion for prospective cognition is with human mental time travel. Indeed, it is because we recognize the importance of this capacity in many behaviors that seem to be uniquely human that we might ask whether foresight represents a discontinuity between our species and other animals. However, there is a risk that this approach is too modular, and that it would be more beneficial to concentrate on individual processes contributing to foresight, rather than the entire cognitive system.

The mental time travel hypothesis lays out an array of cognitive prerequisites for the mental time travel seen in humans, including mental attribution, meta-representation, and the ability to dissociate oneself from the context of the present (Suddendorf & Corballis, 1997). The evolution of human foresight could be a case of "punctuated equilibrium" (Gould & Eldredge, 1977), in which these component processes emerged rapidly in the short time since our split from the chimpanzee lineage. What would be left is a cognitive chasm between humans and the rest of the animal kingdom—a difference in kind rather than degree. If this is not the case, then we must consider the possibility that these processes emerged gradually, with some degree of continuity between species. Either way, it is important to ask whether the component processes of mental time travel facilitate future-oriented action on their own, or whether they are co-opted for that purpose following the emergence of other components. Put simply, does an incomplete mental time travel system—relative to our own—help an animal to act for the future? Where there is a degree of phylogenetic continuity in the processes required for mental time travel, we must consider how each process contributes to future-oriented action in the species endowed with them. Doing so could allow us to relate species differences to differences in ecological demands, and so shed light on the selective forces that shaped the evolution of our own minds.

Corballis (2013, 2014) argues that continuity in hippocampal processing indicates that species differences in foresight are indeed a matter of degree rather than kind. The hippocampus has been implicated as critical in human prospective mental time travel, from neuroimaging studies (e.g., Addis et al., 2007), and from behavioral research on patients with hippocampal damage (Klein et al., 2002; Rosenbaum et al., 2005). The hippocampus is also important for spatial processing, containing "place cells" that fire in response to being in a particular location (O'Keefe & Nadel, 1978). Indeed, it has been argued that a key function of the hippocampus in prospection is to provide a spatial framework for imagined future scenes (Hassabis et al., 2007a; Hassabis et al., 2007b; Hassabis & Maguire, 2007). The evidence for continuity comes from activity in rat hippocampal place cells outside the spatial context to which they refer (Wilson & McNaughton, 1994). This activity can correspond to routes that had not been previously navigated (Gupta et al., 2010), and even to routes that the animal takes subsequently (Pfeiffer & Foster, 2013). We agree that this activity is at least consistent with some component of foresight, and is a plausible building block for the development of human mental time travel (Thom & Clayton, 2014).

## 8. FUTURE DIRECTIONS

It is our belief that comparative research has been excessively focused on fulfillment of a behavioral criterion for foresight. Any such criterion can only ever provide us with an arbitrary cutoff point, with some species being deemed capable and others not. This is both unsatisfactory and uninformative. There will always be those who will view any criterion that admits non-human planning as too lax, as there will be those who view dismissal of some such planning as too strict. A single criterion for prospective mental time travel is unlikely to satisfy neither camp. Indeed, Suddendorf and Corballis (1997, 2007; see also Suddendorf, 2013) present mental time travel as a concoction of various contributory mechanisms. This variety in theory must be reflected in the focus of empirical work.

More important, however, the single criterion approach cannot actually tell us much about the evolution of cognition. A single criterion necessarily dismisses subtle variations in prospective cognition between species. Small variations in processes that are continuous between species give us the opportunity to ask the evolutionary questions we began this chapter with. We can compare these variations to differences in species' ecologies in order to discern their associated adaptive advantages. Similarly, we can ask whether different starting points in cases of convergent evolution offer different constraints on the subsequent evolution of individual processes. It is these processes that should, in our opinion, be the focus of attention in comparative research into prospective mental time travel. Our suggested approach should take its cues from recent neurobiological work, which has made great strides in elucidating the material causes of episodic foresight.

We finish with an example. A defining feature of mental time travel, when compared with semantic memory and non-episodic planning, is that past and future events are experienced in the mind's eye (e.g. Suddendorf & Corballis, 1997; Tulving, 2002). It has been argued that this feature depends on constructive processes—supported by medial temporal brain areas—that establish a spatial framework for the event (Hassabis & Maguire, 2007). This is consistent with work demonstrating hippocampal activity during episodic recall and prospective simulation of the future (e.g., Addis et al., 2007), and also in imagination of an event without any specified time (Hassabis et al., 2007a). Indeed, patients with hippocampal amnesia appear to show impairments in imagination of spatially coherent scenes, but not of single objects (Hassabis et al., 2007b). The processes underpinning "scene construction" form a small but vital component of mental time travel. In order to shed light on their evolutionary history, we need to know how they contribute to planning behavior across species, as well as any role they play in behavior that is not obviously connected with mental time travel.

Fortunately for the purposes of comparative research, scene construction has been linked to a measurable behavioral phenomenon: boundary extension. Boundary extension is the phenomenon whereby people typically report seeing more of a scene than they actually did, resulting in a more "zoomed out" view (Intraub & Richardson, 1989). In other words, the boundaries of the scene have been extended beyond their natural limits. The link between boundary extension and hippocampal construction rests on two planks. First, boundary extension has been associated with hippocampal activity during initial exposure to a scene (Chadwick et al., 2013). Second, the tendency toward boundary extension in healthy individuals produces a paradoxical advantage to recall in patients with hippocampal amnesia (Mullally et al., 2012). Boundary extension in a "match-to-sample" task might therefore be taken as a useful indicator of "scene construction-like" processes, though oriented to the past rather than the future.

Evidence of this kind should not be taken as evidence that an animal is capable of foresight. As we have argued in this chapter, such binary distinctions are unproductive and potentially misleading. Indeed, the processes responsible for boundary extension might be found in an animal that really cannot think about the future (though we are unlikely to ever know this fact with certainty). The hippocampal activity associated with boundary extension was observed primarily during online perception of a scene, leading the authors to suggest that its function might be to provide predictions about the world outside the subject's field of view. This function could be beneficial for navigation, even in the absence of an ability to plan the route ahead. Scene construction–like processes might then have been co-opted to support memory and prospection in some species, but not others.

Our highly speculative account of scene construction–like processes is not intended as a "just-so" story of the evolution of prospective cognition, but as an illustration of the potential benefits of focusing on processes rather than systems. Identifiying variations in these processes and assessing their association with environmental demands could offer insight into the selective pressures that shaped our minds, and could potentially offer scope for inference about ancestral forms of human cognition.

## 9. CONCLUSIONS

We began with Tinbergen's two ultimate questions to be asked of prospective mental time travel: What are the adaptive advantages conferred by a capacity for foresight? And how did our earlier evolutionary history influence the form that capacity takes? As a first step toward answering these questions, it is essential to identify the extent to which other, non-human animals are able to plan for the future with the future in mind.

The Bischof-Köhler hypothesis has been the main null hypothesis in this endeavor, articulating the weakly justifiable preference for "lower" level explanations over more cognitive accounts of behavior. Several studies claim to falsify the central tenet of the Bischof-Köhler hypothesis—that non-human animals cannot anticipate their future needs. The most convincing examples are of tool transportation in great apes (Osvath & Osvath, 2008), food selection in squirrel monkeys (Naqshbandi & Roberts, 2006), and food-caching in corvids (Correia et al., 2007; Cheke & Clayton, 2012). These studies are too few in number to be absolutely conclusive, but they do suggest that it is time to stop saying that non-human animals are "stuck in the present" (a description that has been applied to some humans with hippocampal amnesia; Corkin, 2013; Wearing, 2005).

Some of the criticisms leveled at comparative studies of prospective cognition appear to come down to limitations in Bischof-Köhler hypothesis itself. While the Bischof-Köhler hypothesis has undoubtedly provided a solid foundation for research in this field, it has also been constricting. Its emphasis on anticipation of future needs as distinct from understanding the affordances of a task seems arbitrary. Determining the means of action can tax our ability to imagine and plan, just as identifying our goals can. Similarly, the requirement that present and future rewards be qualitatively distinct is vague, but would seem to exclude, for example, all choices that people make about money.

We believe that the problems seen in the Bischof-Köhler hypothesis are inevitable if any single criterion for a particular cognitive faculty is insisted upon. We have instead argued for an approach that focuses on individual component processes of prospective mental time travel, even if those processes do not contribute to planning in the subject animal. It is these processes that will provide the variation needed to make meaningful comparisons across phylogeny, and to answer the ultimate questions about our own capacity for foresight. It is time to start seeing the trees for the forest.

## ACKNOWLEDGMENTS

We would like to thank Lucy Cheke for her comments on this chapter.

## NOTE

1. There is a necessary ambiguity in the definition of mental time travel as applied to non-human animals, as some degree of difference between species is expected even were we to find great similarities in the nature of prospective thought. Where we use the term *prospective mental time travel*, we refer to the system described by Suddendorf and Corballis (1997, 2007), which has formed the basis for the majority of subsequent comparative research into the topic. We also employ terms such as *foresight*, *prospective cognition*, and *future-thinking*. We use these terms interchangeably to refer to the processes necessary to represent a future event. These processes may or may not qualify as truly episodic in nature, but without access to a non-human animal's subjective experience, this seems unknowable. Our neutrality with respect to the experiential aspects of prospective mental time travel should be seen as analogous to that of Clayton and Dickinson (1998) in their definition of "episodic-like" memory as an analogue of human episodic memory.

## REFERENCES

Addis, D. R., Wong, A. T., & Schacter, D. L. (2007). Remembering the past and imagining the future: common and distinct neural substrates during event construction and elaboration. *Neuropsychologia*, *45*(7), 1363–1377.

Bird, C. D., & Emery, N. J. (2009). Insightful problem solving and creative tool modification by captive nontool-using rooks. *Proceedings of the National Academy of Sciences of the U S A*, *106*(25), 10370–10375.

Bugnyar, T., & Heinrich, B. (2005). Ravens, Corvus corax, differentiate between knowledgeable and ignorant competitors. *Proceedings of the Royal Society of London B: Biological Sciences*, *272*(1573), 1641–1646.

Chadwick, M. J., Mullally, S. L., & Maguire, E. A. (2013). The hippocampus extrapolates beyond the view in scenes: an fMRI study of boundary extension. *Cortex*, *49*(8), 2067–2079.

Cheke, L. G., & Clayton, N. S. (2010). Mental time travel in animals. *Wiley Interdisciplinary Reviews: Cognitive Science*, *1*(6), 915–930.

Cheke, L. G., & Clayton, N. S. (2012). Eurasian jays (Garrulus glandarius) overcome their current desires to anticipate two distinct future needs and plan for them appropriately. *Biology Letters*, *8*(2), 171–175.

Clayton, N., Correia, S., Raby, C., Alexis, D., Emery, N., & Dickinson, A. (2008). Response to Suddendorf & Corballis (2008): in defence of animal foresight. *Animal Behaviour*, *76*(4), e9–e11.

Clayton, N. S., Bussey, T. J., & Dickinson, A. (2003). Can animals recall the past and plan for the future? *Nature Reviews Neuroscience*, *4*(8), 685–691.

Clayton, N. S., & Dickinson, A. (1998). Episodic-like memory during cache recovery by scrub jays. *Nature*, *395*(6699), 272–274.

Clayton, N. S., & Dickinson, A. (1999). Motivational control of caching behaviour in the scrub jay, Aphelocoma coerulescens. *Animal Behaviour*, *57*(2), 435–444.

Corballis, M. C. (2013). Mental time travel: a case for evolutionary continuity. *Trends in Cognitive Sciences*, *17*(1), 5–6.

Corballis, M. C. (2014). Mental time travel: how the mind escapes from the present. *Cosmology*, *18*, 139–145.

Corkin, S. (2013). *Permanent present tense: The man with no memory, and what he taught the world*. London: Penguin UK.

Correia, S. P., Dickinson, A., & Clayton, N. S. (2007). Western scrub jays anticipate future needs independently of their current motivational state. *Current Biology*, *17*(10), 856–861.

de Kort, S. R., Correia, S. P. C., Alexis, D. M., Dickinson, A., & Clayton, N. S. (2007). The control of food-caching behavior by Western scrub-jays (Aphelocoma californica). *Journal of Experimental Psychology-Animal Behavior Processes*, *33*(4), 361–370.

De Waal, F. (1982). *Chimpanzee politics: sex and power among apes*. London: Jonathan Cape.

Dekleva, M., van den Berg, L., Spruijt, B. M., & Sterck, E. H. (2012). Take it or leave it: transport of tools for future use by long-tailed macaques (Macaca fascicularis). *Behavioural Processes*, *90*(3), 392–401.

Dufour, V., & Sterck, E. (2008). Chimpanzees fail to plan in an exchange task but succeed in a tool-using procedure. *Behavioural Processes*, *79*(1), 19–27.

Emery, N. J., & Clayton, N. S. (2001). Effects of experience and social context on prospective caching strategies by scrub jays. *Nature*, *414*(6862), 443–446.

Emery, N. J., & Clayton, N. S. (2004). The mentality of crows: convergent evolution of intelligence in corvids and apes. *Science*, *306*(5703), 1903–1907.

Fitzpatrick, S. (2008). Doing away with Morgan's Canon. *Mind and Language*, *23*(2), 224–246.

Goodall, J. (1986). *The chimpanzees of Gombe: patterns of behavior*. Cambridge, MA: Harvard University Press.

Gould, S. J., & Eldredge, N. (1977). Punctuated equilibria: the tempo and mode of evolution reconsidered. *Paleobiology*, *3*(2), 115–151.

Gupta, A. S., van der Meer, M. A., Touretzky, D. S., & Redish, A. D. (2010). Hippocampal replay is not a simple function of experience. *Neuron*, *65*(5), 695–705.

Hassabis, D., Kumaran, D. & Maguire, E. A. (2007a). Using imagination to understand the neural basis of episodic memory. *The Journal of Neuroscience*, *27*(52), 14365–14374.

Hassabis, D., Kumaran, D., Vann, S. D., & Maguire, E. A. (2007b). Patients with hippocampal amnesia cannot imagine new experiences. *Proceedings of the National Academy of Sciences of the U S A*, *104*(5), 1726–1731.

Hassabis, D., & Maguire, E. A. (2007). Deconstructing episodic memory with construction. *Trends in Cognitive Sciences*, *11*(7), 299–306.

Holliday, M. A. (1986). Body composition and energy needs during growth. In Frank Falkner, James M. Tanner (Eds.), *Postnatal Growth Neurobiology* (pp. 101–117). New York: Springer.

Intraub, H., & Richardson, M. (1989). Wide-angle memories of close-up scenes. *Journal of Experimental Psychology: Learning, Memory, and Cognition, 15*(2), 179–187.

Janmaat, K. R., Polansky, L., Ban, S. D., & Boesch, C. (2014). Wild chimpanzees plan their breakfast time, type, and location. *Proceedings of the National Academy of Sciences, 111*(46), 16343–16348.

Jerison, H. J. (1973). *Evolution of the brain and intelligence.* New York: Academic Press.

Klein, S. B., Loftus, J., & Kihlstrom, J. F. (2002). Memory and temporal experience: the effects of episodic memory loss on an amnesic patient's ability to remember the past and imagine the future. *Social Cognition, 20*(5), 353–379.

Land, M., & Nilsson, D. (2002). *Animal eyes.* New York: Oxford University Press.

Lett, B. T. (1975). Long delay learning in the T-maze. *Learning and Motivation, 6*(1), 80–90.

Mink, J. W., Blumenschine, R. J., & Adams, D. B. (1981). Ratio of central nervous system to body metabolism in vertebrates: its constancy and functional basis. *American Journal of Physiology: Regulatory, Integrative and Comparative Physiology, 241*(3), R203–R212.

Morgan, C. L. (1903). *An introduction to comparative psychology.* London: W. Scott.

Mulcahy, N. J., & Call, J. (2006). Apes save tools for future use. *Science, 312*(5776), 1038–1040.

Mullally, S. L., Intraub, H., & Maguire, E. A. (2012). Attenuated boundary extension produces a paradoxical memory advantage in amnesic patients. *Current Biology, 22*(4), 261–268.

Naqshbandi, M., & Roberts, W. A. (2006). Anticipation of future events in squirrel monkeys (Saimiri sciureus) and rats (Rattus norvegicus): tests of the Bischof-Kohler hypothesis. *Journal of Comparative Psychology, 120*(4), 345.

Nisbett, R. E., & Kanouse, D. E. (1969). Obesity, food deprivation, and supermarket shopping behavior. *Journal of Personality and Social Psychology, 12*(4), 289–294.

O'Keefe, J., & Nadel, L. (1978). *The hippocampus as a cognitive map* (Vol. 3). Oxford: Clarendon Press.

Orenstein, R. I. (1972). Tool-use by the New Caledonian crow (Corvus moneduloides). *The Auk, 89*(3), 674–676.

Ostojic, L., Legg, E. W., Shaw, R. C., Cheke, L. G., Mendl, M., & Clayton, N. S. (2014). Can male Eurasian jays disengage from their own current desire to feed the female what she wants? *Biology Letters, 10*(3), 20140042.

Ostojic, L., Shaw, R. C., Cheke, L. G. & Clayton, N. S. (2013). Evidence suggesting that desire-state attribution may govern food sharing in Eurasian jays. *Proceedings of the National Academy of Sciences U S A, 110*(10), 4123–4128.

Osvath, M. (2009). Spontaneous planning for future stone throwing by a male chimpanzee. *Current Biology, 19*(5), R190–R191.

Osvath, M. (2010). Great ape foresight is looking great. *Animal Cognition, 13*(5), 777–781.

Osvath, M., & Karvonen, E. (2012). Spontaneous innovation for future deception in a male chimpanzee. *PloS One, 7*(5), e36782.

Osvath, M., & Osvath, H. (2008). Chimpanzee (Pan troglodytes) and orangutan (Pongo abelii) forethought: self-control and pre-experience in the face of future tool use. *Animal Cognition, 11*(4), 661–674.

Paxton, R., & Hampton, R. R. (2009). Tests of planning and the Bischof-Köhler hypothesis in rhesus monkeys (Macaca mulatta). *Behavioural Processes, 80*(3), 238–246.

Petrovich, G. D., Ross, C. A., Gallagher, M., & Holland, P. C. (2007). Learned contextual cue potentiates eating in rats. *Physiology and Behavior, 90*(2), 362–367.

Pfeiffer, B. E., & Foster, D. J. (2013). Hippocampal place-cell sequences depict future paths to remembered goals. *Nature, 497*(7447), 74–79.

Raby, C. R., Alexis, D. M., Dickinson, A., & Clayton, N. S. (2007). Planning for the future by western scrub-jays. *Nature, 445*(7130), 919–921.

Rosenbaum, R. S., Kohler, S., Schacter, D. L., Moscovitch, M., Westmacott, R., Black, S. E., Gao, F., & Tulving, E. (2005). The case of K.C.: contributions of a memory-impaired person to memory theory. *Neuropsychologia, 43*(7), 989–1021.

Seed, A. M., Clayton, N. S., & Emery, N. J. (2008). Cooperative problem solving in rooks (Corvus frugilegus). *Proceedings of the Royal Society B: Biological Sciences, 275*(1641), 1421–1429.

Sober, E. (2005). Comparative psychology meets evolutionary biology. In L. Daston & G. Mitman (eds.), *Thinking with animals: new perspectives on anthropomorphism* (pp. 85–99). New York: Columbia University Press.

Suddendorf, T. (2006). Foresight and evolution of the human mind. *Science, 312*(5776), 1006–1007.

Suddendorf, T. (2013). Mental time travel: continuities and discontinuities. *Trends in Cognitive Sciences, 17*(4), 151–152.

Suddendorf, T., & Corballis, M. C. (1997). Mental time travel and the evolution of the human mind. *Genetic, Social, and General Psychology Monographs, 123*(2), 133–167.

Suddendorf, T., & Corballis, M. C. (2007). The evolution of foresight: What is mental time travel, and is it unique to humans? *The Behavioral and Brain Sciences, 30*(3), 299–313; discussion 313–251.

Suddendorf, T., & Corballis, M. C. (2008). New evidence for animal foresight? *Animal Behaviour, 75*, E1–E3.

Suddendorf, T., & Corballis, M. C. (2010). Behavioural evidence for mental time travel in nonhuman animals. *Behavioural Brain Research, 215*(2), 292–298.

Suddendorf, T., Corballis, M. C., & Collier-Baker, E. (2009). How great is great ape foresight? *Animal Cognition, 12*(5), 751–754.

Taylor, A. H., Hunt, G. R., Holzhaider, J. C., & Gray, R. D. (2007). Spontaneous metatool use by new Caledonian crows. *Current Biology, 17*(17), 1504–1507.

Thom, J., & Clayton, N. S. (2015). Route-planning and the comparative study of future-thinking. *Frontiers in Psychology, 6*, 144.

Thom, J. M., & Clayton, N. S. (2014). Continuity in hippocampal function as a constraint on the convergent evolution of episodic-like cognition. *Cosmology, 18*, 461–465.

Tinbergen, N. (1963). On aims and methods in ethology. *Zeitschrift für Tierpsychologie, 20*, 410–433.

Tulving, E. (2002). Episodic memory: from mind to brain. *Annual Review of Psychology, 53*, 1–25.

Van Lawick-Goodall, J. (1970). Tool-using in primates and other vertebrates. *Advances in the Study of Behavior, 3*, 195–249.

Vander Wall, S. B. (1990). *Food hoarding in animals.* Chicago: University of Chicago Press.

Wearing, D. (2005). *Forever today: a memoir of love and amnesia.* London: Random House.

Weir, A. A. S., & Kacelnik, A. (2006). A New Caledonian crow (Corvus moneduloides) creatively re-designs tools by bending or unbending aluminium strips. *Animal Cognition, 9*(4), 317–334.

Wilson, M. A., & McNaughton, B. L. (1994). Reactivation of hippocampal ensemble memories during sleep. *Science, 265*(5172), 676–679.

# 15

# With the Future in Mind

## *Toward a Comprehensive Understanding of the Evolution of Future-Oriented Cognition*

GEMA MARTIN-ORDAS ■

From an early age, humans—at least, in Western societies—are bombarded with questions about what we want to be when we grow up, where we are going to go for our next holidays, what we are going to do next weekend, or what we would like to eat tomorrow. We rapidly learn that the future matters and that our actions in the present have inevitable consequences in the future. There is no doubt that thinking and imagining the future is an extraordinary ability that has astonishing adaptive advantages (e.g., buying food on Friday will prevent me from "starvation" during the weekend).

The images that we construct to think about the future heavily rely on past experiences (Buckner & Carroll, 2007; Dudai & Carruthers, 2005; Klein 2013a; Schacter, Addis, & Buckner, 2007; Suddendorf & Corballis, 1997, 2007; Tulving, 2005). Humans are able to reminisce about the past and to use such recollections to transcend the present and wander the upcoming times. For example, when I try to imagine what my next lecture to anthropology students is going to be like, I might retrieve memories of other past lectures so I can avoid assuming their knowledge in psychology. We mentally travel back (i.e., *episodic memory*) and forward (i.e., *future-oriented cognition*) in time. In fact, from an evolutionary perspective, being able to imagine and plan for one's personal future based on memory of past events(especially being able to plan for future events that do not depend on current needs) is argued to provide an important selective advantage (Klein 2013a; Suddendorf & Corballis 2007).

Are we humans the only animal species that can project backward and forward in time? Or do we share this capacity with other animals? During the past decades there has been an extraordinary increase in the number of studies on future-oriented cognition in both human and non-human animals. The upsurge of this research is reflected in the large number of terms used to describe this ability: *episodic future thought* (Atance & O'Neill, 2001), *prospection* (Buckner & Carroll, 2007; Gilbert & Wilson, 2007), *simulation* (Schacter & Addis, 2007; Schacter, Addis, & Buckner, 2007), *projection* (Okuda et al., 2003), *mental time travel into the future*

(Klein, 2013a, b; Suddendorf & Corballis, 1997), or *episodic foresight* (Suddendorf, 2010). What all these terms have in common is the idea that thinking about the future implies mental travel in time to construct possible future personal episodes or scenarios (Atance & O'Neill, 2001; Szpunar, 2010). Since this capacity seems to rely on subjective experiences, many researchers have argued that mentally traveling in time is uniquely human; and, therefore, only humans are capable of episodic foresight as captured by the preceding definition. However, as argued by Osvath and Martin-Ordas (2014), what is really meant by episodic foresight, or how it is measured, is not straightforward (see also Hudson et al., 2011). For example, whereas animal research has focused upon planning tasks, studies with children and adult humans have mainly examined descriptions of future events (see Martin-Ordas et al., 2014a, for a review). Additionally, different criteria have been used for human and non-human animals. For example, the subjective feeling of recollection has to be proved in animals; however, this is assumed to exist without empirical evidence in children (Suddendorf & Busby, 2003a, b). Therefore, important theoretical and methodological limitations arise from this research.

The aim of this chapter is to describe the cognitive mechanisms argued to underlie future-oriented cognition and how they have been investigated in non-human animals. Since some of the most controversial evidence for future-oriented cognition has come from work with great apes, I will exclusively focus on the research carried out in our closest relatives. I lay the groundwork for the episodic system by reviewing concepts such as autonoetic awareness and noetic awareness. I then provide an overview of the research on episodic foresight in great apes, building on the criteria previously defined in the human research and highlighting methodological limitations. Finally, I will critically review concepts such as episodic memory, semantic memory, and episodic foresight. My ultimate goal is to show that comparative research into future-oriented cognition must be extended not only beyond the current theoretical dichotomies (e.g., memory, foresight) but also beyond the human populations that we have been studying thus far. While the capacity to foresee the future might be universal, it is still an open question whether cultural and environmental factors affect such capacity in ways that make comparisons with our closest relatives more difficult. I argue that if we are to understand which mechanisms affect future cognition, cross-species and cross-cultural studies are needed.

## 1. MENTAL TIME TRAVEL

### 1.1. Noetic Awareness versus Autonoetic Awareness

A few months ago, I visited Zambia for the first time. I landed in Ndola, a small city in the Copperbelt province in the north of the country. I can perfectly remember the feeling of being in a very small airport but still not knowing what to do (e.g., where to get a taxi). My European scripts of how to get around airports seemed to not apply very well to that situation. For me to be able to describe this experience, I have to draw on two memory systems: the episodic and the semantic systems (Tulving, 1972). Tulving (1972) defined semantic memory as our database of knowledge about the world. In contrast, episodic memory was defined as "an information processing system that a) receives and stores information about temporally dated episodes or events, and about temporal-spatial relations among these events, b) retains various aspects of this information, and c) upon instructions transmits specific retained information to other systems, including those responsible for translating it into behavior and conscious awareness" (Tulving, 1972, 385). In the example, knowing that Ndola is a Zambian city in the Copperbelt province is an example of semantic memory. However, when I remember arriving at the airport in Ndola a few months ago, I am drawing on episodic memory.

A few years later, Tulving (1983, 2005) modified his original definition and stated that only episodic memory "makes possible mental time travel through subjective time, from the present to the past, thus allowing one to re-experience, through autonoetic awareness, one's own previous experiences. Its operations require, but go beyond, the semantic memory system" (Tulving, 2005, 5). Autonoetic consciousness is defined as "the capacity that allows adult humans to mentally represent and to become aware of their protracted existence across time" (Wheeler et al., 1997, 335). Thus, the phenomenological experience attached to the content of the memory is crucial in conceiving a memory as episodic. This is so because even though retrieving an episodic memory implies that the individual recollects the spatiotemporal context in which the past event took place, some semantic memories do involve a concrete occurrence (e.g., knowing when you were born). However, the latter does not require any re-experience of the past event (Tulving, 2002). Thus, noetic (as opposed to autonoetic) consciousness is considered the defining property of semantic memory and is expressed without any such self-recollection but simply in awareness of familiarity or knowing (e.g., I *know* when I was born, but I do not *remember* that event).

The episodic/semantic memory distinction has also been applied to future-oriented thinking (Atance & O'Neill, 2001, 2005; Suddendorf & Corballis, 2007). Originally, semantic future thinking was conceived as the capacity to think about facts and context-free general conceptual knowledge (e.g., there are going to be elections next year) (Atance & O'Neill, 2001, 2005). Later, Klein (2002, 2013 a, b, see also Klein et al. 2002) redefined the term, arguing that it can refer to public future events (e.g., the effect of global warming on the planet) but also to personal future events (e.g., I will become a mother) (Rathbone et al., 2011). In contrast, episodic foresight or episodic future thinking[1] is referred to as the capacity to mentally project oneself into the future in order to "pre-experience" (one would need to explain how it is possible to experience something that has not happened yet) a spatiotemporally specific event (e.g., imagining my next trip to Zambia). Episodic memory and episodic foresight have been shown to rely on the same brain areas (Addis, Wong, & Schacter, 2007; Hassabis, Kumaran, Vam, & Maguire, 2007; Klein, Loftus, & Kihlstrom, 2002; Okuda et al., 2003; Szpunar, Watson, & McDermott, 2007; Tulving, 2005). Consequently, it has been argued that past and future events draw on similar information and rely on common underlying processes (Klein 2013b; Schacter & Addis, 2007; Suddendorf & Corballis, 2007). In fact, episodic memory seems to be crucial for imagining future events because it allows for the extraction and combination of episodic details to create future events (Addis et al., 2007, 2009; Suddendorf & Corballis, 2007; Szpunar & McDermott 2008).

## 1.2. Problems for Comparative Psychology

A definition based on purely phenomenological constructs conveys important problems to comparative research. This is so because studies on episodic cognition with non-verbal animals have to rely on behavioral outcomes, which makes it extremely difficult to grasp the difference between performances based on *noetic* awareness and performances based on *autonoetic* awareness. Imagine the following situation. A chimpanzee sees an experimenter hiding a tool in a box. Thirty minutes later, when given access to the room where the box is placed, the chimpanzee retrieves the tool. How does the chimpanzee achieve this task? What does the chimpanzee have in mind when she walks toward the box and finds the tool? One possibility is that the she mentally travels back (*autonoetic awareness*) in time and re-experiences the event in which she saw the experimenter placing the tool inside the box. A second possibility is that the chimpanzee may "simply" know (*noetic awareness*) exactly where the tool is, without having a specific recollection of the episode in which it was placed there. Alternatively, the chimpanzee may know that boxes always contain tools (*associative learning*). Thus, she could always search inside boxes for tools, independently of having witnessed (or not) the experimenter placing the tool inside the box. In all these three cases, the behavioral outcome is the same: the chimpanzee finds the tool. However, what it is in the chimpanzee's mind when she is retrieving the tool differs significantly in each case (see Redshaw, 2014, for similar arguments). If we assume that subjective (*autonoetic*) awareness is the central and crucial component of mental time travel, then demonstrating this capacity in non-verbal animals is going to be a very difficult, if not impossible, task.

While retrieving an episodic memory involves re-experiencing a personal past event, imagining a future event involves pre-experiencing a future situation. However, it has been argued that retrieving episodic information to anticipate an upcoming event might take place without necessarily pre-experiencing (i.e., mentally projecting the *self*) the event (Osvath & Martin-Ordas, 2014; Pillemer, 2003; Szpunar, 2010). For instance, remembering the last time I woke up in the morning and could not have coffee because I ran out of it the day before does not imply that every time I go to the supermarket and buy coffee I do so because I project myself into the (terrible) future event of not having coffee first thing in the morning. Note, though, that the behavioral outcome in both situations (i.e., imagining the specific details of the future event vs. only anticipating the negative consequences of not having coffee) is the same (i.e., I buy coffee). Cheke and Clayton (2010) already have argued that associative learning and future-oriented cognition might suffice for future thinking (i.e., "known future" as termed by Klein, 2013a).[2]

Only recently researchers, who previously argued that animals did not have the capacity to travel in time, have been persuaded by evidence showing that rats' brains have a default mode network (Lu et al., 2012) (see Mantini et al., 2011, and Rilling et al., 2007, for similar results in monkeys and great apes, respectively). Since the hippocampus is a critical region in the default mode network and is also involved in recalling and planning, thus subjective experience might also be present in nonhuman animals (Corballis, 2013). Recently, Allen and Fortin (2013) suggested that the episodic system might have been already present in the common ancestor of mammals, birds, and reptiles. In their review, they identified the existence of the hippocampus, parahippocampal region, and prefrontal cortex (and the equivalent bird brain regions)—the three of them being involved in the episodic system—in mammals and birds. As they suggested, this does not necessarily involve that these animals are able to travel back and forward in time as humans do, but that the episodic system could have evolved from these common brain structures.

Despite the evident limitations that conceptualizing future thinking in terms of autonoetic awareness has for the field of comparative psychology and despite previous attempts to move away from such conceptualizations (e.g., Cheke & Clayton, 2010), the experiential (subjective, phenomenological) aspects are being brought back into play (e.g., Klein, 2015; Redshaw, 2014). In fact, Klein (2015) has recently claimed that

> [u]nfortunately, one cannot remove the experiential aspect of a recollection without expunging the phenomenon under consideration. Memory is an experience, and experience is not reducible to, or knowable from, physical behavior (save for evidence provided by introspective report). While the experiential nature of episodic memory may be a source of methodological inconvenience, this does not justify its exclusion from scientific consideration or empower one to affect definitional emendations designed to compensate for empirical limitations. If our available techniques are unable to capture this core feature of memory, then so much worse for our techniques. (Klein 2015, 23)

With statements like this, the field of comparative psychology is in trouble. In this first section, I have highlighted that the use of concepts based on different levels of consciousness does not allow comparative psychologists to distinguish between actions purely grounded on autonoetic awareness or noetic awareness. Despite the theoretical limitations, the criteria used to identify future-oriented cognition in non-human animals have been based on such dichotomous framework. In the next section, I will examine such criteria and I will review the empirical evidence of future-oriented cognition in great apes.

## 2. HOW TO MEASURE FUTURE-ORIENTED COGNITION

### 2.1. In Theory

In human adult research, the phenomenological experience associated with the capacity to mentally travel in time has been mainly investigated with the use of verbal tasks. In these tasks, participants are asked to describe the content of a memory and the subjective experience associated with the recollection of a past event (e.g., Buckner & Carroll, 2007; Schacter & Addis, 2007). However, the use of verbal tasks to investigate the content and the phenomenological component of recollection poses an obvious problem for the comparative study of mental time travel, since no other animals besides humans can verbally communicate their re-experience or pre-experience of an episode. For this reason, Suddendorf and Corballis (2007) proposed one of the most commonly used criteria in the field of comparative psychology. They suggested that having the future in mind should, at least, provide *flexibility* to the owner of such a behavior. One should also rely on a single *unique experience* (i.e., one trial experience) and there should not be any *cue/stimuli* in the present environment that could trigger such a response. Additionally, some researchers have argued for the importance of the organism planning for a need that it is not being experienced at the time when the future outcome is imagined (e.g., Bischof, 1985; Bischof-Köhler, 1985; Suddendorf & Corballis, 1997, 2007). According to the Bischof-Köhler hypothesis, "... animals other than humans cannot anticipate future needs and drive states and are therefore bound to a present that is defined by their current motivational state" (Suddendorf & Corballis, 1997, 150). The Bischof-Köhler hypothesis has become a cornerstone in contemporary comparative studies of future-oriented cognition (Crystal, 2012; Suddendorf & Busby, 2003a, b; Suddendorf & Corballis, 2007, 2008; Suddendorf, Corballis, & Collier-Baker, 2009; see Boyer, 2008, for a critical view on this issue). While having these criteria certainly help identify future-oriented behaviors (e.g., Suddendorf et al. 2011), there are some weaknesses that need to be addressed. Next I will describe each of these criteria and I will identify some of their limitations.

### 2.1.1. Flexibility

The flexibility criterion is quite intuitive. Being able to interchange and combine features of past events to imagine a future one "frees" us from the current spatio-temporal context, so we can think of situations not experienced before in a non-stereotypical way. Being able to demonstrate flexibility would imply that neither innate responses nor associatively learned responses are driving future-oriented behaviors. However, one can also combine previous knowledge—obviously acquired through personal experiences—to create new knowledge. For example, I know that I should not add mustard to a cheesecake because I *know* that both foods together will not taste good. There is no need for me to re-experience a past event to imagine a future event so I can avoid mixing mustard and cake. The semantic system also provides us with enough *flexibility* to make rational decisions about the future (e.g., to not mix cake and mustard) (Osvath & Martin-Ordas, 2014; Pillemer 2003; Szpunar 2010). There is also no doubt that we can apply a skill (or objects) to new problems that are different from the problems that served as the basis for training/learning (e.g., stimulus generalization). Then, we are left with the open question of what flexibility really involves. One could argue that the more novel the combinations of past impressions, the more flexibly creative or future oriented a behavior/response becomes. And conversely, the more novel the event, the more necessary the combination of past events could be (Osvath & Martin-Ordas, 2014).

### 2.1.2. Single Trials

The unique trial exposure criterion also seems straightforward. This is so because it avoids repeated exposure to the same stimulus, and consequently establishing positive associations between certain stimuli and responses. However, is it true that because learning has taken place, an explanation in terms of the episodic system is unsuitable? Suddendorf and colleagues (2007, 2011) have argued that a single exposure to an event will also allow demonstrating memory of specific events. Three important issues are raised here. First, most memory systems need repeated exposure to a stimulus for it to be retained (Morris, 2001). Moreover, a single trial exposure does not necessarily rule out associative learning, as certain conditioning procedures have demonstrated (e.g., Garcia & Koelling, 1966). For example, in a conditioned taste aversion experimental paradigm, when rats are given only one opportunity to drink flavored water and later they are made sick by giving them a toxic substance, they will avoid the previously given flavored water for a long time. The adaptive value of this fast learning experience is crucial for survival. However, remembering the event in which rats learned such an association might not be as valuable as remembering the stimuli that made them ill. Finally, memories about our past do not necessarily refer to *unique* events but to *general* events (i.e., repeated events in time). Thus, the criterion of single exposure would not necessarily allow us to distinguish between learning experience and episodic events. Additionally, it would not take into account a very frequent type of memories (e.g., general events) that could also be involved in the imagination of future events (see section 3).

#### 2.1.3. Contextual Detachment (Lack of Contextual Cues/Stimuli)

The cue/stimuli independency criterion refers to the lack of cues in the current temporal and spatial contexts for the future-directed action to take place. This is a surprising criterion since the opposite has been postulated for the retrieval of memories for past events. In fact, as mentioned earlier, memories are more likely to be triggered when the context at time of encoding matches the context at retrieval (Tulving & Thompson, 1973). However, as Osvath and Martin-Ordas (2014) have suggested, this idea does not apply to future-oriented behaviors. How, then, is a future-oriented behavior initiated? As Osvath and Martin-Ordas (2014) suggest, cueing is also crucial for episodic foresight and that there must be something in the current situation (e.g., current goals, mental state, physiological state, stimuli in the environment) that facilitates the thought of the future event. In the case of the developmental studies, researchers ask children to choose items (e.g., a tool) to take to a different room. This is an obvious way to elicit and cue planning in children by an adult—even if developmental psychologists do not openly recognize this limitation, one cannot help but wonder to what extent children are planning when they are asked to plan by an adult in such situation. If this is true, the idea of a complete contextual detachment, as suggested by Suddendorf and Corballis (2007), does not seem to provide us with a realistic criterion to distinguish between episodic foresight and associated behaviors.

In the next section I will provide an overview of the research on future-oriented cognition in great apes. I will also describe how researchers have dealt with the criteria just described and the main criticisms that the comparative work has received.

### 2.2. In Practice

#### 2.2.1. Studies in Captivity

Almost 100 years ago, Köhler (1921) observed how chimpanzees seemed to be able to anticipate and plan for future events. Now, almost 100 years later, we have experimental evidence that great apes can plan for future events. One of the most commonly used paradigms to assess future thinking in non-human animals is illustrated by Tulving's (2005) "spoon test." This test is based on the following scenario: a young girl dreams she is at a party where all the guests are being served a delicious chocolate pudding. However, to eat the pudding, guests must have their own spoon, and the young girl does not. That night, she falls asleep again while holding a spoon in her hand because she wants to avoid making the same mistake again. Tulving argues that non-human animals should have difficulty passing any task that draws on the logic of this scenario.

Based on the spoon-test paradigm, Mulcahy and Call (2006) presented orangutans and bonobos with an out-of-reach reward and with a set of useful and useless tools, which they could take into a waiting room. To obtain the reward, subjects had to return to the room where the out-of reach reward was placed, carrying the useful tool, either 1 or 24 hours after having seen the reward (i.e., *contextual detachment* criterion). Mulcahy and Call showed that great apes were capable of saving tools needed in a distant future. However, Suddendorf and colleagues (Suddendorf, 2006; Suddendorf & Corballis, 2007) argued that the subjects could have potentially experienced a desire for the reward throughout, and thus the experiment did not directly address the Bischof-Köhler hypothesis. In addition, Suddendorf and colleagues argued that simple associative learning could explain their results. This is because over the trials, subjects could have learned an association between the tool and the reward; therefore no foresight might have been involved in their decision to take and transport the tool.

More recently, Osvath and Osvath (2008) addressed some of the criticisms aimed at Mulcahy and Call's (2006) study. Osvath and Osvath (2008) also presented chimpanzees and orangutans with a tool-use task, in which the selection, saving, and transportation of a tool was also required to succeed. Note that in all of their four experiments the reward apparatus was out of sight and at a different location from where the tools could be selected (i.e., *contextual detachment* criterion). Osvath and Osvath (2008) showed (1) that apes would disregard an immediate small, but highly liked, reward in favor of the appropriate tool that would give access to the future larger reward (i.e., Bischof-Köhler hypothesis); (2) that the functional tool was not selected merely on the basis of being an associatively learned stimulus, but rather on its functionality; (3) that apes selected tools functional in the apparatus, but which they had never encountered before (i.e., equivalent to *single trial* criterion and matching a novel problem to a memorized problem). Altogether they concluded that the results from their experiments suggest that great apes were engaging in planning behaviors for the future by outcompeting current drives and mentally pre-experiencing an upcoming event. Importantly, associative learning cannot account for Osvath and Osvath's results (see Suddendorf et al., 2009, for a critical review; see Osvath, 2010, for a response).

In an attempt to show that apes can plan besides the tool-use context (i.e., *flexibility* criterion), Dufour and Sterck (2008) tested whether chimpanzees could prepare for future exchanges with a human. While chimpanzees failed at selecting the correct tokens, transporting them and exchanging them at a later time, the same subjects succeeded in a tool-use planning task. Authors concluded that chimpanzees' planning skills could be limited to the context of tool use. Another study using the same exchange paradigm showed that apes could indeed succeed in such a task (Osvath & Persson 2013). Authors suggested that the different results in both studies could be a consequence of two different captive populations with differing long-term experiences of humans and, therefore, different semantic knowledge about the world (see Bourjade et al., 2014 for negative results with chimpanzees).

Another compelling piece of evidence of future-oriented cognition comes from the spontaneous behavior of a single male chimpanzee in captivity described to gather stones hours in advance, before any zoo visitors were present, for later throwing at the visitors (Osvath 2009). More recently, authors observed the chimpanzee preparing for deception by hiding stones under mounds of hay and inhibiting his normally aggressive display behavior when appearing before the visitors. Even though further experimental evidence is needed, these observations suggest that complex planning skills might be present in chimpanzees (Osvath & Karvonen 2012).

One of the first attempts to address the role of memory in planning was carried out by Martin-Ordas et al. (2014b). They investigated how the encoding of information, either intentionally or incidentally, affects subjects' performance in a tool-use task. Chimpanzees, orangutans, bonobos, and human children were all tested under a highly similar setup. The results show that when information relevant for the tool-use task was encoded incidentally, subjects performed worse than when they knew the information's relevance (intentional encoding). This suggests that the type of memory available in a simple planning task is crucial for great apes. Along the same lines, another study with chimpanzees and orangutans investigated the contribution of long-term memory to a simple tool-use planning task in a cue-recall paradigm (Martin-Ordas et al., 2013). The results showed that when presented with the right cues (i.e., same experimenter, same setup, same context), chimpanzees and orangutans could remember a tool-finding event that took place on four occasions (i.e., general event) 3 years earlier, as well as a unique tool-finding event that happened 2 weeks before. Next I describe some of the research carried out in the wild.

### 2.2.2. Studies in the Wild

Evidence of future-oriented cognition comes not only from the lab. Even if scarce, examples of future thinking in the wild also have been reported. Chimpanzees in Tai forest have been shown to carry tools to nut-cracking sites even when the site is out of their view (Boesch & Boesch, 1984; Goodall, 1964). Similarly, chimpanzees in Goualougo Triangle have been observed not only to transport tools from one termite nest to another but also to transport the specific tool for the specific task (Sanz et al., 2009). Orangutans in the wild have also been reported to display future-oriented behaviors (van Schaik et al., 2013). The study tested whether males' long calls are indicative of future travel direction. The results suggest that males not only decided the travel route in advance, but also communicated it to other members of the group. A more recent study also showed evidence of "route planning" in female chimpanzees at the Tai forest (Janmaat et al., 2014). Janmaat and colleagues found not only that chimpanzees woke up earlier when their breakfast was a high-value fruit (e.g., figs) but also that they were more likely to build their nest en route to breakfast when planning to have figs in the morning—which, as a result of high competition, are available for a shorter amount time than other fruits. Although all of these reports are suggestive, due to a lack of a complete control over possible interfering variables, it is not always straightforward to interpret them as evidence of future-oriented cognition (see also Thom & Clayton, 2015, for similar arguments).

### 2.2.3. Meeting the Criteria

Taken together, the studies described thus far provide evidence for the existence of an episodic system in our closest relatives. These pieces of evidence also highlight some of the limitations of the criteria proposed by Suddendorf and colleagues. As mentioned earlier, the idea of pure contextual detachment does not really exist. Even if we use two different rooms and the spatial context in which subjects choose the tools is different from the spatial context in which the task is presented, subjects might be *cued* by their current goals or motivations (Osvath & Osvath, 2008). Similarly, the idea that a single trial is necessary to test episodic events was questioned by Martin-Ordas et al. (2013). In their study, apes could act on events that happened 2 weeks ago but only once (i.e., single trial exposure) and on events that took place a few times 3 years ago (i.e., general events in section 3). Finally, evidence from flexible planning comes from studies investigating future object-exchange (e.g., Osvath & Persson, 2013).

To end, most would agree that what studies in the wild and in captivity can certainly prove is that when a chimpanzee is grabbing a tool she already has a plan of which task she is going to be facing and which tools are necessary to obtain the food (e.g., Martin-Ordas et al. 2013; Sanz et al. 2009) or which type of food she is anticipating to have for breakfast (e.g., Janmat et al., 2014); however, the extent to which the chimpanzee is *mentally simulating* the specific *details* of the *future* event (*sensu* Addis et al. 2007; Suddendorf & Corballis, 2007) remains an open question. This issue leads to my next two points: first, the extent to which it is possible to distinguish between responses based exclusively on episodic and semantic features—at least at the behavioral level—and whether this distinction is a valid one. Second, since foresights are images constructed from memories of past experiences, I will also examine the extent to which memory and foresight are distinct concepts.

## 3. DRAWING LINES TO OPERATIONALIZE (SUBJECTIVE) CONCEPTS

### 3.1. Semantic versus Episodic

As mentioned earlier, episodic memory involves the recollection of personal past events, and it encompasses specific times and places related to individual experiences. In contrast, semantic memory refers to knowledge about the world, such as concepts, names, and features of objects, knowledge about history, relations between concepts, and understanding of causes and effects, among other elements (Binder & Dessai 2011). However, semantic memory also refers to context-free personal facts (e.g., memories about high school) and public events (e.g., the winter Olympics in 2014 took place in Sochi) (e.g., Patterson & Hodges, 2000).

Even though neuropsychological studies provide the strongest evidence for the distinction between episodic and semantic memory (Conway & Fthenaki, 2000; Kapur, 1999; Wheeler & McMillan, 2001), evidence for a clear-cut division between these two types of memories is somewhat controversial (Squire et al., 2004; Tulving, 2005). Can memories be conceptualized as either purely episodic or purely semantic? Let us go back to my memory of my trip to Zambia. When I remember getting off the plane and feeling the warm temperature, I am retrieving episodic details because they refer to vivid contextual information related to that particular event; however, facts such as Ndola is a city in Zambia reflect semantic knowledge. Thus, my recollection illustrates that semantic facts intertwine with episodic details (Balota &

Coane, 2008; Binder & Dessai 2011; Piolino et al. 2009; Race et al. 2011). Now imagine that I have been to Zambia several times over the last few years. The type of personal memory I would be retrieving seems not to have a specific temporal component, or to be associated with any specific event or autonoetic conscious experience (i.e., I might remember several events associated with those trips). However, these memories can be rich regarding other contextual details (e.g., I remember the small restaurant at the airport). Would those memories still be considered episodic in nature? These types of memories are not explicitly addressed in Tulving's classic model, but they fit Neisser's (1981) idea of the merging of memories for past events into one representative event, Barsalou's (1988) concept of summarized or extended events, or Conway's (2001) "general events" level of autobiographical knowledge. What all these concepts have in common is the idea that episodic and semantic features intertwine. Finally, neuropsychological research has also demonstrated such an interdependent relation (see Greenberg & Verfaellie, 2010, for a review).

Given such overlap, Tulving (1995) redefined the relation between episodic and semantic memory. In his new (hierarchical) model (termed SPI: *s*erial encoding, *p*arallel storage, and *i*ndependent retrieval), episodic memory was considered to be a subsystem of semantic memory and to be dependent upon the integrity of semantic knowledge (Tulving, 1995); that is, while information can be encoded into semantic memory independently of episodic memory, information can only be encoded into episodic memory through semantic memory. Nonetheless, Tulving's SPI model has been empirically questioned (e.g., Graham et al., 2000; Squire & Zola, 1998). For example, Squire and Zola have argued that the encoding and retrieval of semantic memory depend on the acquisition of the episode in which such information was experienced.

Thus, the contents of episodic memory involve semantic features, and there is a clear interdependency between these two systems. However, the field of comparative psychology has been stuck in the semantic-episodic dichotomy for the last decades. We should contemplate the possibility that these two concepts are the two extremes of a continuum and should move away from dichotomous conceptions that do not allow the field to move forward. From this, it would follow then that episodic foresight might also contain semantic elements. Therefore, memories and foresights will vary in the degree or number of episodic and semantic components (Osvath & Martin-Ordas, 2014; Szpunar et al., 2014). The extent to which a foresight is integrated by more episodic than semantic components and whether this might be more adaptive and/or might provide us with more flexibility is still an empirical question that has to be addressed. Is the difference between a memory and a foresight as blurry as the difference between semantic and episodic memory? I will address this issue in the next section.

### 3.2. Memory versus Foresight

There is no doubt that whereas the episodic memory is oriented toward the past, foresight is oriented toward the future. Nonetheless, memories and foresights do share features. I have mentioned earlier that episodic memory is crucial for the construction of future events (Addis & Schacter, 2008, 2012; Suddendorf & Corballis, 2007; Szpunar & McDermott, 2008). I have also mentioned that episodic and semantic memories are intertwined memory systems. However, compared to episodic memory, the contribution of semantic memory to episodic foresight has received less empirical attention (but see D'Argembeau, Renaud, & Van der Linden, 2011; Klein, 2013a,b; Martin-Ordas et al., 2012; Martin-Ordas et al., 2014a; Szpunar 2010). Most of the evidence for the importance of semantic memory in episodic foresight has come from neuropsychological research (e.g., Irish & Piguet, 2013; Irish et al., 2012; Duval et al., 2012). Importantly, these results suggest that semantic memory may be necessary for the construction of future events.

Additionally, memories and foresights rely on common underlying processes (Klein, 2013a,b; Schacter & Addis, 2007; Suddendorf & Corballis, 2007). For instance, patients with impaired episodic memory fail not only at retrieving episodic memories but also at imagining personal future events (Addis et al., 2009; Gamboz et al., 2010; Hassabis et al., 2007; Kwan et al., 2010; Race et al., 2011; Tulving, 1985). Finally, some authors have argued that memories themselves are for the future and that simply retrieving memories per se has little fitness value (Klein 2013a; Suddendorf & Corballis 2007). For example, the use of past experiences (e.g., my previous lecture to anthropology students) allows us to anticipate possible future happenings (e.g., assume students' knowledge in psychology), which clearly has an important adaptive value (e.g., I will not lecture about consciousness). Osvath and Martin-Ordas (2014) argued that memories might ensure that one acts appropriately wherever the current situation resembles a past situation. They also suggested that some type of (*internal* [e.g., goal] or *external* [e.g., spatial context]) cueing in the present situation might be at play in order to facilitate the retrieval of a previously experienced event, which is what allows us to behave in ways that affect a future outcome. Boyer (2008) argued that episodic representations could work as a motivational break on any previous motivational state because the construct that evokes the emotion is outcompeting the earlier motivation. For example, in a delay of gratification task, a choice between an immediate and delayed reward would be a choice between two (current) motivations (Osvath & Martin-Ordas, 2014). This line of argument takes us again to the idea that the line between a memory and a foresight is as blurry as the line between semantic and episodic memory. Similar to what others have previously argued, it seems more plausible to think of them as a continuum rather than as dichotomous concepts (Osvath & Martin-Ordas, 2014; Szpunar et al., 2014).

## 3.3. Where Do We Go Next?

Despite the theoretical limitations just described, the field of future-oriented cognition in non-human animals, in general, and in great apes, in particular, has been grounded on this dichotomous framework (e.g., semantic vs. episodic) that does not allow us to disentangle which cognitive mechanisms underlie certain behavioral responses. It is remarkable, though, that comparative research has mainly been criticized because it does not necessarily reflect self-projection and because of the type of memory required to solve the episodic foresight tasks (Suddendorf, 2006; Suddendorf & Corballis, 2008). In the light of what I have argued in previous sections, this criticism does not seem constructive. It is true that we cannot assess the subjective experience of apes' behavior, but that does not mean that we can deny it. Also, even if learning accounts for apes' knowledge about the suitability of the tools, it does not necessarily rule out the possibility that subjects are thinking about the future situation in order to take the appropriate tool. As I argued earlier, semantic and episodic memories are interwoven; therefore continuing to search for responses based on purely episodic memories and foresights does not seem to be the correct path to follow. Finally, even if we were to assume that only a *semantic* prospection is behind apes' performance, no one would deny that this type of prospection requires knowledge that there is a future (Raby & Clayton, 2009).

So where do we go next? If we are to abandon the dichotomous framework, then it becomes difficult to argue that we are the only animals who can mentally travel in time. However, it is not my intention to claim that all animals do so with the same level of complexity. Importantly, the differences in complexity might come from the extent to which memories might be integrated by semantic features and from how this semantic knowledge is built up and transmitted across generations, rather than from the capacity to re-experience or pre-experience an event. In the next section, I will further develop this idea and argue that identifying how cultural differences that affect future-oriented cognition might provide us with better grounds to understand the evolution of the episodic system.

## 4. TOWARD A MORE COMPREHENSIVE UNDERSTANDING OF FUTURE-ORIENTED COGNITION

So far I have described what the episodic system is and how it has been operationalized. I have suggested some limitations associated with the definition and the criteria proposed to identify future-oriented behaviors in non-human animals. I have critically reviewed the literature and also have provided empirical evidence that questions the idea that future-oriented cognition is uniquely human. My intention for the remaining part of this chapter is to introduce a new (old) idea that might help us broaden the way we think about and conceive the field of future-oriented cognition. In order to do that, I will borrow an example from Osvath and Martin-Ordas's (2014) proposal and expand on the idea of why it is believed that human future-oriented cognition surpasses that of other animals.

Osvath and Martin-Ordas (2014) suggested that our (as in humans from Western societies) future-oriented skills would be poor without our extensive semantic knowledge, which is transferred through language and stored in text and in the brains of many (i.e., cumulative culture [see Vale et al., 2012, for a similar take on the issue]). In fact, one of the most extraordinary human planning achievements was to go to the moon (see, e.g., Suddendorf et al., 2009). Osvath and Martin-Ordas (2014) argued that this was indeed a perfect example of what we can do with our *cumulative* semantic knowledge. The only way in which humans could organize such an enterprise was not by episodically simulating past events, but by sharing semantic knowledge. Our brain would not be able to cope with so many combinations of past events and mental simulations of future ones. Osvath and Martin-Ordas (2014) also argued that "we perform long-term planning stretching over years and decades, we most likely depend on our semantic system; we know about things like retirement or academic diploma and what not" (4). This argument points out that the way we foresee the future might not be a human universal. More specifically, humans might be able to imagine and think about the future, but the way (e.g., how far into the future we can think) we do it might differ depending on the physical, social, and historical environment. Clearly, having access to different devices that allow us to foresee and represent a distant future might allow us to think further into the future than if we lacked such devices. Would other humans with different cultures be able to do the same? As I mentioned earlier, different groups of captive chimpanzees seem to display different planning skills depending on the semantic knowledge they have about a particular task (Dufour & Sterck, 2008; Osvath & Persson, 2013; Boujarde et al., 2014). If we find such variety across different populations of captive populations of chimpanzees, obviously the same could be true for different human cultures.

For example, in developmental research, one variable that has been hypothesized to affect how future-oriented cognition develops is parenting style. Quon and Atance (2010) suggested that one of the reasons that preschoolers might not be that good at foreseeing the future is because they do not have control over most of their future events (e.g., going to visit grandma). This is so because parents decide what their children will be doing during the day. Quon and Atance (2010) also argued that there are most likely events for which children have higher levels of control (e.g., what they would do at their next playtime) than others (what they would do at bedtime). These authors found that children provided more accurate responses to high-control events than to low-control events. Is this feature universal? Parenting style varies not only among industrialized but also among non-industrialized societies (Hewlett and Lamb, 2005; Keller et al., 2006). For the purpose of my argument, I will focus on two different populations from the Central African Republic: the Aka and the Ngandu. The Aka population is an egalitarian hunter-gatherer society, and the Ngandu population is a farmer/horticulturalist society (e.g., Boyette, unpublished thesis; Hewlett et al., 2011). Interestingly, Aka parents, as opposed to Ngandu parents, focus on the development of children's autonomy, and they tend to not actively intervene or to be less direct in their children's learning experiences (Hewlett & Lamb, 2005). Even if this is an oversimplified view of parenting style in these two populations, it serves for the purpose of my argument. If having little control over future events affects (e.g., delays) the development of future-oriented cognition, then we will expect to see differences between these two populations in the onset of the development of future-oriented cognition. Grasping this source of variation is crucial not only to understand the ontogeny of the episodic system but also to understand its evolution. Results from this type of research would have crucial implications for comparative research. The combination of comparative and cross-cultural studies regarding future-oriented cognition will provide us with a comprehensive understanding of what processes underlie the evolution of such a capacity. The similarities and differences in future-oriented cognition between great apes species (including humans) will shed light on the (biological) evolutionary history of this cognitive skill. On the other hand, similarities and differences across different human cultures will indicate the extent to which contextual factors influence future-oriented cognition.

As I mentioned earlier, this is a new/old idea. It is old because comparative psychologists and cultural anthropologists, among others, have already highlighted some of the limitations regarding the study of social cognition by only focusing on Western societies and great apes (see Barrett et al., 2012; Boesch, 2007, 2008; Henrich et al., 2005; Henrich et al., 2010; Tomasello & Call 2008). It is new because, so far, the study of the evolution of future-oriented cognition has focused only on the comparison between Western children or adults and great apes. As such I would like to push the idea a bit further by suggesting two future lines of research. The first one will examine how being surrounded by devices that allow us to physically represent and keep track of time (e.g., agendas, calendars) might, on the one hand, facilitate foreseeing a further future and, on the other, release the human brain from episodic simulation and semantic load. I argue that this might have tremendous consequences on the way we—Western humans—think about the future, as opposed to, for example, the way chimpanzees do. While the former populations might be able to imagine what they will be doing in 10 years (e.g., buying a house), the latter might only be able to foresee a closer future (e.g., where to build the nest tonight). In this case, as Osvath and Martin-Ordas (2014) argued, the differences might not rely on the episodic system but on the semantic knowledge. Conducting studies in human populations with no access or with different levels of access to devices that allow keeping track of time might shed light on the extent to which being surrounded by and using such devices affect our future-oriented cognition.

Another interesting question is the extent to which differences in ecology affect future-oriented cognition. Woodburn (1982) described mobile, egalitarian hunter-gatherer groups (such as the Aka mentioned earlier) as "immediate-return" societies, and sedentary, "complex" hunter-gatherers (like those who lived along the northwestern coast of the United States), as well as farmers, as "delayed return" societies. Whereas the delayed-return societies are based on systems in which individuals wait, for example, up to 6 months for a return of an asset, in immediate-return societies the return of the investment takes place almost immediately after the action (e.g., hunting). The cognitive consequences of living in a society in which one is always looking for food to eat within the next hours/days, as opposed to a society in which one is planning which crop to plant next year, might be tremendous on the episodic system. If these ecological factors affect the way humans think about the future, then we might be doing a feeble favor to the field of comparative psychology by not addressing the (possible) wide variety of future-oriented cognition across different (human) populations.

To end, these patterns of variation across human and non-human populations are crucial to understand the interaction between inherited predispositions, culture, and ecology. Cross-species and cross-cultural studies will allow us to have a more comprehensive view of what future-oriented cognition involves and to carry out finer and better comparisons across species. Only by understanding human variation will we be able to understand the evolution of future-oriented cognition.

## 5. CONCLUSIONS

There is no doubt that in science the theoretical framework determines the empirical approach to a particular problem. In the case of the episodic system, Tulving's conceptualization of episodic and semantic memory has had a tremendous influence in the way comparative psychologists have addressed the study of episodic memory and future-oriented cognition in non-human animals. In this chapter, the empirical and theoretical validity of such a dichotomous approach has been brought into question. Research conducted in the last years has provided enough evidence that humans are not the only animals able to mentally travel in time. I have also suggested that if we are to move the field forward, it is time to address the differences across different human populations. It is crucial to learn how future-oriented cognition varies depending on physical, historical, and social factors. The capacity to mentally travel in time might be widespread in the animal kingdom; however, what makes our episodic system unique might depend on factors, such as our impressive capacity to share and accumulate semantic knowledge or ecological factors (e.g., foraging strategies). This is still an open question that should be addressed in the future.

## ACKNOWLEDGMENTS

I am very thankful to Tom Smulders and Adam Boyette for their insightful comments and suggestions.

## NOTES

1. I will be referring to *episodic foresight* or *episodic future thinking* as terms involving self-projection or pre-experience of a future event and to *future-oriented cognition* as a broader term independent of the mechanism underlying it.
2. I would like to highlight that this is not only a common problem in comparative psychology. Developmental psychologists face similar issues, but the urge to show *autonoetic consciousness* in developmental research is not pressing, not even necessary; it is implicitly assumed in children's performance. For example, Suddendorf et al. (2011) developed a task in which children were presented with a box with no key to open it in one room. Then children waited 15 minutes in another room and after that delay, they were told to return to the first room. Before going back, they were presented with several options—one of which (i.e., a key) could be used to open the box and retrieve a reward (e.g., a sticker). In this study, 4-year-olds were able to successfully choose the key, and Suddendorf et al. (2011) concluded that this experiment was evidence for memory of a problem and foresight of a solution. Even though Suddendorf et al. (2011) argued that their study fulfills some of the criteria for future-thinking (e.g., the use of single trials, novel problems, different temporal and spatial contexts)—I will return to these criteria in section 2—what their experiment does not show is whether children are re-experiencing and pre-experiencing (i.e., *autonoetic awareness*) the problem. In fact, the same two other alternative explanations I mentioned earlier for the chimpanzee retrieving the tool could apply to children's performance in Suddendorf et al. (2011)'s research. It seems, though, that no one would deny—or even question—the necessity of having *autonoetic awareness* in order to succeed in such a task. This is in stark contrast with what comparative researchers are required to demonstrate in their studies.

# REFERENCES


Addis, D. R., & Schacter, D. L. (2008). Constructive episodic simulation: temporal distance and detail of past and future events modulate hippocampal engagement *Hippocampus, 18*, 227–237.

Addis, D. R., & Schacter, D.L. (2012). The hippocampus and imagining the future: Where do we stand? *Frontiers in Human Neuroscience, 5*, 173.

Addis, D. R., Wong, A.T, & Schacter, D. L. (2007). Remembering the past and imagining the future: common and distinct neural substrates during event construction and elaboration. *Neuropsychologia, 45*, 1363–1377.

Addis, D. R., Sacchetti, D. C., Ally, B. A., Budson, A. E., & Schacter, D. L. (2009). Episodic simulation of future events is impaired in mild Alzheimer's disease. *Neuropsychologia, 47*, 2660–2671.

Allen, T. A., & Fortin, N. J. (2013). The evolution of episodic memory. *Proceedings of the National Academy of Sciences, 110*(Suppl 2), 10379–10386.

Atance, C. M., & O'Neill, D. K. (2001). Episodic future thinking. *Trends in Cognitive Sciences, 5*, 533–539.

Atance, C. M., & O'Neill, D. K. (2005). Preschoolers' talk about future situations. *First Language, 25*, 5–18.

Balota, D. A., & Coane, J. H. (2008). Semantic memory. In J. H. Byrne, H. Eichenbaum, R. Menzel, H. L. Roediger III, & D. Sweatt (eds.), *Handbook of learning and memory: a comprehensive reference* (pp. 512–531). Amsterdam: Elsevier.

Barsalou, L. W. (1988). The content and organization of autobiographical memories. In U. Neisser & E. Winograd (eds.), *Remembering reconsidered: ecological and traditional approaches to the study of memory* (pp. 193–243) New York: Cambridge University Press.

Barrett, H. C., Stich, S., & Laurence, S. (2012). Should the study of *Homo Sapiens* be part of cognitive science? *Topics in Cognitive Sciences, 4*, 379–386.

Binder, J. R., & Desai, R. (2011). The neurobiology of semantic memory. *Trends in Cognitive Sciences, 15*, 527–536.

Bischof, N. (1985). *Das ratzel odipus*. Munich: Piper.

Bischof- Köhler, D. (1985). Zur Phylogenese menschlicher Motivation [On the phylogeny of human motivation]. In L. H. Eckensberger & E. D. Lantermann (eds.), *Emotion und Reflexivitaet* (pp. 3–47). Urban & Schwarzenberg.

Boesch, C. (2007). What makes us human (Homo sapiens)? The challenge of cognitive cross-species comparison. *Journal of Comparative Psychology, 121*, 227–240.

Boesch, C. (2008). Taking development and ecology seriously when comparing cognition: reply to Tomasello and Call (2008). *Journal of Comparative Psychology, 122*, 453–455.

Boesch, C., & Boesch, H. (1984). Mental map in wild chimpanzees: an analysis of hammer transports for nut cracking. *Primates, 25*, 160–170.

Bourjade, M., Call, J., Pelé, M., Maumy, M., & Dufour, V. (2014). Bonobos and orangutans, but not chimpanzees, flexibly plan for the future in a token-exchange task. *Animal Cognition, 17*, 1329–1340. doi: 10.1007/s10071-014-0768-6.

Boyer, P. (2008). Evolutionary economics of mental time travel? *Trends in Cognitive Sciences, 12*, 219–224.

Boyette, A. H. (2013). Social learning during middle childhood among Aka foragers and Ngandu farmers of the Central African Republic. (Unpublished doctoral thesis). Washsington State University, Vancouver, Canada.

Buckner, R. L., & Carroll, D. C. (2007). Self-projection and the brain. *Trends in Cognitive Sciences, 11*, 49–57.

Cheke, L. G., & Clayton, N. S. (2010). Mental time travel in animals. *Wiley Interdisciplinary Reviews: Cognitive Science*, 1, 1–16.

Conway, M. A. (2001). Sensory-perceptual episodic memory and its context: autobiographical memory. *Philosophical Transactions of the Royal Society of London B*, 356, 1375–1384.

Conway, M. A., & Fthenaki, A. (2000). Disruption and loss of autobiographical memory. In F. Boller and Grafman (eds.), *Handbook of neuropsychology: memory and its disorders* (pp. 281–312). 2nd ed. Amsterdam: Elsevier Science B.V.

Corballis, M. C. (2013). Mental time travel: a case for evolutionary continuity. *Trends in Cognitive Sciences*, 17, 5–6. doi: 10.1016/j.tics.2012.10.009.

Crystal, J. D. (2012). Prospective cognition in rats. *Learning and Motivation, 43*, 181–191.

D'Argembeau, A., Renaud, O., & Van der Linden, M. (2011). Frequency, characteristics, and functions of future-oriented thoughts in daily life. *Applied Cognitive Psychology, 35*, 96–103.

Dudai, Y., & Carruthers, M. (2005). The Janus face of Mnemosyne. *Nature*, 434, 567.

Dufour, V., & Sterck, E. H. M. (2008). Chimpanzees fail to plan in an exchange task but succeed in a tool-using procedure. *Behavioral Processes, 79*, 19–27.

Duval, C., Desgranges, B., De La Sayette, V., Belliard, S. Eustache, F., & Piolino, P. (2012). What happens to personal identity when semantic knowledge degrades? A study of the self and autobiographical memory in semantic dementia. *Neuropsychologia, 50*, 254–256.

Gamboz, N., De Vito, S., Bradimonte, M. A., Pappalardo, S. G., Galeone, F., Iavarone, A., et al. (2010). Episodic future thinking in amnesic mild cognitive impairment. *Neuropsychologia*, 48, 2091–2097.

Garcia, J., & Koelling, R. A. (1966). Relation of cue to consequence in avoidance learning. *Psychonomic Science, 4*, 123–124.

Gilbert, D. T., & Wilson, T. D. (2007). Prospection: experiencing the future. *Science, 317*, 1351–1354.

Goodall, J. (1964). Tool-use and aimed throwing in a community of free-ranging chimpanzees. *Nature, 201*, 1264–1266.

Graham, K. S., Simons, J. S., Pratt, K. H., Patterson, K., & Hodges, J. R. (2000). Insights from semantic dementia on the relationship between episodic and semantic memory. *Neuropsychologia*, 38, 313–324. doi: 10.1016/S0028-3932(99)00073-1.

Greenberg, D. L., & Verfaellie, M. (2010). Interdependence of episodic and semantic memory: evidence from neuropsychology. *Journal of the International Neuropsychological Society*, 16, 748–753.

Hassabis, D., Kumaran, D., Vann, S. D., & Maguire, E. A. (2007). Patients with hippocampal amnesia cannot imagine new experiences. *Proceedings of the National Academy of Sciences U S A, 104*, 1726–1731.

Henrich, J., Heine, S. J., & Norenzayan, A. (2010). The weirdest people in the world? *Behavioral and Brain Sciences, 33*, 61–135.

Henrich, J., Boyd, R., Bowles, S., Camerer, C. F., Fehr, E., Gintis, H., McElreath, R., Alvard, M., Barr, A., Ensminger, J., Henrich, N. S., Hill, K., Gil-White, F., Gurven, M., Marlowe, F. W., Patton, J. Q., & Tracer, D. (2005). "Economic man" in cross-cultural perspective: Behavioral experiments in 15 small-scale societies. *Behavioral and Brain Sciences, 28*, 795–815.

Hewlett, B. S., Fouts, H. N., Boyette, A. H., & Hewlett, B. L. (2011). Social learning among Congo Basin hunter-gatherers. *Philosophical Transactions of the Royal Society of London B, 366*, 1168–1178.

Hewlett, B. S., & Lamb, M. E. (2005). Hunter-gatherer childhoods: evolutionary, developmental and cultural perspectives. New Brunswick, NJ: Transaction/Aldine.

Hudson, J. A, Mayhew, E. M., & Prabhakar, J. (2011). The development of episodic foresight: emerging concepts and methods. *Advances in Child Development and Behavior, 40*, 95–137.

Irish, M., Addis, D. R., Hodges, J. R., & Piguet, O. (2012) Considering the role of semantic memory in episodic future thinking: evidence from semantic dementia. *Brain, 135*, 2178–2191.

Irish, M., & Piguet, O. (2013). The pivotal role of semantic memory in remembering the past and imagining the future. *Frontiers in Behavioral Neuroscience, 7*(27). doi: 10.3389/fnbeh.2013.00027.

Janmaat, K. R., Polansky, L., Ban, S. D., & Boesch, C. (2014). Wild chimpanzees plan their breakfast time, type, and location. *Proceedings of the National Academy of Sciences U S A, 111*, 16343–16348.

Kapur, N. (1999). Syndromes of retrograde amnesia: a conceptual and empirical synthesis. *Psychological Bulletin, 125*, 800–825.

Keller, H., Lamm, B., Abels, M., Yovsi, R. D., Borke, J., Jensen, H., et al. (2006). Cultural models, socialization goals, and parenting ethnotheories: a multi-cultural analysis. *Journal of Cross-Cultural Psychology, 37*, 155–172.

Köhler, W. (1921) Zur psychologie des schimpansen. *Psychologische Forschung, 1*, 2–46.

Klein, S. B. (2013a). The complex act of projecting oneself into the future. *Wiley Interdisciplinary Reviews: Cognitive Science, 4*, 63–79.

Klein, S. B. (2013b). Making the case that episodic recollection is attributable to operations occurring at retrieval rather than to content stored in a dedicated subsystem of long-term memory. *Frontiers in Behavioral Neuroscience, 7*, 3.

Klein, S. B. (2015). What memory is. *Wiley Interdisciplinary Reviews: Cognitive Science, 6*, 1–38.

Klein, S. B., Loftus, J., & Kihlstrom, J. F. (2002). Memory and temporal experience: the effects of episodic memory loss on an amnesic patient's ability to remember the past and imagine the future. *Social Cognition, 20*, 353–379.

Kwan, D., Carson, N., Addis, D. R., & Rosenbaum, R. S. (2010). Deficits in past remembering extend to future imagining in a case of developmental amnesia. *Neuropsychologia, 48*, 3179–3186.

Lu, H. B., Zou, Q., Gu, H., Raichle, M. E., Stein, E. A., & Yang, Y. (2012). Rat brains also have a default mode network. *Proceedings of the National Academy of Sciences U S A, 109*, 3979–3984.

Mantini, D., Gerits, A., Nelissen, K., Durand, J. B., Joly, O., Simone, L., Sawamura, H., Wardak, C., Orban, G. A., Buckner, R. L., & Vanduffel, W. (2011). Default mode of brain function in monkeys. *Journal of Neuroscience, 31*, 12954–12962.

Martin-Ordas, G., Atance, C. M., & Louw, A. (2012). The role of episodic and semantic memory in episodic foresight. *Learning and Motivation, 43*, 209–219.

Martin-Ordas, G., Atance, C. M., & Caza, J. S. (2014a). How do episodic and semantic memory contribute to episodic foresight in young children? *Frontiers in Psychology, 5*, 1–11.

Martin-Ordas, G., Atance, C. M., & Call, J., (2014b) Remembering in tool-use tasks in children: the role of the information at encoding. *Memory, 22*, 129–44.

Martin-Ordas, G., Berntsen, D., & Call, J. (2013). Memory for distant past events in chimpanzees and orangutans. *Current Biology, 23*, 1438–1441.

Morris, R. G. M. (2001). Episodic-like memory in animals: psychological criteria, neural mechanisms and the value of episodic-like tasks to investigate animal models of neurodegenerative disease. *Philosophical Transactions of the Royal Society of London B, 356*, 1453–1465.

Mulcahy, N. J., & Call, J. (2006). Apes save tools for future use. *Science, 312*, 1038–1040.

Neisser, U. (1981). John Dean's memory: a case study. *Cognition*, 9, 1–22.

Okuda, J., Fujii, T., Ohtake, H., Tsukiura, T., Tanji, K., Suzuki, K., Kawashima, R., Fukuda, H., Itoh, M., & Yamadori, A. (2003). Thinking of the future and past: the roles of the frontal pole and the medial temporal lobes. *Neuroimage, 19*, 1369–1380.

Osvath, M. (2009). Spontaneous planning for stone throwing by a male chimpanzee. *Current Biology, 5*, R191–R192.

Osvath, M. (2010). Great ape foresight is looking great. *Animal Cognition, 13*, 777–781.

Osvath, M., & Karvonen, E. (2012), Spontaneous innovation for future deception in a male chimpanzee. *PloS One*, 7, e36782.

Osvath, M., & Martin-Ordas, G. (2014). The future of future-oriented cognition in non-humans: theory and the empirical case of the great apes. *Philosophical Transactions of the Royal Society of London B, 369*, 1655.

Osvath, M., & Osvath, H. (2008). Chimpanzee (*Pan troglodytes*) and orangutan (*Pongo abelii*) forethought: self-control and pre-experience in the face of future tool-use. *Animal Cognition, 11*, 661–674.

Osvath, M., & Persson, T. (2013). Great apes can defer exchange: a replication with different results suggesting future oriented behaviour. *Frontiers in Comparative Psychology, 4*(698). doi: 10.3389/fpsyg.2013.00698.

Patterson, K., & Hodges, J. R. (2000). Semantic dementia: one window on the structure and organization of semantic memory. In L. Cermak (ed.), *Revised handbook of neuropsychology: memory and its disorders* (pp. 313–333). Amsterdam: Elsevier Science.

Pillemer, D. (2003). Directive functions of autobiographical memory: the guiding power of specific episodes. *Memory, 11*, 193–202.

Piolino P., Desgranges B., & Eustache F. (2009). Episodic autobiographical memories over the course of time: cognitive, neuropsychological and neuroimaging findings. *Neuropsychologia, 47*, 2314–2329.

Quon, E., & Atance, C. M. (2010). A comparison of preschoolers' memory, knowledge, and anticipation of events. *Journal of Cognition and Development, 11*, 37–60.

Raby, C. R., & Clayton, N. S. (2009). Prospective cognition in animals. *Behavioural Processes, 80*, 314–324.

Race, E., Keane, M. M., & Verfael-lie, M. (2011). Medial temporal lobe damage causes deficits in episodic memory and episodic memory and episodic future thinking not attributable to deficits in narrative construction. *Journal of Neuroscience, 31*, 10262–10269.

Rathbone, C. J., Conway, M. A., & Moulin, C. J. A. (2011). Remembering and imagining: the role of the self. *Consciousness and Cognition, 20*, 1175–1182.

Redshaw, J. (2014). Does metarepresentation make human mental time travel unique? *Wiley Interdisciplinary Reviews: Cognitive Science, 5*, 519–531.

Rilling, J. K., Barks, S. K., Parr, L. A., Preuss, T. M., Faber, T. L., Pagnoni, G., Bremner, J. D., & Votaw, J. R. (2007). A comparison of resting-state brain activity in humans and chimpanzees. *Proceedings of the National Academy of Sciences U S A, 104,* 17146–17151.

Sanz, C., & Morgan, D. (2009). Flexible and persistent tool-using strategies in honey-gathering by wild chimpanzees. *International Journal of Primatology, 30,* 411–427.

Schacter, D. L., & Addis, D. R. (2007). The ghosts of past and future. *Nature, 445,* 27.

Schacter, D. L., Addis, D. R., & Buckner, R. L. (2007). The prospective brain: remembering the past to imagine the future. *Nature Review Neuroscience, 8,* 657–661.

Squire, L. R., & Zola, S. M. (1998). Episodic memory, semantic memory and amnesia. *Hippocampus, 8,* 205–211.

Squire, L. R., Stark, C. E. L., & Clark, R. E. (2004). The medial temporal lobe. *Annual Review of Neuroscience, 27,* 279–306.

Suddendorf, T. (2010). Episodic memory versus episodic foresight: Similarities and differences. *Wiley Interdisciplinary Reviews Cognitive Science, 1,* 99–107.

Suddendorf, T. (2006). Foresight and evolution of the human mind. *Science, 312,* 1006–1007.

Suddendorf, T., & Busby, J. (2003a). Mental time travel in animals? *Trends in Cognitive Sciences, 7,* 391–396.

Suddendorf, T., & Busby, J. (2003b). Like it or not: the mental time travel debate. *Trends in Cognitive Sciences, 7,* 437–438.

Suddendorf, T., & Corballis, M. C. (1997). Mental time travel and the evolution of the human mind. *Genetic, Social, and General Psychology Monographs, 123,* 133–167.

Suddendorf, T., & Corballis, M. C. (2007). The evolution of foresight: what is mental time travel and is it unique to humans? *Behavioral and Brain Sciences, 30,* 299–313.

Suddendorf, T., & Corballis, M. C. (2008). New evidence for animal foresight? *Animal Behaviour, 75,* 1–3.

Suddendorf, T., & Corballis, M. C. (2010). Behavioural evidence for mental time travel in nonhuman animals. *Behavioural Brain Research, 215,* 292–298.

Suddendorf, T., Corballis, M. C., & Collier-Baker, E. (2009). How great is great ape foresight? *Animal Cognition, 12,* 751–754.

Suddendorf, T., Nielsen, M., & von Gehlen, R. (2011). Children's capacity to remember a novel problem and to secure its future solution. *Developmental Science, 14,* 26–33. doi: 10.1111/j.1467-7687.2010.00950.

Szpunar, K. K. (2010). Episodic future thought: an emerging concept. *Perspective on Psychological Science, 5,* 142–162.

Szpunar, K. K., & McDermott, K. B. (2008). Episodic future thought and its relation to remembering: evidence from ratings of subjective experience. *Consciousness and Cognition, 17,* 330–334.

Szpunar, K. K., Spreng, R. N., & Schacter, D. L. (2014). A taxonomy of prospection: introducing an organizational framework for future-oriented cognition. *Proceedings of the National Academy of Sciences U S A, 111,* 18414–18421.

Szpunar, K. K., Watson, J. M., & McDermott, K. B. (2007). Neural substrates of envisioning the future. *Proceedings of the National Academy of Sciences U S A, 104,* 642–647.

Thom, J. M., & Clayton N. S. (2015). Route-planning and the comparative study of future-thinking. *Frontiers in Psychology, 6*(144). doi: 10.3389/fpsyg.2015.00144.

Tomasello, M., &. Call, J. (2008). Assessing the validity of ape-human comparisons: a reply to Boesch (2007). *Journal of Comparative Psychology, 122,* 449–452.

Tulving E. (1972). Episodic and semantic memory. In E. Tulving, & W. Donaldson (eds.), *Organization of Memory* (pp. 381–403). New York: Academic Press.

Tulving, E. (1983). *Elements of episodic memory.* Oxford: Clarendon Press.

Tulving, E. (1985). Memory and consciousness. *Canadian Psychology,* 26, 1–12.

Tulving, E. (1995). Organization of memory: quo vadis? In M. S. Gazzaniga (ed.), *The cognitive neuroscience* (pp. 839–847). Cambridge, MA: MIT Press.

Tulving, E. (2002). Episodic memory: From mind to brain. *Annual Review of Psychology, 53,* 1–25.

Tulving, E. (2005). Episodic memory and autonoesis: uniquely human? In H. S. Terrace & J. Metcalfe J. (eds.), *The missing link in cognition* (pp. 3–56). Oxford: Oxford University Press.

Tulving, E., & Thompson, D. (1973). Encoding specificity and retrieval processes in episodic memory. *Psychological Review, 8,* 352–373.

Vale, G. L., Flynn, E. G., & Kendal, R. L. (2012). Cumulative culture and future thinking : is mental time travel a prerequisite to cumulative cultural evolution? *Learning and Motivation, 43,* 220–230.

van Schaik, C. P., Damerius, L., & Isler, K. (2013). Wild orangutan males plan and communicate their travel direction one day in advance. *PLoS One 8,* e74896.

Wheeler, M. A., & McMillan, C.T. (2001). Focal retrograde amnesia and the episodic-semantic distinction. *Cognitive, Affective and Behavioral Neuroscience, 1,* 22–36.

Wheeler, M. A., Stuss, D. T., & Tulving, E. (1997). Toward a theory of episodic memory: the frontal lobes and autonoetic consciousness. *Psychological Bulletin, 121,* 331–354.

Woodburn, J. (1982). Egalitarian societies. *Man* (N.S.), *17,* 431–451.

16

# The Future of Memory, Mental Time Travel, and Mind Wandering

MICHAEL C. CORBALLIS ■

The future begins with the past. It has long been recognized that memory systems evolved as adaptive mechanisms for future survival and reproductive success, and that there are different memory systems for different functions, ranging from imprinting in birds to memory of events in humans (Sherry & Schacter, 1987). Even in humans, there are at least three different systems (Squire, 2004). Our capacity to survive depends on the skills we acquire, the knowledge we accumulate, and the experiences we have had. The acquisition of skill depends on what is called procedural memory, and is non-declarative—or inaccessible to consciousness or language. We learn to type, ride a bike, and even talk, without any awareness of the process of learning itself. Our knowledge about the world is semantic memory, and is declarative in the sense that it can be brought into consciousness when required, and can be articulated. Memory of our individual experiences is known as episodic memory (Tulving, 2002), and allows us to replay past events in our lives. It, too, is conscious and declarative.

The adaptiveness of memory lies in its usefulness rather than its accuracy or its completeness. We almost certainly forget much more than we remember; we may have the illusion of good memory, but of course we don't know how much we have forgotten precisely because we have forgotten it! Even our skills decline with lack of use. Semantic memory is less robust; most of us have forgotten much of what we learned in school, or in Physics 101, but I discovered I could still ride a bike after many years off the saddle. Episodic memory, though, is extremely fragile and often fleeting, and probably holds only a tiny fraction of our experiences. With perhaps slight exaggeration, the émigré Czech writer Milan Kundera, in his novel *Ignorance* (2002), put it like this:

> The fundamental given is the ratio between the amount of time in the lived life and the amount of time from that life that is stored in memory. No one has ever tried to calculate this ratio, and in fact there exists no technique for doing so; yet without much risk of error I could assume that the memory retains no more than a millionth, a hundred-millionth, in short an utterly

> infinitesimal bit of the lived life. That fact too is part of the essence of man. If someone could retain in his memory everything he had experienced, if he could at any time call up any fragment of his past, he would be nothing like human beings: neither his loves nor his friendships nor his angers nor his capacity to forgive or avenge would resemble ours. (pp. 122–123)

It is indeed an impediment to remember too much. The best-known case of super-memory is Solomon Shereshevskii (Luria, 1968), who was able to repeat back long lists of digits, nonsense syllables, poetry in unknown languages, even elaborate scientific formulae, and to do so perfectly even years after they were presented to him. He could even repeat them in reverse order. His technique was to convert words into graphic images, and distribute them in different mental locations, typically Gorky Street in Moscow. He could then retrieve them by repeating his mental walk—in either direction. But his remarkable memory was also a burden. He had difficulty reading even simple sentences, because images would appear and crowd in on each other. Reading stories was next to impossible, because images would surface from other things he had read. He did find one solution, though, which was to imagine information he encountered written on a blackboard, and then imagine erasing it!

Memories are also often false. Even our basic knowledge can warp over time, but episodic memory seems especially susceptible to suggestion. Loftus and Ketcham (1994) recount studies in which people gave detailed answers when asked to describe being lost in a mall, or being taken for a ride in a hot-air balloon, or being nearly drowned and rescued by a lifeguard, even though these events never actually happened. In experimental studies, Roediger and McDermott (1995) showed how people could be induced to remember specific words in lists, even though those words were not actually presented. For example, people heard lists of words like *table, sit, legs, seat, soft, desk, arm, sofa, wood, cushion*, and were later tested for recall. Many were adamant that they had heard the word *chair*, even though it was not in the list.

The sheer unreliability of memory, especially memory for specific events, is a clue to its adaptability and flexibility (Schacter, 2012; Schacter & Addis, 2007). We use our memories not so much to visit the past as to mentally construct possible futures, perhaps rearranging the people, places, and things of our episodic memories to create new combinations for some planned event, such as a dinner party or a picnic with friends. Such constructions depend also on semantic memory—our knowledge of how the world works—so that we can set our future imaginings in known real-world contexts (Klein, 2013). The flexibility of memory allows us to generate different options, so that we might choose the most suitable or appropriate course of action. Memory also allows us to generate purely imaginary and often unrealistic scenarios—a point I return to later.

## 1. MENTAL TIME TRAVEL AND THE HIPPOCAMPUS

The ability to mentally project ourselves into both past and future, albeit incompletely, has been dubbed *mental time travel* (Suddendorf & Corballis, 1997, 2007; Tulving, 1985), and depends critically on the hippocampus. Destruction of the hippocampus in humans causes severe loss of episodic memory, as well as an inability to imagine or construct possible future events. One classic example is Henry Molaison, better known in the literature as H. M., whose hippocampus was destroyed as a result of temporal lobe surgery for the relief of intractable epilepsy. His condition is captured in a book appropriately entitled *Permanent Present Tense*, by Corkin (2013). A similar case is the English musician Clive Wearing, who suffered hippocampal loss as result of a viral infection. His condition is described in a book entitled *Forever Today*, written by his wife Deborah Wearing (2005). Tulving (1985) described another well-studied patient, known as N. N., who was also unable to recall past events or imagine future ones. These individuals, though, retained basic skills and semantic memory. All could speak coherently, and Clive Wearing retains an ability to pay the piano and conduct a choir. But each of them is stuck in the present—or at least in a narrow window of time in which access to memory of events that occurred more than a few seconds ago is lost. They can project a few seconds into the future in order to finish a sentence, or perhaps plan a new one, and so maintain the semblance of conversation, but the thread is quickly lost.

The role of the hippocampus is further supported by evidence from neuroimaging. Functional magnetic resonance imaging (fMRI) shows that the hippocampus is active when people recall past events, and also when asked to imagine future events based on new combinations of the elements of past episodes (Addis et al., 2007; Martin et al., 2011). Different regions of the hippocampus appear to be involved in the different aspects of retrieving, constructing, and encoding events, and hippocampal activity is stronger when people construct future episodes than when they re-experience past ones, in part because imagining future events requires extra hippocampal involvement in combining disparate details into a coherent scenario not previously encountered (Gaesser et al., 2014).

The role of the hippocampus in mental time travel, and indeed mental time travel itself, may not be restricted to humans. Individual neurons in the rat hippocampus fire when the animal is in particular locations in an environment, such as a maze, suggesting that the hippocampus is involved in the construction and activation of cognitive maps of the environment (O'Keefe & Nadel, 1978). This activity occurs while the animal is exploring a maze, but also occurs in sharp-wave ripples (SWRs) sometime after the animal has actually been in the maze, either during slow-wave sleep (Wilson & McNaughton, 1994) or when the animal is awake but immobile (Karlsson & Frank, 2009). The paths indicated by the replay need not correspond to the actual paths taken in the maze. Sometimes they correspond to the reverse of those actually taken (Foster & Wilson, 2006), sometimes to paths the animal will take in the future (Pfeiffer & Foster, 2013), and sometimes to paths the animal never takes at all (Gupta et al., 2010). The ripples are typically compressed in time, just as our own mental travels are generally much shorter than actual ones. Summarizing these and other findings, Dragoi and Tonegawa (2014) write:

> The existence of temporally compressed neuronal sequences that are independent of the recent experience of the animal could support the existence of the ability to travel mentally into its past as well as into the future, a sophisticated process that may underlie higher cognitive functions like memory

> recollection, navigational planning, imagining, cognitive map formation, and schema-based rapid learning. (6)

And that is in the rat. This suggests, contrary to my earlier view (Corballis, 2011; Suddendorf & Corballis, 2007), that at least a rudimentary ability to travel mentally in time and space goes far back in evolution (Corballis, 2013).

As far as we know, remembered or imagined trajectories in a spatial environment in rats probably do not have the complexity of human episodes, which include objects, people, animals, emotions, happenings, along with locations in space and time, and their combinations. Nevertheless, it seems likely that mental time travel did not emerge abruptly in humans, but evolved over the tens of millions of years that separate us from our common ancestry with the other mammals, and perhaps goes further back to common ancestry with birds, which also possess a hippocampal analogue (Macphail, 2002). In creatures that move freely over the surface of the earth—and beyond—a mechanism for mental as well as physical exploration in time and space may have been very early requisites for successful adaptation.

## 2. MIND WANDERING

Mental time travel is also part of what has been termed *mind wandering*, in which our thoughts drift away from the present or from the task at hand. Mind wandering is estimated to occupy nearly 50% of our waking hours (Killingsworth & Gilbert, 2010). It activates a subset of the extensive network in the brain known as the default-mode network, first identified as such by Raichle et al. (2001). Its core regions include medial prefrontal cortex, posterior cingulate and retrosplenial cortex, inferior parietal lobule, lateral temporal cortex, and hippocampus (Buckner et al., 2008). It is activated not only by mental travels in time and space, but also by what is known as *theory of mind*—the ability to imagine being in the mental state of others. As Buckner et al. (2008, 1) put it, "the default network is active when individuals are engaged in internally focused tasks including autobiographical memory retrieval, envisioning the future, and conceiving the perspectives of others."

A large-scale survey by Killingsworth and Gilbert (2010) also revealed that people were less happy when their minds were wandering than when they were "on task." Indeed, mind wandering is said to impact negatively on a wide range of activities and dispositions, including comprehension, daily life activities, driving, memory, mood, and even intelligence, as documented in a review by Mooneyham and Schooler (2013). This compendium of woe may have encouraged the pursuit of "mindfulness," a therapeutic technique derived from Buddhist meditational practice. Mindfulness is in many respects the antithesis of mind wandering, and urges a focus on the present and on one's own body (e.g., Siegel, 2007), and is said to alleviate stress. But stress and indeed unhappiness can also be adaptive; brooding over past mistakes can help one avoid similar mistakes in the future. And although mind wandering can impede our "on-line" activities in many ways, it can also provide the sources of innovation and creativity, as well as relief from boredom, not to mention the sheer enjoyment of a pleasant daydream (Singer, 1966).

## 3. THEORY OF MIND

As suggested in the preceding section, mind wandering includes imagined travels not only in space and time, but also into the minds of others. Frith and Frith (2003) identified a network involving medial prefrontal cortex, temporal poles, and superior temporal sulcus, which is responsible for what they term *mentalizing*—the attribution of intentional states to others. This overlaps with the default-mode network (Buckner et al., 2008), and with the areas activated by autobiographical memory. Within the temporal lobes, autobiographical memory and future thinking produce greater activation in the medial portion, including the hippocampus, while theory-of-mind tasks produce greater activation in the lateral portion (Spreng & Grady, 2009). But such distinctions seem to depend also on how well you know the person whose mind you project into; hippocampal involvement is greater when individuals project into the minds of those they know, probably because they draw to some extent on semantic and episodic memories (Rabin & Rosenbaum, 2012).

Theory of mind can sometimes persist following loss of autobiographical memory. As a result of severe traumatic brain injury, two cases known as K. C. and M. L. both lost the ability to consciously recollect personal events from their own lives, but scored normally on routine theory-of-mind tests. K. C. in particular showed clear appreciation of humor and sarcasm, both considered dependent on theory of mind (Rosenbaum et al. 2007). Such cases are rare, but they do suggest a degree of dissociation between theory of mind and autobiographical memory. Conversely, patients with semantic dementia, which affects semantic memory but leaves episodic memory largely intact, show deficits in theory of mind (Duval et al., 2012).

There remains doubt as to whether non-human animals can project into the minds of others. In 1978, David Premack and Guy Woodruff published a seminal paper asking the question "Does the chimpanzee have a theory of mind?" This paper stimulated great interest, but even after 30 years the issue remained unresolved. In an article entitled "Darwin's Mistake: Explaining the Discontinuity Between Human and Nonhuman Minds," Penn et al. (2008) concluded that even chimpanzees, our closest non-human relatives, have no theory of mind—and that Darwin was wrong in declaring the difference between humans and non-humans to be one of degree but not of kind. On the other hand, Call and Tomasello (2008) concluded that 30 years of research have shown chimpanzees to have some understanding of the goals, intentions, perceptions, and knowledge of others, but no understanding of others' beliefs or desires. It may well be that animals are able to read the minds of other animals, but there are often simpler behavioral explanations. We may never really know what goes on in the minds of species other than ourselves.

In spite of the conflicting opinions, it seems fair to conclude that humans have a greater capacity for projecting into the minds of others than do other species. This gives a further dimension to mind wandering, as we project ourselves into the thoughts of others as well as into different places and different times. This extended capacity probably evolved during the Pleistocene, dating from some 2.6 million years ago until around 12,000 years ago, and marked by a tripling of brain size (Wood & Collard, 1999), the development of tools, and the establishment of what has been called the "cognitive niche" as a means of survival in the hostile environment of the African savanna (Tooby & DeVore, 1987).

Theory of mind is especially important in social contexts, critical to empathy and cooperation, but reveals its darker side in Machiavellian intelligence (Byrne & Whiten, 1988). If we know what is in the minds of others, we can then act to deceive them, and it can be especially pleasurable to deceive deceivers. Theory of mind is recursive, adding depth and complexity to social intrigue, as in Premack and Woodruff's (2007) comment on human sexual politics: "Women think that men think that they think that men think that women's orgasm is different" (13,865). Humphrey (1976) described the Machiavellian loop as a "self-winding watch to increase the general intellectual standing of the species" (311).

Although mental time travel may be within the repertoire of non-human species, the addition of theory of mind in humans no doubt enriches the nature and scope of our mind wanderings. We mentally wander not only in space and time, but also into other people's lives. This is also the basis of fiction, which takes us into the lives of imaginary people. We watch plays and soap operas, read novels, compose fantasy worlds of heroes and monsters. In this respect, our mental travels are not simply a means of planning future events; they are also a means of stretching the bounds of possibility, so that we are better prepared for the unexpected.

## 4. LANGUAGE

Theory of mind enriches our mind wandering in another way. It is a necessary component of language, allowing us to share our travels and so benefit from the experiences of others. As we have seen, non-human animals may well have the capacity to imagine future or past events, but what they appear to lack is the ability to communicate about them to others. Language may have evolved precisely because of the adaptive consequences in a highly social species to share knowledge and experience. One of the critical features of language is displacement, the ability to refer to the non-present. Indeed, it may have been this very property that drove its evolution (Bickerton, 2010; Corballis, 2011; Gärdenfors & Osvath, 2010). But it was theory of mind that provided the extra ingredient that made language possible.

The importance of theory of mind in language can be credited to the philosopher Paul Grice (1975), who held that language depends on inference rather than explicit decoding. Animal communication, like computer languages, is generally unambiguous, whereas human language, despite its apparent richness, is characteristically ambiguous and imprecise. In order to converse, individuals must understand what is going on in each other's minds, so that each can infer what the other means. Noam Chomsky (2007) also seems to convey much the same idea:

> Communication is not a matter of producing some mind-external entity that the hearer picks out of the world, the way a natural scientist could. Rather, communication is a more-or-less affair, in which the speaker produces external events and hearers seek to match them as best they can to their own internal resources. (10)

By "communication" here he means language, which he contrasts with animal communication.[1]

The theme has been pursued by Dan Sperber (e.g., Sperber & Origgi, 2010; Sperber & Wilson, 2002). Sperber and Wilson (2002) give an example of how complex mutual understanding can build up prior to the utterance of any words:

> Suppose that Peter and Mary are walking in the park. They are engaged in conversation; there are trees, flowers, birds, and people all around them. Still, when Peter sees their acquaintance John in a group of people coming towards them, he correctly predicts that Mary will notice John, remember that he moved to Australia three months earlier, infer that there must be some reason he is back in London, and conclude that it would be appropriate to ask him about this. Peter predicts Mary's train of thought so easily, and in such a familiar way, that it is not always appreciated how remarkable this is from a cognitive point of view. (15)

So Peter might say to Mary something like "Hmm. Wonder what brought him back so soon." They both know exactly what he means, and know that the other knows it, too.

Through language, then, we can share our mind wanderings, and so increase our knowledge and range of experiences, and enhance social understanding beyond that gained through theory of mind. To a large extent this is done through stories, whether of past experiences or of the purely imaginary. The literary scholar John Niles (2010) suggests that our species should be renamed *Homo narrans*—the storytellers. The elaboration of mental time travels into narratives probably dates from the Pleistocene, dating from some 1.6 million years ago, when our hunger-gatherer forebears foraged over increasingly wide terrain for food. It would have been increasingly adaptive to relay the foraging experiences to others. Children, too, would have benefited from the tales told by the adult hunters and foragers, gaining knowledge about food sources and hunting techniques before themselves graduating to the hunt. Sugiyama (2011) suggests that the importance of stories may explain the prolonged juvenile phase in humans. The juvenile phase began to increase with the earliest members of our genus, *Homo habilis*, early in the Pleistocene, reaching a peak with *Homo sapiens* (Locke & Bogin, 2006).

## 5. CREATIVITY

Our mind wanderings, and our stories, often take us beyond actual experience into the realms of fantasy, stretching the imagination beyond actual experience, perhaps to test the boundaries of the possible, or create the myths of heroic deeds that bind societies together. This imaginative component also underlies creativity. The distinguished psychologist and epistemologist Donald T. Campbell (1960) once described the essence of creativity as "blind variation and selective retention" (380). Blind variation is captured in the very notion of wandering, whether ambulatory or mental—straying from a set path into unknown territory. What we find there then depends to a large extent on chance. It is the randomness of our wanderings, then, that supplies the spark of creativity, although when we do stumble across something new and important we need to recognize it as such and retain it—what Campbell called "selective retention."

The random component of mind wandering is well illustrated by dreams, which are essentially a random jumble of experiences based on past memories. Although many have sought to find coherence and meaning in dreams (e.g., Freud, 1900; Revonsuo, 2000), their sheer unpredictability seems to defy rational explanation. Some have nevertheless claimed a dream as a source of creativity. As the result of a dream, Robert Louis Stevenson is said to have developed the plot of *Dr. Jekyll and Mr. Hyde*, and Jack Nicklaus to have corrected his golf swing. But dreams are probably too unstructured to be a reliable source of creativity, and in any case most dreams are forgotten.

Daydreaming is a more likely source of creativity, since they can be partly under conscious control, and maintain a degree of real-world coherence. The positive effects of daydreaming were well documented in the work of Jerome L. Singer in the 1950s and 1960s (e.g., Singer, 1966), now somewhat neglected in the recent surge of interest (McMillan et al., 2013). Daydreaming can often stimulate creativity and problem-solving indirectly, through what is known as "incubation," in which ideas are developed while one is thinking of something else. Its effectiveness appears to be highest when people work on some undemanding task—higher than when they work on a demanding task or on no task at all (Baird et al., 2012). Doing nothing at all may result in the daydream descending into the chaos of a real dream, while engaging in a demanding task may inhibit the scope of our mental wandering. An undemanding task, such as washing the dishes, may be sufficient to maintain a sense of reality while allowing the mind to roam with freedom. The mathematician Poincaré had a moment of major inspiration while stepping onto a bus, and Archimedes reportedly had his "eureka" discovery when he noticed that the water level rose when he stepped into a bath.

People do seem to differ in their tendency to daydream—every schoolteacher knows and often despairs of the dreamy child in class. People who report that they daydream often do in fact daydream more than others do (Antrobus et al., 1967). People who imagined past or future episodes in greater detail also score higher on tests of creativity (Addis et al., 2016), suggesting that individual differences in creativity may depend on the capacity to provide rich descriptions of actual experience. Although we seem to be calibrated to daydream around 50% of the time on average, there is also natural variation—necessary perhaps for a balanced society that includes poets, painters, and novelists, along with accountants, engineers, and experimental physicists. Theoretical physicists may daydream more often.

## 6. CONCLUSIONS

It is over 40 years ago that Endel Tulving (1972) introduced the distinction between episodic and semantic memory. Although his views on episodic memory have changed somewhat since that time, his view of episodic memory as a reliving of the past was clear:

> A person's episodic memories are located in and refer to his own personal past. Most, if not all, episodic memory claims a person makes can be translated into

> the form: "I did such and such, at such and such a place, at such and such a time." Thus an integral part of the representation of a remembered experience in episodic memory is its reference to the rememberer's knowledge of his personal identity. (Tulving, 1972, 389)

Tulving also pointed out that although the term *episodic memory* was new, the concept of episodic memory was essentially the same as William James's concept of memory itself: "Memory requires more than the mere dating of the fact in the past. It must be dated in *my* past" (James, 1890, 650).

Since Tulving first wrote about episodic memory, the concept was expanded to include reference to future as well as to past events, and Tulving himself seems to have been the first to coin the phrase *mental time travel* (Tulving, 1985). Indeed, precedence must go to future thinking rather than mere recapitulation of the past, since it is future adaptation and future fitness that ensure survival beyond the present; Klein et al. (2010) describe memory itself as "an evolved system for planning future acts" (13). In this chapter, I suggest that our mental travels go even beyond the future, and into the minds of others, and into imaginary worlds that serve to stretch our understanding and powers of adaptation, and to encourage creativity. But perhaps most important of all is our capacity, perhaps unique to our species, to share our mental wanderings, through the invention of language. Through language, the experiences of others can be added to our own, vastly increasing our mental resources for planning and adaptation to the social and physical worlds, along with adventures into the world of imagination.

Let's leave the last word, then, to William Shakespeare:

> The poet's eye, in a fine frenzy rolling,
> Doth glance from heaven to earth, from earth to heaven,
> And as imagination bodies forth
> The forms of things unknown, the poet's pen
> Turns them to shape, and gives to airy nothing
> A local habitation and a name.
>
> —*A Midsummer Night's Dream*, Act V, Scene 3

## NOTE

1. To Chomsky, language is primarily a symbolic mode of thought, not a means of communication. Chomsky calls it *internal language*, or I-language. Some use the term *external language*, or E-language, to refer to the communicative aspect.

## REFERENCES

Addis, D. R., Pan, L., Musicaro, R. & Schacter, D. L. (2016). Divergent thinking and constructing episodic simulations. *Memory, 24*, 84–97.

Addis, D. R., Wong, A. T. & Schacter, D. L. (2007). Remembering the past and imagining the future: common and distinct neural substrates during event construction and elaboration. *Neuropsychologia, 45*, 1363–1377.

Antrobus, J. S., Coleman, R., and Singer, J. L. (1967). Signal-detection performance by subjects differing in predisposition today dreaming. *Journal of Consulting Psychology, 31*, 487–491.

Baird, B., Smallwood, J., Mrazek, M. D., Kam, J. W. Y., Franklin, M. S., & Schooler, J. W. (2012). Inspired by distraction: mind wandering facilitates creative incubation. *Psychological Science, 23*, 1117–1122.

Bickerton, D. (2010). On two incompatible theories of language evolution. In R. K. Larson, V. Déprez, & H. Yamakido (eds.), *The evolution of human language* (pp. 199–210). Cambridge, UK: Cambridge University Press.

Byrne, R. W., & Whiten, A. (1988). *Machiavellian intelligence*. Oxford: Oxford University Press.

Buckner, R. L., Andrews-Hanna, J. R., & Schacter, D. L. (2008). The brain's default network: anatomy, function, and relevance to disease. *Annals of the New York Academy of Sciences, 1124*, 1–38.

Call, J., & Tomasello, M. (2008). Does the chimpanzee have a theory of mind? 30 years later. *Trends in Cognitive Sciences, 12*, 187–192.

Campbell, D. T. (1960). Blind variation and selective retention in creative thought as in other knowledge processes. *Psychological Review, 67*, 380–400.

Chomsky, N. (2007). Biolinguistic explorations: design, development, evolution. *International Journal of Philosophical Studies, 15*, 1–21.

Corballis, M. C. (2011). *The recursive mind: the origins of language, thought, and civilization*. Princeton, NJ: Princeton University Press.

Corballis, M. C. (2013). Mental time travel: the case for evolutionary continuity. *Trends in Cognitive Sciences, 17*, 5–6.

Corkin, S. *Permanent present tense: the man with no memory, and what he taught the world*. London: Allen Lane, 2013.

Dragoi, G., & Tonegawa, S. (2013). Selection of preconfigured cell assemblies for representation of novel spatial experiences. *Philosophical Transactions of the Royal Society of London B Biological Sciences, 369*, 20120522. http://dx.doi.org/10.1098/rstb.2012.0522.

Duval, C., Bejanin, A., Piolino, P., Laisney, M., de La Sayette, V., Belliard, S., Eustache, F. & Desgranges, B. (2012). Theory of mind impairments in patients with semantic dementia. *Brain, 135*, 228–241. doi: 10.1093/brain/awr309.

Foster, D. J., & Wilson, M. A. (2006). Reverse replay of behavioural sequences in hippocampal place cells during the awake state. *Nature, 440*, 680–683.

Freud, S. (1900). *The interpretation of dreams*. New York: Macmillan.

Frith, U., & Frith, C. D. (2003). Development and neurophysiology of mentalizing. *Philosophical Transactions of the Royal Society of London B, 358*, 459–473. doi: 10.1098/rstb.2002.1218

Gaesser, B., Spreng, N., McLelland, V. C., Addis, D. R., & Schacter, D. L. (2014). Imagining the future: evidence for a hippocampal contribution to constructive processing. *Hippocampus, 23*, 1150–1161.

Gärdenfors, P., & Osvath, M. (2010). Prospection as a cognitive precursor to symbolic communication. In R. K. Larson, V. Déprez, & H. Yamakido (eds.), *The evolution of human language* (pp. 103–114). Cambridge, UK: Cambridge University Press.

Grice, H. P. (1975). Logic and conversation. In P. Cole & J. Morgan (eds.), *Syntax and semantics*, Vol. 3, *Speech acts* (pp. 43–58). New York: Academic Press.

Gupta, A. S., van der Meer, M. A. A., Touretzky, D. S., & Redich, A. D. (2010). Hippocampal replay is not a simple function of experience. *Neuron, 65*, 695–705.

Humphrey, N. (1976). The social function of intellect. In P. P. G. Bateson and R. A. Hinde (eds.), *Growing points in ethology* (pp. 303–317). Cambridge, UK: Cambridge University Press.

James, W. (1890). *Principles of psychology.* New York: Henry Holt.

Karlsson, M. P., & Frank, L. M. (2009). Awake replay of remote experiences in the hippocampus. *Nature Neuroscience, 12*, 913–918.

Killingsworth, M. A., & Gilbert, D. T. (2010). A wandering mind is an unhappy mind. *Science, 330*, 932.

Klein, S. B. (2013). The complex act of projecting oneself into the future. *WIREs Cognitive Science, 4*, 63–79.

Klein, S. B., Robertson, T. E., & Delton, A. W. (2010). Facing the future: memory as an evolved system for planning future acts. *Memory and Cognition, 38*, 13–22.

Kundera, M. (2002). *Ignorance.* Translated by L. Asher. New York: HarperCollins.

Locke, J. L., & Bogin, B. (2006). Language and life history: a new perspective on the development and evolution of human language. *Behavioral & Brain Sciences, 29*, 259–325.

Loftus, E., & Ketcham, K. (1994). *The myth of repressed memory: false memories and allegations of sexual abuse.* New York: St. Martin's Press.

Luria, A. R. (1968). *The mind of a mnemonist.* New York: Basic Books.

Machiavelli, N. (1961). *The Prince.* London: Penguin. (Translated from the 1532 Italian original by G. Bull).

Macphail, E. M. (2002). The role of the avian hippocampus in spatial memory. *Psicologica, 23*, 93–108.

McMillan, R. L., Kaufman, S. B., & Singer, J. L. (2013). Ode to positive constructive daydreaming. *Frontiers in Psychology, 4*, article 626.

Martin, V. C., Schacter, D. L., Corballis, M. C. & Addis, D. R. (2011). A role for the hippocampus in encoding simulations of future events. *Proceedings of the National Academy of Sciences U S A, 108*, 13858–13863.

Mooneyham, B. W. & Schooler, J. W. (2013). The costs and benefits of mind-wandering: a review. *Canadian Journal of Experimental Psychology, 67*, 11–18.

Niles, J. D. (2010). *Homo narrans: the poetics and anthropology of oral literature.* Philadelphia: University of Pennsylvania Press.

O'Keefe, J., & Nadel, L. (1978). *The hippocampus as a cognitive map.* Oxford: Clarendon Press.

Penn, D. C., Holyoak, K. J., & Povinelli, D. J. (2008). Darwin's mistake: explaining the discontinuity between human and nonhuman minds. *Behavioral & Brain Sciences, 31*, 108–178.

Pfeiffer, B. E., & Foster, D. J. (2013). Hippocampal place-cell sequences depict future paths to remembered goals. *Nature, 497*, 74–80.

Premack, D., & Woodruff, G. (1978). Does the chimpanzee have a theory of mind? *Behavioral and Brain Sciences, 4*, 515–526.

Rabin, J. S., & Rosenbaum, R. S. (2012). Familiarity modulates the functional relationship between theory of mind and autobiographical memory. *NeuroImage, 62*, 520–529.

Raichle, M. E., MacLeod, A. M., Snyder, A. Z., Powers, W. J. Gusnard, D. A., & Shulman, G. L. (2001). A default mode of brain function. *Proceedings of the National Academy of Sciences U S A, 109*, 3979–3984.

Revonsuo, A. (2000). The reinterpretation of dreams: an evolutionary hypothesis of the function of dreaming. *Behavioral and Brain Sciences, 23*, 877–901.

Roediger, H. L., & McDermott, K. B. (1995). Creating false memories: remembering words not presented in lists. *Journal of Experimental Psychology: Learning Memory and Cognition*, *21*, 803–814.

Rosenbaum, R. S., Stuss, D. T., Levine, B. & Tulving, E. (2007). Theory of mind is independent of episodic memory. *Science*, *318*, 1257.

Schacter, D. L. (2012). Adaptive constructive processes and the future of memory. *American Psychologist*, *67*, 603–613. doi: 10.1037/a0029869.

Schacter, D. L., & Addis, D. R. (2007). The optimistic brain. *Nature Neuroscience*, *10*, 1345–1347. doi: 10.1038/nn1107-1345.

Sherry, D. F., & Schacter, D. L. (1987). The evolution of multiple memory systems. *Psychological Review*, *94*, 439–454.

Siegel, D. J. (2007). Mindfulness and neural integration: differentiation of distinct streams of awareness and the cultivation of well-being. *Social, Cognitive & Affective Neuroscience*, *2*, 259–263.

Singer, J. L. (1966). *Daydreaming: an introduction to the experimental study of inner experience*. New York: Random House.

Sperber, D., & Origgi, G. (2010). A pragmatic perspective on the evolution of language. In R. K. Larson, V. Déprez, & H. Yamakido (eds.), *The evolution of human language* (pp. 124–131). Cambridge, UK: Cambridge University Press.

Sperber, D., & Wilson, D. (2002). Pragmatics, modularity and mind-reading. *Mind and Language*, *17*, 3–23.

Spreng, R. N., & Grady, S. L. (2009). Patterns of brain activity supporting autobiographical memory, prospection, and theory of mind, and their relationship to the default mode network. *Journal of Cognitive Neuroscience*, *22*, 1112–1123.

Squire, L. R. (2004). Memory systems of the brain: a brief history and current perspective. *Neurobiology of Learning & Memory*, 82, 171–177.

Suddendorf, T., & Corballis, M. C. (1997). Mental time travel and the evolution of the human mind. *Genetic, Social, and General Psychology Monographs*, *123*, 133–167.

Suddendorf, T., & Corballis, M. C. (2007). The evolution of foresight: what is mental time travel, and is it unique to humans? *Behavioral and Brain Sciences*, *30*, 299–351.

Sugiyama, M. S. (2011). The forager oral tradition and the evolution of prolonged juvenility. *Frontiers in Psychology*, *2*, Article 133.

Tooby, J., & DeVore, I. (1987). The reconstruction of hominid evolution through strategic modeling. In W. G. Kinzey (ed.), *The evolution of human behavior: primate models* (pp. 183–237). Albany: State University of New York Press.

Tulving, E. (1972). Episodic and semantic memory. In E. Tulving and W. Donaldson (eds.), *Organization of memory* (pp. 381–403). New York: Academic Press.

Tulving, E. (1985). Memory and consciousness. *Canadian Psychology*, *26*, 1–12.

Tulving, E. (2002). Episodic memory: from mind to brain. *Annual Review of Psychology*, *53*, 1–25.

Wearing, D. *Forever today*. New York: Doubleday, 2005.

Wilson, M. A., & McNaughton, B. L. (1994). Reactivation of hippocampal ensemble memories during sleep. *Science*, *265*, 676–679.

Wood, B., & Collard, M. (1999). The human genus. *Science*, *284*, 65–71.

# PART VII

# Developmental Perspectives

17

# Shaping One's Future Self

## *The Development of Deliberate Practice*

THOMAS SUDDENDORF, MELISSA BRINUMS, AND KANA IMUTA ■

What we hope ever to do with ease, we must learn first to do with diligence.

—*Samuel Johnson*

A key adaptive advantage of future-oriented mental time travel is that it allows humans not only to shape and control the world of the future, but also to intentionally change their future selves. People can set out to acquire skills and knowledge, and so deliberately alter their future being—which in turn influences their chances of future survival and reproduction. One powerful way in which one can acquire capacities is through practice. Humans regularly spend thousands of hours practicing skills that they (or others) believe to be worthwhile improving, and this is essential to any explanation of the extraordinary diversity of expertise that characterizes human societies (Suddendorf, 2013a). Yet, surprisingly little research has examined the nature and development of deliberate practice.

Deliberate practice can be regarded as repeated actions driven by the goal to improve one's future capacities and, so defined, it presupposes the ability to think about oneself in the future. We can imagine future situations in which it would be beneficial for our future self to have certain skills (or detrimental if we did not have them), and so we plan and pursue courses of action that we hope will let us acquire and improve specific capacities to a desired level. This is typically effortful, involving prolonged repetitions of certain actions. It often benefits from careful monitoring of progress and selection of what precise aspect of the skill ought to be trained next, and tends to progress from a focus on copying a model to a focus on outcome (e.g., Ericsson, 2008; Rossano, 2003). Deliberate practice allows us to learn relatively simple skills, such as how to write a letter of the alphabet, and with dedication, enables us to master extremely complex skills such as the exquisite playing of a musical instrument.

Not all skills, however, require deliberate practice. Certain skills mature in a quasi-universal manner. Infants learn to walk and talk without having to deliberately set out with such a goal in mind. Some capacities are learned incidentally. For instance, children learn to calibrate complex movements in the wake of playing throwing games, but are engaging in the repetitive actions because they are enjoying them in the present moment, rather than because they are aiming to improve hand-eye coordination. Of course, there can be many reasons that one might engage in repetitive actions that are beneficial for skill acquisition, and these need not be mutually exclusive. We can deliberately engage in, say, regular games of football because we enjoy the activity, because it keeps us fit, as well as because we want to get better at the game.

Deliberate practice does not need to rely on external triggers, but can be autocued (e.g., Donald, 1999). Indeed, it has been argued to depend on conscious, voluntary control (Rossano, 2003). We can choose to spend time honing any number of skills, and individuals differ vastly in what they decide to pursue and how much time and effort they are willing to dedicate. Nonetheless, much practice behavior is instigated or guided by others. Parents and teachers employ diverse techniques to get children to practice and improve, whether or not they want to improve themselves (more on how teaching and practice are intertwined further below). The focus of this chapter, however, is on our powerful ability to self-initiate practice with the goal of improving our future skills in mind. Furthermore, although deliberate practice may be helpful in improving a diversity of capacities (e.g., memorizing information through verbal rehearsal), here we will focus on the acquisition of motor skills.

## 1. LEARNING MOTOR SKILLS

Repeated practice can lead to neuronal changes that allow individuals to accomplish motor tasks in a faster and more accurate manner. Motor skill learning is a multilayered process that appears to be hierarchically organized, comprising action selection and action execution processes (e.g., Diedrichsen & Kornysheva, 2015). Action selection refers to the task of comparing alternative response options and selecting the most appropriate one based on expected rewards, costs, and goals. Once initiated, the action execution processes control the coordinated muscle activity in a replicable spatiotemporal pattern. Learning can take place at the execution level, at the selection level, or at intermediate levels. A novel action sequence requires demanding and time-consuming action selection and execution processes that, with repeated practice, can become increasingly fast and automatized. Elementary movements can be bound into chunks and executed faster and more accurately than when individually triggered by action selection processes (Rosenbaum, 1983). Hierarchical encoding allows for great flexibility as it enables individuals to combine and recombine motor primitives to generate novel action sequences.

Research on the role of practice in motor learning has identified various factors that play a role, such as the type of practice (e.g., Jeannerod, 1994), whether practice is massed or distributed (e.g., Lee & Genovese, 1989), sleep between practice sessions (e.g., Walker, 2009; Wilhelm, Metzkow-Mészàros, Knapp, & Born, 2012), and the timing, type, and frequency of feedback involved (e.g., Schmidt, 1991; Sidaway,

Bates, Occhiogrosso, Schlagenhaufer, & Wilkes, 2012). There is some evidence, for instance, that self-initiated feedback is more effective than feedback controlled by others (Kaefer, Chiviacowsky, Meira Jr., & Tani, 2014; Post, Fairbrother, & Barros, 2011). Though it may be questioned whether specific findings from research on simple motor skills can be generalized to more complex skill learning (Wulf & Shea, 2002), feedback certainly allows one in general to monitor progress and adjust practice accordingly, such that one may focus on those components of a sequence that are not yet acquired to one's satisfaction. Football players, for instance, may benefit from spending more time practicing kicks with their weaker foot than from spending equal time practicing with each foot (Coughlan, Williams, McRobert, & Ford, 2014).

Once elements are accomplished, they can intentionally be assembled into novel sequences. Though motor skills are generally regarded as part of procedural control systems, deliberately engaging in such practice behavior demonstrates that declarative, even episodic, processes can direct and shape its development. Although debates about the relationship between foresight and episodic and semantic processes have long featured prominently (e.g., Martin-Ordas, Atance, & Caza, 2014), the effect that foresight can have on procedural processes has been largely overlooked in the mental time travel literature. Yet, only through imaging a future self with improved skills may we be able to motivate, plan, and execute the honing of skills through deliberate practice.

Indeed, we can even choose to practice through deliberate mental simulation alone, rather than repeating actions physically, and our motor skills can improve as a result (e.g., Cahn, 2008; Feltz & Landers, 1983; Zimmermann-Schlatter, Schuster, Puhan, Siekierka, & Steurer, 2008). Without immediate reinforcers, practicing, whether mentally or physically, is effortful and costly (at least in terms of opportunity). It therefore requires motivation derived from a commitment to the cause, even in the face of more immediately rewarding temptations. Indeed, it has been argued that to achieve outstanding expertise—consistently performing at superior levels in a domain—one has to engage in years of deliberate effort aimed at improving performance (e.g., Ericsson, 2008).

## 2. MASTERY OF COMPLEX SKILLS

Unsurprisingly, there has been considerable research on how people acquire expert performance levels in complex skills associated with desirable arts, crafts, and professions. Much of this research has suggested that deliberate practice is the most important factor in determining expertise. Ericsson and colleagues (1993) championed this idea and proposed that individual differences in performance across many different domains can largely be accounted for by differential amounts of past and present practice. They define deliberate practice much more narrowly than we do here, limiting it to highly structured activities that were designed for the express purpose of improving performance through feedback, and that are demanding and not inherently enjoyable.[1] Exceptional skills are proposed to be the result of very long periods of deliberate practice, and the amount of time spent on this is suggested to be strongly and monotonically related to performance. In support of these assertions, they report, for instance, that highly skilled musicians practiced for longer periods of time and had accumulated more practice hours over their lifetime than musicians of lesser skill. Expert violinists and pianists had accumulated on average over 10,000 hours of deliberate practice.

With large amounts of practice, motor skills develop gradually from sensory-driven responses to integrated patterns of behavior relying on anticipation. This is associated with changes in the structural and functional organization of cortical motor areas (Dayan & Cohen, 2011). For instance, the volume of M1 and of premotor areas is larger in professional musicians compared to amateurs or non-musicians (Gaser & Schlaug, 2003) and the relevant sensory and motor representations of performing body parts are enlarged in professional musicians (Elbert, Pantev, Wienbruch, Rockstroh, & Taub, 1995).

While Ericsson and colleagues (1993) concede that other factors, such as parental support and the age a person begins practicing, are related to expertise, they assert that the role of these factors is fully mediated by deliberate practice. This view has garnered considerable support (e.g., Ericsson, 2008), and has been popularized in many books (e.g., Colvin, 2008; Coyle, 2009; Levitin, 2007; Shenk, 2010). Part of its popular appeal, no doubt, derives from the implication that anyone can become brilliant at whatever they would like, as long as they put in the requisite hard work.

Alas, as many critics have pointed out, this is simply not the case (e.g., Sternberg, 1996). Many people fail to reach expert levels in spite of copious amounts of practice, and comparable levels of elite expertise can be achieved with different amounts of effort (see Baker, Côté, & Deakin, 2005; Campitelli & Gobet, 2011; Hambrick et al., 2014; Hodges, Kerr, Starkes, Weir, & Nananidou, 2004; Macnamara, Hambrick, & Oswald, 2014). Studies suggest that, although deliberate practice is necessary, it is not sufficient to explain individual differences in performance. Additional factors, such as starting age (Gobet & Campitelli, 2007), working memory capacity (Meinz & Hambrick, 2010), and inheritance (Hambrick & Tucker-Drob, 2015), also contribute to the development of exceptional expertise. Hambrick and Tucker-Drob (2015), for instance, found genetic effects on musical accomplishment in 800 pairs of twins, with a quarter of this explained by genetic effects on music practice. A recent meta-analysis on studies on chess and music found that about one-third of the reliable variance in performance in both of these areas of expertise could be explained by deliberate practice (Hambrick et al., 2014).

Ericsson (2014) retorted by arguing that data on large samples of beginners may not be extrapolated to samples of elite performers. However, there are also data on top performers that question his claim about the supremacy of deliberate practice. For instance, Hornig, Aust, and Güllich (2014) found that the most successful professional football (soccer) players did not engage in more organized training practice, but in more non-organized football in childhood and more engagement in other sports in adolescence than other footballers, suggesting that incidental rather than deliberate practice was critical to their outstanding success. But be that as it may, here we are not so much interested in how individuals achieve the most outstanding performance levels in one specific field, but rather in how virtually all humans achieve some success in a variety of areas through deliberate practice. The bottom line is that what these studies have in common is that deliberate practice is an important factor in honing one's skills.

## 3. THE NATURE OF DELIBERATE PRACTICE

We want to examine the most fundamental capacity that allows humans to attempt to become an expert, whether they succeed or not. People regularly employ deliberate practice to acquire any number of mundane skills, from tying one's shoelaces to learning a particular dance, and this allows humans to adapt their skills to their individual desires and anticipated demands of future situations. It enables people to acquire any number of different skills and gives human groups supreme diversity in skill patterns.

Other animals also learn skills through practice. Mammalian play typically involves repetitive actions that resemble real fight, hunt, or flight patterns, and as such improve critical survival skills—natural selection favoring those who acquired better skills when the serious struggle for existence arrives. However, Donald (1999) argued that non-human animals do not spontaneously rehearse and refine their movement patterns. We know of no obvious evidence that other animals deliberately set out to improve particular future skills. Nor do we know of evidence for any of the consequences of having such a capacity. For instance, there does not appear to be a great diversity of skills in groups of the same species that could be attributed to differential deliberate practice histories. But we need to be cautious here, as this perception may reflect absence of evidence, rather than evidence of absence. We cannot rule out that a young cat really sets out to become a better hunter when it repeatedly "play-pounces" on a fluff ball. And if one does not demand that deliberate practice requires the goal of future skill improvement, then various cases for animal "deliberate" practice may be made (Helton, 2005, 2008).

However, given that, according to our definition, thinking about future skill improvement is essential, a capacity for mental time travel is a prerequisite for deliberate practice. Without such a capacity, organisms may not be mentally confronted with the problems of future skills. The claim that mental time travel into the future is a distinctly human capacity (Suddendorf & Corballis, 1997, 2007) continues to have some currency, even though it has been challenged by a range of studies over the last two decades. Purported evidence comes from studies in which animals appear to take future hunger or thirst into account, for instance, by caching food (Correia, Dickinson, & Clayton, 2007; Naqshbandi & Roberts, 2006; Raby, Alexis, Dickinson, & Clayton, 2007), and from studies in which apes secure tools for future use (Mulcahy & Call, 2006; Osvath, 2009; Osvath & Osvath, 2008). However, these studies have been criticized, and alternative explanations have been put forward (e.g., Suddendorf, 2006; Suddendorf & Corballis, 2008). The most recent challenge comes from work on hippocampal place cells in rats, whose activity suggest that the animals trace paths in a maze even when they are not in it (e.g., Gupta, van der Meer, Touretzky, & Redish, 2010), leading one of the original proponents of the uniqueness claim to change his mind (Corballis, 2013; but also see Suddendorf, 2013b).

Even if one takes this available evidence at face value, and we are not able to review it here, none of the limited number of cases points to flexible, long-term foresight in non-human animals. Consider, for instance, the capacity to delay gratification (Mischel & Metzner, 1962). When given the choice between a small reward now and a larger reward later, our closest animal relatives, the great apes, can wait much longer than most other animals—but they still only wait for a few minutes

(Beran, 2002; Dufour, Pelé, Sterck, & Thierry, 2007; Rosati, Stevens, Hare, & Hauser, 2007). Non-human animal foresight may be severely limited for a number of reasons (Suddendorf & Corballis, 2007). But putting skeptical critiques to one side, the evidence suggests that apes may consider events hours or, at the very best, a day in advance (e.g., Janmaat, Polansky, Ban, & Boesch, 2014; Mulcahy & Call, 2006). Deliberate practice for future skill acquisition, however, is by its very nature a long-term endeavor that often only leads to expertise days, weeks, or even years later.

Without being able to foresee the remote future and the desirability of relevant future skills, one may not be able to derive and follow a long-term plan toward changing one's capacities through deliberate practice. Immediate rewards may lead to behavior that improves skills, and human trainers can deliberately use standard conditioning techniques to shape diverse expert animal performances, such as turning dogs into herders, guards, hunters, helpers, retrievers, or drug detectors. However, in the absence of such immediate reinforcement, non-human animals may not autocue repeated actions to improve their skills. Humans, on the other hand, willingly forego present rewards in order to follow dreams of future accomplishment. As we saw, many spend thousands of hours honing their skills. This begs the question of when this peculiar trait became part of our ancestors' repertoire.

Though we cannot directly examine deliberate practice in ancient hominins, there are some clues in the archaeological record that are suggestive. It has been argued that the earliest hard evidence for mental time travel into the future comes from *Homo erectus* some 1.8 million years ago: Acheulean tools (e.g., Suddendorf & Corballis, 2007). This tool set includes symmetrical tools such as bifacial hand-axes and cleavers. Production of these tools requires considerable knowledge about rocks and skill at working them. Hand-axes are handy tear-drop-shaped tools that have many potential uses, and represent perhaps the most enduringly popular tools ever made: they were manufactured for over a million years. At some sites, such as the site of Olorgesailie in Kenya, the ground is still littered with them. This indicates that *Homo erectus* may have been practicing manufacture because they were making new tools even though plenty of usable tools were around them (Rossano, 2003). Acheulean tools are complex to produce but were made in a uniform manner, suggesting that they were made according to a deliberate plan. Furthermore, in addition to their manufacture, tool use indicates foresight because there is evidence that *Homo erectus* carried these tools to put them to use repeatedly (e.g., Hallos, 2005). So *Homo erectus* was likely making them with the future in mind and spent considerable amounts of time learning how to make them properly. We therefore suggest that they probably had the capacity to practice deliberately (cf., Rossano, 2003). Indeed, it is difficult to imagine that they could have become proficient at making these tools in any other way.

This is not to say that each individual learned how to make bifaces independently. In fact, the uniformity of the tools strongly suggests that the method was passed on from one generation to the next through social learning. Imitation and teaching are usually hailed as the two pillars of cultural transmission, our second inheritance system, allowing information to be passed on, with either the intent of the knowledgeable or the ignorant (e.g., Whiten, 2005). A recent study suggests that already the making of earlier Oldowan tools may have been fostered by instruction (Morgan et al., 2015; though also see Balter 2015). But even when one has observed how the tool is made, or has been instructed on how to do it, ultimately one needs to practice the behavior repeatedly to train the motor skills at the selection and execution levels.

This need not mean that *Homo erectus* could already draw on all the hallmarks of modern human culture. One peculiar aspect of the Acheulean tool tradition is that it changed so little for over a million years. There is no evidence of what has been called the ratchet effect (e.g., Tomasello, 1999), through which individuals built upon the inventions of others over time. This effect has long been considered central to distinguishing modern human cumulative culture from non-human social traditions (Dennett, 1995; Sterelny, 2003; Tomasello, 1999). By retaining previous inventions with high accuracy, new generations can draw on the accumulated wisdom of those who went before. Consequently, a lot of recent research has focused on the importance of "over-imitation" (e.g., Horner & Whiten, 2005; Nielsen, 2006; Whiten, McGuigan, Marshall-Pescini, & Hopper, 2009) and the role of teaching (e.g., Hoppitt et al., 2008).

Yet, the transfer of information is only adaptive when someone has something useful to transfer, and inventions worth retaining are not just single ideas or designs that individuals stumbled upon. To make an invention work, one typically has to master novel skills. For instance, the invention of weapons that can hurt at a distance are only useful insofar as one can become proficient enough to use them effectively (many involve the acquisition of complex skills, as one can easily demonstrate to oneself by practicing them—you may have tried a bow and arrow, for instance). Through deliberate practice, these skills can be honed; only when mastered to a certain level are the tools beneficial. A spear-thrower is only better than a thrown spear if one can acquire the motor skills to use this powerful tool effectively. And by the same token, bow and arrow are only advantageous over a spear-thrower when one has sufficiently mastered this weapon. Incidentally, chimpanzees may make spears and thrust them into a tree hollow to kill a bushbaby (Pruetz & Bertolani, 2007), but we know of no observation that they have practiced thrusting, let alone aimed throwing, with this tool. Without a capacity for deliberate practice, they could not benefit from the invention of a spear thrower.

Deliberate practice is closely linked to transmission mechanisms. Learning from others, whether via imitation or teaching, often involves deliberate practice. Observation and practice can powerfully combine in motor skill learning (e.g., Wulf, Shea, & Lewthwaite, 2010). And in a sense, deliberate practice is teaching oneself rather than others. Teaching oneself and observing progress in self and others, in turn, can provide insights into how best to teach others. Indeed, as we noted earlier, teaching others often involves guided practice and instructions on how to go about deliberate practice. Teaching and deliberate practice are intricately linked and arguably represent two sides of the same coin (Premack, 2007). Both are flexible tools that we employ to shape our own and others' behavior in deliberate ways.

Acquisition and transmission of skills are essential to human cultural learning and appear only to exist as functional analogues in other animals. Some species, including meerkats and ants, can be said to engage in functional teaching in the sense that they will modify their behavior in a way that aids another individual to acquire knowledge or skills without obtaining an immediate benefit themselves (Caro & Hauser, 1992; Hoppitt et al., 2008). There is no sign, however, that teaching in non-human animals is a deliberate, flexible act, in which the teacher plans the acquisition of future knowledge (Premack, 2007). Likewise, cases may be made for functional or incidental practice in some non-human animals, but as yet, there appears to be no sign of deliberate practice (Suddendorf, 2013a).

With the emergence of deliberate practice and teaching, our ancestors became increasingly capable of *flexibly specializing*. That is, instead of being a *generalist* or a *specialist* species adapted to specific environmental challenges, humans have the generalist capacity to rapidly (i.e., within a generation) specialize to changing demands. We can train ourselves to become diverse experts, specializing in what is needed at present and future points in time. This makes us extremely adaptable (and, of course, we have also adapted our environment increasingly to our future needs). Cooperative groups that had the flexibility to create complementary skills as required would no doubt have out-competed others. Given how much time is involved in becoming a master in certain skills, it is clear that individuals cannot achieve elite levels in all, or even in very many, domains. However, each individual does seem to have the flexible capacity for deliberate practice and, within the constraint of his or her natural talent, can therefore become competent in a selection of a great variety of skills. In this way, humans could flexibly open up new niches. More research is needed on how flexible specialization evolved, and how cultural groups increasingly used cooperative teaching and deliberate practice to establish a successful mix of expertise.

What is clear, though, is that deliberate practice is important for the extensive variety of human skills we see today. Without it, we would not have the technology, civilization, and colorful diversity that we take for granted—and that we continue to build on for the future (Suddendorf, 2013a). It is therefore important, we argue, that more research is devoted to understanding this corollary of mental time travel into the future. For a comprehensive understanding of deliberate practice, we need to find out much more about its evolution, function, mechanisms, and development (cf. Tinbergen, 1963). When exactly did it first evolve, and in what contexts (just because stone tools survive the test of time does not mean the capacity emerged in that context)? What were its original functions? What are the neurological, cognitive, and social mechanisms of deliberate practice? And how does the capacity develop in children? Here, for the remainder of this chapter, we will focus on this last question.

## 4. THE DEVELOPMENT OF DELIBERATE PRACTICE

When do children begin to engage in repeated action with the goal of improving their future skills? As already noted, to conceive of the goal of improving one's future performance levels, one has to be able to think about future situations in the first place. It is hence appropriate to first consider what is currently known about the development of mental time travel into the future. Children also need to have some appreciation of the link between practice and skill acquisition to pursue this goal. Further factors may play important roles in determining whether a child actually engages in deliberate practice, such as executive capacities necessary to maintain practice behaviors in the face of more immediate urges and temptations. We will discuss the development of these capacities before describing a first study directly aimed at examining deliberate practice in children.

The development of mental time travel into the future, or episodic foresight, has been of interest to researchers for almost 20 years (Atance & O'Neill, 2001; Suddendorf & Corballis, 1997), but the last few years have witnessed a considerable increase in studies (e.g., McCormack & Atance, 2011; Suddendorf & Moore, 2011). The simplest way of examining children's thinking about the future is to document their future-directed vocabulary (e.g., Busby Grant & Suddendorf, 2010; Harner, 1975), and to directly ask them (e.g., Friedman, 2005; Hudson, Shapiro, & Sosa, 1995). When questioned about events that will happen tomorrow or about plans for hypothetical future events, 3-year-olds are already capable of reporting some basic information, but older pre-schoolers provide significantly more events and details (Busby Grant & Suddendorf, 2005; Hayne, Gross, McNamee, Fitzgibbon, & Tustin, 2011; Quon & Atance, 2010; Suddendorf, 2010). Verbal reports may be misleading, however, because children may understand without having the means of expression, and they may use words appropriately without properly understanding their meaning (e.g., Lyon & Flavell, 1994). Thus, paradigms focusing on future-oriented behavior have been called for (Suddendorf & Busby, 2005).

Fortunately, the results of behavioral tasks are largely in line with the verbal evidence (see Suddendorf & Redshaw, 2013, for a review). One recent study, for instance, carefully followed strict criteria to rule out alternative explanations for children's behavior and found that 4-year-olds, though not 3-year-olds, were able to remember a novel past problem sufficiently enough to prepare for its future solution by securing an appropriate tool at a different time and place (Suddendorf, Nielsen, & von Gehlen, 2011; see also Atance, Louw, & Clayton, 2015; Payne, Taylor, Hayne, & Scarf, 2014; Redshaw & Suddendorf, 2013; Scarf, Gross, Colombo, & Hayne, 2013). In these tasks, children were presented with problems in one room, and later were given the opportunity in another room to obtain tools for the potential future solution upon return to the first room. As we will see, one can adapt this basic paradigm to let children select a behavior to practice and so give them an opportunity to carry potentially improved skills (rather than tools) back to the first room.

Although a major shift occurs during the preschool years, children's capacity to reason about the future and prudently guide their behavior continues to develop throughout childhood (Lagattuta, 2014; Suddendorf & Redshaw, 2013). One reason for this is that future-directed mental time travel is not an isolated module but depends on a range of components that tend to develop at quite diverse rates. Working memory capacity, for instance, an essential component for staging mental simulations (Suddendorf & Corballis, 2007), increases linearly between the ages of 4 and 11 (Alloway, Gathercole, & Pickering, 2006). Whereas other components, such as recursive capacities, which are, for instance, necessary for embedding elements of scenarios into larger narratives (Suddendorf & Corballis, 2007), emerge much more rapidly around the end of the preschool years (e.g., Corballis, 2014; Wells, 1985). Together, the developmental pattern of the various components enables distinct capacities and limits for children's episodic foresight (Suddendorf & Redshaw, 2013). Next we will briefly review the development of some of those purported components that may influence the capacity for deliberate practice, not only via their effects on mental time travel, but also more directly.

Episodic memory is intimately linked in mind and brain to episodic foresight (e.g., Dudai & Carruthers, 2005; Klein, 2013; Schacter & Addis, 2007; Suddendorf & Corballis, 1997; Szpunar, Watson, & McDermott, 2007). There are commonalities also in their development, such that there are links, for instance, between children's capacity to report events from yesterday and tomorrow (e.g., Busby Grant & Suddendorf, 2005). Episodic memory provides a database from which children can construct future events by combining and recombining basic elements (Schacter & Addis, 2007; Suddendorf & Corballis, 2007). This is also critical for deliberate practice, in that a child who remembers a past situation in which performance was poor may be able to imagine a future scenario in which it is better, motivating skill acquisition through deliberate practice. By around 5 years of age, children have a robust understanding and expectation that a negative experience from the past is likely to reoccur (Lagattuta, 2007). Remembering what worked and what did not work in previous attempts is also critical to monitoring and fine-tuning effective practice. The development of episodic memory may be traced to the earliest declarative memories of toddlers, as evident in deferred imitation tasks (e.g., Barr, Dowden, & Hayne, 1996; Herbert, Gross, & Hayne, 2007; Meltzoff, 1988) and matures into more coherent accounts of past events over the preschool years (Bauer, 2007; Hayne et al., 2011; Levine, 2004; Suddendorf & Corballis, 1997).

To drive deliberate practice, one has to do more than re-experience past events, however. One must also imagine and compare alternative situations. Through entertaining thoughts on what could have happened instead, and with this, feeling regret for not having been better prepared in the past (e.g., Weisberg & Beck, 2010), one may deliberately practice for more desirable outcomes in the future. Similarly, by simulating what would happen if one did and if one did not have a particular skill at a future occasion, one can create motivation to practice. Counterfactual thinking may emerge around age 3 to 4 years (Harris, German, & Mills, 1996), though some researchers argue that it may emerge only as late as around age 6 (Rafetseder, Cristi-Vargas, & Perner, 2010; Rafetseder & Perner, 2010). The case for late development is in line with research that has found that children represent multiple potential outcomes of a single event only from around that age (Beck, Robinson, Carroll, & Apperly, 2006). To stage and compare alternative mental scenarios, one may need considerable working memory capacity, which, as we saw earlier, only expands gradually between age 4 and 11.

To drive deliberate practice, one may also need some level of theory of mind. After all, one needs to evaluate how one will feel in different future situations, and one needs some understanding of how feelings, skills, and knowledge can change. Theory of mind has long been linked to mental time travel (Moore & Lemmon, 2001; Suddendorf & Corballis, 1997; but see Hanson, Atance, & Paluck, 2014). Indeed, William Hazlitt already observed that "[t]he imagination . . . must carry me out of myself into the feelings of others by one and the same process by which I am thrown forward as it were into my future being, and interested in it" (1805, 1). The findings of extensive research in recent years suggest that theory of mind develops gradually and through specific stages over the preschool years (e.g., Peterson, Wellman, & Slaughter, 2012; Shahaeian, Peterson, Slaughter, & Wellman, 2011; Wellman & Liu, 2004). A critical step in this is generally considered to be the understanding of false beliefs around age 3 to 4 (e.g., Perner, 1991; Wellman, Cross, & Watson, 2001), as this is thought to demonstrate a representational understanding of mind. That is, at this stage, children can meta-represent how others (and oneself) represent the world (e.g., Perner, 1991). This allows children to reflect on the relation between representations and what they represent—including the fact that a representation may misrepresent the world (e.g., Wimmer & Perner, 1983). This reflection is essential for assessing one's own cognitive abilities and processes (Schneider & Lockl, 2008; Suddendorf, 2013a).

From around the time children understand false beliefs in others, they also attribute former false beliefs to self (Gopnik & Astington, 1988), they begin to appreciate how perception can lead to knowledge (e.g., Gopnik & Graf, 1988; Perner & Ruffman, 1995), and they recognize that they have not always known what in fact they have just learned today (e.g., Taylor, Esbensen, & Bennett, 1994). From this age onward, they can also distinguish between someone who acquired knowledge yesterday and someone who is going to acquire it tomorrow—such that only the former has access to the knowledge now (e.g., Busby Grant & Suddendorf, 2010). When one knows that one does not know, one can plot paths toward finding out what is required in the future. A first study of such deliberate information seeking found that 5-, but not 4-year-olds, appropriately sought information they needed to solve a future problem (Redshaw & Suddendorf, 2015). Such reflection on and pursuit of knowledge may well extend to skills. It stands to reason that, in order to monitor what one knows and can do, and what one still has to learn, one needs some common capacity for metacognition.

Through reflection, children may gain some measure of executive control about what future-directed actions to pursue. Such control is critical to allowing thought about the future to drive current action, such as initiating and sustaining deliberate practice over a prolonged period of time and in the face of competing drives. The term *executive function* usually refers to a range of processes that help to monitor and regulate thought and behavior and to guide goal-directed behavior (Beck, Schaefer, Pang, & Carlson, 2011; Carlson, Davis, & Leach, 2005; Carlson, Zelazo, & Faja, 2013; Suddendorf & Corballis, 2007). Even toddlers can demonstrate basic forms of inhibition, such as withholding an automatic response (e.g., Kochanska, Tjebkes, & Forman, 1998). However, young children often have difficulties disengaging from previously rewarding actions and so commit perseveration errors (e.g., Suddendorf, 2003). They gradually acquire capacities to coordinate inhibition and activation prudently. Four-year-olds, for instance, have some capacity—though it often deteriorates over trials—to follow a rule that runs against a pre-potent tendency, as in a Stroop task (Gerstadt, Hong, & Diamond, 1994) or in card sorting first by shape and then by color (Zelazo, Muller, & Frye, 2003). However, they continue to struggle with simple inhibition tasks, such as Simon Says (Carlson, 2005), and with tasks in which they have to self-direct control, such as when asked to sort cards in a new way (e.g., Jacques & Zelazo, 2001). Significant developments in this domain continue to be observed well into adolescence (Luna, Garver, Urban, Lazar, & Sweeney, 2004). One can distinguish three major transitions in the development of control: from perseveration to externally triggered responding; from reactive control to proactive control; and from environmental triggers to autocuing (Munakata, Snyder, & Chatham, 2012). To self-initiate deliberate practice, the final transition needs to be taken.

The fact that executive control continues to develop into late childhood and beyond means that children may often struggle in implementing actions aimed at imagined future goals, such as sustained practice in the absence of more immediate reinforcers. To diligently pursue deliberate practice, one must resist the temptation to follow more pleasant and more immediately rewarding courses of action. In wait-based delay of gratification tasks, for instance, children have to wait until an experimenter returns with a larger reward. Preschoolers gradually improve in the length of time they are willing and able to wait, with most 4-year-olds only able to resist temptation for a smaller immediate reward for about 5 minutes (e.g., Atance & Jackson, 2009; Mischel & Ebbesen, 1970). Better control in the preschool years is associated with various advantages later in life (e.g., Ayduk et al., 2000; Casey et al., 2012; Mischel, Shoda, & Peake, 1988; Schlam, Wilson, Shoda, Mischel, & Ayduk, 2013).

In addition to a basic capacity to imagine future events, one must be able to organize current and future actions with longer-term goals in mind, in order to engage in consistent deliberate practice that can shape future skill levels. Similar to children's performance on the wait-based delay of gratification task, children's propensity to make prudent choices to receive a larger reward later over a small reward now increases over the preschool years (e.g., Hongwanishkul, Happaney, Lee, & Zelazo, 2005; Imuta, Hayne, & Scarf, 2014; Lemmon & Moore, 2007; Moore & Lemmon, 2001; Prencipe & Zelazo, 2005). Through the development in ability to connect the behaviors of the present self with potential outcomes that would be experienced by the future self (recall the Hazlitt quote earlier in this chapter), children gain the self-control to forgo immediate gratification and make future-oriented decisions, such as engaging in deliberate practice.

Another critical capacity for autocuing deliberate practice is the formation of future-directed intentions to practice—and to remember such intentions. The latter is known as *prospective memory*, and research on its development points to significant changes over the later preschool years and beyond (Kliegel, Mackinlay, & Jäger, 2008; Kvavilashvili, Kyle, & Messer, 2012). Note though, that in virtually all prospective memory paradigms, the experimenter instructs the participants as to what they are expected to do (but see Lyons, Henry, Rendell, Corballis, & Suddendorf, 2014, for a new approach), and so we know next to nothing about when children begin to form such future-directed intentions themselves—which would be essential for self-initiating deliberate practice.

No doubt, most children are introduced to deliberate practice by adults, such as parents and teachers, who initially take advantage of incidental practice in play to facilitate specific skill development (Bloom, 1985). Many parents opt for specific skills being trained in a more dedicated fashion, enrolling their children in music, craft, or sport classes given by experienced teachers and coaches. Social pressures encourage children to keep practicing their skills outside formal classes to foster improvement. Many children learn the phrase "practice makes perfect" by heart. Immediate rewards (especially praise) are frequently deployed to reinforce incremental steps along the way. Children may or may not eventually adopt their parents' goals as their own and pursue subsequent skill development on their own accord.

By the age of 6 or 7, most children in countries with compulsory education begin formal schooling (UNISCO Institute of Statistics, 2014), in which practice is systematically encouraged by teachers in academic, physical, and practical education. For instance, children are encouraged to engage in verbal rehearsal in order to memorize information (Bebko & McKinnon, 1990; Bebko, McMorris, Metcalfe, Ricciuti, & Goldstein, 2014; Flavell, Beach, & Chinsky, 1966). This strategy is largely dependent on language proficiency (Bebko et al., 2014), and children with poorer language skills (e.g., deaf children) may not begin to rehearse information until years later (Bebko & McKinnon, 1990). In school, children learn the role that effort plays in achieving success (Gipps & Tunstall, 1998) and are taught strategies to help them regulate their own learning (e.g., Coffman, Ornstein, McCall, & Curran, 2008), but the many effects that formal teaching has on children's development are beyond the scope of this chapter. Suffice to say that deliberate practice is an essential part of any schooling.

## 5. A FIRST STUDY ON THE DEVELOPMENT OF DELIBERATE PRACTICE

The review of the development of related capacities suggests that the essential building blocks for deliberate practice are only in place around the time that formal schooling tends to start. This is probably no coincidence. Yet, there is a lack of formal research on the development of this critical capacity. We know of only one study that has directly aimed at examining this (Davis, Cullen, & Suddendorf, in press). The researchers conducted two experiments on deliberate practice. In the first, they adopted a version of the above-mentioned rooms-task (Suddendorf et al., 2011) to test 3-, 4-, and 5-year-olds' ability to selectively practice motor task games in one room in preparation for doing such a task in the other room later. In the first room, children were shown four slightly different versions of a motor game (e.g., they had to guide a loop around a wire without touching it), and were given a turn at each version to demonstrate that these were difficult tasks. One version of the game was singled out as a game they would play later to win a sticker reward. Children were then taken to the second room, where they could select one version of the game to play with (i.e., to practice on). They then returned to the first room and were given the opportunity to win stickers by playing the game that was singled out. Four- and 5-year-olds, but not 3-year-olds, selected to play the most relevant game in the second room in apparent preparation for a return to the first. It remained unclear, however, whether children made their selection because they wanted to practice for the future, or because this game was singled out earlier without them understanding the relationship between repetition and improved skills.

Experiment 2 therefore examined children's explicit understanding of deliberate practice, while also including one version of the paradigm pioneered in the first experiment. To assess their understanding, children were given a story-based task in which they had to rate which of two puppets was more likely to win a particular competition: one who tried the relevant behavior just once, or one who tried the behavior in question every day for a long period. Following two versions of this task, children were also asked directly what they should do if they wanted to get better at something. Whereas 4-year-olds did not perform above chance on the story tasks, 5-year-olds selected the character who had practiced as the one likely to win a competition over a character who had only done the critical actions once. Furthermore, of the children who selected the correct puppet across both age groups, 83% reported that practice or repetition was the reason that the character would win the competition. When asked directly what they should do to get better at something, two-thirds of 5-year-olds mentioned "practice" or "repetition," whereas merely 10% of 4-year-olds did so.

Perhaps most important, the performances on the tasks were associated. Those children who referred to practice and repetition were also more likely to pick the practicing character in the story tasks, as well as to pick the relevant motor game in their own practice task. This suggests that, by 5 years of age, many children have an explicit understanding of the link between deliberate practice and skill acquisition. They are able to select to practice a task relevant to a future situation, to predict skill change in others based on practice behaviors, and to recognize an association between practice or repetition of behavior and skill improvement. This study provides the first direct indication of when understanding of deliberate practice may emerge in young children, and the results are in line with the developmental data reviewed earlier. Nonetheless, given that this is a first study, we caution that it certainly requires replication before firm conclusions can be drawn.

## 6. CONCLUSIONS

Many basic questions remain to be answered. How does understanding of deliberate practice change over time? For instance, when do children appreciate that although practice is effective in some contexts, it has little use in others? Informal observations suggest that many children generalize the idea that practice makes you better even to domains where practice, in fact, makes no difference at all (such as in the ability to throw dice). How do the developmental changes in the related capacities we discussed earlier influence children's early understanding and use of deliberate practice? What are the long-term consequences for children when parents do, or do not, encourage intensive deliberate practice? We envisage that future work can unearth both theoretically and practically significant information about this essential human capacity.

In the literature, the adaptive significance of mental time travel into the future has been largely discussed in terms of our ability to flexibly anticipate and prepare (e.g., Suddendorf, 2006), and in terms of our capacity to shape the future world to our design (Dong, Collier-Baker, & Suddendorf, in press). Here we hope to have demonstrated that the human capacity to shape one's own future self through deliberate practice deserves considerably more attention than it has received. Thoughts about our future selves tend to trigger emotions and drive many of our actions. Considerable research has documented that we tend to exaggerate the intensity and duration of future feelings (e.g., Gilbert, 2006). One ultimate reason for such a bias may be that it helps us motivate future-directed actions in the face of more immediate urges (Miloyan & Suddendorf, 2015). Forgoing current pleasures in order to practice a skill that we anticipate will give us more pleasure at a future point in time may be a case in point. After all, with diligent practice, we can learn to do very complex things with apparent ease, as Samuel Johnson observed. Without it, we would have considerably fewer skills, less diversity, and less power to control our own destiny.

## NOTE

1. Others argue that deliberate practice further requires knowledge of the correct mechanisms involved (Colvin, 2008). We maintain that engaging in repeated actions with the goal of achieving a change in skill levels qualifies as deliberate practice, even if the mechanisms involved are not quite understood, or the activity is enjoyable.

## REFERENCES


Alloway, T. P., Gathercole, S. E., & Pickering, S. J. (2006). Verbal and visuospatial short-term and working memory in children: are they separable? *Child Development*, *77*(6), 1698–1716. doi: 10.1111/j.1467-8624.2006.00968.x.

Atance, C. M., & Jackson, L. K. (2009). The development and coherence of future-oriented behaviors during the preschool years. *Journal of Experimental Child Psychology*, *102*(4), 379–391. doi: 10.1016/j.jecp.2009.01.001.

Atance, C. M., Louw, A., & Clayton, N. S. (2015). Thinking ahead about where something is needed: new insights about episodic foresight in preschoolers. *Journal of Experimental Child Psychology*, *129*, 98–109. doi: 10.1016/j.jecp.2014.09.001.

Atance, C. M., & O'Neill, D. K. (2001). Episodic future thinking. *Trends in Cognitive Sciences*, *5*(12), 533–539. doi: 10.1016/S1364-6613(00)01804-0.

Ayduk, O., Mendoza-Denton, R., Mischel, W., Downey, G., Peake, P. K., & Rodriguez, M. (2000). Regulating the interpersonal self: strategic self-regulation for coping with rejection sensitivity. *Journal of Personality and Social Psychology*, *79*(5), 776–792. doi: 10.1037/0022-3514.79.5.776.

Baker, J., Côté, J., & Deakin, J. (2005). Expertise in ultra-endurance triathletes early sport involvement, training structure, and the theory of deliberate practice. *Journal of Applied Sport Psychology*, *17*(1), 64–78. doi: 10.1080/10413200590907577.

Balter, M. (2015). Humans language may have evolved to help our ancestors make tools. *Science, News 13 January*. doi: 10.1126/science.aaa6332.

Barr, R., Dowden, A., & Hayne, H. (1996). Developmental changes in deferred imitation by 6- to 24-month-old infants. *Infant Behavior and Development*, *19*(2), 159–170. doi: 10.1016/S0163-6383(96)90015-6.

Bauer, P. J. (2007). *Remembering the times of our lives: memory in infancy and beyond*. Mahwah, NJ: Lawrence Erlbaum Associates.

Bebko, J. M., & McKinnon, E. E. (1990). The language experience of deaf children: its relation to spontaneous rehearsal in a memory task. *Child Development*, *61*(6), 1744–1752. doi: 10.2307/1130835.

Bebko, J. M., McMorris, C. A., Metcalfe, A., Ricciuti, C., & Goldstein, G. (2014). Language proficiency and metacognition as predictors of spontaneous rehearsal in children. *Canadian Journal of Experimental Psychology*, *68*(1), 46–58. doi: 10.1037/cep0000013.

Beck, D. M., Schaefer, C., Pang, K., & Carlson, S. M. (2011). Executive function in preschool children: test-retest reliability. *Journal of Cognition and Development*, *12*(2), 169–193. doi: 10.1080/15248372.2011.563485.

Beck, S. R., Robinson, E. J., Carroll, D. J., & Apperly, I. A. (2006). Children's thinking about counterfactuals and future hypotheticals as possibilities. *Child Development*, *77*(2), 413–426. doi: 10.1111/j.1467-8624.2006.00879.x.

Beran, M. (2002). Maintenance of self-imposed delay of gratification by four chimpanzees (Pan troglodytes) and an orangutan (Pongo pygmaeus). *The Journal of General Psychology*, *129*(1), 49–66. doi: 10.1080/00221300209602032.

Bloom, B. S. (1985). Generalizations about talent development. In B. S. Bloom & L. A. Sosniak (eds.), *Developing talent in young people* (pp. 507–549). New York: Ballantine Books.

Busby Grant, J., & Suddendorf, T. (2005). Recalling yesterday and predicting tomorrow. *Cognitive Development*, *20*(3), 362–372. doi: 10.1016/j.cogdev.2005.05.002.

Busby Grant, J., & Suddendorf, T. (2010). Young children's ability to distinguish past and future changes in physical and mental states. *The British Journal of Developmental Psychology, 28*(4), 853–870. doi: 10.1348/026151009X482930.

Cahn, D. (2008). The effects of varying ratios of physical and mental practice, and task difficulty on performance of a tonal pattern. *Psychology of Music, 36*(2), 179–191. doi: 10.1177/0305735607085011.

Campitelli, G., & Gobet, F. (2011). Deliberate practice: necessary but not sufficient. *Current Directions in Psychological Science, 20*(5), 280–285. doi: 10.1177/0963721411421922.

Carlson, S. M. (2005). Developmentally sensitive measures of executive function in preschool children. *Developmental Neuropsychology, 28*(2), 595–616. doi: 10.1207/s15326942dn2802_3.

Carlson, S. M., Davis, A. C., & Leach, J. G. (2005). Less is more: executive function and symbolic representation in preschool children. *Psychological Science, 16*(8), 609–616. doi: 10.1111/j.1467-9280.2005.01583.x.

Carlson, S. M., Zelazo, P. D., & Faja, S. (2013). Executive function. In P. D. Zelazo (ed.), *The Oxford handbook of developmental psychology* (Vol. 1, pp. 706–743). New York: Oxford University Press

Caro, T. M., & Hauser, M. D. (1992). Is there teaching in nonhuman animals? *The Quarterly Review of Biology, 67*(2), 151–174. doi: 10.1086/417553.

Casey, B. J., Teslovich, T., Glover, G., Zayas, V., Mischel, W., Shodae, Y., . . . Wilson, N. L. (2012). Behavioral and neural correlates of delay of gratification 40 years later. *Annals of Neurosciences, 19*(1), 27–28. doi: 10.5214/ans.0972.7531.180407.

Coffman, J. L., Ornstein, P. A., McCall, L. E., & Curran, P. J. (2008). Linking teachers' memory-relevant language and the development of children's memory skills. *Developmental Psychology, 44*(6), 1640–1654. doi: 10.1037/a0013859.

Colvin, G. (2008). *Talent is overrated: what really separates world-class performers from everybody else.* New York: Portfolio.

Corballis, M. C. (2013). Mental time travel: a case for evolutionary continuity. *Trends in Cognitive Sciences, 17*(1), 5–6. doi: 10.1016/j.tics.2012.10.009.

Corballis, M. C. (2014). Recursive cognition as a prelude to language. In F. Lowenthal & L. Lefebvre (eds.), *Language and recursion* (pp. 27–36). New York: Springer.

Correia, S. P. C., Dickinson, A., & Clayton, Nicola S. (2007). Western scrub-jays anticipate future needs independently of their current motivational state. *Current Biology, 17*(10), 856–861. doi: 10.1016/j.cub.2007.03.063.

Coughlan, E. K., Williams, A. M., McRobert, A. P., & Ford, P. R. (2014). How experts practice: a novel test of deliberate practice theory. *Journal of Experimental Psychology: Learning, Memory, and Cognition, 40*(2), 449–458. doi: 10.1037/a0034302.

Coyle, D. (2009). *The talent code.* New York: Bantam.

Davis, J. T. M., Cullen, E., & Suddendorf, T. (in press). *Understanding deliberate practice in preschool aged children. The Quarterly Journal of Experimental Psychology.*

Dayan, E., & Cohen, L. G. (2011). Neuroplasticity subserving motor skill learning. *Neuron, 72*(3), 443–454. doi: 10.1016/j.neuron.2011.10.008.

Dennett, D. C. (1995). *Darwin's dangerous idea: evolution and the meanings of life.* New York: Simon & Schuster.

Diedrichsen, J., & Kornysheva, K. (2015). Motor skill learning between selection and execution. *Trends in Cognitive Sciences, 19*(4), 227–233. doi: 10.1016/j.tics.2015.02.003.

Donald, M. (1999). Preconditions for the evolution of protolanguage. In M. C. Corballis & S. E. G. Lea (eds.), *The descent of mind: psychological perspectives on hominid evolution* (pp. 138–154). Oxford: Oxford University Press

Dong, A., Collier-Baker, E., & Suddendorf, T. (in press). Building blocks of human design thinking in animals. *International Journal of Design Creativity and Innovation*. doi: 10.1080/21650349.2015.1011700.

Dudai, Y., & Carruthers, M. (2005). The Janus face of Mnemosyne. *Nature, 434*(7033), 567–567. doi: 10.1038/434567a.

Dufour, V., Pelé, M., Sterck, E. H. M., & Thierry, B. (2007). Chimpanzee (Pan troglodytes) anticipation of food return: coping with waiting time in an exchange task. *Journal of Comparative Psychology, 121*(2), 145–155. doi: 10.1037/0735-7036.121.2.145.

Elbert, T., Pantev, C., Wienbruch, C., Rockstroh, B., & Taub, E. (1995). Increased cortical representation of the fingers of the left hand in string players. *Science, 270*(5234), 305–307. doi: 10.1126/science.270.5234.305.

Ericsson, K. A. (2008). Deliberate practice and acquisition of expert performance: a general overview. *Academic Emergency Medicine, 15*(11), 988–994. doi: 10.1111/j.1553-2712.2008.00227.x.

Ericsson, K. A. (2014). Why expert performance is special and cannot be extrapolated from studies of performance in the general population: a response to criticisms. *Intelligence, 45*(1), 81–103. doi: 10.1016/j.intell.2013.12.001.

Ericsson, K. A., Krampe, R. T., & Tesch-Römer, C. (1993). The role of deliberate practice in the acquisition of expert performance. *Psychological Review, 100*(3), 363–406. doi: 10.1037/0033-295X.100.3.363.

Feltz, D. L., & Landers, D. M. (1983). The effects of mental practice on motor skill learning and performance: a meta-analysis. *Journal of Sport Psychology, 5*(1), 25–57.

Flavell, J. H., Beach, D. R., & Chinsky, J. M. (1966). Spontaneous verbal rehearsal in a memory task as a function of age. *Child Development, 37*(2), 283–299. doi: 10.2307/1126804.

Friedman, W. J. (2005). Developmental and cognitive perspectives on humans' sense of the times of past and future events. *Learning and Motivation, 36*(2), 145–158. doi: 10.1016/j.lmot.2005.02.005.

Gaser, C., & Schlaug, G. (2003). Brain structures differ between musicians and non-musicians. *Journal of Neuroscience, 23*(27), 9240–9245. http://www.jneurosci.org

Gerstadt, C. L., Hong, Y. J., & Diamond, A. (1994). The relationship between cognition and action: performance of children 3½–7 years old on a stroop-like day-night test. *Cognition, 53*(2), 129–153. doi: 10.1016/0010-0277(94)90068-X.

Gilbert, D. T. (2006). *Stumbling on happiness*. New York: A. A. Knopf.

Gipps, C., & Tunstall, P. (1998). Effort, ability and the teacher: young children's explanations for success and failure. *Oxford Review of Education, 24*(2), 149–165. doi: 10.1080/0305498980240201.

Gobet, F., & Campitelli, G. (2007). The role of domain-specific practice, handedness, and starting age in chess. *Developmental Psychology, 43*(1), 159–172. doi: 10.1037/0012-1649.43.1.159.

Gopnik, A., & Astington, J. W. (1988). Children's understanding of representational change and its relation to the understanding of false belief and the appearance-reality distinction. *Child Development, 59*(1), 26–37. doi: 10.2307/1130386.

Gopnik, A., & Graf, P. (1988). Knowing how you know: young children's ability to identify and remember the sources of their beliefs. *Child Development, 59*(5), 1366–1371. doi: 10.1111/j.1467-8624.1988.tb01505.x.

Gupta, A. S., van der Meer, M. A. A., Touretzky, D. S., & Redish, A. D. (2010). Hippocampal replay is not a simple function of experience. *Neuron*, *65*(5), 695–705. doi: 10.1016/j.neuron.2010.01.034.

Hallos, J. (2005). "15 minutes of fame": exploring the temporal dimension of middle pleistocene lithic technology. *Journal of Human Evolution*, *49*(2), 155–179. doi: 10.1016/j.jhevol.2005.03.002.

Hambrick, D. Z., Oswald, F. L., Altmann, E. M., Meinz, E. J., Gobet, F., & Campitelli, G. (2014). Deliberate practice: is that all it takes to become an expert? *Intelligence*, *45*(1), 34–45. doi: 10.1016/j.intell.2013.04.001.

Hambrick, D. Z., & Tucker-Drob, E. M. (2015). The genetics of music accomplishment: Evidence for gene–environment correlation and interaction. *Psychonomic Bulletin & Review*, *22*(1), 112–120. doi: 10.3758/s13423-014-0671-9.

Hanson, L. K., Atance, C. M., & Paluck, S. W. (2014). Is thinking about the future related to theory of mind and executive function? Not in preschoolers. *Journal of Experimental Child Psychology*, *128*, 120–137. doi: 10.1016/j.jecp.2014.07.006

Harner, L. (1975). Yesterday and tomorrow: development of early understanding of the terms. *Developmental Psychology*, *11*(6), 864–865. doi: 10.1037/0012-1649.11.6.864.

Harris, P. L., German, T., & Mills, P. (1996). Children's use of counterfactual thinking in causal reasoning. *Cognition*, *61*(3), 233–259. doi: 10.1016/S0010-0277(96)00715-9.

Hayne, H., Gross, J., McNamee, S., Fitzgibbon, O., & Tustin, K. (2011). Episodic memory and episodic foresight in 3- and 5-year-old children. *Cognitive Development*, *26*(4), 343–355. doi: 10.1016/j.cogdev.2011.09.006.

Hazlitt, W. (1805). *Essay on the principles of human action and some remarks on the systems of Hartley and Helvetius*. London: J. Johnson.

Helton, W. S. (2005). Animal expertise, conscious or not. *Animal Cognition*, *8*(2), 67–74. doi: 10.1007/s10071-004-0234-y.

Helton, W. S. (2008). Expertise acquisition as sustained learning in humans and other animals: commonalities across species. *Animal Cognition*, *11*(1), 99–107. doi: 10.1007/s10071-007-0093-4.

Herbert, J., Gross, J., & Hayne, H. (2007). Crawling is associated with more flexible memory retrieval by 9-month-old infants. *Developmental Science*, *10*(2), 183–189. doi: 10.1111/j.1467-7687.2007.00548.x.

Hodges, N. J., Kerr, T., Starkes, J. L., Weir, P. L., & Nananidou, A. (2004). Predicting performance times from deliberate practice hours for triathletes and swimmers: what, when, and where is practice important? *Journal of Experimental Psychology: Applied*, *10*(4), 219–237. doi: 10.1037/1076-898X.10.4.219.

Hongwanishkul, D., Happaney, K. R., Lee, W. S. C., & Zelazo, P. D. (2005). Assessment of hot and cool executive function in young children: age-related changes and individual differences. *Developmental Neuropsychology*, *28*(2), 617–644. doi: 10.1207/s15326942dn2802_4.

Hoppitt, W. J. E., Brown, G. R., Kendal, R., Rendell, L., Thornton, A., Webster, M. M., & Laland, K. N. (2008). Lessons from animal teaching. *Trends in Ecology & Evolution*, *23*(9), 486–493. doi: 10.1016/j.tree.2008.05.008.

Horner, V., & Whiten, A. (2005). Causal knowledge and imitation/emulation switching in chimpanzees (Pan troglodytes) and children (Homo sapiens). *Animal Cognition*, *8*(3), 164–181. doi: 10.1007/s10071-004-0239-6.

Hornig, M., Aust, F., & Güllich, A. (2014). Practice and play in the development of German top-level professional football players. *European Journal of Sport Science*, *2*, 1–10. doi: 10.1080/17461391.2014.982204.

Hudson, J. A., Shapiro, L. R., & Sosa, B. B. (1995). Planning in the real world: preschool children's scripts and plans for familiar events. *Child Development*, *66*(4), 984–998. doi: 10.1111/j.1467-8624.1995.tb00917.x.

Imuta, K., Hayne, H., & Scarf, D. (2014). I want it all and I want it now: delay of gratification in preschool children. *Developmental Psychobiology*, *56*(7), 1541–1552. doi: 10.1002/dev.21249.

Jacques, S., & Zelazo, P. D. (2001). The Flexible Item Selection Task (FIST): a measure of executive function in preschoolers. *Developmental Neuropsychology*, *20*(3), 573–591. doi: 10.1207/875656401753549807.

Janmaat, K. R. L., Polansky, L., Ban, S. D., & Boesch, C. (2014). Wild chimpanzees plan their breakfast time, type, and location. *Proceedings of the National Academy of Sciences*, *111*(46), 16343–16348. doi: 10.1073/pnas.1407524111.

Jeannerod, M. (1994). The representing brain: neural correlates of motor intention and imagery. *Behavioral and Brain Sciences*, *17*(2), 187–202. doi: 10.1017/S0140525X00034026.

Kaefer, A., Chiviacowsky, S., Meira Jr., C. d. M., & Tani, G. (2014). Self-controlled practice enhances motor learning in introverts and extroverts. *Research Quarterly for Exercise and Sport*, *85*(2), 226–233. doi: 10.1080/02701367.2014.893051

Klein, S. B. (2013). The temporal orientation of memory: it's time for a change. *Journal of Applied Research in Memory and Cognition*, *2*(4), 222–234. doi: 10.1016/j.jarmac.2013.08.001.

Kliegel, M., Mackinlay, R., & Jäger, T. (2008). Complex prospective memory: development across the lifespan and the role of task interruption. *Developmental Psychology*, *44*(2), 612–617. doi: 10.1037/0012-1649.44.2.612.

Kochanska, G., Tjebkes, T. L., & Forman, D. R. (1998). Children's emerging regulation of conduct: restraint, compliance, and internalization from infancy to the second year. *Child Development*, *69*(5), 1378–1389. doi: 10.2307/1132272.

Kvavilashvili, L., Kyle, F. E., & Messer, D. J. (2012). The development of prospective memory in children: methodological issues, empirical findings, and future directions. In M. Kliegel, M. A. McDaniel, & G. O. Einstein (eds.), *Prospective memory: cognitive, neuroscience, developmental, and applied perspectives* (pp. 115–140). New York: Taylor and Francis.

Lagattuta, K. H. (2007). Thinking about the future because of the past: young children's knowledge about the causes of worry and preventative decisions. *Child Development*, *78*(5), 1492–1509. doi: 10.1111/j.1467-8624.2007.01079.x.

Lagattuta, K. H. (2014). Linking past, present, and future: children's ability to connect mental states and emotions across time. *Child Development Perspectives*, *8*(2), 90–95. doi: 10.1111/cdep.12065.

Lee, T. D., & Genovese, E. D. (1989). Distribution of practice in motor skill acquisition: different effects for discrete and continuous tasks. *Research Quarterly for Exercise and Sport*, *60*(1), 59–65. doi: 10.1080/02701367.1989.10607414.

Lemmon, K., & Moore, C. (2007). The development of prudence in the face of varying future rewards. *Developmental Science*, *10*(4), 502–511. doi: 10.1111/j.1467-7687.2007.00603.x.

Levine, B. (2004). Autobiographical memory and the self in time: brain lesion effects, functional neuroanatomy, and lifespan development. *Brain and Cognition*, *55*(1), 54–68. doi: 10.1016/S0278-2626(03)00280-X.

Levitin, D. J. (2007). *This is your brain on music: understanding a human obsession.* London: Atlantic.

Luna, B., Garver, K. E., Urban, T. A., Lazar, N. A., & Sweeney, J. A. (2004). Maturation of cognitive processes from late childhood to adulthood. *Child Development, 75*(5), 1357–1372. doi: 10.1111/j.1467-8624.2004.00745.x.

Lyon, T. D., & Flavell, J. H. (1994). Young children's understanding of 'remember' and 'forget.' *Child Development, 65*(5), 1357–1371. doi: 10.2307/1131503.

Lyons, A. D., Henry, J. D., Rendell, P. G., Corballis, M. C., & Suddendorf, T. (2014). Episodic foresight and aging. *Psychology and aging, 29*(4), 873.

Macnamara, B. N., Hambrick, D. Z., & Oswald, F. L. (2014). Deliberate practice and performance in music, games, sports, education, and professions: a meta-analysis. *Psychological Science, 25*(8), 1608–1618. doi: 10.1177/0956797614535810.

Martin-Ordas, G., Atance, C. M., & Caza, J. S. (2014). How do episodic and semantic memory contribute to episodic foresight in young children? *Frontiers in Psychology, 5*, 732. doi: 10.3389/fpsyg.2014.00732.

McCormack, T., & Atance, C. M. (2011). Planning in young children: a review and synthesis. *Developmental Review, 31*(1), 1–31. doi: 10.1016/j.dr.2011.02.002.

Meinz, E. J., & Hambrick, D. Z. (2010). Deliberate practice is necessary but not sufficient to explain individual differences in piano sight-reading skill: the role of working memory capacity. *Psychological Science, 21*(7), 914–919. doi: 10.1177/0956797610373933.

Meltzoff, A. N. (1988). Infant imitation and memory: nine-month-olds in immediate and deferred tests. *Child Development, 59*(1), 217–225. doi: 10.1111/j.1467-8624.1988.tb03210.x.

Miloyan, B., & Suddendorf, T. (2015). Feelings of the future. *Trends in Cognitive Sciences, 19*(4), 196–200. doi: 10.1016/j.tics.2015.01.008.

Mischel, W., & Ebbesen, E. B. (1970). Attention in delay of gratification. *Journal of Personality and Social Psychology, 16*(2), 329–337. doi: 10.1037/h0029815.

Mischel, W., & Metzner, R. (1962). Preference for delayed reward as a function of age, intelligence, and length of delay interval. *The Journal of Abnormal and Social Psychology, 64*(6), 425–431. doi: 10.1037/h0045046.

Mischel, W., Shoda, Y., & Peake, P. K. (1988). The nature of adolescent competencies predicted by preschool delay of gratification. *Journal of Personality and Social Psychology, 54*(4), 687–696. doi: 10.1037/0022-3514.54.4.687.

Moore, C., & Lemmon, K. (2001). *The self in time: developmental perspectives.* Mahwah, NJ: Lawrence Erlbaum.

Morgan, T. J. H., Uomini, N. T., Rendell, L. E., Chouinard-Thuly, L., Street, S. E., Lewis, H. M., . . . Laland, K. N. (2015). Experimental evidence for the co-evolution of hominin tool-making teaching and language. *Nature Communications, 6*, 6029.

Mulcahy, N. J., & Call, J. (2006). Apes save tools for future use. *Science, 312*(5776), 1038–1040. doi: 10.1126/science.1125456.

Munakata, Y., Snyder, H. R., & Chatham, C. H. (2012). Developing cognitive control: three key transitions. *Current Directions in Psychological Science, 21*(2), 71–77. doi: 10.1177/0963721412436807.

Naqshbandi, M., & Roberts, W. A. (2006). Anticipation of future events in squirrel monkeys (Saimiri sciureus) and rats (Rattus norvegicus): tests of the Bischof-Kohler Hypothesis. *Journal of Comparative Psychology, 120*(4), 345–357. doi: 10.1037/0735-7036.120.4.345.

Nielsen, M. (2006). Copying actions and copying outcomes: social learning through the second year. *Developmental Psychology, 42*(3), 555–565. doi: 10.1037/0012-1649.42.3.555.

Osvath, M. (2009). Spontaneous planning for future stone throwing by a male chimpanzee. *Current Biology, 19*(5), R190–R191. doi: 10.1016/j.cub.2009.01.010.

Osvath, M., & Osvath, H. (2008). Chimpanzee (Pan troglodytes) and orangutan (Pongo abelii) forethought: self-control and pre-experience in the face of future tool use. *Animal Cognition, 11*(4), 661–674. doi: 10.1007/s10071-008-0157-0.

Payne, G., Taylor, R., Hayne, H., & Scarf, D. (2014). Mental time travel for self and other in three- and four-year-old children. *Memory, 23*(5), 675–682. doi: 10.1080/09658211.2014.921310.

Perner, J. (1991). *Understanding the representational mind.* Cambridge, MA: MIT Press.

Perner, J., & Ruffman, T. (1995). Episodic memory and autonoetic consciousness: developmental evidence and a theory of childhood amnesia. *Journal of Experimental Child Psychology, 59*(3), 516. doi: 10.1006/jecp.1995.1024.

Peterson, C. C., Wellman, H. M., & Slaughter, V. (2012). The mind behind the message: advancing theory-of-mind scales for typically developing children, and those with deafness, autism, or asperger syndrome. *Child Development, 83*(2), 469–485. doi: 10.1111/j.1467-8624.2011.01728.x.

Post, P. G., Fairbrother, J. T., & Barros, J. A. C. (2011). Self-controlled amount of practice benefits learning of a motor skill. *Research Quarterly for Exercise and Sport, 82*(3), 474–481. doi: 10.1080/02701367.2011.10599780.

Premack, D. (2007). Human and animal cognition: continuity and discontinuity. *Proceedings of the National Academy of Sciences U S A, 104*(35), 13861–13867. doi: 10.1073/pnas.0706147104.

Prencipe, A., & Zelazo, P. D. (2005). Development of affective decision making for self and other: evidence for the integration of first- and third-person perspectives. *Psychological Science, 16*(7), 501–505. doi: 10.1111/j.0956-7976.2005.01564.x.

Pruetz, J. D., & Bertolani, P. (2007). Savanna chimpanzees, Pan troglodytes verus, hunt with tools. *Current Biology, 17*(5), 412–417. doi: 10.1016/j.cub.2006.12.042.

Quon, E., & Atance, C. (2010). A comparison of preschoolers' memory, knowledge, and anticipation of events. *Journal of Cognition and Development, 11*(1), 37–60. doi: 10.1080/15248370903453576.

Raby, C., Alexis, D., Dickinson, A., & Clayton, N. (2007). Empirical evaluation of mental time travel. *Behavioral and Brain Sciences, 30*(3), 330–331. doi: 10.1017/S0140525X07002166.

Rafetseder, E., Cristi-Vargas, R., & Perner, J. (2010). Counterfactual reasoning: developing a sense of nearest possible world. *Child Development, 81*(1), 376–389. doi: 10.1111/j.1467-8624.2009.01401.x.

Rafetseder, E., & Perner, J. (2010). Is reasoning from counterfactual antecedents evidence for counterfactual reasoning? *Thinking & Reasoning, 16*(2), 131–155. doi: 10.1080/13546783.2010.488074.

Redshaw, J., & Suddendorf, T. (2013). Foresight beyond the very next event: four-year-olds can link past and deferred future episodes. *Frontiers in Psychology, 4*, 404. doi: 10.3389/fpsyg.2013.00404.

Redshaw, J., & Suddendorf, T. (2015). *The development of future-oriented information seeking and encoding.* Manuscript submitted for publication.

Rosati, A. G., Stevens, J. R., Hare, B., & Hauser, M. D. (2007). The evolutionary origins of human patience: temporal preferences in chimpanzees, bonobos, and human adults. *Current Biology, 17*(19), 1663–1668. doi: 10.1016/j.cub.2007.08.033.

Rosenbaum, D. A. (1983). Hierarchical versus nonhierarchical models of movement sequence control: a reply to Klein. *Journal of Experimental Psychology: Human Perception and Performance*, *9*(5), 837–839. doi: 10.1037/0096-1523.9.5.837.

Rossano, M. J. (2003). Expertise and the evolution of consciousness. *Cognition*, *89*(3), 207–236. doi: 10.1016/S0010-0277(03)00120-3.

Scarf, D., Gross, J., Colombo, M., & Hayne, H. (2013). To have and to hold: episodic memory in 3- and 4-year-old children. *Developmental Psychobiology*, *55*(2), 125–132. doi: 10.1002/dev.21004.

Schacter, D. L., & Addis, D. R. (2007). The cognitive neuroscience of constructive memory: remembering the past and imagining the future. *Philosophical Transactions of the Royal Society of London B: Biological Sciences*, *362*(1481), 773–786. doi: 10.1098/rstb.2007.2087.

Schlam, T. R., Wilson, N. L., Shoda, Y., Mischel, W., & Ayduk, O. (2013). Preschoolers' delay of gratification predicts their body mass 30 years later. *Journal of Pediatrics*, *162*(1), 90–93. doi: 10.1016/j.jpeds.2012.06.049.

Schmidt, R. A. (1991). Frequent augmented feedback can degrade learning: evidence and interpretations. In J. Requin & G. E. Stelmach (eds.), *Tutorials in motor neuroscience* (Vol. 62, pp. 59–75). Netherlands: Springer.

Schneider, W., & Lockl, K. (2008). Procedural metacognition in children: evidence for developmental trends. In J. Dunlosky & R. A. Bjork (eds.), *Handbook of metamemory and memory* (pp. 391–410). New York: Psychology Press.

Shahaeian, A., Peterson, C. C., Slaughter, V., & Wellman, H. M. (2011). Culture and the sequence of steps in theory of mind development. *Developmental Psychology*, *47*(5), 1239–1247. doi: 10.1037/a0023899.

Shenk, D. (2010). *The genius in all of us: why everything you've been told about genetics, talent, and IQ is wrong.* New York: Anchor Books.

Sidaway, B., Bates, J., Occhiogrosso, B., Schlagenhaufer, J., & Wilkes, D. (2012). Interaction of feedback frequency and task difficulty in children's motor skill learning. *Physical Therapy*, *92*(7), 948–957. doi: 10.2522/ptj.20110378.

Sterelny, K. (2003). *Thought in a hostile world: the evolution of human cognition.* Malden, MA: Blackwell.

Sternberg, R. J. (1996). Costs of expertise. In K. A. Ericsson (ed.), *The road to excellence: the acquisition of expert performance in the arts and sciences, sports, and games* (pp. 347–354). Mahway, NJ: Lawrence Erlbaum.

Suddendorf, T. (2003). Early representational insight: twenty-four-month-olds can use a photo to find an object in the world. *Child Development*, *74*(3), 896–904. doi: 10.1111/1467-8624.00574.

Suddendorf, T. (2006). Foresight and evolution of the human mind. *Science*, *312*(5776), 1006–1007. doi: 10.1126/science.1129217.

Suddendorf, T. (2010). Linking yesterday and tomorrow: preschoolers' ability to report temporally displaced events. *British Journal of Developmental Psychology*, *28*(2), 491–498. doi: 10.1348/026151009X479169

Suddendorf, T. (2013a). *The gap: the science of what separates us from other animals.* New York: Basic Books.

Suddendorf, T. (2013b). Mental time travel: continuities and discontinuities. *Trends in Cognitive Sciences*, *17*(4), 151. doi: 10.1016/j.tics.2013.01.011.

Suddendorf, T., & Busby, J. (2005). Making decisions with the future in mind: developmental and comparative identification of mental time travel. *Learning and Motivation*, *36*(2), 110–125. doi: 10.1016/j.lmot.2005.02.010.

Suddendorf, T., & Corballis, M. C. (1997). Mental time travel and the evolution of the human mind. *Genetic Social and General Psychology Monographs, 123*(2), 133–167.

Suddendorf, T., & Corballis, M. C. (2007). The evolution of foresight: what is mental time travel, and is it unique to humans? *Behavioral and Brain Sciences, 30*(3), 299–313. doi: 10.1017/S0140525X07001975.

Suddendorf, T., & Corballis, M. C. (2008). New evidence for animal foresight? *Animal Behaviour, 75*(5), e1–e3. doi: 10.1016/j.anbehav.2008.01.006.

Suddendorf, T., & Moore, C. (2011). Introduction to the special issue: the development of episodic foresight. *Cognitive Development, 26*(4), 295–298. doi: 10.1016/j.cogdev.2011.09.001.

Suddendorf, T., Nielsen, M., & von Gehlen, R. (2011). Children's capacity to remember a novel problem and to secure its future solution. *Developmental Science, 14*(1), 26–33. doi: 10.1111/j.1467-7687.2010.00950.x.

Suddendorf, T., & Redshaw, J. (2013). The development of mental scenario building and episodic foresight. *Annals of the New York Academy of Sciences, 1296*(1), 135–153. doi: 10.1111/nyas.12189.

Szpunar, K. K., Watson, J. M., & McDermott, K. B. (2007). Neural substrates of envisioning the future. *Proceedings of the National Academy of Sciences U S A, 104*(2), 642–647. doi: 10.1073/pnas.0610082104.

Taylor, M., Esbensen, B. M., & Bennett, R. T. (1994). Children's understanding of knowledge acquisition: the tendency for children to report that they have always known what they have just learned. *Child Development, 65*(6), 1581–1604. doi: 10.1111/j.1467-8624.1994.tb00837.x.

Tinbergen, N. (1963). On aims and methods of ethology. *Zeitschrift für Tierpsychologie, 20*(4), 410–433. doi: 10.1111/j.1439-0310.1963.tb01161.x.

Tomasello, M. (1999). The human adaptation for culture. *Annual Review of Anthropology, 28*(1), 509–529. doi: 10.1146/annurev.anthro.28.1.509.

UNISCO Institute of Statistics. (2014). Education: official entrance age by level of education. Retrieved April 16, 2014, from http://data.uis.unesco.org/index.aspx?queryname=218.

Walker, M. P. (2009). Sleep-dependent memory processing. In R. Stickgold & M. P. Walker (eds.), *The neuroscience of sleep* (pp. 230–240). London: Academic Press Elsevier.

Weisberg, D. P., & Beck, S. R. (2010). Children's thinking about their own and others' regret and relief. *Journal of Experimental Child Psychology, 106*(2), 184–191. doi: 10.1016/j.jecp.2010.02.005.

Wellman, H. M., Cross, D., & Watson, J. (2001). Meta-analysis of theory-of-mind development: the truth about false belief. *Child Development, 72*(3), 655–684. doi: 10.1111/1467-8624.00304.

Wellman, H. M., & Liu, D. (2004). Scaling of theory-of-mind tasks. *Child Development, 75*(2), 523–541. doi: 10.1111/j.1467-8624.2004.00691.x.

Wells, G. (1985). *Language development in the preschool years* (Vol. 2). Cambridge, UK: Cambridge University Press.

Whiten, A. (2005). The second inheritance system of chimpanzees and humans. *Nature, 437*(7055), 52–55. doi: 10.1038/nature04023.

Whiten, A., McGuigan, N., Marshall-Pescini, S., & Hopper, L. M. (2009). Emulation, imitation, over-imitation and the scope of culture for child and chimpanzee. *Philosophical Transactions of the Royal Society of London B: Biological Sciences, 364*(1528), 2417–2428. doi: 10.1098/rstb.2009.0069.

Wilhelm, I., Metzkow-Mészàros, M., Knapp, S., & Born, J. (2012). Sleep-dependent consolidation of procedural motor memories in children and adults: the pre-sleep level of performance matters. *Developmental Science, 15*(4), 506–515. doi: 10.1111/j.1467-7687.2012.01146.x.

Wimmer, H., & Perner, J. (1983). Beliefs about beliefs: representation and constraining function of wrong beliefs in young children's understanding of deception. *Cognition, 13*(1), 103–128. doi: 10.1016/0010-0277(83)90004-5.

Wulf, G., Shea, C., & Lewthwaite, R. (2010). Motor skill learning and performance: a review of influential factors. *Medical Education, 44*(1), 75–84. doi: 10.1111/j.1365-2923.2009.03421.x.

Wulf, G., & Shea, C. H. (2002). Principles derived from the study of simple skills do not generalize to complex skill learning. *Psychonomic Bulletin & Review, 9*(2), 185–211. doi: 10.3758/BF03196276.

Zelazo, P. D., Muller, U., & Frye, D. (2003). The development of executive function in early childhood. *Monographs of the Society for Research in Child Development, 68*(3), 1–151. doi: 10.1111/j.1540-5834.2003.06803002.x.

Zimmermann-Schlatter, A., Schuster, C., Puhan, M. A., Siekierka, E., & Steurer, J. (2008). Efficacy of motor imagery in post-stroke rehabilitation: a systematic review. *Journal of NeuroEngineering and Rehabilitation, 5*(1), 8–8. doi: 10.1186/1743-0003-5-8.

18

# Episodic Future Thinking in Children

## *Methodological and Theoretical Approaches*

CRISTINA M. ATANCE AND CAITLIN E. V. MAHY ■

The focus of this chapter is the development of episodic future thinking, defined as the capacity to mentally project the self into the future to pre-experience events (Atance & O'Neill, 2001). This capacity can be thought of as the "future-directed" counterpart to episodic memory—though it is important to note that, from early on, Tulving (1985) conceived of episodic memory as allowing for mental time travel into *both* the past and the future. Episodic future thinking is argued to be distinct from what could be termed *semantic future thinking*—or, more "knowledge-based" predictions about the future that do not necessarily pertain to the self (e.g., Atance & O'Neill, 2001; Suddendorf & Corballis, 1997, 2007; Tulving, 2005). Episodic future thinking is typically assessed in adults by asking them to describe personal events that will occur at different time points in the future (e.g., next week, next year, etc.) (e.g., Addis & Schacter, 2008). Researchers may also present participants with cue words (e.g., birthday) and ask them to elaborate on a birthday party that they will attend in the future (e.g., Szpunar, Watson, & McDermott, 2007). Using such methods has resulted in a number of important insights about the brain regions involved in thinking about the future, how these overlap with people's capacities to remember the past (e.g., Addis, Wong, & Schacter, 2007; Okuda et al., 2003; Szpunar et al., 2007), and hence theories about how memory plays a critical role in people's ability to simulate the future (e.g., Schacter & Addis, 2007; Suddendorf & Corballis, 2007).

An implicit assumption of these approaches is that adults can think about the future and often do so in a rich and detailed way. In contrast, we cannot make these assumptions when studying the development of children's capacity to conceptualize the future. Rather, questions such as "Can children think about the future?" "How do we measure their future thinking ability?" and "What is the nature of the representations that underlie children's future thinking?" are at the fore of what is being explored. Accordingly, the starting point of developmental research about episodic future thinking is very different from the starting point with adults. This, in turn, has led to different methodological approaches to the study of future thinking, as well as differences in the particular aspects of future thinking ability under study.

The overarching aim of this chapter is to provide an overview of the methodological and theoretical approaches to the development of children's episodic future thinking. We begin by carefully considering the developmental trajectory of future thinking ability from the preschool years into early/middle childhood. Although the majority of the work on episodic future thinking has been conducted with preschoolers, research about this topic beyond the preschool years is a burgeoning area of study that we will also touch upon. In considering developmental trajectories, we place a strong focus on delineating/highlighting how future thinking changes, along with the methodologies used to detect such changes. Thus, rather than arguing that children either "do" or "do not" have the capacity to mentally project themselves into the future, we highlight the important developments occurring at each age, and how these are subsequently built upon, resulting in more sophisticated forms of future thought. We also touch upon what we know about how related cognitive factors, including memory, theory of mind, and executive functioning, contribute to developments in episodic future thinking in preschoolers and older children. Finally, we conclude our chapter by suggesting some important distinctions that researchers may wish to consider in the study of episodic future thinking. These include the notion of "conflict" between present and future, and potential differences between thinking about the "future self" versus a "future other."

## 1. THE DEVELOPMENT OF EPISODIC FUTURE THINKING

Although "primitive" forms of future orientation emerge in infancy (e.g., Atance & Hanson, 2011; Haith, 1997; McCormack & Atance, 2011), it is unlikely that prior to about 2 years of age the necessary cognitive abilities, and understanding of "self," are sufficiently developed to allow children to mentally project themselves into the future. Accordingly, the first efforts to test episodic future thinking in development have targeted children between the ages of 3 and 5 (i.e., the preschool years) and have mostly examined children's talk about the future, using what we will refer to as "verbal" methodologies. Such methodologies have provided a useful starting point for the study of episodic future thinking, and have also served to highlight some of the contexts/manipulations that may render talking about (and conceptualizing) the future easier for young children. With these points in mind, we now turn to the research that has examined children's talk about the future.

### 1.1. Verbal Methodologies

Although there is substantially more research that has been devoted to children's memory than to their thought about the future, some researchers within the memory literature have nonetheless considered the emergence of future-oriented talk. For example, Nelson (1989) detailed the "bedtime" talk of a young child, Emily who, at 28 months, constructed extremely sophisticated future narratives of the following sort:

> We are gonna . . .
> at the ocean.
> Ocean is a little far away.
> baw, baw, buh (etc.)
> far away . . .
> I think it's . . .
> couple blocks . . . away.
> Maybe it's down, downtown,
> and across the ocean,
> and down the river,
> and maybe it's in,
> the hot dogs will be in a fridge,
> and the fridge (would) be in the water over by a shore,
> and then we could go in,
> and get a hot dog and bring it out to the river,
> and then sharks go in the river and bite me,
> in the ocean. . . . (66)

Some of Emily's narrative appears to be based on details from a prior conversation with her father (e.g., that they would go to the ocean and that there would be hot dogs), coupled with her own predictions about what might occur (e.g., that the hot dogs would be in the fridge). Yet, Emily also included events that seemed quite unlikely to occur in the future (e.g., "and then sharks go in the river and bite me").

The inclusion of such "unlikely" events (at least as verified by parents) has since been reported in larger samples of young children who are asked to predict what they will do "tomorrow" (e.g., Busby & Suddendorf, 2005), or the next time they will go to the park, for example (e.g., Quon & Atance, 2010). In several experiments conducted by Busby and Suddendorf (2005), children were asked to predict something that they would do "tomorrow." Parents were also asked to rate the accuracy of these predictions. Whereas the 4- and 5-year-olds' responses were relatively accurate (69% and 63%, respectively, Experiment 2), those of the 3-year-olds were significantly less so (31%), suggesting that children's capacity to talk about future events improves between ages 3 and 5 (see also Atance & Meltzoff, 2005).

Importantly, it was not that the 3-year-olds were unable to provide a verbal response—69% of them did so—rather, they appeared limited in their ability to provide a response that parents judged as likely to occur the next day (similar to Emily, who predicts being bitten by a shark). In a more recent study, Quon and Atance (2010) asked 3-, 4-, and 5-year-olds about specific events (e.g., "What are you going to eat for breakfast tomorrow?"), rather than the more open-ended "tomorrow." Higher levels of accuracy were reported overall, with the parents of 3-, 4-, and 5-year-olds deeming that the events that their children described were likely to occur 60%, 69%, and 81% of the time, respectively. Five-year-olds' responses were significantly more accurate than those of either the 3- or 4-year-olds', whose performance did not differ. Yet, as seemed to be the case in Busby and Suddendorf (2005), children did not appear to be limited by their ability to provide a specific verbal response, per se. Indeed, Quon and Atance (2010) reported that 89%, 89%, and 95% of the 3-, 4-, and 5-year-olds, respectively, produced a verbal response, and these percentages did not significantly differ. Together, these findings suggest that children younger than 5 are not limited by verbal ability, per se, or by the ability to generate an event, but, rather, by their ability to generate an event that is *likely* to occur the next day. If so, this implies that these younger children may be capable of "imagining" numerous future possibilities but may subsequently have difficulty screening out/eliminating those that are unlikely to occur.

Although speculative at this point, we theorize that neither developments in imaginative capacity nor verbal ability can fully account for the developments we see in children's future thinking ability during the preschool years. In fact, one could even argue that younger children are considering more possibilities than older children and adults (and thus may, in a sense, be even more "imaginative"), but then have difficulty constraining their prediction to an event that is likely to occur. This is consistent with a study by Hayne, Gross, McNamee, Fitzgibbon, and Tustin (2011) in which children were also asked to talk about future events but, in this case, the event was generated by the parent—thus removing the need for children, themselves, to generate a plausible event. For example, a child might be told the following, "Your mum said that you're going to the farmer's market tomorrow. What can you tell me about this?" The researchers found that the content of children's reports was very accurate (90%) and, importantly, that accuracy did not vary as a function of age; though 5-year-olds provided more information overall than 3-year-olds. Thus, it is interesting that, once the event is generated for children, even 3-year-olds can quite accurately elaborate on it.

Relatedly, when children are asked about events for which they have more control—as rated by parents—(e.g., what they will eat for breakfast *tomorrow*, or what they will do at playtime *tomorrow*), accuracy levels are higher than for events for which they have little control (e.g., what they will eat for supper *tonight*, or what they will do at bedtime *tonight*) (Quon & Atance, 2010). Thus, part of children's difficulty making accurate predictions about the future may be because they do not often have control over how the future will unfold. However, as Quon and Atance have demonstrated, when the level of control increases, so too does accuracy. Nonetheless, it is important to note that both Hayne et al.'s (2011) and Quon and Atance's (2010) approaches may serve to scaffold or "cue" children's future talk. In the context of episodic memory, some have argued that the fact that it is difficult to elicit memory talk from young children without such cues or external prompts indicates that their episodic memory system may not be fully developed (Wheeler, Stuss, & Tulving, 1997).

Although much of the work on children's episodic future thinking has focused on the development of this ability during the preschool years (e.g., Atance & Meltzoff, 2005; Hudson, Mayhew, & Prabhakar, 2011; Suddendorf, Nielsen, & von Gehlen, 2011), some recent research has targeted its development during middle childhood. Most studies have entailed the type of verbal methodologies used with adults, and have also directly compared children's thinking about the future with their memory for the past (we will describe the results from such comparisons in more detail in our section entitled "Links Between Episodic Future Thinking and Memory in Children"). In addition, some studies have examined how temporal distance (i.e., near vs. distant future) impacts children's episodic future thinking.

Accordingly, most studies with older children entail asking them about different events in order to examine richness, level of detail, and fluency of both past and future events that are near (e.g., think of a time next week that the word "family" makes you think of) or far (e.g., think of a time next year that the word "school" makes you think of) in time.

Coughlin, Lyons, and Ghetti (2014) studied 5-, 7-, and 9-year-olds' narratives of near and distant past and future events. Children were given cue words (e.g., book, cake, family, etc.) and were asked to describe eight personal events (two events from last week, last year, next week, and next year). Once a narrative was provided (which was scored for episodicity), children were asked to rate their phenomenological experience bringing these events to mind, including: the clarity of the event, whether they imagined it from a first- or third-person perspective, and how easy it was to bring to mind. Researchers contacted parents by phone after each session to rate the thematic and temporal veracity of children's narrative responses. Results showed that compared to episodic memory for past events, the development of prospection for future events was more protracted developmentally; older children had higher episodicity scores than younger children. The authors suggest that this may be due to the additional skills above and beyond recollection required for prospection, including imagining possible futures. Finally, temporally near events were rated as clearer and easier to bring to mind than temporally distant events.

Similarly, Wang, Capous, Koh, and Hou (2014) interviewed 7- to 10-year-olds about temporally near (e.g., going to the science museum) or distant (e.g., getting married) past and future events. Children were asked to describe two specific past events and two specific future events. There were no developmental changes in the amount of specific (i.e., episodic information relevant to the event) or general (non-episodic, external information) details provided in this age range, suggesting that measures that rely on the amount of detail provided in children's narratives about the future might not be sensitive to developmental changes. However, children did provide more specific details for temporally near versus distant events, suggesting that they were better at mentally projecting themselves into events closer in time.

In an older sample, Gott and Lah (2014) gave children (aged 8–10) and adolescents (aged 14–16) the Child Autobiographical Interview, which required them to generate two past and two future personally relevant events (temporal distance was not manipulated/studied). Participants were given a list of autobiographical events at the start of the interview (e.g., birthday party, Christmas) and were asked to recall/imagine the event (either one from the list or one that was personally generated) that they had been/would be personally involved in, was specific to time and place, and occurred over the course of one day. After children gave their responses, the experimenter provided general and then specific probes to encourage them to produce more details. Results showed a developmental increase in the level of richness (i.e., more episodic and semantic details mentioned) used to describe future events, suggesting that adolescents, compared to children, are better able to episodically project themselves into these events (or, at least, are better able to describe these events richly).

Based on the results of verbal methodologies that have been used from early to later childhood, we can conclude that when asked relatively open-ended questions about the future (e.g., "what are you going to do tomorrow?") 3-year-olds have difficulty providing an accurate response, and this does not appear to be fully explained by language or imaginative deficits, per se. However, when children are given more guidance about the future event in question, then 3-year-olds' accuracy appears to improve. As children move into middle childhood and adolescence, their accounts of the future tend to become richer and more detailed. There also seems to be some effect of temporal distance, such that at least in one study (Wang et al., 2014), children provided more detail about near versus distant future events. This finding is thus consistent with the research with adults showing that distant events are constructed more abstractly than proximal ones (which, in turn, are constructed more concretely than their distant counterparts) (e.g., D'Argembeau & Van der Linden, 2004; Liberman & Trope, 1998). Relatedly, Friedman (2000) has found that between ages 4 and 10 there is improvement in the ability to differentiate future distances of events, with 5-year-olds, for example, being able to distinguish between events occurring in the coming weeks versus the coming months.

## 1.2. Behavioral Methodologies

Although the research that has assessed children's ability to talk about future events has provided a wealth of information to this relatively new area of study, like any approach, it has its limitations. For example, it is difficult (if not impossible) to ask children to talk about the future without using specific temporal terms (e.g., tomorrow, next time, later, when you're bigger, etc.) that signal the point in time that they are being asked about. Yet, we know that preschoolers (i.e., children aged 3 to 5) have limitations in their understanding of various temporal terms (e.g., Busby Grant & Suddendorf, 2011). As such, the use of these terms when asking children about the future may partly mask their underlying representations. To address this (and other related) issues, researchers have devised methods that rely less on children's ability to formulate a verbal response and more on their ability to behave in a way that can be considered future-oriented. This "behavioral" approach to measuring young children's episodic future thinking abilities is very different from the approaches used with adults—and much more similar to those used by researchers who study non-human animals (e.g., Mulcahy & Call, 2006; Raby, Alexis, Dickinson, & Clayton, 2007).

Central to such behavioral approaches has been to identify various criteria that need be fulfilled to conclude that children are showing foresight. These are best illustrated by Tulving's (2005; see also Suddendorf & Busby, 2005) example drawn from an Estonian children's story. In this story, a young girl dreams about attending a party where a delicious pudding is being served but, not having brought a spoon, she is unable to have any. Prior to going to bed the next night, the little girl gets a spoon in anticipation of eating the pudding, and avoiding further disappointment. According to Tulving, the little girl's behavior reflects her ability to mentally travel into both the past and the future. Interestingly, Tulving also makes the prediction that this little girl is at least 4 years of age. In what follows, we describe some of the research based on the "spoon test." In doing so, we highlight what appear to be limiting factors for young children who fail such a test, but also the particular factors that are in place, and that likely form the building blocks of episodic future thinking ability.

Most paradigms inspired by the spoon test have required children to spend several minutes in one room/location in which a critical item to solve a future problem is missing. For example, Suddendorf, Nielsen, and von Gehlen (2011) showed 3- and 4-year-olds a locked box with a triangular keyhole. The experimenter then showed children that the box could be unlocked using a triangular key to retrieve a sticker. Children were also given the opportunity (on two trials) to use the key to obtain a reward. Children were then shown a box with a *square* keyhole, and it was emphasized to them that the triangular key could no longer be used to obtain a reward. Children in the "instant" condition were then led across the room and, with their backs to the box, were asked to select one key among three—circle, star, and square—to take back to the box. In contrast, in the "delay" condition, children were led to another room where they played games for 15 minutes. After this time had elapsed, they were similarly presented with the three keys and asked to select one to bring back to the first room. In the instant condition, both age groups were significantly above chance (and practically at ceiling) in selecting the correct square key. Although the 4-year-olds' performance was also significantly above chance in the delayed condition (roughly 60% correct), it was significantly worse than in the instant condition. This was also true of the 3-year-olds, but their performance in this condition was no better than chance (roughly 35% correct). Several additional conditions in this study confirmed that children's difficulty in the delayed condition was not due to a spatial displacement (i.e., being brought to another room) but, rather, a temporal one (i.e., the delay between being presented with the problem and being given the opportunity to solve it).

These results are consistent with the claim that around age 4, children can remember a past problem and use this as a basis for formulating an appropriate future-oriented response. However, children's significantly superior performance in the instant condition compared to the delayed condition also suggests that memory—and not foresight, per se—may be an important factor that is driving children's errors. Indeed, several other studies that have involved children selecting an item to address a future problem (e.g., Atance & Sommerville, 2014; Scarf, Gross, Colombo, & Hayne, 2013) are also consistent with the argument that 3-year-olds' difficulty solving the task is rooted in their inability to remember the critical aspects of the problem (or memory failures).

If we want to conclude that children's limitation in episodic future thinking tasks is at least partly rooted in their inability to think about the future (rather than remember the past), then it is important to show that, when there is no delay between being presented with the problem and being given the opportunity to solve it, children still make an incorrect future-oriented choice. Similarly, if it can be shown that children remember the critical information even after a delay, but then still make an incorrect future-oriented choice, it is likely that the limiting factor is foresight, not memory. Several existing studies are relevant in this respect. In Russell, Alexis, and Clayton (2010), 3-, 4-, and 5-year olds learned how to play a "blow football" game. In this game, the child and the experimenter stood on either side of a table and needed to use straws to blow a ball into the other person's goal. Children played the game on the "red" side of the table, while the experimenter played from the "blue" side. They were then told that they would return the next day to play the game from the blue side. Children were told that they could save two items (from a group of six) for "tomorrow." Choices included several distracter items (e.g., cardboard referee and team badge), as well as one essential item: a straw with which to blow the football. In addition, to successfully play the game from the experimenter's side of the table (the blue side), children needed a box to stand on because this side of the table—unlike the red side they had just played on—did not have a box to allow children to reach the tabletop. Only 5-year-olds selected the box and straw significantly more often than chance.

Because there was virtually no delay between the time that children played the blow football game and the time at which they needed to make the critical choice for the next day, it seems unlikely that memory was a limiting factor (the items were also visible to children when they made their selection). Moreover, children performed better in this task when they were told that their choice was for "right now" rather than "tomorrow" (though it is important to note that even in the "right now" condition, children solved the task only 33% of the time). As such, it seems plausible that children's difficulty with this task centered on their inability to conceive of the future time in which they would play the game. The fact that the 4-year-olds in the study also performed better in a version in which they had to imagine another child playing the game "tomorrow" suggests that difficulties understanding this particular temporal term also cannot fully explain children's failure.

Using a different design based on a study conducted on planning ability in scrub jays (Raby et al., 2007), Atance, Louw, and Clayton (2015) had children visit two separate rooms, two consecutive times: one that contained toys, and one that did not. One of the rooms was labeled as "Big Bird's" and one as "Ernie's"; which of these rooms had toys, as well as which room children visited first, were both counterbalanced. If children first entered the toy room, they were told that Big Bird/Ernie had toys for them to play with. A 3-minute timer was then set and children played with the toys while the Experimenter stated that she had work to do and thus did not interact with the child. Next, children visited the room with no toys, where they were told that Big Bird/Ernie did not have toys and thus the child had nothing to do during the 3-minute stay. As in the toy room, the Experimenter stated that she had work to do. After this sequence was repeated one more time, children were brought to an outside hallway and were queried about their memory for which room had toys and which room did not. They were then asked the "foresight" question. Rather than ask children about "tomorrow," as was the case in Russell et al. (2010), they were asked about a point in time when they would be one year older than they were that day. For example, 3-year-olds were told: "You're going to get to come back here when you're 4 years old to visit Ernie's room and Big Bird's room." This wording was chosen because similar references to time, such as, "when I was little" and "when I get bigger," appear to be fairly well understood even by 3-year-olds (Busby Grant & Suddendorf, 2011). Children were then shown several additional toys and were asked, "Where would you like to put these for next time?" All three age groups performed significantly above chance on the two memory questions (though 5-year-olds performed significantly better than 3-year-olds), but only the 4- and 5-year-olds' performance was above chance on the foresight question. In addition, even those 3-year-olds who responded to both memory questions correctly did not perform above chance on the foresight question. These findings are thus consistent with the claim that the 3-year-olds' below-chance performance on the foresight question was not due to limitations in memory but, rather, limitations in foresight.

The previous studies demonstrate that although memory limitations may account for children's failure in some episodic future thinking tasks, this is not true for all tasks (e.g., Atance et al., 2015; Russell et al., 2010). Perhaps those tasks that are most susceptible to memory failures are ones for which children have to remember which particular item from a set must be used to address a previously encountered problem. When the items in the set are quite similar or conceptually related (e.g., four almost-identical keys with only one shape opening the box in question), as in Suddendorf et al. (2011), this may be especially taxing on young children's memory.

If memory is indeed a limiting factor on some of the future thinking tasks we have discussed, then it follows that boosting children's memory should also boost their future thinking. Indeed, there is now evidence that making the past event that forms the basis of an adaptive future-oriented choice particularly distinctive leads to improved task performance. For example, Payne et al. (2014) used an item-choice/spoon test in which children and the experimenter, together, engaged in a task that required them to build two separate musical instruments (one for the child, one for the experimenter). However, in doing so, they discovered that the last piece of each instrument was missing (e.g., a bell for the jingle bells). The child and the experimenter then went outside to a sandbox and were told that a pirate had visited earlier in the day and had buried something in the sandbox. Three locked treasure chests were then uncovered and all of the children recognized that a key was needed to open them. Children were then given one colored key that corresponded to the color of one of the treasure chests, only to discover that it was empty. The experimenter then told children that she had no more keys but invited them to shake the two remaining chests to see if anything was inside. One chest made the sound of the instrument that the child had earlier constructed, whereas the other made the sound of the instrument that the experimenter had constructed. After an approximately 20-minute delay in another location, children were told that they would be returning to the sandbox and were presented with the key that had opened the empty chest, two additional keys (one that corresponded to the chest that made the sound of the child's instrument, and one that corresponded to the chest that made the sound of the experimenter's instrument) and a distracter item (ball). Children were asked to select the items that both they and the experimenter should take to the sandbox. Results indicated that both the 3- and 4-year-olds selected the correct keys significantly more often than would be expected by chance. The authors argue that the use of a distinctive event (digging up treasure chests, etc.), as well as one that was motivating to children, may have contributed to the relatively strong performance of the 3-year-olds. This suggests that context and, more specifically, such factors as event distinctiveness and motivation play an important role in children's ability to solve item-choice/spoon tests. More broadly, these findings fit with the theory that memory is an important limiting factor in spoon tests and that boosting children's memory increases their future-oriented behavior.

The kinds of behavioral paradigms that we have discussed thus far have been some of the most popular means to test episodic future thinking in young children, yet, to date, have not been used with older children. This is because, in most cases, performance on these sorts of tasks is almost at ceiling by age 5. However, it might be possible to adapt such tasks for older children. Specifically, one interesting observation about verbal and behavioral methodologies is that, in the former, experimenters have varied the extent to which children are required to spontaneously generate a future prediction. With verbal paradigms, we saw that when children are asked about a relatively "open-ended" future (e.g., "What are you going to do tomorrow?"), they perform more poorly (or less "accurately") than when they are asked about events that are more specific (e.g., Quon & Atance, 2010), or ones that are generated for them (e.g., Hayne et al., 2011). As such, we see a notable developmental progression reflecting the fact that younger children need more support when required to provide verbal responses. Interestingly, this similar distinction has not, until recently, been tested in the behavioral realm. That is, in virtually all of the "behavioral" paradigms (though see Metcalf & Atance, 2011) that have been used to measure children's episodic future thinking, children are presented with "forced-choice" options. For example, they are presented with items from which to choose, or they must select in which of two rooms to place a resource for the future (e.g., Atance et al., 2015). Thus, we know very little about the extent to which children "spontaneously" generate solutions to future-oriented problems.

However, in a recent study by Mahy, Atance, Moses, and Kopp (2015), children were given a version of the spoon test (adapted from Suddendorf et al., 2011) but, prior to being given a forced-choice question about which item they should bring to the first room they had visited, children were asked to more spontaneously generate this choice (e.g., "What would be a good thing to bring to the first room?"). Whereas only 20% of children responded correctly to this open-ended question, 76% responded correctly to the forced-choice question. Moreover, children's performance on neither question showed age-related differences. It is somewhat surprising that the 5-year-olds were not better than they were at spontaneously generating a response (and also no better at this than the younger children). Although more data and different variations to create more "spontaneous" spoon tasks are needed, these preliminary results suggest that even older preschoolers' ability to independently generate future-oriented responses is quite limited. This is not a trivial point given that people must often make such spontaneous (i.e., in the absence of prompts) decisions for the future and, in fact, this is exactly what the little girl in Tulving's (2005) story was described as doing.

Earlier in this section, we argued that memory is a limiting factor in children's future-oriented thinking. Another factor/context that also appears to pose difficulty for younger and older children alike (and even adults) is when their future prediction pertains to physiological states. For example, Atance and Meltzoff (2006) introduced a task that pitted a current state against a future preference using the so-called pretzel task. In this task, 3- to 5-year-old children were asked to indicate a baseline preference for water or pretzels. Not surprisingly, most children reported preferring pretzels to water. Then, children were given the opportunity to eat pretzels (creating a state of thirst) and were asked to predict what they would like the next day: pretzels to eat or water to drink. Despite children's initial preference for pretzels, after consuming them, the majority of children indicated that they would prefer water the next day. Children in a control condition, who did not eat pretzels, continued to prefer pretzels for their future self, showing that current physiological states have a powerful influence on children's future thinking. Interestingly, there was no developmental increase in performance on this task between ages 3 and 5. Accordingly, to further examine the developmental trajectory of this task, Mahy et al. (2015) tested 3- to 7-year-olds and similarly found that after consuming pretzels, children mostly switched their future preference to water. Moreover,

7-year-olds performed no better on this task than 3-year-olds, confirming earlier findings of no developmental increase between ages 3 and 7 (Mahy, Grass, Wagner, & Kliegel, 2014). Seven-year-olds were nowhere near ceiling on this task, which is in fact similar to research with adults that shows that physiological states powerfully influence thinking about the future (Nisbett & Kanouse, 1969). Because physiological states (such as thirst) may be especially difficult to overcome when reasoning about the future from early childhood into adulthood, it may be best to consider this as an especially challenging form of future prediction.

So far, we have reviewed some of the developments in episodic future thinking during the preschool years and beyond, with an eye to what task factors may pose difficulty for children, as well as what specific competencies in this domain younger (e.g., 3-year-old) children may have. We shift our focus in the next section to discuss what we know about how cognitive factors such as memory, theory of mind (ToM), and executive function (EF) are related to future thinking. With respect to memory, there is much evidence from studies with adults that there is important cognitive and neural overlap between the abilities to remember the past and think about the future (e.g., Addis et al., 2007). One prominent theory, the "constructive episodic simulation hypothesis" (Schacter & Addis, 2007), posits that our memories of the past provide a database from which we flexibly recombine elements to construct possible future scenarios. There are now data to support links between memory and future thinking in younger and older children as well, and this is the topic to which we turn next.

## 1.3. Links Between Episodic Future Thinking and Memory in Children

We have already outlined research that suggests that preschoolers' inability to remember critical past information limits their performance on certain future thinking tasks. In this section we now specifically discuss research showing that younger and older children's ability to talk about past events is related to their ability to predict future events, and that performance on certain memory tasks (e.g., relational memory) predicts children's ability to construct future events. Several studies with preschoolers have shown that their ability to accurately respond to questions such as "What did you do yesterday?" are positively correlated to their responses to questions such as "What are you going to do tomorrow?" (e.g., Busby & Suddendorf, 2005; Hayne et al., 2011). In addition, there is evidence that relational memory (or, the capacity to flexibly recombine details of past events) is positively related to 3- to 5-year-olds' ability to mentally project into the future (Richmond & Pan, 2013).

Similar to the work with younger children, most studies examining episodic future thinking in older children show that this ability is correlated with episodic memory. For example, Gott and Lah (2014) found that relational memory, measured by a word-pairs task, was positively related to the number of episodic but not semantic details that 8- to 10-year-olds and 14- to 16-year-olds provided on the Child Autobiographical Interview (described earlier). This finding suggests that, by later childhood, relational memory contributes to children's ability to construct future episodic, but not semantic, details. Similarly, in 5- to 9-year-olds, past episodic memory predicted prospection into the future (Coughlin et al., 2014) further supporting the role of episodic memory for the past as an important factor influencing future thinking.

In addition to such correlational approaches, Lagattuta and her colleagues have assessed children's understanding that thinking about the future can be impacted by events that have happened in the past. For example, Lagattuta (2007) told 3- to 7-year-olds stories about a character that experiences a negative event, and then at a point in the future felt worried, or changed his behavior when he saw the item associated with the event (e.g., a red-haired boy takes David's teddy bear, David sees *the* red-haired boy or *a* red-haired boy and feels worried/hides his toys behind a tree). Between ages 3 and 5 (but not between ages 5 and 7), there was an increase in the frequency with which children reported that the character's reaction was caused by the anticipation of a negative event (e.g., the red-haired boy stealing David's toys). These results suggest that, during the preschool years, children develop the ability to anticipate negative future outcomes based on past experiences.

As is true of the adult research, most of the research with young children that examines factors that relate to episodic future thinking has focused on memory. However, researchers have also been interested in possible links between children's ability to think about the future and theory of mind (ToM) and executive function (EF). Some of this interest has been fueled by theoretical accounts about why these abilities may be related (e.g., Buckner & Carroll, 2007; Suddendorf & Corballis, 2007). For example, with respect to ToM, in particular, Buckner and Carroll (2007) argue that thinking about the future, remembering the past, taking the perspective of others (i.e., ToM), and some forms of spatial navigation all involve the capacity to "self-project" out of one's current perspective. Because of this common requirement, it is conceivable that these various forms of thought emerge together in development. Suddendorf and Corballis (2007) have also stressed the importance of executive functioning to thinking about the future. For example, inhibitory control, in particular, may be required to allow an individual to forgo instant rewards in the interest of longer-term gains (see also Suddendorf & Redshaw, 2013, for a more comprehensive account of the various component abilities required for the emergence of episodic future thinking).

### 1.4. Links Between Episodic Future Thinking and Theory of Mind and Executive Function in Children

We are aware of only one study that has examined the links between episodic future thinking, ToM, and EF in younger children. Hanson, Atance, and Paluck (2014) gave 3-, 4-, and 5-year-olds batteries of episodic future thinking, ToM, and EF tasks. The future thinking tasks included ones that assess planning for an event that pertains to the self (e.g., "Can you tell me a plan for going to the beach?") (Hudson, Shapiro, & Sosa, 1995); reasoning about the temporal order of events—an ability argued to be involved in constructing future scenarios (McColgan & McCormack, 2008); a temporal sequencing task that requires children to make predictions about when various life events will occur (Busby Grant & Suddendorf, 2009); and, finally, two tasks specifically designed to tap into children's ability to mentally project the self into the future: a picture book task that requires children to anticipate and reason about future states of the self (Atance & Meltzoff, 2005); and Busby and Suddendorf's (2005) task, described earlier in this chapter, that requires children to predict what they will do "tomorrow." ToM tasks included those measuring false belief and desire understanding, and EF tasks included those measuring inhibition, working memory, planning, cognitive flexibility, and generativity.

As expected, most of the individual ToM tasks were inter-correlated, as were most of the individual EF tasks. In contrast, the episodic future thinking tasks were not; though four of the five showed significant age-related improvement. This lack of correlation is theoretically interesting because it suggests that these tasks were measuring different aspects of children's foresight—an important point to which we return. The second relevant finding is that there were virtually no significant correlations between the individual episodic future thinking tasks and either the ToM or EF batteries. As such, these findings do not support the claim that, at least during the preschool years, there are strong links between children's developing capacity to think about the future and their theory of mind and executive functioning abilities.

To our knowledge, there is no research that examines the link between episodic future thinking and ToM in older children. However, there is some evidence that, unlike in younger children, the capacity to think about the future is related to some aspects of executive functioning. Most notably, Gott and Lah (2014) found that executive functioning, measured by digit span and a trail-making task, were positively related to the number of episodic, but not semantic, details that 8- to 10-year-olds and 14- to 16-year-olds provided on the Child Autobiographical Interview. This particular finding appears to suggest that by later childhood, executive functions contribute to children's ability to construct future episodic, but not semantic, details. This finding is interesting because it suggests that there are important differences between the processes required to think about the *self* in the future, and those required to think about more generic knowledge about the future.

Although it appears that cognitive abilities such as memory, theory of mind, and executive functioning influence episodic future thinking ability, the research in these areas (especially with respect to ToM and EF) is still quite limited, making it difficult to draw firm conclusions. Perhaps the most unequivocal finding from the data that we have discussed thus far is that, for younger and older children alike, memory plays a key role in the ability to think about the future. Although we have mostly highlighted the role of episodic memory in this respect, it is important to recognize that different types of memory (including semantic) contribute—and may in fact be necessary—to an individual's capacity for episodic future thinking (see Klein, 2013, for a comprehensive review). As such, a potentially important contribution of developmental work is to explore how children's episodic future thinking changes as a function of the development of different types of memory abilities over early and middle childhood. The roles of ToM and EF—at least in early childhood—are less clear and thus merit further attention, with a particular eye to which specific aspects of future thinking ability are related to which particular aspects of ToM and/or EF. In older children, at least, it appears that the extent to which one is able to provide episodic future details is related to several aspects of EF (Gott & Lah, 2014). More broadly, as research in this area starts to grow, there are other potential factors that could be explored in relation to children's developing capacities for future thought. For example, there may be certain personality characteristics that may make an individual better or at least more prone to thinking about the future, such as conscientiousness, anxiety level, or a future-oriented focus (vs. a present-oriented focus).

## 2. THEORETICAL AND EMPIRICAL CHALLENGES

In this final section of the chapter, we outline what we perceive as some theoretical and empirical challenges in the study of children's episodic future thinking. The first is that episodic future thinking has been measured using a variety of different tasks in early childhood and, as highlighted in the study by Hanson et al. (2014), children's performance on these tasks is not significantly correlated. Although Hanson et al. did not include any of the prototypical behavioral tasks described earlier in this chapter (e.g., "spoon tests"), their findings nonetheless suggest that these tasks are measuring different aspects of future orientation. Perhaps this is unsurprising given that episodic future thinking, or "prospection," likely encompasses many different components (as recently highlighted by Szpunar, Spreng, & Schacter, 2014). Moreover, because different researchers have approached the topic from different theoretical and empirical perspectives, many of the tasks that have been developed differ on a number of dimensions. This fact may also partly account for Hanson et al.'s (2014) failure to find significant correlations between episodic future thinking ability and either ToM or EF. Accordingly, a useful future direction in this area of study is to shed light on what some of these key dimensions may be.

We believe that one of these dimensions is the extent to which thinking about the future requires overcoming conflict between what is currently true, desirable, and so on, and what will be true, desirable, and so on, in the future. At one end of this continuum are scenarios in which one's current state directly conflicts with a state that will be experienced in the future. Of the tasks that we have discussed in this chapter, the pretzel task (e.g., Atance & Meltzoff, 2006) best characterizes this end of the continuum. And, as we saw, this task seems to pose equal difficulty for younger and older children alike. Although, to our knowledge, this task has not been run with adults, the fact that they too struggle with accurately predicting their future states (e.g., Gilbert, Gill, & Wilson, 2002; Nisbett & Kanouse, 1969; Wilson & Gilbert, 2003) suggests that performance will be poor. In addition, it is likely that having to overcome a salient, *physiological* state (e.g., thirst) makes this task especially difficult. Some evidence to support this idea is that, by age 5, children are able to acknowledge that a preference that they currently hold (e.g., liking Kool-Aid more than coffee) will likely change in the future (i.e., liking coffee more than Kool-Aid) (Bélanger, Atance, Varghese, Nguyen, & Vendetti, 2014). Although this kind of reasoning seems quite similar to the reasoning required to pass the pretzel task, it is not centered on a physiological state but, rather, one than is more mental/emotional.

At the other end of the continuum are tasks in which what is currently true, desirable, and so on, does not differ appreciably from what will be true, desirable, and so on, in the future. Although such tasks may still require that children imagine what will happen in the future (e.g., thinking about what they will do "tomorrow") and do indeed show age-related improvement between ages 3 and 5, doing so may not require a strong suppression of one's current state. We would argue that most of the behavioral tasks that have been developed for use with young children could be characterized in this way. For example, selecting a key to address the problem of opening a locked box, or putting toys in the room where there are none, may require children to think ahead but do not necessarily involve children thinking about a future that is appreciably different from the present. Moreover, as discussed earlier in this chapter, by about age 5, children no longer have difficulty with this sort of task. Interestingly, some preliminary data from our labs suggest that certain types of "conflict" episodic future thinking tasks show a more protracted development than "no-conflict" tasks (Mahy et al., 2015).

Another important distinction that we believe merits further attention in this area of research is the extent to which thinking about one's own future is similar to or different from thinking about the future of another person. Because both of these abilities require that an individual "self-project" out of his or her current perspective to consider a future one (Buckner & Carroll, 2007; Payne et al., 2014), we might expect a fairly strong relation between the two. In contrast, integral to many definitions of mental time travel is the assertion that it is the *self* that is mentally projected in time (e.g., Atance & O'Neill, 2001; Suddendorf & Corballis, 2007; Tulving, 1985). As such, we might expect that mentally projecting *someone else's self* requires different cognitive mechanisms. There are now a handful of developmental studies that have included questions that pertain to both the child participant's future, as well as the future of another child. Interestingly, on those tasks in which there is more conflict between children's current and future states, there appears to be some benefit to adopting another person's future perspective. For example, Bélanger et al. (2014) found that children were more accurate at predicting that another child would prefer coffee to Kool-Aid when he or she was all grown up, than they were at making this same prediction for themselves. In contrast, on tasks that arguably entail less conflict between current and future states (e.g., Payne et al., 2014), children performed similarly when asked to think about their own and another person's future. These potential differences merit further study because they suggest that one method to improve children's ability to accurately think about the future—at least when current-future conflict is high—is to encourage them to consider what someone else should do.

## 3. CONCLUSIONS

Over the past few years, there has been an increase in research devoted to young children's episodic future thinking. As a result, we are learning more about the contexts that enhance as well as hinder this ability. We have outlined some of these factors by tracing the developmental trajectory and roots of episodic future thinking ability, along with what partial knowledge/conceptual understanding may look like across development. With respect to talk about the future, we have argued that young children may have the necessary linguistic and imaginative capacities to consider the future, but are less skilled than older children at formulating future scenarios that will actually play themselves out. The challenges that young preschoolers face with behavioral methodologies—at least those used to date—seem less centered on their inability to envision "realistic" future scenarios but, rather, on their inability to remember past information that is critical to formulating adaptive future-oriented solutions. Nonetheless, even when memorial requirements are minimized (e.g., Atance et al., 2015; Russell et al., 2010), older preschoolers outperform their younger counterparts, suggesting that capacities specific to "foresight" also play a role. Whether these capacities include and/or rely upon theory of mind and/or executive function is still an open question (though data with preschoolers do no support this claim), as are questions about key dimensions that impact children's ability to think about the future.

## REFERENCES


Addis, D. R., & Schacter, D. L. (2008). Constructive episodic simulation: temporal distance and detail of past and future events modulate hippocampal engagement. *Hippocampus, 18*, 227–237. doi: 10.1002/hipo.20405.

Addis, D. R., Wong, A. T., & Schacter, D. L. (2007). Remembering the past and imagining the future: common and distinct neural substrates during event construction and elaboration. *Neuropsychologia, 45*, 1363–1377. doi: 10.1016/j.neuropsychologia.2006.10.016.

Atance, C. M., & Hanson, L. K. (2011). Making predictions: a developmental perspective. In *Predictions in the brain: using our past to generate a future* (pp. 311–324). New York: Oxford University Press.

Atance, C. M., Louw, A., & Clayton, N. S. (2015). Thinking ahead about where something is needed: new insights about episodic foresight in preschoolers. *Journal of Experimental Child Psychology, 129*, 98–109. doi: 10.1016/j.jecp.2014.09.001.

Atance, C. M., & Meltzoff, A. N. (2005). My future self: young children's ability to anticipate and explain future states. *Cognitive Development, 20*, 341–361. doi: 10.1016/j.cogdev.2005.05.001.

Atance, C. M., & Meltzoff, A. N. (2006). Preschoolers' current desires warp their choices for the future. *Psychological Science, 17*(7), 583–587. doi: 10.1111/j.1467-9280.2006.01748.x.

Atance, C. M., & O'Neill, D. K. (2001). Episodic future thinking. *Trends in Cognitive Sciences, 5*, 533–539. doi: 10.1016/S1364-6613(00)01804-0

Atance, C. M., & Sommerville, J. A. (2014). Assessing the role of memory in preschoolers' performance on episodic foresight tasks. *Memory, 22*, 118–128. doi: 10.1080/09658211.2013.820324.

Bélanger, M. J., Atance, C. M., Varghese, A. L., Nguyen, V., & Vendetti, C. (2014). What will I like best when I'm all grown up? Preschoolers' understanding of future preferences. *Child Development, 85*, 2419–2431. doi: 10.1111/cdev.12282.

Buckner, R. L., & Carroll, D. C. (2007). Self-projection and the brain. *Trends in Cognitive Sciences, 11*, 49–57. doi: 10.1016/j.tics.2006.11.004.

Busby Grant, J., & Suddendorf, T. (2009). Preschoolers begin to differentiate the times of events from throughout the lifespan. *European Journal of Developmental Psychology, 6*, 746–762. doi: 10.1080/17405620802102947.

Busby Grant, J., & Suddendorf, T. (2011). Production of temporal terms by 3-, 4-, and 5-year-old children. *Early Childhood Research Quarterly, 26*, 87–95. doi: 10.1016/j.ecresq.2010.05.002.

Busby, J., & Suddendorf, T. (2005). Recalling yesterday and predicting tomorrow. *Cognitive Development, 20*, 362–372. doi: 10.1016/j.cogdev.2005.05.002.

Coughlin, C., Lyons, K. E., & Ghetti, S. (2014). Remembering the past to envision the future in middle childhood: developmental linkages between prospection and episodic memory. *Cognitive Development, 30*, 96–110. doi: 10.1016/j.cogdev.2014.02.001.

D'Argembeau, A., & Van der Linden, M. (2004). Phenomenal characteristics associated with projecting oneself back into the past and forward into the future: influence of valence and temporal distance. *Consciousness and Cognition, 13*, 844–858. doi: 10.1016/j.concog.2004.07.007.

Friedman, W. J. (2000). The development of children's knowledge of the times of future events. *Child Development, 71*, 913–932.

Gilbert, D. T., Gill, M. J., & Wilson, T. D. (2002). The future is now: temporal correction in affective forecasting. *Organizational Behavior and Human Decision Processes, 88*, 430–444. doi: 10.1006/obhd.2001.2982.

Gott, C., & Lah, S. (2014). Episodic future thinking in children compared to adolescents. *Child Neuropsychology, 20*, 625–640. doi: 10.1080/09297049.2013.840362.

Haith, M. (1997). The development of future thinking as essential for the emergence of skill in planning. In S. L. Friedman & E. Kofsky-Scholnick (eds.), *The developmental psychology of planning: why, how, and when do we plan* (pp. 25–42). Mahwah, NJ: Lawrence Erlbaum Associates.

Hanson, L. K., Atance, C. M., & Paluck, S. W. (2014). Is thinking about the future related to theory of mind and executive function? Not in preschoolers. *Journal of Experimental Child Psychology, 128*, 120–137. doi: 10.1016/j.jecp.2014.07.006.

Hayne, H., Gross, J., McNamee, S., Fitzgibbon, O., & Tustin, K. (2011). Episodic memory and episodic foresight in 3- and 5-year-old children. *Cognitive Development, 26*, 343–355. doi: 10.1016/j.cogdev.2011.09.006.

Hudson, J. A., Mayhew, E. M. Y., & Prabhakar, J. (2011). The development of episodic foresight: emerging concepts and methods. In Janette B. Benson (ed.), *Advances in child development and behavior* (Vol. 40, pp. 95–137). Boston: Elsevier. http://www.sciencedirect.com/science/article/pii/B9780123864918000037.

Hudson, J. A., Shapiro, L. R., & Sosa, B. B. (1995). Planning in the real world: preschool children's scripts and plans for familiar events. *Child Development, 66*, 984. doi: 10.2307/1131793.

Klein, S. B. (2013). The complex act of projecting oneself into the future. *Wires Cognitive Science, 4*, 63–69. doi: 10.1002/wcs.1210.

Lagattuta, K. H. (2007). Thinking about the future because of the past: young children's knowledge about the causes of worry and preventative decisions. *Child Development, 78*, 1492–1509. doi: 10.1111/j.1467-8624.2007.01079.x.

Liberman, N., & Trope, Y. (1998). The role of feasibility and desirability considerations in near and distant future decisions: a test of temporal construal theory. *Journal of Personality and Social Psychology, 75*, 5–18. doi: 10.1037/0022-3514.75.1.5.

Mahy, C. E. V., Atance, C. M., Moses, L. J., & Kopp, L. (2015). Developmental trajectories of episodic foresight for self and other in early childhood. Manuscript submitted for publication.

Mahy, C. E. V., Grass, J., Wagner, S., & Kliegel, M. (2014). These pretzels are going to make me thirsty tomorrow: differential development of hot and cool episodic foresight in early childhood? *British Journal of Developmental Psychology, 32*, 65–77. doi: 10.1111/bjdp.12023.

McColgan, K. L., & McCormack, T. (2008). Searching and planning: young children's reasoning about past and future event sequences. *Child Development*, *79*, 1477–1497. doi: 10.1111/j.1467-8624.2008.01200.x.

McCormack, T., & Atance, C. M. (2011). Planning in young children: a review and synthesis. *Developmental Review*, *31*, 1–31. doi: 10.1016/j.dr.2011.02.002.

Metcalf, J. L., & Atance, C. M. (2011). Do preschoolers save to benefit their future selves? *Cognitive Development*, *26*, 371–382. doi: 10.1016/j.cogdev.2011.09.003.

Mulcahy, N. J., & Call, J. (2006). Apes save tools for future use. *Science*, *312*, 1038–1040. doi: 10.1126/science.1125456.

Nelson, K. (1989). *Narratives from the crib*. Cambridge, MA: Harvard University Press.

Nisbett, R. E., & Kanouse, D. E. (1969). Obesity, food deprivation, and supermarket shopping behavior. *Journal of Personality and Social Psychology*, *12*, 289.

Okuda, J., Fujii, T., Ohtake, H., Tsukiura, T., Tanji, K., Suzuki, K., . . . Yamadori, A. (2003). Thinking of the future and past: the roles of the frontal pole and the medial temporal lobes. *NeuroImage*, *19*, 1369–1380. doi: 10.1016/S1053-8119(03)00179-4.

Payne, G., Taylor, R., Hayne, H., & Scarf, D. (2014). Mental time travel for self and other in three- and four-year-old children. *Memory*, 1–8. doi: 10.1080/09658211.2014.921310.

Quon, E., & Atance, C. M. (2010). A comparison of preschoolers' memory, knowledge, and anticipation of events. *Journal of Cognition and Development*, *11*, 37–60. doi: 10.1080/15248370903453576.

Raby, C. R., Alexis, D. M., Dickinson, A., & Clayton, N. S. (2007). Planning for the future by western scrub-jays. *Nature*, *445*, 919–921. doi: 10.1038/nature05575.

Richmond, J. L., & Pan, R. (2013). Thinking about the future early in life: the role of relational memory. *Journal of Experimental Child Psychology*, *114*, 510–521. doi: 10.1016/j.jecp.2012.11.002.

Russell, J., Alexis, D., & Clayton, N. (2010). Episodic future thinking in 3- to 5-year-old children: the ability to think of what will be needed from a different point of view. *Cognition*, *114*, 56–71. doi: 10.1016/j.cognition.2009.08.013.

Scarf, D., Gross, J., Colombo, M., & Hayne, H. (2013). To have and to hold: episodic memory in 3- and 4-year-old children. *Developmental Psychobiology*, *55*, 125–132. doi: 10.1002/dev.21004.

Schacter, D. L., & Addis, D. R. (2007). The cognitive neuroscience of constructive memory: remembering the past and imagining the future. *Philosophical Transactions of the Royal Society of London B: Biological Sciences*, *362*, 773–786. doi: 10.1098/rstb.2007.2087.

Suddendorf, T., & Busby, J. (2005). Making decisions with the future in mind: Developmental and comparative identification of mental time travel. *Learning and Motivation*, *36*, 110–125. doi:10.1016/j.lmot.2005.02.010

Suddendorf, T., & Corballis, M. C. (1997). Mental time travel and the evolution of the human mind. *Genetic, Social, and General Psychology Monographs*, *123*, 133–167.

Suddendorf, T., & Corballis, M. C. (2007). The evolution of foresight: what is mental time travel, and is it unique to humans? *Behavioral and Brain Sciences*, *30*, 299–351. doi: 10.1017/S0140525X07001975.

Suddendorf, T., Nielsen, M., & von Gehlen, R. (2011). Children's capacity to remember a novel problem and to secure its future solution. *Developmental Science*, *14*, 26–33. doi: 10.1111/j.1467-7687.2010.00950.x.

Suddendorf, T., & Redshaw, J. (2013). The development of mental scenario building and episodic foresight. *Annals of the New York Academy of Sciences*, *1296*, 135–153. http://doi.org/10.1111/nyas.12189.

Szpunar, K. K., Spreng, R. N., & Schacter, D. L. (2014). A taxonomy of prospection: introducing an organizational framework for future-oriented cognition. *Proceedings of the National Academy of Sciences, 11*, 18414–18421. doi: 10.1073/pnas.1417144111.

Szpunar, K. K., Watson, J. M., & McDermott, K. B. (2007). Neural substrates of envisioning the future. *Proceedings of the National Academy of Sciences, 104*, 642–647.

Tulving, E. (1985). Memory and consciousness. *Canadian Psychology/Psychologie canadienne, 26*, 1–12. doi: 10.1037/h0080017.

Tulving, E. (2005). Episodic memory and autonoesis: uniquely human? In H. S. Terrace & J. Metcalfe (eds.), *The missing link in cognition: origins of self-reflective consciousness* (pp. 3–56). New York: Oxford University Press.

Wang, Q., Capous, D., Koh, J. B. K., & Hou, Y. (2014). Past and future episodic thinking in middle childhood. *Journal of Cognition and Development, 15*, 625–643. doi: 10.1080/15248372.2013.784977.

Wheeler, M. A., Stuss, D. T., & Tulving, E. (1997). Toward a theory of episodic memory: the frontal lobes and autonoetic consciousness. *Psychological Bulletin, 121*, 331–354. doi: 10.1037/0033-2909.121.3.331.

Wilson, T. D., & Gilbert, D. T. (2003). Affective forecasting. *Advances in Experimental Social Psychology, 35*, 345–411.

# PART VIII

# Clinical Perspectives

19

# Semantic Memory as the Essential Scaffold for Future-Oriented Mental Time Travel

MUIREANN IRISH

The ability to mentally project oneself into the future allows humans to engage in complex and evolutionary adaptive future-oriented behaviors such as simulating possible events, predicting the outcome of actions, forming intentions, affective forecasting, and planning. Future mental time travel represents a fascinating aspect of human cognition and is the focus of increasing attention within the scientific literature. Over the last decade, research in this field has witnessed a dramatic surge, with efforts directed to explicating the cognitive and neural architecture of this uniquely human capacity. It is becoming increasingly apparent, however, that the conceptualization and, by extension, study of future-oriented mental time travel has to date been couched with reference to the episodic memory system. As will be argued in this chapter, the undifferentiated and flexible nature of semantic memory renders it equally well suited to support complex acts of future-oriented mental time travel. Although largely neglected within the discourse on future thinking, the idea that semantic memory may provide a fundamental framework for the expression of many complex cognitive functions is gradually gaining traction within the field of cognitive neuroscience. Here, I will provide an overview of the semantic memory system, beginning with its conceptualization as a time-invariant repository of conceptual knowledge. I will then discuss mounting evidence demonstrating the interdependence between episodic and semantic memory, and the movement toward a conceptual continuum, rather than a strict dichotomy, of these memory systems. I will pay particular attention to studies of future thinking in clinical populations, as a complement to the extant functional neuroimaging literature, which have proved revelatory for clarifying the contribution of semantic memory to future mental time travel. Finally, I will expand upon the idea of *semantic scaffolding*, addressing current inconsistencies in the literature and highlighting areas that warrant further attention. In this manner, the objective of this chapter is not to argue that semantic memory takes precedence above episodic memory in the service of future simulation, but rather to emphasize the necessary confluence of episodic and semantic memory in the genesis of complex acts of future-oriented mental time travel.

## 1. WHAT IS SEMANTIC MEMORY?

Semantic memory is typically conceptualized as an individual's repository of acquired knowledge of the world (Binder & Desai, 2011). As semantic constructs are posited to be divorced from any specific experience, the content of semantic memory is said to be "conceptual" (Binder & Desai, 2011). In many ways, conceptual knowledge can be thought of as a multimodal "semantic database" that imparts meaning to all stimuli and experiences that we encounter in our daily lives (Lambon Ralph, 2014). In this regard, semantic memory is integral to all aspects of human cognition, as it brings meaning to verbal and non-verbal information, thus providing the foundation for a host of complex cognitive endeavors (Binder & Desai, 2011).

### 1.1. Models of Semantic Memory

Perhaps the most prominent model of semantic memory in the literature is that modality-specific regions in the brain are connected to, and communicate through, a common or shared amodal "hub" (Patterson, Nestor, & Rogers, 2007). This hub-and-spoke model of semantic memory holds that modality-specific associations are processed in each respective modality, but are drawn together into coherent concepts via an amodal hub (Lambon Ralph, Sage, Jones, & Mayberry, 2010; Patterson et al., 2007). The hub thus permits the transmodal representation of information, which in turn facilitates flexible acts of cognition such as semantic-based generalization to new or changing concepts (reviewed by Lambon Ralph, 2014). From a neuroanatomical perspective, this shared amodal hub is proposed to reside in the anterior temporal lobes of the brain (see Figure 19.1). By this view, damage to the semantic hub is predicted to produce a global and transmodal semantic impairment, as is the case in the disease syndrome of semantic dementia (Patterson et al., 2007).

More recently, however, an alternative position has been put forward emphasizing two separate, but interacting, routes of processing within the temporal lobes (Mesulam et al., 2013). This two-route account argues that distributed cross-network interactions between the language and object networks in the brain support conceptual processing, without the need for a single domain-independent hub. As such, a left lateralized temporosylvian language network specialized for verbal concepts is proposed to interact with a right-sided or potentially bilateral inferotemporal/fusiform object recognition network (Mesulam et al., 2013).

### 1.2. Neuroanatomy of the Semantic Memory System

Neuroimaging studies of healthy individuals have clarified the neuroanatomical signature of semantic memory. A recent meta-analysis of 120 functional

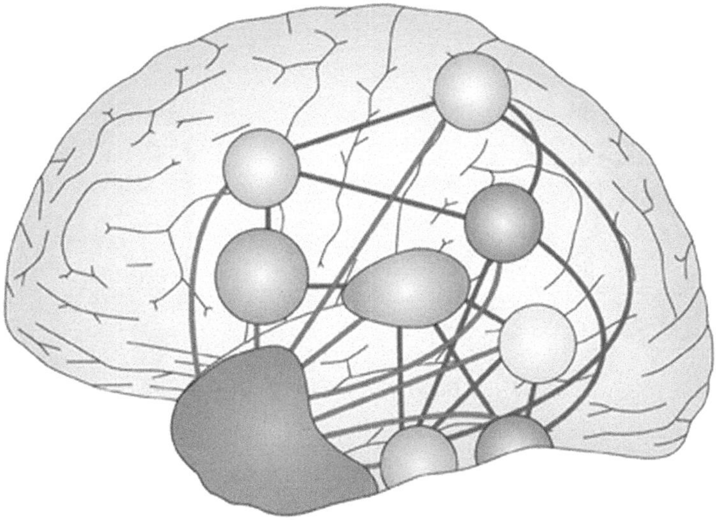

**Figure 19.1.** The hub-and-spoke model of semantic memory proposed by Patterson et al. (2007).
Reprinted by permission from Macmillan Publishers: Patterson, K., Nestor, P. J., Rogers, T. T., Where do you know what you know? the representation of semantic knowledge in the human brain, *Nature Reviews Neuroscience, 8*(12), 976–987, © 2007. Adapted from Figure 1B, 977.

neuroimaging studies investigating semantic processing revealed remarkably consistent patterns of activation in a left-lateralized network comprising temporal and parietal heteromodal cortex, as well as prefrontal and posterior parietal regions (Figure 19.2; Binder, Desai, Graves, & Conant, 2009). This finding has led to the proposal of a large-scale neural model of semantic processing encompassing modality-specific and supramodal representations (Binder & Desai, 2011). Accordingly, high-level convergence zones located in the inferior parietal, lateral temporal, and ventral temporal association cortices are preferentially recruited in the service of undifferentiated conceptual processing (Binder & Desai, 2011; Binder et al., 2009). These supramodal convergences of information are argued to provide the requisite framework for an array of conceptually driven functions including language, object recognition, social cognition, and the capacity to remember the past and imagine the future (Binder & Desai, 2011).

## 2. TEMPORALITY AND THE EPISODIC-SEMANTIC DISTINCTION

The treatment of episodic and semantic memory as largely dissociable systems can be traced back to Tulving's highly influential theory of declarative memory. Tulving originally proposed that episodic memory consists of "temporally dated episodes or events, and the temporal-spatial relations" between these events (Tulving, 1972, 385). On the other hand, semantic memory was conceptualized as knowledge

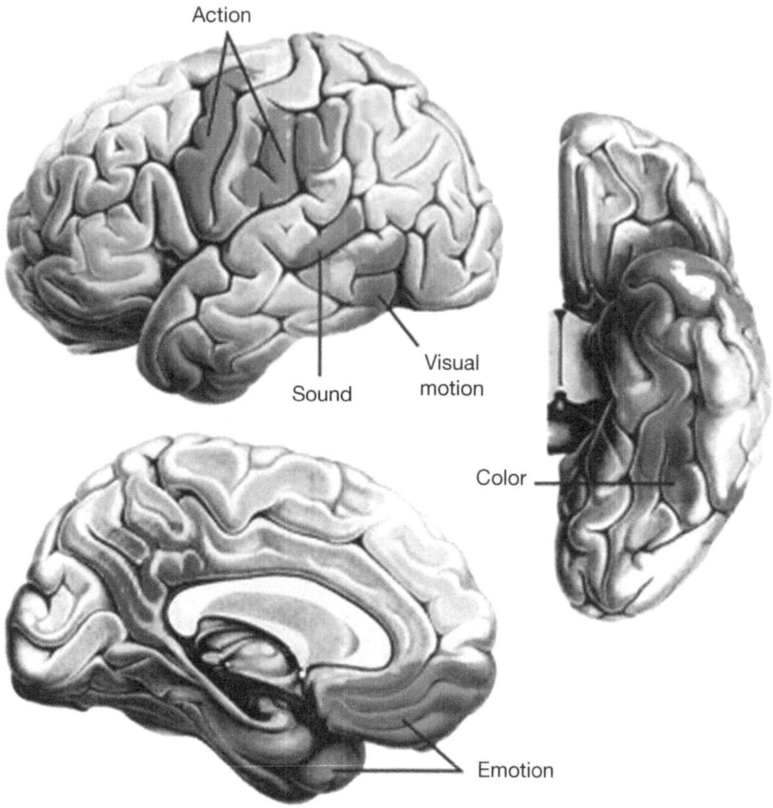


**Figure 19.2.** Binder and Desai's (2009) neuroanatomical model of semantic processing. Reprinted from Binder, J. H, Desai, R. H., The neurobiology of semantic memory, *Trends in Cognitive Sciences, 15*(11), 531, © 2011, with permission from Elsevier.

experienced as independent from the original source of acquisition (Tulving, 1972, 1983). Taken further, the episodic-semantic heuristic holds that different forms of temporal phenomenology are associated with the two forms of memory (Klein, 2015). Whereas episodic memory gives rise to an event being experienced in the past or future, the temporal experience of semantic memory is typically confined to the "here and now," and this form of memory is not accompanied by a feeling of re-experiencing the original act of acquisition (reviewed by Klein, 2015).

By this view, episodic memory, via the accompanying autonoetic experience, provides the essential foundation for mental time travel across past and future contexts (Tulving, 1985; Wheeler, Stuss, & Tulving, 1997). Autonoetic awareness is held as the prerequisite for mental time travel as it incorporates a personally meaningful relationship to time (Szpunar & Tulving, 2011). Semantic memory, by contrast, is mediated by noetic ("knowing") consciousness, which permits the retrieval of conceptual information "in the absence of a feeling of re-experiencing the past" (Szpunar, 2010, 144). As such, the accompanying temporal experience of semantic memory is typically viewed as confined to the present moment (reviewed by Klein, 2015), precluding the ability to re-experience the past or to pre-experience the future (Szpunar & Tulving, 2011; Tulving, 1985). Semantic memory has thus been regarded as a temporally unconstrained system, which lacks the temporal resources fundamental for future-oriented thinking (Suddendorf & Corballis, 2007; Tulving, 1985, 2002).

## 3. INTERACTIONS BETWEEN THE EPISODIC AND SEMANTIC MEMORY SYSTEMS

Advances in functional neuroimaging and neuropsychological research have begun to reveal significant interplay between episodic and semantic memory, moving beyond the long-held idea that these memory systems are, by definition, orthogonal. For example, the distributed semantic memory system identified by Binder and colleagues shows striking overlap with the core network that subtends autobiographical memory retrieval (Binder et al., 2009; Maguire, 2001; Svoboda, McKinnon, & Levine, 2006), resonating with the proposal that autobiographical memories, by necessity, encompass a large proportion of semantic concepts (Binder et al., 2009; Greenberg & Verfaellie, 2010). Mounting evidence from the functional neuroimaging literature corroborates the interdependence of episodic and semantic memory, underscoring the unclear boundaries that currently exist between these systems (see Burianova, McIntosh, & Grady, 2010). This overlap intuitively stands to reason if we consider the type of information that is invariably called to mind during the process of autobiographical memory retrieval. For example, when we remember our wedding day (e.g., "I got married in 2011 in a small church in Ireland"), interleaved within this description will be semantic elements that are pivotal to convey meaning to the overall narrative (e.g., "This was the church where my parents were married"). Semantic memory, therefore, is viewed as integral to the autobiographical memory trace, a position that is supported by numerous studies in clinical populations (Graham, Patterson, & Hodges, 1999; Irish et al., 2011; McKinnon, Black, Miller, Moscovitch, & Levine, 2006; Westmacott, Leach, Freedman, & Moscovitch, 2001).

The core constituents of episodic memory, namely the "what," "where," and "when" (Tulving, 1985), also reside within semantic memory (Klein, 2015). For example, we can know specific event, spatial, and temporal information, without recollecting the original source of acquisition of this information (Grilli & Verfaellie, 2014; Martinelli, Sperduti, & Piolino, 2013). Such instances of retrieval further confound the heretofore strict separation of episodic and semantic processes. Indeed, it has recently been proposed that mental time travel can be conceptualized in *episodic-semantic neutral* terms, with self-projection across subjective time occurring regardless of the episodic or semantic memory content (Stocker, 2012). This conceptualization supports the prevailing view that simulation of future events relies upon the integrity of retrieval from episodic memory (Schacter & Addis, 2007a; Schacter, Addis, & Buckner, 2007) but also resonates with demonstrations of semantic forms of future-oriented mental time travel (Irish, Addis, Hodges, & Piguet, 2012a; Klein, Loftus, & Kihlstrom, 2002; Race, Keane, & Verfaellie, 2013). Thus, the conception of semantic memory as a system orthogonal to that of episodic memory is gradually shifting toward a unified account in which semantic processing is viewed as playing a central role in all aspects of internal mentation (Binder & Desai, 2011; Irish & Piguet, 2013).

## 4. THE LATE ARRIVAL OF SEMANTIC MEMORY

Semantic memory has without question experienced something of a late arrival to the field, becoming a topic of formal investigation in its own right only relatively recently (Binder et al., 2009). Traditionally, studies of conceptual knowledge tended to be restricted to the domain of object knowledge, failing to embrace the many other ways in which humans occupy their time, such as internally driven forms of cognition including remembering, imagining, planning, and reasoning (Binder & Desai, 2011). Where future-oriented mental time travel is concerned, the late arrival of semantic memory is particularly palpable, the focus to date resting almost exclusively upon contributions from episodic memory (Irish & Piguet, 2013; Klein, 2013a; Schacter et al., 2012). If semantic memories are "the basic material from which complex and detailed episodic memories are constructed" (Greenberg & Verfaellie, 2010, 750), it follows that the same principle should hold with regard to future-oriented cognition. In the next section, I will present the case that the very nature of semantic representations renders them ideally suited to engaging in complex acts of future mental time travel.

### 4.1. Semantic Representations Are Ideally Suited to Future Mental Time Travel

By definition, semantic representations are derived from the accumulation of, and abstraction from, multiple experiences, and thus can be generalized appropriately from one exemplar to another (Binder & Desai, 2011; Lambon Ralph, Pobric, & Jefferies, 2009). In this regard, semantic representations are arguably more flexible than episodic memories, as their undifferentiated and atemporal nature facilitates generalization and extrapolation across contexts without recourse to any specific experience (Hegdé, 2007; Lambon Ralph et al., 2010). As such, we can harness our conceptual store of knowledge to predict and simulate events that have not yet occurred and of which we have no prior personal experience. For example, one can imagine unlikely (e.g., travelling to Antarctica) and even implausible (e.g., walking on the moon) experiences simply by drawing upon and combining abstracted representations from one's conceptual store. Related to this, semantic memory has been shown to underlie acts of conceptual expansion in which existing concepts are widened to include additional features and attributes (Rutter et al., 2012), thus facilitating expressions of creativity. Accordingly, semantic memory may afford significantly more flexibility where imaginative endeavors are concerned. It has been argued that episodic memory is in fact the least compatible form of memory for future-oriented cognition, in that "what one can predict about the future will be limited by one's episodic memory" (Hegdé, 2007, 324). Given this hypothetical "computational bottleneck" of episodic memory (Hegdé, 2007), the modulating role of other forms of memory for acts of future thinking is gradually gaining acceptance, with the proposal that semantic memory may not only facilitate but may be *essential* for future-oriented forms of mental time travel (Irish & Piguet, 2013; Klein, 2013a). Incidentally, this proposal is not new, as Ingvar (1985) made the prescient observations that semantic memory "is mainly involved in the cognitive or future consequences and meaning of events," whereas episodic memory "pertains to actual sensory and cognitive percepts in the past and present" (Ingvar, 1985, 129).

### 4.2. Semantic Contributions to Future Thinking: Evidence from Neuroimaging Studies

While the episodic-semantic heuristic has persisted into the future-thinking domain, a number of anomalies challenge the prevailing view that episodic memory exclusively provides the building blocks for acts of future simulation. For instance, an fMRI study of past and future modes of thinking across personal (episodic) and non-personal (semantic) contexts revealed significant recruitment of semantic processing regions, including the inferior temporal gyrus and the temporal poles, regardless of temporal context (past/future) or self-referential condition (Abraham, Schubotz, & von Cramon, 2008). Similarly, an fMRI study of healthy older adults revealed dissociations between the neural substrates of remembered personal events and envisioned future plans (Viard et al., 2011). Specifically, the envisaging of future plans recruited the inferior frontal and lateral temporal gyri, regions classically implicated in semantic memory (Viard et al., 2011). It is notable that lateral temporal regions have been implicated in many of the functional neuroimaging studies on future simulation in healthy individuals (reviewed by Bubić & Abraham, 2014), which is suggestive of a fundamental foundation for semantic processing in future-oriented mental time travel.

## 5. INSIGHTS FROM AMNESIC PATIENTS

The most compelling evidence for a core contribution of semantic memory to future-oriented thinking, however, comes from the study of clinical populations in which divergent damage to the episodic and semantic memory systems is present. Patient D. B. suffered a global loss of anterograde and retrograde episodic memory function as a result of an anoxic episode following a heart attack (Klein et al., 2002). Using a simple questionnaire, Klein and colleagues distinguished between the experience of *lived time*, mediated by episodic memory, versus that of *known time*, taken to reflect the experience of time mediated by semantic memory (Klein et al., 2002). The authors specifically included the *known time* component in recognition of the fact that it is possible to envisage a non-personal future without placing oneself squarely within the simulation, for example knowledge of public events and world issues (Klein et al., 2002). In keeping with his severe episodic impairment, D. B. displayed significant deficits on *lived time* conditions across past and future contexts. As such, he was unable to remember previously experienced events within his past, or to project himself forward in subjective time to envisage possible events he might experience in the future (e.g., "When will be the next time you visit a doctor?"). Despite these profound impairments in *lived time,* D. B. could engage in impersonal or non-episodic forms of mental time travel. He demonstrated relatively intact retrieval of non-personal events (e.g., important issues facing his community over the last 10 years), and he was surprisingly capable of envisaging non-episodic future issues (e.g., issues facing the environment in the next 10 years). This case study elegantly demonstrates that the tendency to conceptualize mental time travel in terms of one form of temporal subjectivity may be overly restrictive, in that amnesic patients appear capable of certain forms of impersonal temporal projection (Klein, 2013a; Klein et al., 2002; Race et al., 2013). A recent study investigating the capacity for semantic prospection in episodic amnesic patients is noteworthy in this regard (Race et al., 2013). While patients appeared adept at describing potential future issues in the public domain (e.g., foreign policy, the environment), marked difficulties were evident when participants were asked to elaborate on these issues. Notably, this semantic elaboration impairment occurred in the context of relatively intact semantic processing. The authors proposed that while generic and routinized information can be retrieved independently of the medial temporal lobe (see also Steinvorth, Levine, & Corkin, 2005), elaboration of semantic information during prospection likely draws upon mnemonic processes supported by the hippocampus. As such, the extent to which semantic forms of future thinking can be achieved in amnesic populations may hinge critically upon demands placed on core elaborative processes (Race et al., 2013).

## 5.1. The Representation of Time in Amnesia

It is becoming clear that amnesic patients are not necessarily "trapped in time," but may avail of non-episodic forms of temporal consciousness (Craver, Kwan, Steindam, & Rosenbaum, 2014; Klein, 2013a). For example, it was suggested that the amnesic patient D. B. could accomplish acts of semantic prospection via his preserved comprehension of the language of time (Klein, 2013b). Temporal consciousness in this respect is defined as "a form of consciousness that allows individuals to remember their personal past, to be oriented in their present world, and to predict their personal futures" (Dalla Barba & La Corte, 2013, 102). A recent study of patient K. C. is notable in this respect. K. C. represents one of the most studied amnesic patients, who presents with perhaps the most severe episodic amnesia, due to extensive brain damage sustained during a motorcycle accident (see Tulving, 1985). Little doubt exists regarding the profound episodic memory impairment exhibited by K. C., attributable to significant volume loss in the bilateral hippocampus and parahippocampal gyrus (Rosenbaum et al., 2005; Tulving, 1985). Initial investigations of K. C.'s perception of subjective time (Tulving, 1985) led to the pervasive assumption that episodic amnesic patients lack "temporal consciousness" and hence are stuck in time (e.g., Corkin, 2013; Dalla Barba & La Corte, 2013). Craver and colleagues, however, argue that temporal consciousness is in fact relatively spared in patient K. C., supported by intact retrograde semantic memories, personal semantics, and his spared semantic knowledge of time (Craver et al., 2014). Across a series of carefully chosen tasks, they demonstrated that K. C. patently maintains semantic knowledge of discrete time series and can meaningfully order personal events from his life on a timeline. He displays temporal discounting and decision-making styles similar to healthy controls and lacks hedonistic and fatalistic attitudes that might be expected if one could not appreciate the future (Suddendorf & Corballis, 2007). In stark contrast, a recent case study of a patient with semantic dementia, in which semantic memory is irrevocably degraded (see later discussion in this chapter), reveals that impaired future prospection is closely related to feelings of emptiness, decreased sense of being human, and suicidal behavior (Hsiao, Kaiser, Fong, & Mendez, 2013). In the case of K. C., however, his profound episodic deficit does not place the future out of his reach. Instead, he maintains an awareness of time and temporal consciousness that is largely driven by his relatively intact semantic representations (Craver et al., 2014).

## 6. LOSS OF SEMANTIC MEMORY: THE SYNDROME OF SEMANTIC DEMENTIA

Arguably the most revelatory insights regarding the contribution of semantic memory to future-oriented mental time travel stem from the study of patients who have pathological insult to the semantic memory system. In this regard, the syndrome of semantic dementia, a temporal variant of frontotemporal dementia and one of the primary progressive aphasias (Gorno-Tempini et al., 2011), has proved most illuminating. Semantic dementia offers a compelling view of the neurocognitive architecture of the brain in relation to the systematic and coordinated degeneration of the semantic memory system (Irish, Piguet, & Hodges, 2012). The most prominent feature of this syndrome is anomia, whereby the individual fails to name concepts, objects, and even people (Hodges & Patterson, 2007). Importantly, this anomia reflects the progressive and amodal loss of conceptual knowledge. One of the striking features of semantic dementia is that this loss of world knowledge occurs irrespective of modality, and has thus been attributed to the degradation of the central amodal semantic hub (Patterson et al., 2007). From a neural standpoint, semantic dementia is characterized by the progressive deterioration of the anterior temporal lobes, most severe on the ventral surface, and encompassing the anterior fusiform gyrus and perirhinal cortex, as well as medial temporal lobe structures including the hippocampus (Chan et al., 2001; Galton et al., 2001; Mion et al., 2010). This pattern of atrophy, while ultimately bilateral, is typically lateralized to the left hemisphere (see Figure 19.3), producing a selective impairment in semantic processing (Hodges & Patterson, 2007). Despite these profound semantic impairments, semantic dementia patients display a range of relatively intact cognitive functions, notably that of episodic memory, particularly when non-verbal tasks are employed (Bozeat, Gregory, Ralph, & Hodges, 2000; Crutch & Warrington, 2002; Maguire, Kumaran, Hassabis, & Kopelman, 2010). As such, semantic dementia represents a unique lesion model to clarify the contribution of semantic processes to future-oriented mental time travel (Irish & Piguet, 2013).

### 6.1. Patients with Semantic Dementia Cannot Imagine The Future

If semantic memory is critical for the apprehension of time, and provides a foundation for the generalization and expansion of concepts across contexts, disruption to the semantic memory system should preclude the ability to engage in future-oriented forms of mental time travel, irrespective of content. This is precisely the pattern of results we see in the case of semantic dementia. The first study of future-oriented thought in semantic dementia revealed marked impairments in the capacity to form self-representations at a future time point in the context of a relatively

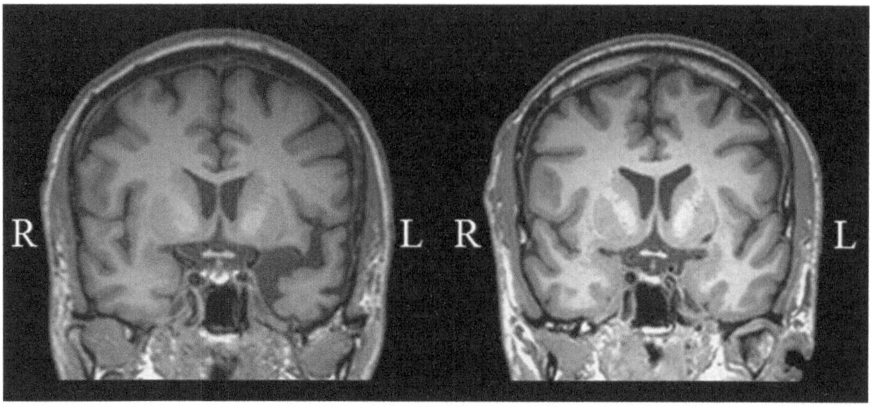


**Figure 19.3.** Structural MRI displaying characteristic pattern of atrophy in semantic dementia (*left*) versus age-matched healthy older control individual (*right*). The semantic dementia patient presents with predominantly left-lateralized atrophy in the anterior temporal cortex and temporal pole.
L = left; R = right.
Credit: Muireann Irish.

spared ability to retrieve self-representations from the past (Duval et al., 2012). Despite intact past retrieval, however, the patients were unable to construct future self-images or to generate relevant contextual details to support their conception of their future selves. The authors concluded that personal semantic information must therefore make a fundamental contribution to future mental time travel (Duval et al., 2012). Work from our group has replicated this asymmetric impairment of future-oriented relative to past-oriented thinking in semantic dementia and has clarified the neural underpinnings of the future-thinking deficit (Irish, Addis, et al., 2012a). Using a modified version of the Past-Future task (Addis, Wong, & Schacter, 2008), we revealed that patients with semantic dementia could remember and describe in rich contextual detail episodic events from their recent past, yet displayed marked impairments when attempting to envisage possible events in the future. Correlation analyses revealed that future thinking impairments were significantly related to disrupted performance on semantic memory tasks. This relationship was confirmed on the neuroanatomical level by voxel-based morphometry analyses of structural neuroimaging scans, which revealed that future thinking deficits related exclusively to gray matter intensity decrease in the left anterior inferior temporal gyrus and the bilateral temporal poles (Figure 19.4; Irish, Addis, et al., 2012a), regions specialized for semantic processing (Mion et al., 2010). Crucially, our findings reveal that despite a relatively intact store of episodic details, which ostensibly can be used to populate future simulations, the act of future thinking is impeded in semantic dementia (reviewed by Irish & Piolino, 2015).

In addition, we used the *known time* subscale developed by Klein and colleagues to probe forms of non-episodic mental time travel in semantic dementia and revealed gross impairments across past and future contexts for the retrieval and envisaging of non-personal events (e.g., community affairs, advances in technology). Importantly, our neuroimaging analyses underscored the importance of the

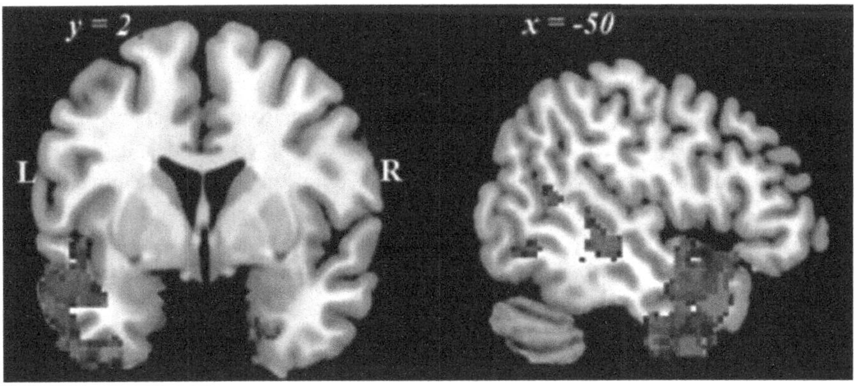


**Figure 19.4.** Voxel-based morphometry analyses displaying significant associations between regions of gray matter intensity decrease and impairments in personal/episodic (red) and non-personal/semantic (blue) forms of future-oriented mental time travel in semantic dementia. The common region implicated in both forms of prospection resides in the left anterior inferior temporal cortex. Taken from Irish et al. (2012a). (See color insert). L = left; R = right.
Figure adapted from Irish, M., Addis, D. R., Hodges, J. R., Piguet, O. (2012), Considering the role of semantic memory in episodic future thinking: evidence from semantic dementia, *Brain, 135*, 2187.

left anterior inferior temporal gyrus of the semantic memory network in mediating past and future non-episodic forms of mental time travel (Irish, Addis, et al., 2012a). These studies provide converging behavioral and neuroimaging evidence in favor of a core role for semantic memory in mediating acts of future mental time travel, irrespective of content, broadening the current impetus of research beyond the contribution of the episodic memory system (see also Andelman, Hoofien, Goldberg, Aizenstein, & Neufeld, 2010; Hassabis & Maguire, 2007).

## 6.2. The Phenomenon of Recasting in Semantic Dementia

One issue within the literature on semantic memory concerns the process by which multimodal experiences are brought together and reconfigured into coherent and generalizable representations (Lambon Ralph et al., 2010). Indeed, the machinations behind such recombination processes are inherently complex, requiring the convergence of multimodal information drawn from different points in time in a unique fashion (Lambon Ralph et al., 2010). A noteworthy observation in this regard concerns the novelty of future events generated by patients with semantic dementia. Our initial exploration of future-oriented thinking in this cohort reveals that semantic dementia patients are particularly hampered in the generation of novel future events (Irish, Addis, et al., 2012a). We found that 80% of the future events generated by these patients had in fact occurred previously in the past, suggesting that the patients merely recast the past event to occur at a future time point. Importantly, this recapitulation of past events occurred despite explicit task instructions requiring participants to create an event not previously experienced.

As such, it appears that the generation of novel events heavily taxes the semantic memory system and hinges on the flexibility and generalizability of semantic representations. This flexibility renders them ideally suited to association in unique and original ways, for example in the case of conceptual expansion for acts of creativity (Rutter et al., 2012). When the conceptual knowledge base degrades, as occurs in semantic dementia, flexible and malleable representations critical for novel event generation can no longer be harnessed, leading to an overreliance upon the relatively intact episodic memory store. As such, we see a preponderance of recast future simulations that more accurately represent retrieval of episodes that have previously transpired (Irish & Piguet, 2013).

## 7. ACCESSIBILITY OF DETAILS DURING FUTURE MENTAL TIME TRAVEL

A second study by our group supports the pivotal role of semantic memory for the construction of novel future events (Irish, Addis, Hodges, & Piguet, 2012b). We conducted a fine-grained analysis of the contextual details provided by semantic dementia patients across past and future temporal contexts to address the question of whether discrete details are differentially compromised as the patients attempt to project themselves into the future. Interestingly, internal event details (i.e., episodic details that convey the crux of the event) displayed an asymmetric profile, which we labeled a Past > Future effect. Semantic dementia patients scored in line with healthy controls for their provision of episodic event details in the past, yet demonstrated marked impairments for these details during future simulation. In contrast, when we analyzed the external event details (i.e., details reflecting non-episodic content), the converse profile was revealed, whereby a Future > Past effect was found (Irish, Addis, et al., 2012b). This concurrent decline of internal and inflation of external details has previously been reported in healthy aging during autobiographical memory retrieval (Levine, Svoboda, Hay, Winocur, & Moscovitch, 2002), and the envisaging of future events (Gaesser, Sacchetti, Addis, & Schacter, 2011), but appears particularly pronounced in patients with semantic dementia. From a theoretical standpoint, this finding is notable in two respects. First, the capacity to envisage a richly detailed future event appears doomed to fail in semantic dementia, despite a readily available store of intact episodic event details. Second, the overproduction of off-target external details seems to reflect the relatively spared retrieval from disparate and unrelated past events. As such, the external information provided by these patients is largely episodic, albeit unrelated to the future event being simulated (Irish & Piguet, 2013). This finding resonates with the proposition that the content of future simulation inextricably relies upon the accessibility of information and is not necessarily restricted to the episodic domain (Szpunar, 2010). By this view, Szpunar proposes that healthy individuals will draw upon abstracted information in the service of future thinking, as such representations should be more accessible than the relatively constrained focus of episodic memory (i.e., once-off, unique occurrences located within a discrete spatiotemporal framework). Evidence from the study of semantic dementia provides unequivocal support for this view whereby the contents of patients' future simulations invariably reflect the most efficient and accessible mode of information (Irish, Addis, et al., 2012a, 2012b; Szpunar, 2010).

## 8. THE SEMANTIC SCAFFOLDING HYPOTHESIS

The studies reviewed here provide compelling evidence for the crucial role for semantic memory in supporting complex acts of future-oriented mental time travel. Therefore, it is becoming increasingly evident that semantic memory may not only facilitate prospection, but may in fact provide the crucial framework for this ability. In recognition of this pivotal role, the semantic scaffolding hypothesis has been advanced to provide a conceptual account of how semantic memory likely underpins all forms of internally driven mental time travel (Irish, Addis, et al., 2012a; Irish & Piguet, 2013). Accordingly, semantic knowledge provides the requisite framework that enables (re)construction of the past and simulation of the future. Thus when we attempt to project forward in subjective time, we invoke a representative semantic schema or foundation to provide the necessary structure and meaning to the overall event. For example, when envisaging a possible trip to Paris, the requisite semantic framework would involve semantic details about travel, Paris, and France, which collectively impart meaning and structure to guide the simulation (e.g., the French language, French cuisine, the Eiffel tower). Once this broad framework is in place, relevant sensory-perceptual details from episodic and semantic memory can then be harnessed and assimilated into the schema. This model dovetails with current construction metaphors in the field that postulate that the building blocks of future simulations are the domain of episodic memory (Schacter & Addis, 2007a, 2007b) and that the generation of a spatially contiguous scene provides the foundation for constructive endeavors (Mullally & Maguire, 2014). The semantic scaffolding hypothesis thus offers a means of accounting for previous findings of impaired mental time travel in patients with semantic dementia (Duval et al., 2012; Irish, Addis, et al., 2012a, 2012b) but also relatively spared prospection in patients with intact semantic knowledge (Craver et al., 2014; Klein et al., 2002). Across all studies, we see the importance of abstract semantic representations in providing the necessary framework within which to ground future-oriented projection.

In addition, this theory can account for recent findings of seemingly intact future prospection in semantic dementia despite the profound semantic impairment that is characteristic of this syndrome. Viard and colleagues conducted an fMRI study of future simulation in which patients with semantic dementia were instructed to simulate events that were pre-selected by family and related to planned events that would occur in the future (Viard et al., 2014). Patients demonstrated an intact capacity for prospection, which was modulated by the functional integrity of the anterior hippocampus and the bilateral superior medial prefrontal gyri.

At first glance, this finding seems at odds with the fundamental role of semantic memory for future-oriented thinking. The methodological protocol used in this study holds the key, however, as patients were presented with event descriptors for the pre-defined events prior to event elaboration. Thus, patients were not required to generate an appropriate semantic framework, lessening the demands placed upon the conceptual knowledge store, in contrast with previous studies in which generative paradigms have been used (Duval et al., 2012; Irish, Addis, et al., 2012a, 2012b). It appears, therefore, that under certain conditions it is possible for the semantic deficit in semantic dementia to be circumvented to enable successful future-oriented mental time travel. When the essential semantic scaffold for the event in question is explicitly provided, semantic dementia patients succeed in populating this framework with contents that can be sampled from their relatively intact episodic memory store. Thus the Viard et al. (2014) study provides a counterpoint to previous work on prospection in semantic dementia by highlighting the fundamental role for semantic frameworks during event simulation. Finally, it is important to note that while the semantic scaffolding hypothesis argues for a central role for semantic memory, it acknowledges that complex acts of future-oriented mental time travel represent the confluence of the episodic and semantic memory systems, moving toward a synergistic perspective that is gaining traction within the field.

## 9. A SEMANTIC-EPISODIC MEMORY CONTINUUM

As has been demonstrated, converging evidence from the neuroimaging and neuropsychological literature highlights the overly restrictive reliance on the classic episodic-semantic heuristic where complex acts of memory and prospection are concerned. The construct of personal semantics is an excellent example of the considerable overlap between the different forms of declarative memory, and has recently been posited to represent something of a crossroads between episodic and semantic memory (Renoult, Davidson, Palombo, Moscovitch, & Levine, 2012). Where prospection is concerned, activation in semantic processing regions of the brain is consistently reported in functional neuroimaging studies of future-oriented thinking (reviewed by Bubić & Abraham, 2014). In addition, patients with gross episodic memory dysfunction can achieve certain forms of future-oriented mental time travel, which appear to be mediated by relatively spared semantic memory (Craver et al., 2014; Klein et al., 2002). Similarly, hippocampal amnesic patients have been shown to display, in some instances, a preserved capacity to construct future scenarios (Hurley, Maguire, & Vargha-Khadem, 2011; Maguire, Vargha-Khadem, & Hassabis, 2010; Mullally, Hassabis, & Maguire, 2012) with this preservation suggested to rely, in part, upon the integrity of semantic knowledge (Hurley et al., 2011). Finally, the disintegration of the semantic knowledge base results in marked impairments in envisaging the future, despite relatively intact episodic memory (Duval et al., 2012; Irish, Addis, et al., 2012a, 2012b). To this end, it has been suggested that episodic and semantic contributions to prospection can be viewed on a continuum, rather than in terms of a strict dichotomy, with differential contributions of each memory system contingent on current task demands and goals (Szpunar, Spreng, & Schacter, 2014).

Figure 2.1 in Chapter 2 displays a new taxonomy of prospection that places episodic and semantic memory on a continuum along which modes of future thinking can be conceptualized as relying more or less upon each form of memory (Szpunar et al., 2014). This model is significant, as it allows for considerable flexibility and interplay between the episodic and semantic memory systems in the service of complex future acts. In addition, however, Szpunar and colleagues acknowledge that prospection does not necessarily have to occur within the episodic domain, arguing that episodic and semantic forms of prospection can equally occur, as well as hybrid forms of future-oriented cognition in which significant interdependence between the memory systems is present (see also Greenberg & Verfaellie, 2010; Irish & Piguet, 2013; Klein, 2013a). As such, this new taxonomy represents a useful compass with which to reorient research on future-oriented thinking toward a more synergistic viewpoint in which the episodic and semantic memory systems are viewed as complementary and facilitatory systems, rather than mutually exclusive constructs.

## 10. OUTSTANDING CHALLENGES AND FUTURE DIRECTIONS

This chapter concludes at an exciting juncture in research on prospection, with parallel lines of study all directed to explicating the complex underpinnings of future-oriented mental time travel. A number of outstanding challenges remain, however, which will be important for future studies to clarify. One such issue is the relationship between autonoetic consciousness and semantic memory in facilitating the expression of future-oriented mental time travel. It has been suggested that it is autonoetic consciousness, not the memory content per se, which serves as the fundamental platform from which we project ourselves across subjective time to re-experience the past and pre-experience the future (Klein, 2015; Wheeler et al., 1997). As such, the content of simulations is viewed as essentially atemporal, and thus the crucial determinant of temporality is the nature of the accompanying conscious experience that occurs at the point of retrieval/simulation (Klein, 2015). In this light, Klein proposes that studies of patients with impaired access to content but intact autonoetic consciousness will prove particularly revelatory in clarifying the component processes of prospection. Notably, this profile of intact autonoetic consciousness and loss of content is present in patients with semantic dementia (see Piolino et al., 2003), suggesting that this syndrome holds the key to many of the conceptual debates surrounding future-oriented cognition. In contrast, Craver and colleagues (2014) argue that temporality is conferred by the semantic apprehension of time, pointing to a further area of study in which semantic memory and consciousness may interact to enable mental time travel. Finally, the precise mechanisms by which contextual details are reconfigured within semantic scaffolding to generate future events remain to be elucidated, although candidate processes such as hippocampally dependent recombination (Addis, Cheng, Roberts, & Schacter, 2011), scene construction (Mullally & Maguire, 2014), and conceptual expansion (Abraham et al., 2012) all potentially interact to ensure the flexible manipulation, association, and reconfiguring of contextual and abstracted representations during future simulation.

## 11. CONCLUSIONS

The purpose of this chapter is to highlight the fundamental contribution of semantic memory for many, if not all, forms of future-oriented mental time travel and to reorient research efforts to include semantic memory within the discourse on future thinking. Semantic memory has long been neglected within the memory literature; it is time to adopt a more synergistic approach whereby semantic memory is integrated as a core feature within current theoretical models. Work on the semantic contributions to prospection remains in its infancy; however, it is hoped that discussions of this nature will stimulate future research to ultimately explicate the cognitive and neural machinery that support the fascinating capacity for future-oriented mental time travel.

## ACKNOWLEDGMENTS

MI is supported by an Australian Research Council Discovery Early Career Researcher Award (DE130100463).

## REFERENCES

Abraham, A., Pieritz, K., Thybusch, K., Rutter, B., Kroger, S., Schweckendiek, J., . . . Hermann, C. (2012). Creativity and the brain: uncovering the neural signature of conceptual expansion. *Neuropsychologia, 50*(8), 1906–1917. doi: S0028-3932(12)00172-8 [pii] 10.1016/j.neuropsychologia.2012.04.015 [doi].

Abraham, A., Schubotz, R. I., & von Cramon, D. Y. (2008). Thinking about the future versus the past in personal and non-personal contexts. *Brain Research, 1233*, 106–119. doi: S0006-8993(08)01824-6 [pii] 10.1016/j.brainres.2008.07.084 [doi].

Addis, D. R., Cheng, T., Roberts, R. P., & Schacter, D. L. (2011). Hippocampal contributions to the episodic simulation of specific and general future events. *Hippocampus, 21*(10), 1045–1052. doi: 10.1002/hipo.20870 [doi].

Addis, D. R., Wong, A. T., & Schacter, D. L. (2008). Age-related changes in the episodic simulation of future events. *Psychological Science, 19*(1), 33–41.

Andelman, F., Hoofien, D., Goldberg, I., Aizenstein, O., & Neufeld, M. (2010). Bilateral hippocampal lesion and a selective impairment of the ability for mental time travel. *Neurocase, 16*(5), 426–435.

Binder, J. R., & Desai, R. H. (2011). The neurobiology of semantic memory. *Trends in Cognitive Sciences, 15*(11), 527–536. doi: S1364-6613(11)00214-2 [pii] 10.1016/j.tics.2011.10.001 [doi].

Binder, J. R., Desai, R. H., Graves, W. W., & Conant, L. L. (2009). Where is the semantic system? A critical review and meta-analysis of 120 functional neuroimaging studies. *Cerebral Cortex, 19*(12), 2767–2796. doi: bhp055 [pii] 10.1093/cercor/bhp055 [doi].

Bozeat, S., Gregory, C. A., Ralph, M. A., & Hodges, J. R. (2000). Which neuropsychiatric and behavioural features distinguish frontal and temporal variants of frontotemporal dementia from Alzheimer's disease? *Journal of Neurology, Neurosurgery and Psychiatry, 69*(2), 178–186.

Bubić, A., & Abraham, A. (2014). Neurocognitive bases of future-oriented cognition. *Review of Psychology, 21*(1), 3–15.

Burianova, H., McIntosh, A. R., & Grady, C. L. (2010). A common functional brain network for autobiographical, episodic, and semantic memory retrieval. *Neuroimage, 49*(1), 865–874. doi: S1053-8119(09)00978-1 [pii] 10.1016/j.neuroimage.2009.08.066 [doi].

Chan, D., Fox, N., Scahill, R., Crum, W., Whitwell, J., Leschziner, G., . . . Rossor, M. (2001). Patterns of temporal lobe atrophy in semantic dementia and Alzheimer's disease. *Annals of Neurology, 49*(4), 433–442.

Corkin, S. (2013). *Permanent present tense: the unforgettable life of the amnesic patient H.M.* New York: Basic Books.

Craver, C. F., Kwan, D., Steindam, C., & Rosenbaum, R. S. (2014). Individuals with episodic amnesia are not stuck in time. *Neuropsychologia, 57*, 191–195. doi: 10.1016/j.neuropsychologia.2014.03.004.

Crutch, S. J., & Warrington, E. K. (2002). Preserved calculation skills in a case of semantic dementia. *Cortex, 38*(3), 389–399.

Dalla Barba, G., & La Corte, V. (2013). The hippocampus, a time machine that makes errors. *Trends in Cognitive Sciences, 17*(3), 102–104. doi: 10.1016/j.tics.2013.01.005.

Duval, C., Desgranges, B., de La Sayette, V., Belliard, S., Eustache, F., & Piolino, P. (2012). What happens to personal identity when semantic knowledge degrades? A study of the self and autobiographical memory in semantic dementia. *Neuropsychologia, 50*(2), 254–265. doi: S0028-3932(11)00530-6 [pii] 10.1016/j.neuropsychologia.2011.11.019 [doi].

Gaesser, B., Sacchetti, D. C., Addis, D. R., & Schacter, D. L. (2011). Characterizing age-related changes in remembering the past and imagining the future. *Psychology and Aging, 26*(1), 80–84. doi: 2010-22974-001 [pii] 10.1037/a0021054 [doi].

Galton, C., Patterson, K., Graham, K., Lambon-Ralph, M., Williams, G., Antoun, N., . . . Hodges, J. (2001). Differing patterns of temporal atrophy in Alzheimer's disease and semantic dementia. *Neurology, 57*(2), 216–225.

Gorno-Tempini, M. L., Hillis, A. E., Weintraub, S., Kertesz, A., Mendez, M., Cappa, S. F., . . . Grossman, M. (2011). Classification of primary progressive aphasia and its variants. *Neurology, 76*(11), 1006–1014. doi: WNL.0b013e31821103e6 [pii] 10.1212/WNL.0b013e31821103e6 [doi].

Graham, K. S., Patterson, K., & Hodges, J. R. (1999). Episodic memory: new insights from the study of semantic dementia. *Current Opinion in Neurobiology, 9*(2), 245–250. doi: S0959-4388(99)80035-X [pii].

Greenberg, D. L., & Verfaellie, M. (2010). Interdependence of episodic and semantic memory: evidence from neuropsychology. *Journal of the International Neuropsychological Society, 16*(5), 748–753. doi: S1355617710000676 [pii] 10.1017/S1355617710000676 [doi].

Grilli, M. D., & Verfaellie, M. (2014). Personal semantic memory: insights from neuropsychological research on amnesia. *Neuropsychologia, 61*, 56–64. doi: 10.1016/j.neuropsychologia.2014.06.012.

Hassabis, D., & Maguire, E. A. (2007). Deconstructing episodic memory with construction. *Trends in Cognitive Sciences, 11*(7), 299–306.

Hegdé, J. (2007). Mental time travel sickness and a Bayesian remedy. *Brain and Behavioral Sciences, 30*, 323–324.

Hodges, J. R., & Patterson, K. (2007). Semantic dementia: a unique clinicopathological syndrome. *The Lancet Neurology, 6*(11), 1004–1014.

Hsiao, J. J., Kaiser, N., Fong, S. S., & Mendez, M. F. (2013). Suicidal behavior and loss of the future self in semantic dementia. *Cognitive and Behavioral Neurology, 26*(2), 85–92.

Hurley, N. C., Maguire, E. A., & Vargha-Khadem, F. (2011). Patient HC with developmental amnesia can construct future scenarios. *Neuropsychologia, 49*(13), 3620–3628. doi: S0028-3932(11)00422-2 [pii] 10.1016/j.neuropsychologia.2011.09.015 [doi].

Ingvar, D. H. (1985). "Memory of the future": an essay on the temporal organization of conscious awareness. *Human Neurobiology, 4*(3), 127–136.

Irish, M., Addis, D. R., Hodges, J. R., & Piguet, O. (2012a). Considering the role of semantic memory in episodic future thinking: evidence from semantic dementia. *Brain, 135*(Pt 7), 2178–2191. doi: aws119 [pii] 10.1093/brain/aws119 [doi].

Irish, M., Addis, D. R., Hodges, J. R., & Piguet, O. (2012b). Exploring the content and quality of episodic future simulations in semantic dementia. *Neuropsychologia, 50*(14), 3488–3495. doi: S0028-3932(12)00384-3 [pii] 10.1016/j.neuropsychologia.2012.09.012 [doi].

Irish, M., Hornberger, M., Lah, S., Miller, L., Pengas, G., Nestor, P. J., . . . Piguet, O. (2011). Profiles of recent autobiographical memory retrieval in semantic dementia, behavioural-variant frontotemporal dementia, and Alzheimer's disease. *Neuropsychologia, 49*(9), 2694–2702. doi: S0028-3932(11)00266-1 [pii] 10.1016/j.neuropsychologia.2011.05.017 [doi].

Irish, M., & Piguet, O. (2013). The pivotal role of semantic memory in remembering the past and imagining the future. *Frontiers in Behavioral Neuroscience, 7*, 27. doi: 10.3389/fnbeh.2013.00027 [doi]

Irish, M., Piguet, O., & Hodges, J. R. (2012). Self-projection and the default network in frontotemporal dementia. *Nature Reviews Neurology, 8*(3), 152–161. doi: nrneurol.2012.11 [pii] 10.1038/nrneurol.2012.11 [doi].

Irish, M., & Piolino, P. (2015). Impaired capacity for prospection in the dementias - Theoretical and clinical implications. *British Journal of Clinical Psychology.* doi: 10.1111/bjc.12090.

Klein, S. B. (2013a). The complex act of projecting oneself into the future. *WIREs Cognitive Science, 4*, 63–79.

Klein, S. B. (2013b). Sameness and the self: philosophical and psychological considerations. *Frontiers in Psychology: Perception Science, 5*, 29. doi: 10.3389/fpsyg.2014.00029.

Klein, S. B. (2015). Autonoetic consciousness: reconsidering the role of episodic memory in future-oriented self-projection. *Quarterly Journal of Experimental Psychology (Hove)*, 1–21. doi: 10.1080/17470218.2015.1007150.

Klein, S. B., Loftus, J., & Kihlstrom, J. F. (2002). Memory and temporal experience: the effects of episodic memory loss on an amnesic patient's ability to remember the past and imagine the future. *Social Cognition, 20*(5), 353–379.

Lambon Ralph, M. A. (2014). Neurocognitive insights on conceptual knowledge and its breakdown. *Philosophical Transactions of the Royal Society of London B: Biological Sciences, 369*(1634), 20120392. doi: 10.1098/rstb.2012.0392.

Lambon Ralph, M. A., Pobric, G., & Jefferies, E. (2009). Conceptual knowledge is underpinned by the temporal pole bilaterally: convergent evidence from rTMS. *Cerebral Cortex, 19*(4), 832–838. doi: 10.1093/cercor/bhn131.

Lambon Ralph, M. A., Sage, K., Jones, R. W., & Mayberry, E. J. (2010). Coherent concepts are computed in the anterior temporal lobes. *Proceedings of the National Academy of Sciences U S A, 107*(6), 2717–2722. doi: 10.1073/pnas.0907307107.

Levine, B., Svoboda, E., Hay, J. F., Winocur, G., & Moscovitch, M. (2002). Aging and autobiographical memory: dissociating episodic from semantic retrieval. *Psychology and Aging, 17*(4), 677–689.

Maguire, E. A. (2001). Neuroimaging studies of autobiographical event memory. *Philosophical Transactions of the Royal Society of London B: Biological Sciences, 356*(1413), 1441–1451.

Maguire, E. A., Kumaran, D., Hassabis, D., & Kopelman, M. D. (2010a). Autobiographical memory in semantic dementia: a longitudinal fMRI study. *Neuropsychologia*, *48*(1), 123–136.

Maguire, E. A., Vargha-Khadem, F., & Hassabis, D. (2010b). Imagining fictitious and future experiences: evidence from developmental amnesia. *Neuropsychologia*, *48*(11), 3187–3192.

Martinelli, P., Sperduti, M., & Piolino, P. (2013). Neural substrates of the self-memory system: new insights from a meta-analysis. *Human Brain Mapping*, *34*(7), 1515–1529. doi: 10.1002/hbm.22008 [doi].

McKinnon, M., Black, S., Miller, B., Moscovitch, M., & Levine, B. (2006). Autobiographical memory in semantic dementia: implications for theories of limbic-neocortical interaction in remote memory. *Neuropsychologia*, *44*(12), 2421–2429.

Mesulam, M. M., Wieneke, C., Hurley, R., Rademaker, A., Thompson, C. K., Weintraub, S., & Rogalski, E. J. (2013). Words and objects at the tip of the left temporal lobe in primary progressive aphasia. *Brain*, *136*(Pt 2), 601–618. doi: aws336 [pii] 10.1093/brain/aws336 [doi].

Mion, M., Patterson, K., Acosta-Cabronero, J., Pengas, G., Izquierdo-Garcia, D., Hong, Y. T., . . . Nestor, P. J. (2010). What the left and right anterior fusiform gyri tell us about semantic memory. *Brain*, *133*(11), 3256–3268. doi: awq272 [pii] 10.1093/brain/awq272 [doi].

Mullally, S. L., Hassabis, D., & Maguire, E. A. (2012). Scene construction in amnesia: an FMRI study. *Journal of Neuroscience*, *32*(16), 5646–5653. doi: 32/16/5646 [pii] 10.1523/JNEUROSCI.5522-11.2012 [doi].

Mullally, S. L., & Maguire, E. A. (2014). Memory, imagination, and predicting the future: a common brain mechanism? *The Neuroscientist*, *20*, 220–234.

Patterson, K., Nestor, P., & Rogers, T. (2007). Where do you know what you know? The representation of semantic knowledge in the human brain. *Nature Reviews Neuroscience*, *8*(12), 976–987.

Piolino, P., Desgranges, B., Belliard, S., Matuszewski, V., Lalevee, C., De La Sayette, V., & Eustache, F. (2003). Autobiographical memory and autonoetic consciousness: triple dissociation in neurodegenerative diseases. *Brain*, *126*(10), 2203–2219.

Race, E., Keane, M. M., & Verfaellie, M. (2013). Losing sight of the future: impaired semantic prospection following medial temporal lobe lesions. *Hippocampus*, *23*(4), 268–277. doi: 10.1002/hipo.22084 [doi].

Renoult, L., Davidson, P. S., Palombo, D. J., Moscovitch, M., & Levine, B. (2012). Personal semantics: at the crossroads of semantic and episodic memory. *Trends in Cognitive Science*, *16*(11), 550–558. doi: 10.1016/j.tics.2012.09.003.

Rosenbaum, R. S., Kohler, S., Schacter, D. L., Moscovitch, M., Westmacott, R., Black, S. E., . . . Tulving, E. (2005). The case of K.C.: contributions of a memory-impaired person to memory theory. *Neuropsychologia*, *43*(7), 989–1021. doi: 10.1016/j.neuropsychologia.2004.10.007.

Rutter, B., Kroger, S., Stark, R., Schweckendiek, J., Windmann, S., Hermann, C., & Abraham, A. (2012). Can clouds dance? Neural correlates of passive conceptual expansion using a metaphor processing task: implications for creative cognition. *Brain and Cognition*, *78*(2), 114–122. doi: 10.1016/j.bandc.2011.11.002.

Schacter, D. L., & Addis, D. R. (2007a). The cognitive neuroscience of constructive memory: remembering the past and imagining the future. *Philosophical Transactions of the Royal Society of London B: Biological Sciences*, *362*(1481), 773–786.

Schacter, D. L., & Addis, D. R. (2007b). On the constructive episodic simulation of past and future events. *Behavioral and Brain Sciences, 30*(03), 331–332.

Schacter, D. L., Addis, D. R., & Buckner, R. L. (2007). Remembering the past to imagine the future: the prospective brain. *Nature Reviews Neuroscience, 8*(9), 657–661.

Schacter, D. L., Addis, D. R., Hassabis, D., Martin, V. C., Spreng, R. N., & Szpunar, K. (2012). The future of memory: remembering, imagining, and the brain. *Neuron, 76*(4), 677–694. doi: S0896-6273(12)00991-9 [pii] 10.1016/j.neuron.2012.11.001 [doi].

Steinvorth, S., Levine, B., & Corkin, S. (2005). Medial temporal lobe structures are needed to re-experience remote autobiographical memories: evidence from H.M. and W.R. *Neuropsychologia, 43*(4), 479–496. doi: S0028-3932(05)00019-9 [pii] 10.1016/j.neuropsychologia.2005.01.001 [doi].

Stocker, K. (2012). The time machine in our mind. *Cognitive Science, 36*(3), 385–420. doi: 10.1111/j.1551-6709.2011.01225.x [doi].

Suddendorf, T., & Corballis, M. (2007). The evolution of foresight: what is mental time travel, and is it unique to humans? *Behavioral and Brain Sciences, 30*(03), 299–313.

Svoboda, E., McKinnon, M. C., & Levine, B. (2006). The functional neuroanatomy of autobiographical memory: a meta-analysis. *Neuropsychologia, 44*(12), 2189–2208. doi: S0028-3932(06)00209-0 [pii] 10.1016/j.neuropsychologia.2006.05.023 [doi].

Szpunar, K. (2010). Episodic future thought. *Perspectives on Psychological Science, 5*(2), 142–162.

Szpunar, K., Spreng, R. N., & Schacter, D. L. (2014). A taxonomy of prospection: introducing an organizational framework for future-oriented cognition. *Proceedings of the National Academy of Sciences U S A, 111*(52), 18414–18421. doi: 10.1073/pnas.1417144111.

Szpunar, K., & Tulving, E. (2011). Varieties of future experience. In M. Bar (ed.), *Predictions and the brain: using our past to generate a future* (pp. 3–12). New York: Oxford University Press.

Tulving, E. (1972). Episodic and semantic memory. In E. Tulving & W. Donaldson (eds.), *Organization of memory* (pp. 381–403). New York: Academic Press.

Tulving, E. (1983). *Elements of episodic memory.* New York: Oxford University Press.

Tulving, E. (1985). Memory and consciousness. *Canadian Psychology/Psychologie canadienne, 26*, 1–12.

Tulving, E. (2002). Episodic memory: from mind to brain. *Annual Review of Psychology, 53*, 1–25.

Viard, A., Chetelat, G., Lebreton, K., Desgranges, B., Landeau, B., de La Sayette, V., . . . Piolino, P. (2011). Mental time travel into the past and the future in healthy aged adults: an fMRI study. *Brain and Cognition, 75*(1), 1–9. doi: S0278-2626(10)00149-1 [pii] 10.1016/j.bandc.2010.10.009 [doi].

Viard, A., Piolino, P., Belliard, S., de La Sayette, V., Desgranges, B., & Eustache, F. (2014). Episodic future thinking in semantic dementia: a cognitive and FMRI study. *PLoS One, 9*(10), e111046. doi: 10.1371/journal.pone.0111046 [doi] PONE-D-14-21880 [pii].

Westmacott, R., Leach, L., Freedman, M., & Moscovitch, M. (2001). Different patterns of autobiographical memory loss in semantic dementia and medial temporal lobe amnesia: a challenge to consolidation theory. *Neurocase, 7*(1), 37–55.

Wheeler, M. A., Stuss, D. T., & Tulving, E. T. (1997). Toward a theory of episodic memory: the frontal lobes and autonoetic consciousness. *Psychological Bulletin, 121*, 331–354.

20

# The Impact of Multiple Sclerosis on Future-Oriented Mental Time Travel

*Neuropsychological and Neuroimagining Approaches*

ALEXANDRA ERNST AND LILIANN MANNING ■

## 1. INTRODUCTION

The study of human memory has aroused the interest of philosophers and researchers throughout the ages. In the fourth century, Saint Augustine posited that remembering our past coexists with the ability to anticipate our own future (see Manning, Cassel, & Cassel, 2013, for a review). Several centuries after Saint Augustine's intuition, this notion was further developed with the concept of mental time travel (Suddendorf & Corballis, 1997; Tulving, 2001, 2002), contributing to the initiation of a large body of theoretical and clinical investigations. Within this theoretical framework, *future-oriented mental time travel* (FMTT) is defined as the ability to imagine or pre-experience personal episodes that may occur in the future, while *autobiographical memory* (AM) corresponds to the ability to mentally re-experience personal past events.

A broad set of studies have established that AM and FMTT share common cognitive and neural mechanisms (see Schacter, Addis, Hassabis, Martin, Spreng, & Szpunar, 2012, for a review). However, converging lines of evidence also have revealed phenomenological and functional differences between AM and FMTT. In particular, past events are generally more richly detailed than future events (D'Argembeau & Van der Linden, 2004; McDonough & Gallo, 2010), or future events are frequently rated as being more positive than memories (Szpunar, Addis, & Schacter, 2012). Neuroimaging studies also revealed differential brain activations associated with the evocation of past or future events, especially greater brain activations in the frontopolar and hippocampal regions when imagining future events (Addis, Wong, & Schacter, 2007). To account for these past/future commonalities and differences, Schacter and Addis (2007) have put forward the *constructive episodic simulation hypothesis*, which suggests that past and future events rely on similar information, stored in episodic memory, which is flexibly recombined into a novel coherent event, but that pre-experiencing an event requires higher constructive processes than re-experiencing an event. Specifically related to the hippocampus, its greater recruitment for FMTT might reflect the extraction and the recombination of details into a coherent novel event, and the additional encoding of the simulation product in memory (Addis & Schacter, 2012).

FMTT has a strong adaptive value. Specifically, FMTT is involved in (1) coping, with a positive influence of the simulation of hypothetical scenario on emotion regulation and problem-solving; (2) goal achievement, with the setting up of a plan of action, taking into account different approaches and options to reach a goal; (3) the implementation of intentions, with an association between the act of forming the intention to achieve an action and the likelihood of doing that action; and (4) the construction and maintenance of a sense of personal continuity over time (Szpunar, 2010; D'Argembeau, Lardi, & Van der Linden, 2012). More generally, a positive association between FMTT and psychological well-being has been put forward.

Based on these different functional benefits, not surprisingly, a growing interest in the study of FMTT in clinical conditions has been observed in the last decade. Studies conducted in single-case amnesic patients (Klein, Loftus, & Kihlstrom, 2002; Tulving, 1985) or group studies in patients with Alzheimer's disease (Addis, Sacchetti, Ally, Budson, & Schacter, 2009b), schizophrenia (D'Argembeau, Raffard, & Van der Linden, 2008), or depression (Williams, Ellis, Tyers, Healy, Rose, & MacLeod, 1996) showed impaired AM and FMTT abilities, with in general a greater deficit for future than past events.

Recently, AM and FMTT impairment has also been documented in multiple sclerosis (MS) patients (Ernst, Blanc, Sellal, De Seze, Chauvin, & Manning, 2014). This finding has opened into a broader line of research that will be discussed in this chapter, which has aimed at investigating the cognitive and neural mechanisms involved in FMTT impairment in MS patients, and exploring the possibility of developing a cognitive rehabilitation program to improve FMTT in this clinical population. A brief overview of the clinical characteristics of MS will be given in the next section, before tackling the investigation itself.

## 2. CLINICAL FEATURES OF MULTIPLE SCLEROSIS

MS is a chronic neurological disease characterized by inflammatory processes, which could appear in the entire central nervous system, the latter conferring to MS its particular feature of multifocal and diffuse brain lesions. These inflammation processes conduct to demyelination in focal areas of the brain or in the spinal cord of patients, in which myelin, myelin-forming cells (oligodendrocytes), and axons are destroyed (Trapp & Nave, 2008). It is now established that MS affects both gray and white matter and can lead to significant brain atrophies and disconnections (Dineen et al., 2009; Filippi & Rocca, 2010). Different disease courses are distinguished, the most frequent being the relapsing-remitting (RR-MS) form, in which patients alternate between phases of relapses and partial or complete remissions, but with no disease progression between relapses. In most of the cases, the diagnosis of MS is established in patients between 20 and 40 years of age—a life period that is in general particularly productive in individuals' professional and personal lives (DeLuca & Nocentini, 2011).

The presence of cognitive impairment in MS patients is well accepted and occurs in 30%–70% of the patients, sometimes even in the early stage of the disease. The most common cognitive impairments observed in MS concern anterograde memory (verbal and visual modalities), information processing speed, attention, working memory, visuospatial abilities, and executive functioning (Chiaravalloti & DeLuca, 2008). Importantly, cognitive impairment in MS patients has an impact on everyday life functioning and quality of life, especially on employment, social activities, and routine household tasks (Chiaravalloti & DeLuca, 2008; Rao, Leo, Ellington, Nauertz, Bernardin, & Unverzagt, 1991).

As previously mentioned, while FMTT is involved in many aspects of daily life functioning and the occurrence of cognitive impairment is now well established, the interest in FMTT in MS patients is very recent. The investigation of FMTT in MS patients could raise issues because of the clinical presentation of the disease, especially for RR-MS patients. Because of the chronicity of the disease and the uncertainty surrounding MS evolution, one could argue that FMTT impairment would likely reflect psychological factors such as an explicit refusal to engage in future projects. Taking the opposite view, we would like to put forward that the clinical features of MS actually call for clinical investigations on FMTT. In particular, the mean age of MS onset coincides with a critical life period (20–40 years) in terms of identity construction and important life events and decisions (Rathbone, Moulin, & Conway, 2008). In addition, the setting up of effective coping and emotion regulation strategies is central to helping patients to better cope with the disease (McCabe, McKern & McDonald, 2004; Taylor et al., 1998). All of these domains are related to FMTT abilities. From this perspective, we will highlight in this chapter that FMTT impairment in MS patients has a cognitive and neural basis and, more important, that this deficit could be alleviated by a *tailor-made* cognitive intervention.

## 3. THE IMPACT OF MS ON FMTT: NEUROPSYCHOLOGICAL FINDINGS

The investigation of FMTT abilities in MS patients originates from studies exploring AM functioning in these patients. Overall, AM studies remain scarce in MS patients, and conflicting results have been reported (Kenealy, Beaumont, Lintern, & Murrell, 2002; Müller et al., 2013; Paul, Blanco, Hames & Beatty, 1997), which seem to reflect methodological discrepancies, such as the sensitivity of the AM test or the heterogeneity of the samples of MS patients. Using the Modified Crovitz test (Graham & Hodges, 1997), Ernst and colleagues (2013) documented AM impairment in non-depressed RR-MS patients who were characterized by a mild degree of functional disability and only mild impairment in other cognitive domains (e.g., anterograde memory, executive functions, attention). This AM impairment covered the entire life span, but with a better recall of memories from early adulthood, which includes memories that benefit from a privileged access (i.e., the reminiscence bump phenomenon; Conway & Rubin, 1993) and that thus involve less retrieval effort. This pattern of AM results, together with the positive association between AM and executive functions performance reported in this group of MS patients and the great occurrence of frontal lesions in MS (Bendfeldt et al., 2009; Benedict et al., 2002), led Ernst and colleagues to suggest a retrieval deficit at the origin of AM impairment in MS patients. Thus, this hypothesis put the frontal regions at the core of AM deficit in MS.

Considering the relationships between AM and FMTT and the particular involvement of executive functions and of the prefrontal regions in FMTT (Addis, Wong, & Schacter, 2007; Berntsen & Bohn, 2010; De Vito, Gamboz, Brandimonte, Barine, Amboni, & Della Salla, 2012), the extension of these findings to FMTT was a natural continuation. In this vein, we conjointly explored AM and FMTT in non-depressed RR-MS patients (Ernst et al., 2014). Here, the use of the Autobiographical Interview (AI; Levine, Svoboda, Hay, Winocur, & Moscovitch, 2002) enabled us to compare MS patients' performance in free recall and retrieval support conditions, using a stringent scoring procedure. Focusing on strictly episodic details (i.e., internal details), relative to healthy controls, results showed impaired AM and FMTT performance in MS patients, with a similar degree of impairment for the two temporal directions. However, in both MS patients and healthy controls, the amount of internal details produced in the future condition was lower than that for the past, reproducing the pattern of performance reported in previous studies (Addis, Wong, & Schacter, 2008; Addis et al., 2009b; D'Argembeau & Van der Linden, 2004). While the retrieval support condition led to better AM/FMTT performance in MS patients, their scores remained inferior to those obtained by healthy controls. The latter result brought supportive findings to the dysexecutive origin of AM and FMTT impairment in MS patients. Indeed, this pattern of results has been reported in clinical populations presenting with retrieval strategy impairment, in the context of prefrontal lesions (Levine et al., 2002; McKinnon, Black, Miller, Moscovitch, & Levine, 2006; Svoboda, Hynes, Campbell, Dade, Moscovitch, & Levine, 2002). Overall, it appears that external details (i.e., episodic details from other incidents, semantic knowledge, repetition, or metacognitive statements) are preserved in MS patients. A summary of the results obtained for the generation of future events in MS patients relative to healthy controls is provided in Figure 20.1.

In relation to the aforementioned concern of the influence of psychological factors on FMTT in MS, we observed in the current sample of patients a positive correlation between AM and FMTT scores. The replication of this robust finding in the literature of the link between AM and FMTT performance (Schacter et al., 2012) supported the idea that FMTT deficit does not likely rely on the presence of psychological factors in MS patients.

As we were particularly interested in the impact of FMTT impairment on everyday life functioning, the AI assessment was completed by a semi-structured interview, which aimed at collecting patients' self-reports on their difficulties associated with FMTT during testing and in their daily life. Based on the Memory Experiences Questionnaire (Sutin & Robins, 2007), this semi-structured interview tackled four key dimensions of FMTT pre-experiencing: the amount of detail, the accessibility, the vividness, and the emotional intensity of future events. As illustrated in the following, the analysis of the content of the semi-structured interview highlighted that MS patients are aware of their FMTT difficulties, which is a critical aspect for the setup of FMTT rehabilitation.

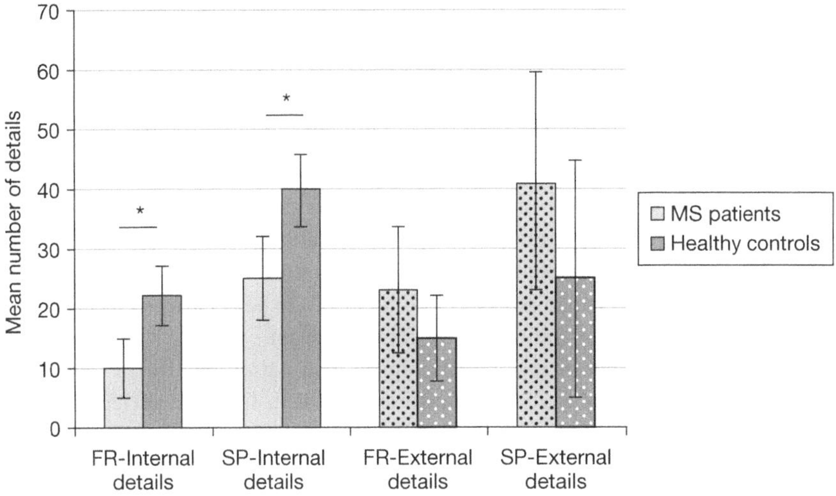


**Figure 20.1.** Mean number of internal and external details provided by MS patients and healthy controls during the free recall (FR) and the specific probe (SP) phases of the future event task (Autobiographical Interview).
Credit: Alexandra Ernst and Liliann Manning.

> "I'd say that, well it's not clear for me because I've really tried to get it, really tried to find the . . . no, there were no details. Let's say that I've just taken bits of what I've lived at some moments and put them in the project." (Patient C. L.)
>
> "It's no longer flashes or images I have; I think it's something in between. It's more static; if there is a little bit of movement it doesn't last long [. . .]. Yes, I was saying things that have already happened; I know I had to dissociate them. . . ." (Patient V. F.)
>
> "It's clear that I'm unable to project myself . . . I just can't, well, it's near future, if I try to think . . . I'll only be able to tell you one . . . but no, no details for me. No, no details, well I'd say it's less than a flash, it's an act, if you know what I mean [. . .] But otherwise, it's for sure I can't project myself, I just can't." (Patient M. F. W.)

These excerpts of comments illustrate MS patients' particular difficulty to depart from past events, together with an impoverishment of details and a low degree of vividness during the testing session, but also in everyday life.

This overlap between objective AI results and qualitative self-reports provided by MS patients indicate that patients are able to convey a clear message about their perceived difficulties. In this context, the introduction of similar questions in routine clinical practice could be informative for clinicians for a better understanding of the impact of FMTT impairment on daily life functioning.

### 3.1. Comparisons of MS Patients' FMTT with Other Clinical Conditions

Considering the extensive brain network and the wide range of cognitive processes involved in FMTT, not surprisingly, impaired FMTT performance has been reported in various clinical conditions. In that vein, Irish and colleagues (2013) have suggested that damage in any key nodes of the FMTT brain network will impinge on the ability to mentally project into the future.

Focusing on patients with damage affecting primarily the fontal lobe, De Vito et al. (2012) reported impaired FMTT in Parkinson's disease patients. De Vito et al.'s aim was to assess FMTT performance in non-amnesic patients and therefore differs from the goal of our Autobiographical Interview of amnesic MS patients. Despite this difference, a central involvement of executive processes at the origin of FMTT difficulties was suggested in both conditions. Interestingly, Parkinson's disease patients also expressed a particular difficulty in departing from their memories and recombining details into a novel coherent event.

A more severe pattern of results was reported in patients with the behavioral variant of frontotemporal dementia (Irish et al., 2013). In these patients, a more general difficulty in producing self-referential information was shown, for both episodic and semantic information.

The role of executive functions in the occurrence of FMTT impairment has also been suggested in psychiatric conditions such as schizophrenia or depression. In both cases, the ability to imagine personal detailed future events was found to be reduced (D'Argembeau, Raffard, & Van der Linden, 2008; King et al., 2011; Sarkohi et al., 2011; Williams et al., 1996).

FMTT impairment has also been described following predominant temporal lobe damage, such as in amnesic patients (Hassabis et al., 2007), Alzheimer's disease patients (Addis et al., 2009b), or semantic dementia (Irish et al., 2012a). While the two first conditions involve mostly the medial temporal regions, including the hippocampus, semantic dementia affects primarily the lateral temporal regions. Again, damage to these different key nodes of the FMTT brain network lead to different clinical profiles, impinging, respectively, the extraction and recombination of details at the hippocampal level or the semantic scaffolding that guides future simulation.

The constructive episodic simulation hypothesis (Schacter & Addis, 2007), suggests that different stages of the future simulation processes are affected across these clinical conditions. Beyond its theoretical contribution, the distinction of these different profiles has important implications for the development of interventions aiming to alleviate FMTT impairment. In the case of MS patients, we suggested that the early stage of the future simulation process is compromised, with impaired strategic abilities that negatively impact the extraction and recombination of details into a novel and coherent event. This hypothesis, suggesting a particular involvement of the frontal regions in the occurrence of FMTT impairment in MS patients, led us to explore more specifically the structural and functional correlates of impaired FMTT in this condition.

## 4. STRUCTURAL AND FUNCTIONAL UNDERPINNINGS OF FMTT IMPAIRMENT IN MS PATIENTS

As previously mentioned, MS is characterized by diffuse brain lesions that could be observed within the entire brain, leading to an important heterogeneity of lesion burden between patients. The group of MS patients included in our FMTT studies presented with significant brain atrophy in the parahippocampal regions, the frontoparietal regions, and the cerebellum (Ernst, Noblet, Gounot, Blanc, De Seze, & Manning, 2015). In this group of MS patients, FMTT performance was positively and extensively correlated with the volume of the left prefrontal regions, as well as with the volume of the lateral temporal region, the inferior parietal lobule, the precuneus, and the cuneus. In comparison with healthy controls, the positive association between FMTT performance and the prefrontal region was particularly critical.

Regarding the functional changes associated with FMTT impairment, using an adapted version of the fMRI paradigm developed by Addis et al. (2007), we explored the cerebral activation changes and the neural response to the difficulty of imagining and to the amount of detail associated with imagining personal future events in MS patients.

As illustrated in Figure 20.2, increased activations were reported in MS patients (relative to healthy controls) within an extended brain network. In accordance with the hypothesis of impaired strategic abilities, enhanced activations were particularly observed in the bilateral prefrontal regions during both the initial building of future events (i.e., the construction phase) and the search of detail associated with future events (i.e., the elaboration phase). In addition, the prefrontal regions were also particularly responsive to the increase in the amount of detail in patients. These findings are consistent with the involvement of the prefrontal regions in the executive demand necessary for the extraction and recombination of an infinite assortment of detail into a novel event and in the integration and maintenance of multimodal information (Addis, Pan, Vu, Laiser, & Schacter, 2009a; Weiler, Suchan, Koch, Schwarz, & Daum, 2011).

Apart from the frontal region, functional activation changes in MS patients have also been observed in other brain regions, such as the lateral and medial temporal regions (i.e., the parahippocampal gyrus). In particular, the lateral temporal region sustains the reactivation of a "semantic scaffolding" to guide future simulations (Irish et al., 2012a; Irish & Piguet, 2013) and its role in the recombination process has been recently discussed (Schacter et al., 2012). Regarding the parahippocampal gyrus, this structure is engaged in contextual processing and plays a pivotal role in the recruitment of the posterior brain regions (Botzung, Denkova, & Manning, 2008; Viard, Desgranges, Eustache, & Piolino, 2012). The greater involvement of the parahippocampal gyrus in association with the difficulty of imagining future events, together with the greater neural activity reported during the elaboration phase in the posterior brain areas (e.g., precuneus, cuneus, occipital regions, and the posterior cingulate gyrus), suggests that visual imagery processes are abnormally recruited in MS patients. A particular increase of the activations in the posterior brain regions was observed in response to the amount of detail in MS patients. In other words, the involvement of visual imagery process in MS patients is modulated by the amount of detail of future events to a greater extent than in healthy controls. This was also the case for brain regions supporting self-referential processes such as the medial frontal region and the anterior cingulate gyrus, for which greater activations relative to healthy controls were consistently observed in MS patients (Northoff, Heinzel, de Greck, Bermpohl, Dobrowolny, & Panksepp, 2006). Overall, it appears that in our patients, it is only when a certain amount of detail is reached that the key processes of visual imagery and self-involvement are recruited, with the requirement of greater neural resources.

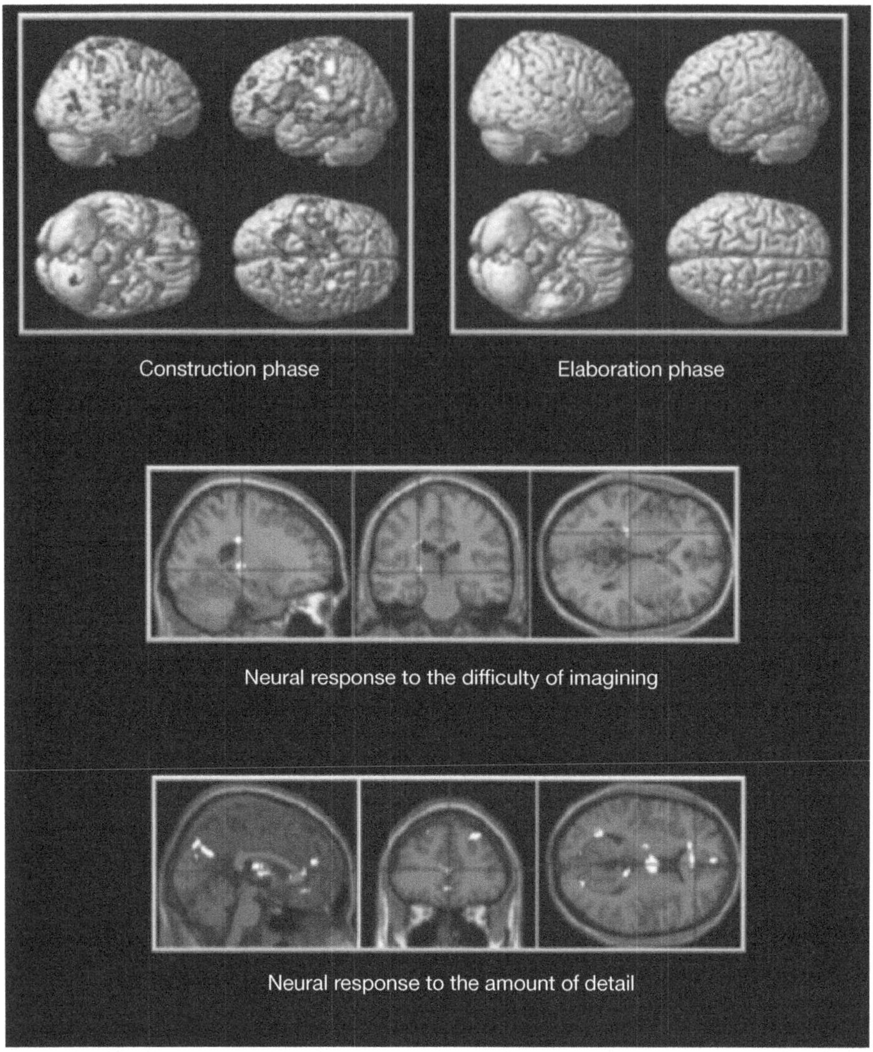


**Figure 20.2.** Brain regions showing an increased activity in MS patients when imagining future events (construction and elaboration) and a greater neural response to the difficulty of imagining and to the amount of detail of future events ($p = 0.005$, $k = 10$ voxels, uncorrected). (See color insert).
Credit: Alexandra Ernst and Liliann Manning.

While a critical role of the hippocampus in FMTT has been described in the literature, no functional change was observed in this brain structure in MS patients. In particular, the hippocampus is involved in the retrieval of details stored in memory, their recombination into a coherent future simulation and the encoding of the simulation product in memory, and it is also particularly responsive to the amount of detail (Addis & Schacter, 2008, 2012). Different hypotheses could be put forward to account for this result in MS patients. Bearing in mind that future events were poorly detailed in patients in comparison with healthy controls, the absence of functional changes in the hippocampus could reflect this result. Alternatively, the significant brain atrophy observed in the adjacent parahippocampal gyrus in our MS patients may have compromised the further involvement of the hippocampus. Finally, as previously mentioned, the executive component necessary for the extraction/recombination of details being sustained by the prefrontal regions, the absence of a differential hippocampal activity between groups could be related to the negative impact of the executive component, which might have affected the extraction/recombination of detail at the hippocampal level, and the further stream of activations within the neocortex. This is consistent with the reciprocal connections shared between the prefrontal regions and the medial temporal lobe, and the neocortex (Maguire, Kumaran, Hassabis, & Kopelman, 2010). A neuroanatomical corroboration of the latter is the main association found between the volume of prefrontal regions and FMTT performance in MS patients, whereas no such association was found for the hippocampus.

Overall, the structural and functional underpinnings of FMTT in MS patients involved a discrete set of brain regions that pertain to the FMTT core network, with a central role of the prefrontal regions. In addition, the partial overlaps observed between the structural and functional correlates of FMTT impairment in MS patients are consistent with previous studies conducted in MS patients in other cognitive domains (e.g., Mainero et al., 2004; Morgen et al., 2007).

Our findings are consistent with the few neuroimaging studies exploring FMTT carried out to date. With a particular focus on clinical conditions involving the frontal lobe, the critical role of the prefrontal regions in FMTT has been suggested in patients with the behavioral variant of frontotemporal dementia, together with an involvement of the temporal regions, especially the hippocampus (Irish et al., 2013). These results highlighted the contribution of disrupted strategic, semantic, and scene construction processes at the origin of FMTT impairment in the behavioral variant of frontotemporal dementia.

The functional integrity of the frontal regions, especially the medial frontal regions, and of the hippocampus appeared critical for FMTT in patients with semantic dementia (Viard et al., 2014). In relation to the primary deterioration of the lateral temporal regions in semantic dementia, FMTT performance was also associated with the structural integrity of this brain region, which is in agreement with the disruption of semantic knowledge as the core of FMTT impairment in this condition (Irish, Addis, Hodges, & Piguet, 2012b).

At the functional level, Hach et al. (2014) have shown that individuals with depression recruited brain regions pertaining to the FMTT core brain network, but with signs of up- or down-regulation in comparison with healthy controls. Interestingly, both depressed and MS patients presented a lack of episodic richness and specificity of their future events at the clinical level, and these similarities remained present at the neural level. In fact, similar to MS patients, increased prefrontal activations were reported in depressed individuals, and both groups showed a decreased contribution of the brain regions sustaining mental visual imagery processes. However, the individuals with depression departed from the MS patients regarding the hippocampus, which showed a reduced activity in depressed individuals. These brain activation changes were also accompanied by functional connectivity changes in depressed individuals.

Taken together, the structural and functional underpinnings of FMTT impairment in MS patients involve different key nodes that have been previously identified in different clinical conditions, such as the behavioral variant of frontotemporal dementia, semantic dementia, or depression, with a central role of the prefrontal regions. The involvement of a widespread brain network in MS patients is consistent with the diffuse nature of lesions characterizing this condition, contrary to the others etiologies studied to date. This finding raises the question of the spatiotemporal dynamic associated with these increased brain activations, and future studies aiming at exploring the functional connectivity changes related to FMTT impairment in MS patients might further our understanding of the neural mechanisms underlying this impairment.

Based on the convergence of behavioral and neuroimaging findings that pointed to a disruption of strategic processes at the origin of FMTT impairment in MS patients, a natural development of this research was to explore the possibility of alleviating this executive-related FMTT impairment, as described in the next section.

## 5. COGNITIVE REHABILITATION OF FMTT IN MS PATIENTS

To the best of our knowledge, the possibility of alleviating FMTT impairment in patients has never been addressed to date. However, numerous studies have suggested different strategies to improve AM functioning, with two main lines of research: the use of an external device such as the SenseCam (e.g., Pauly-Takacs et al., 2011; Woodberry et al., 2015), or training programs (e.g., Moradi et al., 2014; Neshat-Doost et al., 2013). In the case of MS patients, we developed a tailor-made rehabilitation program specifically designed to enhance the strategic abilities necessary for the retrieval of personal memories. Bearing in mind that MS occurs in general in young adults who are in a very active period of their life, our goal was to provide an integrated strategy that could be easily used in everyday life functioning. To that end, we built up a mental visual imagery (MVI)-based facilitation program (Ernst et al., 2013) to reinforce access to memories and to improve their vividness by making use of the cueing role of MVI in AM (Greenberg & Rubin, 2003). Positive outcomes have been obtained in AM performance, with an effective transfer in daily life functioning.

Considering the relationships between AM and FMTT, our objective was to extend these findings to FMTT (Ernst et al., 2015). We conducted a randomized controlled trial, which included 40 non-depressed RR-MS patients presenting with an FMTT impairment, in the context of mild to moderate cognitive impairment in attention and/or executive functions and in the absence of major anterograde memory deficit—the latter inclusion criteria being of importance to guarantee the further good completion of the facilitation program. Patients were then randomly assigned to three groups and were blind to their allocation: (1) the experimental group, who followed the MVI program; (2) the control group, who followed a sham control intervention to control the presence of a "nursing effect"; and (3) the stability group, who underwent the AI assessment twice, but with no intervention in between, in order to control a potential learning effect.

Specifically, the MVI program followed a goal-directed approach (Wilson & Gracey, 2009) and is based on the ability to mentally construct scenes and to pay close attention to details in the mind's eye. It encompasses six 2-hour sessions and is organized in four steps of increasing difficulty. An important aspect is the continuous guidance provided by the neuropsychologist to probe the patient from general aspects to more detailed ones, adopting a "funnel-approach":

1. The *screening step*, based on three subtests from the "Imagery and Perception Battery" (Bourlon et al., 2009), aims to probe basic visual imaging abilities.
2. The *external visualization* included verbal items to imagine and describe in as many details as possible (e.g., shape, color, size, etc.), with the complementary visualization of an action made with the item (e.g., visualize an onion and visualize it again, after being sliced).
3. The *construction phase* consisted in figuring out complex scenes, bringing into play several characters. A first training step (e.g., imagine the hotel of your holidays) and a subsequent mental scene construction, sharing thematic similarities (e.g., imagine the house of your dreams), were proposed, allowing the patient to rely on the training section to construct the next scene.
4. The *self-visualization* followed the same procedure, but here, patients were asked to visualize themselves within a given scenario, to imagine it as though they were actually living the scene, with the description of all kinds of details, sensations, or feelings that came to mind. A first training scene was proposed (e.g., imagine you take part in a magic show), followed by a second scene with a similar theme (e.g., imagine you enter the big cats' cage in a circus).

Regarding the verbal control program, it follows the same structure and procedure as the MVI program, but it focuses on the importance of information organization and narration. Based on a series of texts, patients have to construct a discussion about the topic of text, following different directions depending on the different steps. Briefly, (1) the *external discussion* focuses on the form of the texts (e.g. clarity, vocabulary used); (2) the *discussion construction* consists in discussion about the theme of the text, with a training and a construction step, with two thematically related texts (e.g., a trip to South Africa was followed by a text about a trip to Ireland); and (3) the *self-involved discussion* was similar to the previous step, with the addition of questions about his or her own opinion (e.g., a first text about taxing sodas to reduce their consumption was followed by a second text concerning the usefulness of anti-smoking campaigns).

Before and immediately after facilitation, all the patients underwent the AI and the semi-structured interview to gather information on the perceived difficulties/benefits associated with FMTT functioning. Results obtained by the different groups of patients at the AI are illustrated in Figure 20.3.

Equivalent FMTT performance was observed between groups before facilitation; after facilitation, only the experimental group showed a significant improvement of their AI score, that is, the number of episodic details provided. In addition, patients from the experimental group obtained significantly higher AI scores than the two other groups after facilitation. To obtain a comprehensive picture of the benefits of the MVI program, a follow-up assessment was also conducted 6 months after the

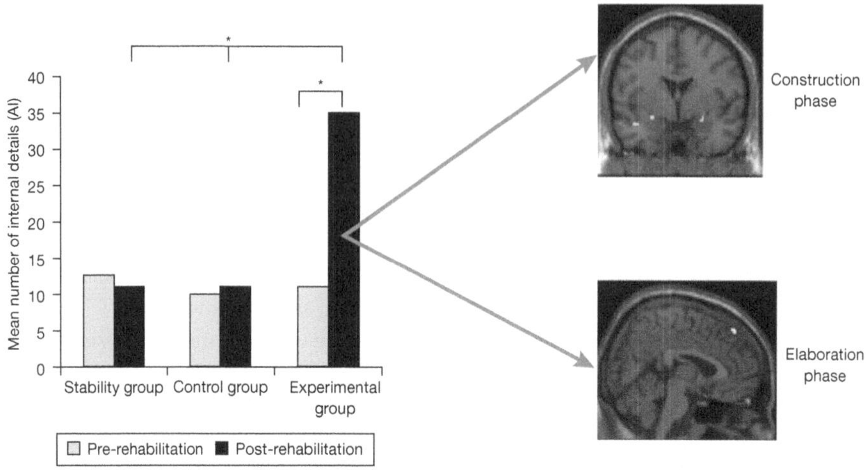


**Figure 20.3.** Clinical and functional brain changes associated with FMTT improvement in MS patients (* significant difference $p < 0.001$; fMRI analysis threshold: $p < 0.001$; $k = 10$ voxels, uncorrected).
Credit: Alexandra Ernst and Liliann Manning.

intervention to test the long-term robustness of treatment effects. While a slight decrease of AI scores was present, FMTT performance remained largely superior to those obtained before facilitation.

At the individual level, nearly all the patients who followed the MVI program showed a clear benefit from the intervention, with a normalization of their AI scores. This was supported by the patients' comments gathered after facilitation, which illustrated that the benefits of the MVI program were efficiently transferred to daily functioning. Examples of comments are provided here:

> "The first time we have done this exercise, it was more like some 'flashes,' whereas now, it is more like films." (Patient S. E.)

> "It's easier to imagine things . . . To project myself. The first time, before the program, it was really hard [. . .]. In fact, it's a mechanism that I have learnt, little by little, that I have integrated. And it was not a 'harsh' learning. I had the time to integrate it. It was spontaneous." (Patient A. S.)

> "Yes, I think that something is set up, in everyday life, for the future. I have the feeling that, as we went along the sessions, I had the feeling that I was actually living the thing. I'm in. I'm seeing things. . . . Yes, the future events, I feel them." (Patient I. B.)

Overall, it appears that the MVI program was efficient in improving FMTT performance, with an effective transfer to daily life functioning and a long-term robustness of clinical benefits, which were not linked to nursing or learning effects.

Over the last decade, the need to provide solid evidence of the efficacy of cognitive interventions led to the inclusion of neuroimaging as a tool to visualize the influence of cognitive rehabilitation on the brain (Strangman et al., 2005). While this line of research is very recent in MS patients, cerebral functional changes have been shown after cognitive interventions aiming to improve, for instance, attention (Penner, Kappos, Rausch, Opwis, & Radü, 2006) or anterograde memory (Chiaravalloti, Wylie, Leavitt, & DeLuca, 2012; Leavitt, Wylie, Girgis, DeLuca, & Chiaravalloti, 2014). This raises the question of the structural and functional brain changes that could be observed following the MVI program and the ensuing FMTT improvement in MS patients.

## 5.1. Brain Plasticity Following FMTT Improvement in MS Patients

As previously discussed in this chapter, increased brain activations have been reported in several key nodes of the FMTT brain network in MS patients presenting with impaired future simulation abilities. An open question is the impact of a cognitive intervention on this "abnormal" brain network, a question that has not yet been addressed, to our knowledge. In the case of MS patients who followed the MVI program, we suggested that brain changes would be observed in the posterior brain regions—considering their role in visual imagery processes (Greenberg & Rubin, 2003), and in key nodes of the FMTT brain network, namely the frontopolar and the hippocampal regions.

From the RR-MS patients included in the previous study (Ernst et al., 2015), 17 patients (10 from the experimental group and 7 from the control group) also underwent neuroimaging sessions, before and after their respective cognitive intervention. Neuroimaging sessions followed strictly the same paradigm as the one previously presented, with each task developed in two versions (randomly assigned to pre- or post-facilitation).

Prior to intervention, no difference was observed between groups regarding the brain network engaged in FMTT. Following the MVI program, increased brain activations were reported in the bilateral parahippocampal regions, the left lateral temporal, and the left frontopolar regions during the construction phase of future events. These enhanced prefrontal activations remained present during the further elaboration of future events, more specifically in the left frontopolar and the right lateral frontal regions (see Figure 20.3). These functional changes reported in this key node of the FMTT brain network could be interpreted in light of its role in the overall integration and maintenance of multimodal information (Addis et al., 2009a). In this context, we suggested that the increased frontopolar activity observed after facilitation in MS patients might reflect the greater cognitive demand of the integration process due to the greater episodic richness of future events. Similarly, the greater activation exhibited by patients after the MVI program in the lateral frontal region could be related to the greater need of prefrontal resources required to bind together richer episodic and semantic information during future simulations (Weiler et al., 2011).

In parallel, the enhanced activity in the lateral temporal region could be associated with the reactivation of personal semantic information to guide future simulations (Irish et al., 2012). Considering both the role of the parahippocampal region and that of MVI in FMTT, through the facilitation program, the greater neural activity in this region could reflect an increased reliance on contextual processing and the greater episodic quality of future events after facilitation. This was accompanied by functional connectivity changes in the posterior default-mode network, especially in the left occipital region and precuneus, the right posterior cingulate gyrus, and the left lateral temporal region. In addition, a positive correlation was obtained between FMTT improvement in post-facilitation and an increase of the gray matter volume in the right prefrontal regions and the posterior cingulate gyrus. This result, beyond reflecting the emphasis made on the role of MVI in FMTT, could be related to the dynamic interactions between the posterior brain regions and the parahippocampal gyrus.

While no FMTT improvement was reported in the control group, slight brain activation changes were observed in this group. In particular, decreased activations were shown in left frontoparietal regions and the right fusiform gyrus during the construction of future events. Similarly, decreased functional connectivity was observed in the anterior default mode network after intervention, especially in the medial frontal region and the right anterior cingulate gyrus. However, no change in gray matter volume was reported in this group. This general pattern of down-regulated neural activity has previously been related to practice effects and habituation to the task in MS patients who underwent a sham cognitive intervention (Chiaravalotti et al., 2012; Leavitt et al., 2014).

To summarize, at the neural level, the clinical benefits of the MVI program are manifested by increased brain activations, higher functional connectivity, as well increased gray matter volume, in key structures of the FMTT brain network that sustain visual imagery, semantic processes, and the integration and maintenance of multimodal information. Conversely, only minor changes, mostly related to practice effects, were observed in the absence of clinical improvement.

## 6. CONCLUSIONS

Taken together, this set of studies exploring FMTT in MS patients demonstrated the occurrence of FMTT impairments in this condition that are associated with cerebral activation changes, especially in the prefrontal regions. Converging behavioral and neuroimaging findings support the idea that the early stage of the simulation process (i.e., strategic abilities) might be compromised in MS patients.

Across these different studies, we have observed that FMTT impairment is very frequent in RR-MS patients, even in the context of relative preservation of general cognitive functioning. The latter underlines the particular sensitivity of FMTT to brain damage and is consistent with the idea that damage to any key nodes of the FMTT brain network will impinge on future simulations (Irish et al., 2013).

From a clinical standpoint, these findings provide new insights into cognitive functioning in MS patients, but most important, argue for the inclusion of FMTT measure in clinical routine, especially since MS patients were generally aware of their difficulties in projecting themselves in the future. This suggestion is reinforced by the long-term benefits on FMTT that we reported after a short cognitive intervention focusing on MVI. This recommendation makes particular sense, considering that FMTT plays important roles in everyday life functioning, for instance in the construction and the sense of continuity of the self, coping with stressful events, or goal achievement—all of these domains being particularly affected by the experience of brain injury, which provokes major disruptions in different realms of everyday life, such as self-perception, social and professional situations, or emotional issues.

The significant structural and functional brain changes that accompanied FMTT improvement in MS patients are supportive arguments of the effectiveness of cognitive rehabilitation, even in a complex cognitive domain like FMTT that involves several cognitive mechanisms.

Since MS is characterized by diffuse brain lesions, with a great heterogeneity between patients, the theoretical contribution of this line of research in the understanding of underlying cognitive and neural mechanisms of FMTT is limited. Nevertheless, these findings suggest that the early stage of future simulation could be efficiently supported by cognitive mechanisms such as visual imagery, the involvement of which has been mainly described in the late stage of future simulation. In this context, our results are consistent with the key role of scene construction in mental time travel (Hassabis & Maguire, 2007).

## 7. FUTURE DIRECTIONS

The growing interest in the study of FMTT in clinical conditions observed in the last years clearly conveys the importance of this field for both clinicians and neuroscientists. In particular, the development of cognitive interventions to alleviate FMTT impairment represents an avenue for future research programs and has the potential to provide significant theoretical and clinical contributions. This is particularly important considering the growing body of research that highlights the critical role of FMTT in everyday life functioning and in well-being. In this vein, the investigation of the relationship between impaired/improved future simulation abilities and domains such as personal identity, emotional regulation, coping strategies, or quality of life would be of clinical relevance.

In addition, the inclusion of different neuroimaging techniques to visualize the structural and functional brain changes following a cognitive intervention could be helpful in disentangling the key neural mechanisms of both impaired and improved FMTT and the similar and different neural signatures associated with different clinical conditions.

## ACKNOWLEDGMENTS

We are grateful to the Fondation pour la Recherche sur la Sclérose en Plaques (ARSEP; Ile de France; grant to L. M.) for research funding and to the Ministry of National Education and Research (A. E.'s PhD grant). A. E. is now a postdoctoral researcher supported by a research funding from the Region Bourgogne (France) accorded to Dr. Chris Moulin and Dr. Céline Souchay (LEAD, CNRS UMR5022, University of Burgundy).

## REFERENCES

Addis, D. R., Pan, L., Vu, M. A., Laiser, N., & Schacter, D. L. (2009a). Constructive episodic simulation of the future and the past: distinct subsystems of a core brain network mediate imagining and remembering. *Neuropsychologia*, *47*, 2222–38.

Addis, D. R., Sacchetti, D. C., Ally, B. A., Budson, A. E., & Schacter, D. L. (2009b). Episodic simulation of future events is impaired in mild Alzheimer's disease. *Neuropsychologia*, *47*, 2660–2671.

Addis, D. R., & Schacter, D. L. (2012). The hippocampus and imagining the future: where do we stand? *Frontiers in Human Neuroscience*, *5*, 1–15.

Addis, D. R., Wong, A. T., & Schacter, D. L. (2007). Remembering the past and imagining the future: common and distinct neural substrates during event construction and elaboration. *Neuropsychologia*, *45*, 1363–1377.

Addis, D. R., Wong, A. T., & Schacter, D. L. (2008). Age-related changes in the episodic simulation of future events. *Psychological Science*, *19*, 33–41.

Bendfeldt, K., Kuster, P., Traud, S., Egger, H., Winklhofer, S., Mueller-Lenke, N., et al. (2009). Association of regional gray matter volume loss and progression of white matter lesions in multiple sclerosis: a longitudinal voxel-based morphometry study. *NeuroImage*, *45*, 60–67.

Benedict, R. H. B., Bakshi, R., Simon, J. H., Priore, R., Miller, C., & Munschauer, F. (2002). Frontal cortex atrophy predicts cognitive impairment in multiple sclerosis. *Journal of Neuropsychiatry and Clinical Neurosciences*, *14*, 44–51.

Berntsen, D., & Bohn, A. (2010). Remembering and forecasting: the relation between autobiographical memory and episodic future thinking. *Memory and Cognition*, *38*, 265–278.

Botzung, A., Denkova, E., & Manning, L. (2008). Experiencing past and future personal events: functional neuroimaging evidence on the neural bases of mental time travel. *Brain and Cognition*, *66*, 202–212.

Bourlon, C., Chokron, S., Bachoud-Lévi, A. C., Coubard, O., Bergeras, I., Moulignier, A., et al. (2009). Presentation of an assessment battery for visual mental imagery and visual perception. *Revue Neurologique*, *165*, 1045–4054.

Chiaravalloti, N. D., & DeLuca, J. (2008). Cognitive impairment in multiple sclerosis. *Lancet Neurology*, *7*, 1139–1151.

Chiaravalloti, N. D., Wylie, G., Leavitt, V., & DeLuca, J. (2012). Increased cerebral activation after behavioral treatment for memory deficits in MS. *Journal of Neurology*, *259*, 1337–1346.

Conway, M. A., & Rubin, D. C. (1993). The structure of autobiographical memory. In A. Collins, S. Gathercole, Martin A. Conway & P. E. Morris (eds.), *Theories of Memory* (pp. 103–137). Lawrence Erlbaum.

D'Argembeau, A., Lardi, C., & Van der Linden, M. (2012). Self-defining future projections: exploring the identity function of thinking about the future. *Memory*, *20*, 110–120.

D'Argembeau, A., Raffard, S., & Van der Linden, M. (2008). Remembering the past and imagining the future in schizophrenia. *Journal of Abnormal Psychology*, *117*, 247–251.

D'Argembeau, A., & Van der Linden, M. (2004). Phenomenal characteristics associated with projecting oneself back into the past and forward into the future: influence of valence and temporal distance. *Consciousness and Cognition*, *13*, 844–858.

DeLuca, J., & Nocentini, U. (2011). Neuropsychological, medical and rehabilitative management of persons with multiple sclerosis. *NeuroRehabilitation*, *29*, 197–219.

De Vito, S., Gamboz, N., Brandimonte, M. A., Barone, P., Amboni, M., & Della Salla, S. (2012). Future thinking in Parkinson's disease: an executive function? *Neuropsychologia*, *50*, 1494–501.

Dineen, R. A., Vilisaar, J., Hlinka, J., Bradshaw, C. M., Morgan, P. S., Constantinescu, C. S., et al. (2009). Disconnection as a mechanism for cognitive dysfunction in multiple sclerosis. *Brain, 132*, 239–249.

Ernst, A., Blanc, F., De Seze, J., & Manning, L. (2015). Using mental visual imagery to improve autobiographical memory and episodic future thinking in relapsing-remitting multiple sclerosis patients: a randomised-controlled trial study. *Restorative Neurology and Neuroscience, 33*, 621–638.

Ernst, A., Blanc, F., Sellal, F., De Seze, J., Chauvin, B., & Manning, L. (2014). Impaired mental simulation of specific past and future personal events in non-depressed multiple sclerosis patients. *Journal of the Neurological Sciences, 345*, 68–74.

Ernst, A., Blanc, F., Voltzenlogel, V., de Seze, J., Chauvin, B., & Manning L. (2013). Autobiographical memory in multiple sclerosis patients: assessment and cognitive facilitation. *Neuropsychological Rehabilitation, 23*, 161–181.

Ernst, A., Noblet, V., Gounot, D., Blanc, F., De Seze, J., & Manning, L. (2015). Neural correlates of episodic future thinking impairment in multiple sclerosis patients. *Journal of Clinical and Experimental Neuropsychology, 37*, 1107–1123.

Filippi, M., & Rocca, M. A. (2010). MR imaging of gray matter involvement in multiple sclerosis: implications for understanding disease pathophysiology and monitoring treatment efficacy. *American Journal of Neuroradiology, 31*, 1171–1177.

Graham, K., & Hodges, J. R. (1997). Differentiating the roles of the hippocampal complex and the neocortex in LTM storage: evidence from the study of semantic dementia and Alzheimer's disease. *Neuropsychology, 11*, 77–89.

Greenberg, D. L., & Rubin, D. C. (2003). The neuropsychology of autobiographical memory. *Cortex, 39*, 687–728.

Hach, S., Tippett, L. J., & Addis, D. R. (2014). Neural changes associated with the generation of specific past and future events in depression. *Neuropsychologia, 65*, 41–55.

Hassabis, D., & Maguire, E. A. (2007). Deconstructing episodic memory with construction. *Trends in Cognitive Sciences, 11*, 299–306.

Hassabis, D., Kumaran, D., Vann, S. D., & Maguire, E. A. (2007). Patients with hippocampal amnesia cannot imagine new experiences. *Proceedings of the National Academy of Sciences, 104*, 1726–1731.

Irish, M., & Piguet, O. (2013). The pivotal role of semantic memory in remembering the past and imagining the future. *Frontiers in Behavioral Neuroscience, 7*, 27.

Irish, M., Addis, D. R., Hodges, J. R., & Piguet, O. (2012a). Exploring the content and quality of episodic future simulations in semantic dementia. *Neuropsychologia, 50*, 3488–3495.

Irish, M., Addis, D. R., Hodges, J. R., & Piguet, O. (2012b). Considering the role of semantic memory in episodic future thinking: evidence from semantic dementia. *Brain, 135*, 2178–2191.

Irish, M., Hodges, J. R., & Piguet, O. (2013). Episodic future thinking is impaired in the behavioural variant of frontotemporal dementia. *Cortex, 49*, 2377–2388.

Kenealy, P., Beaumont, G., Lintern, T., & Murrell, R. C. (2002). Autobiographical memory in advanced multiple sclerosis: assessment of episodic and personal semantic memory across three time spans. *Journal of the International Neuropsychological Society, 8*, 855–860.

King, M. J., Macdougall, A. G., Ferris, S., Herdman, K. A., & McKinnon, M. C. (2011). Episodic simulation of future events is impaired in patients with major depressive disorder. *Psychiatry Research, 187*, 465–467.

Klein, S. B., Loftus, J., & Kihlstrom, J. F. (2002). Memory and temporal experience: the effects of episodic memory loss on an amnesic's patient ability to remember the past and imagine the future. *Social Cognition*, *20*, 353–379.

Leavitt, V. M., Wylie, G. R., Girgis, P. A., DeLuca, J., & Chiaravalloti, N. D. (2014). Increased functional connectivity within memory networks following memory rehabilitation in multiple sclerosis. *Brain Imaging and Behavior*, *8*, 394–402.

Levine, B., Svoboda, E., Hay, J. F., Winocur, G., & Moscovitch, M. (2002). Aging and autobiographical memory: dissociating episodic fromsemantic retrieval. *Psychology and Aging*, *17*, 677–689.

Maguire, E. A., Kumaran, D., Hassabis, D., & Kopelman, M. D. (2010). Autobiographical memory in semantic dementia: a longitudinal fMRI study. *Neuropsychologia*, *48*, 123–136.

Mainero, C., Caramia, F., Pozzilli, C., Pisani, A., Pestalozza, I., Borriello, G., et al. (2004). fMRI evidence of brain reorganization during attention and memory tasks in multiple sclerosis. *NeuroImage*, *21*, 858–867.

Manning, L., Cassel, D., & Cassel, J. C. (2013). St. Augustine's reflections on memory and time and the current concept of subjective time in mental time travel. *Behavioral Sciences*, *3*, 232–243.

McCabe, M. P., McKern, S., & McDonald, E. (2004). Coping and psychological adjustment among people with multiple sclerosis. *Journal of Psychosomatic Research*, *56*, 355–361.

McDonough, I. M., & Gallo, D. A. (2010). Separating past and future autobiographical events in memory: evidence for a reality monitoring asymmetry. *Memory and Cognition*, *38*, 3–12.

McKinnon, M. C., Black, S. E., Miller, B., Moscovitch, M., & Levine, B. (2006). Autobiographical memory in semantic dementia: implications for theories of limbic–neocortical interaction in remote memory. *Neuropsychologia*, *44*, 2421–2429.

Moradi, A. R., Moshirpanahi, S., Parhon, H., Mirzaei, J., Dalgleish, T., & Jobson, L. (2014). A pilot randomized controlled trial investigating the efficacy of memory specificity training in improving symptoms of posttraumatic stress disorder. *Behaviour Research and Therapy*, *56*, 68–74.

Morgen, K., Sammer, G., Courtney, S. M., Wolters, T., Melchior, H., Blecker, C. R., et al. (2007). Distinct mechanisms of altered brain activation in patients with multiple sclerosis. *NeuroImage*, *37*, 937–946.

Müller, S., Saur, R., Greve, B., Melms, A., Hautzinger, M., Fallgatter, A. J., et al. (2013). Similar autobiographical memory impairment in long-term secondary progressive multiple sclerosis and Alzheimer's disease. *Multiple Sclerosis Journal*, *19*, 225–232.

Neshat-Doost, H. T., Dalgleish, T., Yule, W., Kalantari, M., Ahmadi, S. J., Dyregrov, A., et al. (2013). Enhancing autobiographical memory specificity through cognitive training: an intervention for depression translated from basic science. *Clinical Psychological Science*, *1*, 84–92.

Northoff, G., Heinzel, A., de Greck, M., Bermpohl, F., Dobrowolny, H., & Panksepp, J. (2006). Self-referential processing in our brain: a meta-analysis of imaging studies on the self. *NeuroImage*, *3*, 440–457.

Paul, R., Blanco, C., Hames, K., & Beatty, V. W. (1997). Autobiographical memory in multiple sclerosis. *Journal of the International Neuropsychological Society*, *3*, 246–251.

Pauly-Takacs, K., Moulin, C. J. A., & Estlin, E. J. (2011). SenseCam as a rehabilitation tool in a child with anterograde amnesia. *Memory*, *19*, 705–712.

Penner, I. K., Kappos, L., Rausch, M., Opwis, K., & Radü, E. W. (2006). Therapy-induced plasticity of cognitive functions in MS patients: insights from fMRI. *Journal of Physiology, 99*, 455–462.

Rao, S. M., Leo, G. J., Ellington, L., Nauertz, T., Bernardin, L., & Unverzagt, F. (1991). Cognitive dysfunction in multiple sclerosis. II. Impact on employment and social functioning. *Neurology, 41*, 692–696.

Rathbone, C. J., Moulin, C. J. A., & Conway, M. A. (2008). Self-centered memories: the reminiscence bump and the self. *Memory and Cognition, 36*, 1403–1414.

Sarkohi, A., Bjärehed, J., & Andersson, G. (2011). Links between future thinking and autobiographical memory specificity in major depression. *Psychology, 3*, 261–265.

Schacter, D. L., & Addis, D. R. (2007). The cognitive neuroscience of constructive memory: remembering the past and imagining the future. *Philosophical Transactions of the Royal Society of London B: Biological Sciences, 362*, 773–786.

Schacter, D. L., Addis, D. R., Hassabis, D., Martin, V. C., Spreng, N., & Szpunar, K. K. (2012). The future of memory: remembering, imagining, and the brain. *Neuron, 76*, 677–694.

Strangman, G., O'Neil-Pirozzi, T. M., Burke, D., Cristina, D., Goldstein, R., Rauch, S. L., et al. (2005). Functional neuroimaging and cognitive rehabilitation for people with traumatic brain injury. *American Journal of Physical Medicine and Rehabilitation, 84*, 62–75.

Suddendorf, T., & Corballis, M. C. (1997). Mental time travel and the evolution of the human mind. *Genetic, Social, and General Psychology Monographs, 123*, 133–167.

Sutin, A. R., & Robins, R. W. (2007). Phenomenology of autobiographical memories: The Memory Experiences Questionnaire. *Memory, 15*, 390–411.

Svoboda, E., Hynes, C. A., Campbell, A. F., Dade, L. A., Moscovitch, M., & Levine, B. (2002). The frontal lobes and autobiographical memory: differential effects of dorsolateral and ventrolateral prefrontal damage. *Journal of the International Neuropsychological Society, 8*, 275.

Szpunar, K. K. (2010). Episodic Future Thought: An Emerging Concept. *Perspectives on Psychological Science, 5*, 142–162.

Szpunar, K. K., Addis, D. R., & Schacter, D. L. (2012). Memory for emotional simulations: remembering a rosy future. *Psychological Sciences, 23*, 24–29.

Taylor, S. E., Pham, L. B., Rivkin, I. D., & Armor, D. A. (1998). Harnessing the imagination: mental simulation, self-regulation and coping. *American Psychologist, 53*, 429–439.

Trapp, B. D., & Nave, K. A. (2008). Multiple sclerosis: an immune or neurodegenerative disorder? *Annual Review of Neuroscience, 31*, 247–269.

Tulving, E. (1985). Memory and consciousness. *Canadian Psychology/Psychologie canadienne, 26*, 1–12.

Tulving, E. (2001). Episodic memory and common sense: how far apart? *Philosophical Transactions of Royal Society of London B: Biological Sciences, 356*, 1505–1515.

Tulving, E. (2002). Episodic memory: from mind to brain. *Annual Review of Psychology, 53*, 1–25.

Viard, A., Desgranges, B., Eustache, F., & Piolino, P. (2012). Factors affecting medial temporal lobe engagement for past and future episodic events: an ALE meta-analysis of neuroimaging studies. *Brain and Cognition, 80*, 111–125.

Viard, A., Piolino, P., Belliard, S., de La Sayette, V., Desgranges, B., & Eustache, F. (2014). Episodic future thinking in semantic dementia: a cognitive and fMRI study. *Plos One, 9*, 1–22.

Weiler, J., Suchan, B., Koch, B., Schwarz, M., & Daum, I. (2011). Differential impairment of remembering the past and imagining novel events after thalamic lesions. *Journal of Cognitive Neuroscience*, *23*, 3037–3051.

Williams, J. M. G., Ellis, N. C., Tyers, C., Healy, H., Rose, G., & MacLeod, A. K. (1996). The specificity of autobiographical memory and imaginability of the future. *Memory and Cognition*, *24*, 116–125.

Wilson, B. A., & Gracey, F. (2009). Towards a comprehensive model of neuropsychological rehabilitation. In Wilson, B. A., Gracey, F., Evans, J. J., Bateman, A. *Neuropsychological rehabilitation: theory, models, therapy and outcome* (pp. 1–21). Cambridge, UK: Cambridge University Press.

Woodberry, E., Browne, G., Hodges, S., Watson, P., Kapur, N., & Woodberry, K. (2015). The use of a wearable camera improves autobiographical memory in patients with Alzheimer's disease. *Memory*, *23*, 340–349.

# INDEX

The manufacturer's authorised representative in the EU for product
safety is Oxford University Press España S.A. of El Parque Empresarial
San Fernando de Henares, Avenida de Castilla, 2 - 28830 Madrid
(www.oup.es/en or product.safety@oup.com). OUP España S.A. also acts
as importer into Spain of products made by the manufacturer.
Printed and bound by CPI Group (UK) Ltd, Croydon, CR0 4YY
14/09/2026
02216308-0018